Fundamental Principles of
Optical Lithography

Fundamental Principles of Optical Lithography:
The Science of Microfabrication

CHRIS MACK

www.lithoguru.com

John Wiley & Sons, Ltd

To Susan, Sarah and Anna

Contents

Preface

There was a time when I was sure this book would never be finished. Believe it or not, I began working on it about 18 years ago, and for a long time it seemed that each year left me farther from completion (a consequence of working in a rapidly changing field). Several things conspired to finally make this book a reality. Ron Hershel, the closest person to a mentor I have known, gave me sage advice on writing a book: Don't try to write it all at once – instead, write and publish pieces of your book as journal articles over time, then collect them up when you have written enough. That advice has served me well, especially since I have been writing a quarterly column for *Microlithography World* since 1992. That approach helped me finish my first book, *Inside PROLITH: A Comprehensive Guide to Optical Lithography Simulation*, in 1997.

While I enjoyed finishing and publishing *Inside PROLITH*, my ambition was to write a more comprehensive university textbook on the topic of semiconductor lithography. I have been teaching a graduate level class on optical lithography at the University of Texas at Austin since 1991, using handouts in place of a real book. Each year I strove to add more material to make those notes more complete. But I ran into the same problem as before – new material in this quickly changing field was needed at a faster rate than I was writing.

I solved this problem in two ways. First, I began to focus solely on fundamental principles needed to understand the science of lithography. While lithography practice changes quickly, the fundamental principles underlying that practice do not. The result, though, is that this book has very little 'practical' advice, that is, descriptions of best practices in the industry. While such practical descriptions can be very useful, they also become dated very quickly. I hope that by focusing on fundamentals this book might be useful to the reader many years after it is purchased.

Secondly, I quit my job and worked on this book full-time (well, almost full time) for the final year before its completion. I hope that this admission doesn't scare off the earnest would-be author, but the reality was, for me at least, that dedicated effort was required to complete the project. Much of the material contained herein is, of course, a tutorial review of the published literature on lithography and related sciences. But a significant portion is new work, having never before been published. Thus, I hope that this book will contribute to the body of lithography literature mostly as a convenient repository of a useful portion of the collective knowledge in this field, but also as a repository for the information contained in various notebooks, files and scraps of paper scattered around my cluttered office.

While the length of the final book surprised even me, still, there are many fascinating and important topics in lithography that have been excluded for lack of space. In particular, rigorous treatment of electromagnetic scattering through the topography of a photomask is extremely important in lithography today, but goes without even a mention in the book. With respect to photoresist, etch resistance, spin coat rheology, adhesion, shelf life and quality control, resist formulations and defects all receive short shrift, and nothing is mentioned of top surface imaging or the various multilayer and nontraditional resist schemes. For those interested in imaging tools, the topics of geometrical optics, lens design, aberration measurement and tool component functions such as alignment, auto-focus, stage motion, etc., are essentially ignored. The world of mask making is left to other books, as are the topics of chip design and design for manufacturability, despite their obvious impact on the field of lithography. Metrology, especially critical dimension metrology, is extremely important to lithography, since data from these measurements drive much of our knowledge of how lithography behaves. However, even though metrology deserves an entire chapter, I have left it out almost completely. Surprisingly to some, I have chosen not to cover an aspect of lithography very dear to me – lithography simulation. While the use of simulation is illustrated throughout the book, and of course a scientific description of lithography must necessarily serve as the foundation of a lithographic simulator, I have avoided the topic of the numerical solutions to the equations presented in the book as well as to the topics of model speed, accuracy and calibration. Finally, there is no effort to describe research into next generation lithography technologies, and no description of the many lithographic approaches that don't use projection optical imaging. Despite these glaring omissions, I still might be criticized for making the book too long, with too many topics, for use as a university text. I can only say that I have successfully covered most of the information contained in the book in a one semester course, with only the usual amount of grumbling from the affected students.

I am indebted to many, many people for their help with this book. In the 24 years that I have worked in the field of lithography I have been taught by many, many people. I couldn't possible begin to count, let alone recount, the published works that have so greatly contributed to my understanding in this field. In fact, I will here issue a blanket apology to all those whose important works I have relied on but did not include in the references in the book. The lack of proper references is, I think, the biggest failure of this work, though I hope to be forgiven based on its format as a university textbook. I would also like to thank the students at the University of Texas at Austin and at Notre Dame whom I punished with early drafts of this book. Their feedback and experiences, good and bad, helped me to greatly improve the material and make it more suitable for the classroom.

There are many people who helped by reviewing chapters and providing feedback: Gary Bernstein, John Biafore, Robert Bunch, Jeff Byers, John Kulp and John Petersen among others that I am sure I am forgetting. I would especially like to thank Warren Grobman who carefully read every chapter in the book and provided invaluable feedback, and Trey Graves who bravely defied Finman's Law of Mathematics and re-derived many of the equations in Chapters 2 and 3. I am indebted to Chris Sallee and KLA-Tencor for allowing me the use of the lithography simulator PROLITH, which I employed extensively in generating many of the figures found throughout the book.

As a final note, I encourage the interested reader to visit the web page for this book: www.lithoguru.com/textbook. There I will post errata and other information that might be useful to the reader, and information that might prove valuable to the professor interested in using this text as the basis for a university course.

Chris A. Mack
Austin, Texas
June, 2007

1
Introduction to Semiconductor Lithography

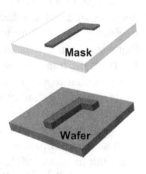

The fabrication of an integrated circuit (IC) involves a great variety of physical and chemical processes performed on a semiconductor (e.g. silicon) substrate. In general, the various processes used to make an IC fall into three categories: film deposition, patterning and semiconductor doping. Films of both conductors (such as polysilicon, aluminum, tungsten and copper) and insulators (various forms of silicon dioxide, silicon nitride and others) are used to connect and isolate transistors and their components. Selective doping of various regions of silicon allows the conductivity of the silicon to be changed with the application of voltage. By creating structures of these various components, millions (or even billions) of transistors can be built and wired together to form the complex circuitry of a modern microelectronic device. Fundamental to all of these processes is lithography, i.e. the formation of three-dimensional (3D) relief images on the substrate for subsequent transfer of the pattern into the substrate.

The word lithography comes from the Greek *lithos*, meaning stones, and *graphia*, meaning to write. It means quite literally writing on stones. In the case of semiconductor lithography, our stones are silicon wafers and our patterns are written with a light-sensitive polymer called a *photoresist*. To build the complex structures that make up a transistor and the many wires that connect the millions of transistors of a circuit, lithography and etch pattern transfer steps are repeated at least 10 times, but more typically 25 to 40 times to make one circuit. Each pattern being printed on the wafer is aligned to the previously formed patterns as slowly the conductors, insulators and selectively doped regions are built up to form the final device.

The importance of lithography can be appreciated in two ways. First, due to the large number of lithography steps needed in IC manufacturing, lithography typically accounts

Fundamental Principles of Optical Lithography: The Science of Microfabrication, Chris Mack.
© 2007 John Wiley & Sons, Ltd.

for about 30 % of the cost of manufacturing a chip. As a result, IC fabrication factories ('fabs') are designed to keep lithography as the throughput bottleneck. Any drop in output of the lithography process is a drop in output for the entire factory. Second, lithography tends to be the technical limiter for further advances in transistor size reduction and thus chip performance and area. Obviously, one must carefully understand the trade-offs between cost and capability when developing a lithography process for manufacturing. Although lithography is certainly not the only technically important and challenging process in the IC manufacturing flow, historically, advances in lithography have gated advances in IC cost and performance.

1.1   Basics of IC Fabrication

A *semiconductor* is not, as its name might imply, a material with properties between an electrical conductor and an insulator. Instead, it is a material whose conductivity can be readily changed by several orders of magnitude. Heat, light, impurity doping and the application of an electric field can all cause fairly dramatic changes in the electrical conductivity of a semiconductor. The last two can be applied locally and form the basis of a *transistor*: by applying an electric field to a doped region of a semiconductor material, that region can be changed from a good to a poor conductor of electricity, or vice versa. In effect, the transistor works as an electrically controlled *switch*, and these switches can be connected together to form digital logic circuits. In addition, semiconductors can be made to *amplify* an electrical signal, thus forming the basis of analog solid-state circuits.

By far the most common semiconductor in use is silicon, due to a number of factors such as cost, formation of a stable native oxide and vast experience (the first silicon IC was built in about 1960). A wafer of single-crystal silicon anywhere from 75–300 mm in diameter and about 0.6–0.8 mm thick serves as the substrate for the fabrication and interconnection of planar transistors into an IC. The most advanced circuits are built on 200- and 300-mm-diameter wafers. The wafers are far larger than the ICs being made so that each wafer holds a few hundred (and up to a few thousand) IC devices. Wafers are processed in lots of about 25 wafers at a time, and large fabs can have throughputs of greater than 10 000 wafers per week. The cycle time for making a chip, from starting bare silicon wafers to a finished wafer ready for dicing and packaging, is typically 30–60 days. Semiconductor processing (or IC fabrication) involves two major tasks:

- Creating small, interconnected 3D structures of insulators and conductors in order to manipulate local electric fields and currents
- Selectively doping regions of the semiconductor (to create p–n junctions and other electrical components) in order to manipulate the local concentration of charge carriers

1.1.1   Patterning

The 3D microstructures are created with a process called *patterning*. The common *subtractive* patterning process (Figure 1.1) involves three steps: (1) deposition of a uniform film of material on the wafer; (2) lithography to create a positive image of the pattern

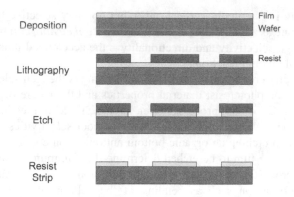

Figure 1.1 *A simple subtractive patterning process*

that is desired in the film; and (3) etch to transfer that pattern into the wafer. An *additive* process (such as electroplating) changes the order of these steps: (1) lithography to create negative image of the pattern that is desired; and (2) selective deposition of material into the areas not protected by the lithographically produced pattern. Copper is often patterned additively using the damascene process (named for a unique decorative metal fill process applied to swords and developed in Damascus about 1000 years ago).

Deposition can use many different technologies dependent on the material and the desired properties of the film: oxide growth (direct oxidation of the silicon), chemical vapor deposition (CVD), physical vapor deposition (PVD), evaporation and sputtering. Common films include insulators (silicon dioxide, silicon nitride, phosphorous-doped glass, etc.) and conductors (aluminum, copper, tungsten, titanium, polycrystalline silicon, etc.). Lithography, of course, is the subject of this book and will be discussed at great length in the pages that follow. Photoresists are classed as positive, where exposure to light causes the resist to be removed, and negative, where exposed patterns remain after development. The goal of the photoresist is to resist etching after it has been patterned so that the pattern can be transferred into the film.

1.1.2   Etching

Etch involves both chemical and mechanical mechanisms for removal of the material not protected by the photoresist. *Wet etch*, perhaps the simplest form of etch, uses an etchant solution such as an acid that chemically attacks the underlying film while leaving the photoresist intact. This form of etching is *isotropic* and thus can lead to undercutting as the film is etched from underneath the photoresist. If *anisotropic* etching is desired (and most always it is), directionality must be induced into the etch process. *Plasma etching* replaces the liquid etchant with a plasma – an ionized gas. Applying an electric field causes the ions to be accelerated downward toward the wafer. The resulting etch is a mix of chemical etching due to reaction of the film with the plasma and physical sputtering due to the directional bombardment of the ions hitting the wafer. The chemical nature of the etch can lead to good *etch selectivity* of the film with respect to the resist (selectivity being defined as the ratio of the film etch rate to the resist etch rate) and with respect to the substrate below the film, but is essentially isotropic. Physical sputtering is very directional (etching is essentially vertical only), but not very selective (the resist etches at about the same rate as the film to be etched). *Reactive ion etching* combines both effects to give good enough selectivity and directionality – the accelerated ions provide energy to drive a chemical etching reaction.

The photoresist property of greatest interest for etching is the etch selectivity, which is dependent both on the photoresist material properties and the nature of the etch process for the specific film. Good etch processes often have etch selectivities in excess of 4 (for example, polysilicon with a novolac resist), whereas poor selectivities can be as low as 1 (for example, when etching an organic bottom antireflection coating). Etch selectivity and the thickness of the film to be etched determine the minimum required resist thickness. Mechanical properties of the resist, such as adhesion to the substrate and resistance to mechanical deformation such as bending of the pattern, also play a role during etching.

For an etch process without perfect selectivity, the shape and size of the final etched pattern will depend on not only the size of the resist pattern but its shape as well. Consider a resist feature whose straight sidewalls make an angle θ with respect to the substrate (Figure 1.2). Given vertical and horizontal etch rates of the resist R_V and R_H, respectively, the rate at which the critical dimension (CD) shrinks will be

$$\frac{\mathrm{d}CD}{\mathrm{d}t} = 2(R_H + R_V \cot \theta) \qquad (1.1)$$

Thus, the rate at which the resist CD changes during the etch is a function of the resist sidewall angle. As the angle approaches 90°, the vertical component ceases to contribute and the rate of CD change is at its minimum. In fact, Equation (1.1) shows three ways to minimize the change in resist CD during the etch: improved etch selectivity (making both R_H and R_V smaller), improved anisotropy (making R_H/R_V smaller) and a sidewall close to vertical (making $\cot \theta$ smaller).

The example above, while simple, shows clearly how resist profile shape can affect pattern transfer. In general, the ideal photoresist shape has perfectly vertical sidewalls. Other nonideal profile shapes, such as rounding of the top of the resist and resist footing, will also affect pattern transfer.

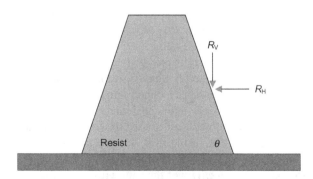

Figure 1.2 *Erosion of a photoresist line during etching, showing the vertical and horizontal etch rate components*

1.1.3 Ion Implantation

Selective doping of certain regions of the semiconductor begins with a patterning step (Figure 1.3). Regions of the semiconductor that are not covered by photoresist are exposed to a dopant impurity. *p-type* dopants like boron have three outer shell electrons and when inserted into the crystal lattice in place of silicon (which has four outer electrons) create mobile *holes* (empty spots in the lattice where an electron could go). *n-type* dopants such as phosphorous, arsenic and antimony have five outer electrons, which create excess mobile electrons when used to dope silicon. The interface between p-type regions and n-type regions of silicon is called a *p–n junction* and is one of the foundational structures in the building of semiconductor devices.

The most common way of doping silicon is with *ion implantation*. The dopant is ionized in a high-vacuum environment and accelerated into the wafer by an electric field (voltages of hundreds of kilovolts are common). The depth of penetration of the ions into the wafer is a function of the ion energy, which is controlled by the electric field. The force of the impact of these ions will destroy the crystal structure of the silicon, which then must be restored by a high-temperature annealing step, which allows the crystal to reform (but also causes diffusion of the dopant). Since the resist must block the ions in the regions where dopants are not desired (that is, in the regions covered by the resist), the resist thickness must exceed the penetration depth of the ions.

Ion implantation penetration depth is often modeled as a Gaussian distribution of depths, i.e. the resulting concentration profile of implanted dopants follows a Gaussian shape. The mean of the distribution (the peak of the concentration profile) occurs at a depth called the *projected range*, R_p. The standard deviation of the depth profile is called the *straggle*, ΔR_p. For photoresists, the projected range varies approximately linearly with implant energy, and inversely with the atomic number of the dopant (Figure 1.4a). A more accurate power-law model (as shown in Figure 1.4a) is described in Table 1.1. The straggle varies approximately as the square root of implant energy, and is about independent of the dopant (Figure 1.4b). For higher energies (1 MeV and above), higher atomic number dopants produce more straggle. When more detailed predictions are needed, Monte Carlo implantation simulators are frequently used.

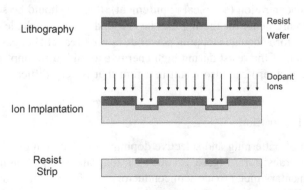

Figure 1.3 Patterning as a means of selective doping using ion implantation

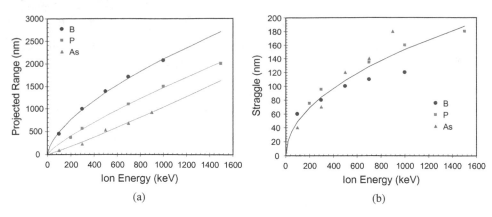

Figure 1.4 *Measured and fitted ion implantation penetration depths for boron, phosphorous and arsenic in AZ 7500 resist: (a) projected range and (b) straggle. Symbols are data[1] and curves are power-law fits to the data as described in Table 1.1. For the straggle data, the empirical model fit is $\Delta R_p = 4.8E^{0.5}$ where E is the ion energy in keV and ΔR_p is the straggle in nm*

Table 1.1 *Empirical model of ion implanted projected range (R_p, in nm) into photoresist versus ion energy (E, in keV) as $R_p = aE^b$*

Dopant	Coefficient a	Power b
Boron	26.9	0.63
Phosphorous	5.8	0.80
Arsenic	0.49	1.11

In order to mask the underlying layers from implant, the resist thickness must be set to at least

$$resist\ thickness \geq R_p + m\Delta R_p \qquad (1.2)$$

where m is set to achieve a certain level of dopant penetration through the resist. For example, if the dopant concentration at the bottom of the resist cannot be more than 10^{-4} times the peak concentration (a typical requirement), then m should be set to 4.3. While the resolution requirements for the implant layers tend not to be challenging, often the thickness required for adequate stopping power does pose real challenges to the lithographer. Carbonization of the resist during high energy and high dose implantation (as well as during plasma etching) can also result in a film that is very difficult to strip away at the end.

1.1.4   Process Integration

The combination of patterning and selective doping allows the buildup of the structures required to make transistors. Figure 1.5 shows a diagrammatical example of a pair of CMOS (complementary metal oxide semiconductor) transistors. Subsequent metal layers (up to 10 metal levels are not uncommon) can connect the many transistors into a full

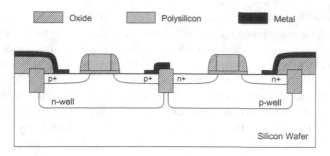

Figure 1.5   *Cross section of a pair of CMOS transistors showing most of the layers through metal 1*

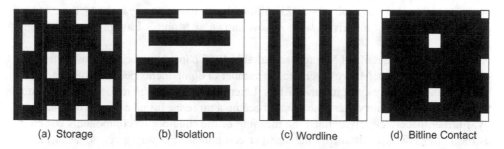

| (a) Storage | (b) Isolation | (c) Wordline | (d) Bitline Contact |

Figure 1.6   *Critical mask level patterns for a 1-Gb DRAM chip[2]. Each pattern repeats in both x and y many times to create the DRAM array*

circuit and the final metal layer will provide connections to the external pins of the device package.

Many lithographic levels are required to fabricate an IC, but about 1/3 of these levels are considered 'critical', meaning that those levels have challenging lithographic requirements. Which levels are critical depends on the process technology (CMOS logic, DRAM, BiCMOS, etc.). The most common critical levels of a CMOS process are active area, shallow trench isolation (STI), polysilicon gate, contact (between metal 1 and poly) and via (between metal layers), and metal 1 (the first or bottom most metal layer). For a large logic chip with 10 layers of metal, the first three will be '1×', meaning the dimensions are at or nearly at the minimum metal 1 dimensions. The next three metal layers will be '2×', with dimensions about twice as big as the 1× metal levels. The next few metal levels will be 4×, with the last few levels as large as 10×. For a DRAM device, some of the critical levels are known as storage, isolation, wordline and bitline contact (see Figure 1.6 for example design patterns for these four levels).

1.2   Moore's Law and the Semiconductor Industry

The impact of semiconductor ICs on modern life is hard to overstate. From computers to communication, entertainment to education, the growth of electronics technology, fueled by advances in semiconductor chips, has been phenomenal. The impact has been so profound that it is now often taken for granted: consumers have come to expect increasingly sophisticated electronics products at ever lower prices, and semiconductor companies expect growth and profits to improve continually. The role of optical lithography in these trends has been, and will continue to be, vital.

The remarkable evolution of semiconductor technology from crude single transistors to billion-transistor microprocessors and memory chips is a fascinating story. One of the first 'reviews' of progress in the semiconductor industry was written by Gordon Moore, a founder of Fairchild Semiconductor and later Intel, for the 35th anniversary issue of *Electronics* magazine in 1965.[3] After only 6 years since the introduction of the first commercial planar transistor in 1959, Moore observed an astounding trend – the number of electrical components per IC chip was doubling every year, reaching about 60 transistors in 1965. Extrapolating this trend for a decade, Moore predicted that chips with 64 000 components would be available by 1975! Although extrapolating any trend by three orders of magnitude can be quite risky, what is now known as Moore's Law proved amazingly accurate.

Some important details of Moore's remarkable 1965 paper have become lost in the lore of Moore's Law. First, Moore described the number of components per IC, which included resistors and capacitors, not just transistors. Later, as the digital age reduced the predominance of analog circuitry, transistor count became a more useful measure of IC complexity. Further, Moore clearly defined the meaning of the 'number of components per chip' as the number which minimized the cost per component. For any given level of manufacturing technology, one can always add more components – the problem being a reduction in yield and thus an increase in the cost per component. As any modern IC manufacturer knows, cramming more components onto ICs only makes sense if the resulting manufacturing yield allows costs that result in more commercially desirable chips. This 'minimum cost per component' concept is in fact the ultimate driving force behind the economics of Moore's Law.

Consider a very simple cost model for chip manufacturing as a function of lithographic feature size. For a given process, the cost of making a chip is proportional to the area of silicon consumed divided by the final yield of the chips. Will shrinking the feature sizes on the chip result in an increase or decrease in cost? The area of silicon consumed will be roughly proportional to the feature size squared. But yield will also be a function of feature size. Assuming that the only yield limiter will be the parametric effects of reduced feature size, a simple yield model might look something like

$$Yield = 1 - e^{-(w-w_0)^2/2\sigma^2}, \ Cost \propto \frac{w^2}{Yield} \qquad (1.3)$$

where w is the feature size (which must be greater than w_0 for this model), w_0 is the ultimate resolution (feature size at which the yield goes to zero), and σ is the sensitivity of yield to feature size. Figure 1.7 shows this yield model and the resulting cost function for arbitrary but reasonable parameters.

But minimizing cost is not really the goal of a semiconductor fab – it is maximizing profit. Feature size affects total profit in two ways other than chip cost. The number of chips per wafer is inversely proportional to the area of each chip, thus increasing the

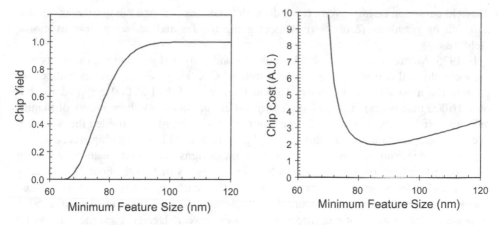

Figure 1.7 *A very simple yield and cost model shows the feature size that minimizes chip cost ($w_0 = 65\,nm$, $\sigma = 10\,nm$). Lowest chip cost occurs, in this case, when $w = 87\,nm$, corresponding to a chip yield of about 90 %*

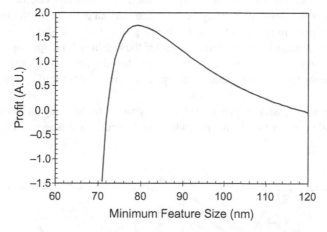

Figure 1.8 *Example fab profit curve using the yield and cost models of Figure 1.7 and assuming the value of the chip is inversely proportional to the minimum feature size. For this example, maximum profit occurs when $w = 80\,nm$, even though the yield is only 65 %*

number of chips that can be sold (that is, the total possible throughput of chips for the fab). Also, the value of each chip is often a function of the feature size. Smaller transistors generally run faster and fit in smaller packages – both desirable features for many applications. Assuming the price that a chip can be sold for is inversely proportional to the minimum feature size on the chip, an example profit model for a fab is shown in Figure 1.8. The most important characteristic is the steep falloff in profit that occurs when trying to use a feature size below the optimum. For the example here, the profit goes to zero if one tries to shrink the feature size by 10 % below its optimum value (unless, of course, the yield curve can be improved). It is a difficult balancing act for a fab to try to maximize its profit by shrinking feature size without going too far and suffering from excessive yield loss.

In 1975, Moore revisited his 1965 prediction and provided some critical insights into the technological drivers of the observed trends.[4] Checking the progress of component growth, the most advanced memory chip at Intel in 1975 had 32 000 components (but only 16 000 transistors). Thus, Moore's original extrapolation by three orders of magnitude was off by only a factor of 2. Even more importantly, Moore divided the advances in circuit complexity among its three principle components: increasing chip area, decreasing feature size, and improved device and circuit designs. Minimum feature sizes were decreasing by about 10 % per year (resulting in transistors that were about 21 % smaller in area, and an increase in transistors per area of 25 % each year). Chip area was increasing by about 20 % each year (Figure 1.9). These two factors alone resulted in a 50 % increase in the number of transistors per chip each year. Design cleverness made up the rest of the improvement (33 %). In other words, the 2× improvement = (1.25)(1.20)(1.33).

Again, there are important details in Moore's second observation that are often lost in the retelling of Moore's Law. How is 'minimum feature size' defined? Moore explained that both the linewidths and the spacewidths used to make the circuits are critical to density. Thus, his density-representing feature size was an average of the minimum linewidth and the minimum spacewidth used in making the circuit. Today, we use the equivalent metric, the minimum pitch divided by 2 (called the minimum half-pitch). Unfortunately, many modern forecasters express the feature size trend using features that do not well represent the density of the circuit. Usually, minimum half-pitch serves this purpose best.

By breaking the density improvement into its three technology drivers, Moore was able to extrapolate each trend into the future and predict a change in the slope of his

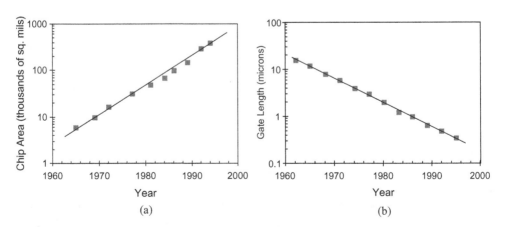

Figure 1.9   *Moore's Law showing (a) an exponential increase (about 15 % per year) in the area of a chip, and (b) an exponential decrease (about 11 % per year) in the minimum feature size on a chip (shown here for DRAM initial introduction)*

observation. Moore saw the progress in lithography allowing continued feature size shrinks to 'one micron or less'. Continued reductions in defect density and increases in wafer size would allow the die area trend to continue. But in looking at the 'device and circuit cleverness' component of density improvement, Moore saw a limit. Although improvements in device isolation and the development of the MOS transistor had contributed to greater packing density, Moore saw the latest circuits as near their design limits. Predicting an end to the design cleverness trend in 4 or 5 years, Moore predicted a change in the slope of his trend from doubling every year, to doubling every 2 years.

Moore's prediction of a slowdown was both too pessimistic and too generous. The slowdown from doubling each year had already begun by 1975 with Intel's 16-Kb memory chip. The 64-Kb DRAM chip, which should have been introduced in 1976 according to the original trend, was not available commercially until 1979. However, Moore's prediction of a slowdown to doubling components every 2 years instead of every year was too pessimistic. The 50% improvement in circuit density each year due to feature size and die size was really closer to 60% (according to Moore's retelling of the story[5]), resulting in a doubling of transistor counts per chip every 18 months or so (Figure 1.10). Offsetting the curve to switch from component counts to transistor counts and beginning with the 64-Kb DRAM in 1979, the industry followed the 'new' Moore's Law trend throughout the 1980s and early 1990s.

After 40 years, extrapolation of Moore's Law now seems less risky. In fact, predictions of future industry performance have reached such a level of acceptance that they have been codified in an industry-sanctioned 'roadmap' of the future. The *National Technology Roadmap for Semiconductors* (NTRS)[6] was first developed by the Semiconductor Industry Association in 1994 to serve as an industry-standard Moore's Law. It extrapolated then current trends to the year 2010, where 70-nm minimum feature sizes were predicted to enable 64-Gb DRAM chip production. This official industry roadmap has been updated many times, going international in 1999 to become the ITRS, the *International Technology Roadmap for Semiconductors*.

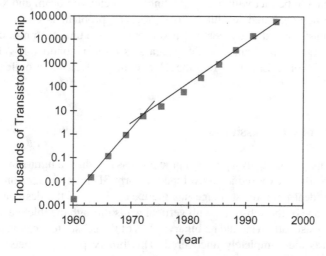

Figure 1.10 *Moore's Law showing an exponential increase in the number of transistors on a semiconductor chip over time (shown here for DRAM initial introduction)*

Ultimately, the drivers for technology development fall into two categories: push and pull. *Push drivers* are technology enablers, those things that make it possible to achieve the technical improvements. Moore described the three push drivers as increasing chip area, decreasing feature size and design cleverness. *Pull drivers* are the economic drivers, those things that make it worthwhile to pursue the technical innovations. Although the two drivers are not independent, it is the economic drivers that always dominate. As Bob Noyce, cofounder of Intel, wrote in 1977 '. . . further miniaturization is less likely to be limited by the laws of physics than by the laws of economics.'[7]

The economic drivers for Moore's Law are extraordinarily compelling. As the dimensions of a transistor shrink, the transistor becomes smaller, faster, consumes less power and in many cases is more reliable. All of these factors make the transistor more desirable for virtually every possible application. But there is more. Historically, the semiconductor industry has been able to manufacture silicon devices at an essentially constant cost per area of processed silicon. Thus, as the devices shrink, they enjoy a shrinking cost per transistor. As many have observed, it is a life without tradeoffs (unless, of course, you consider the stress on the poor engineers trying to make all of this happen year after year). Each step along the roadmap of Moore's Law virtually guarantees economic success. Advances in lithography, and in particular optical lithography, have been critical enablers to the continued rule of Moore's Law.

The death of optical lithography has been predicted so often by industry pundits, incorrectly so far, that it has become a running joke among lithographers. In 1979, conventional wisdom limited optical lithography to 1-μm resolution and a 1983 demise (to be supplanted by electron-beam imaging systems).[8] By 1985, the estimate was revised to 0.5-μm minimum resolution and a 1993 replacement by x-ray lithography.[9] The reality was quite a bit different. In 2006, optical lithography was used for 65-nm production (about 90-nm half-pitch). It seems likely that optical lithography will be able to manufacture devices with 45-nm half-pitch, and experts hedge their bets on future generations. Interestingly, the resolution requirements of current and future lithography processes are not so aggressive that they cannot be met with today's technology – electron beam and x-ray lithography have both demonstrated resolution to spare. The problem is one of cost. Optical lithography is unsurpassed in the cost per pixel (one square unit of minimum resolution) when printing micron-sized and submicron features on semiconductor wafers. To keep the industry on Moore's Law well into the 21st century, advances in optical lithography must continue.

1.3   Lithography Processing

Optical lithography is basically a photographic process by which a light-sensitive polymer, called a photoresist, is exposed and developed to form 3D relief images on the substrate. In general, the ideal photoresist image has the exact shape of the designed or intended pattern in the plane of the substrate, with vertical walls through the thickness of the resist. Thus, the final resist pattern should be binary: parts of the substrate are covered with resist while other parts are completely uncovered. This binary pattern is needed for pattern transfer since the parts of the substrate covered with resist will be protected from etching, ion implantation, or other pattern transfer mechanism.

The general sequence of processing steps for a typical optical lithography process is: substrate preparation, photoresist spin coat, post-apply bake, exposure, post-exposure bake, development and postbake. Metrology and inspection followed by resist strip are the final operations in the lithographic process, after the resist pattern has been transferred into the underlying layer. This sequence is shown diagrammatically in Figure 1.11, and most of these steps are generally performed on several tools linked together into a contiguous unit called a *lithographic cluster* or *cell* (Figure 1.12). A brief discussion of each

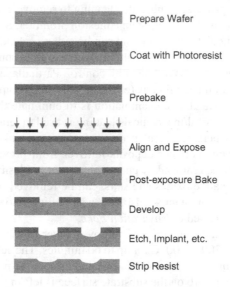

Prepare Wafer

Coat with Photoresist

Prebake

Align and Expose

Post-exposure Bake

Develop

Etch, Implant, etc.

Strip Resist

Figure 1.11 *Example of a typical sequence of lithographic processing steps, illustrated for a positive resist*

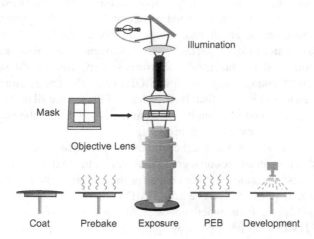

Illumination

Mask

Objective Lens

Coat   Prebake   Exposure   PEB   Development

Figure 1.12 *Iconic representation of the integration of the various lithographic process steps into a photolithography cell. Many steps, such as chill plates after the bake steps, have been omitted*

step is given below, pointing out some of the practical issues involved in photoresist processing. More fundamental and theoretical discussions on these topics will be provided in subsequent chapters.

1.3.1   Substrate Preparation

Substrate preparation is intended to improve the adhesion of the photoresist material to the substrate and provide for a contaminant-free resist film. This is accomplished by one or more of the following processes: substrate cleaning to remove contamination, dehydration bake to remove water and addition of an adhesion promoter. Substrate contamination can take the form of particulates or a film and can be either organic or inorganic. Particulates result in defects in the final resist pattern, whereas film contamination can cause poor adhesion and subsequent loss of linewidth control. Particulates generally come from airborne particles or contaminated liquids (e.g. dirty adhesion promoter). The most effective way of controlling particulate contamination is to eliminate their source. Since this is not always practical, chemical/mechanical cleaning is used to remove particles. Organic films, such as oils or polymers, can come from vacuum pumps and other machinery, body oils and sweat, and various polymer deposits leftover from previous processing steps. These films can generally be removed by chemical, ozone, or plasma stripping. Similarly, inorganic films, such as native oxides and salts, can be removed by chemical or plasma stripping. One type of contaminant – adsorbed water – is removed most readily by a high-temperature process called a *dehydration bake*.

A dehydration bake, as the name implies, removes water from the substrate surface by baking at temperatures of 200 to 400 °C for up to 60 minutes. The substrate is then allowed to cool (preferably in a dry environment) and coated as soon as possible. It is important to note that water will re-adsorb on the substrate surface if left in a humid (nondry) environment. A dehydration bake is also effective in volatilizing organic contaminants, further cleaning the substrate. Often, the normal sequence of processing steps involves some type of high-temperature process immediately before coating with photoresist, for example, thermal oxidation. If the substrate is coated immediately after the high-temperature step, the dehydration bake can be eliminated. A typical dehydration bake, however, does not completely remove water from the surface of silica substrates (including silicon, polysilicon, silicon dioxide and silicon nitride). Surface silicon atoms bond strongly with a monolayer of water forming silanol groups (SiOH) and bake temperatures in excess of 600 °C are required to remove this final layer of water. Further, the silanol quickly reforms when the substrate is cooled in a nondry environment. As a result, the preferred method of removing this silanol is by chemical means.

Adhesion promoters are used to react chemically with surface silanol and replace the —OH group with an organic functional group that, unlike the hydroxyl group, offers good adhesion to photoresist. Silanes are often used for this purpose, the most common being hexamethyl disilizane (HMDS).[10] (As a note, HMDS adhesion promotion was first developed for fiberglass applications, where adhesion of the resin matrix to the glass fibers is important.) The HMDS can be applied by spinning a diluted solution (10–20 % HMDS in cellosolve acetate, xylene, or a fluorocarbon) directly on to the wafer and allowing the HMDS to spin dry (HMDS is quite volatile at room temperature). If the HMDS is not allowed to dry properly, dramatic loss of adhesion will result. Although direct spinning is easy, it is only effective at displacing a small percentage of the silonal groups. By far the preferred method of applying the adhesion promoter is by subjecting the substrate to HMDS vapor at elevated temperatures and reduced pressure. This allows good coating of the substrate without excess HMDS deposition, and the higher temperatures cause more complete reaction with the silanol groups. Once properly treated with HMDS, the substrate can be left for up to several days without significant re-adsorption of water. Performing the dehydration bake and vapor prime in the same oven gives optimum performance. Such vapor prime systems are often integrated into the wafer processing tracks used for the subsequent steps of resist coating and baking.

A simple method for testing for adsorbed water on the wafer surface, and thus the likelihood of resist adhesion failure, is to measure the contact angle of a drop of water. If a drop of water wets the surface (has a low contact angle), the surface is hydrophilic and the resist will be prone to adhesion failure during development. For a very hydrophobic surface, water will have a large contact angle (picture water beading up on a waxed automobile). Contact angles can be easily measured on a primed wafer using a goniometer, and should be in the 50–70° range for good resist adhesion[11] (see Figure 1.13).

1.3.2   Photoresist Coating

A thin, uniform coating of photoresist at a specific, well-controlled thickness is accomplished by the seemingly simple process of *spin coating*. The photoresist, rendered into a liquid form by dissolving the solid components in a solvent, is poured onto the wafer, which is then spun on a turntable at a high speed producing the desired film. (For the case of DNQ/novolac resists, the resist solutions are often supersaturated, making them prone to precipitation.) Stringent requirements for thickness control and uniformity and low defect density call for particular attention to be paid to this process, where a large number of parameters can have significant impact on photoresist thickness uniformity and control. There is the choice between static dispense (wafer stationary while resist is dispensed) or dynamic dispense (wafer spinning while resist is dispensed), spin speeds and times, and accelerations to each of the spin speeds. Also, the volume of the resist dispensed and properties of the resist (such as viscosity, percent solids and solvent composition) and the substrate (substrate material and topography) play an important role in the resist thickness uniformity. Further, practical aspects of the spin operation, such as exhaust, ambient temperature and humidity control, resist temperature, spin cup geometry, point-of-use filtration and spinner cleanliness often have significant effects on the resist film. Figure 1.14a shows a generic photoresist spin coat cycle. At the end of this cycle, a thick, solvent-rich film of photoresist covers the wafer, ready for post-apply bake. By the end

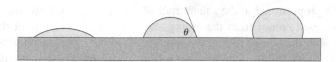

Figure 1.13   *A water droplet on the surface of the wafer indicates the hydrophobicity of the wafer: the left-most drop indicates a hydrophilic surface, the right-most drop shows an extremely hydrophobic surface. The middle case, with a contact angle of 70°, is typically about optimum for resist adhesion*

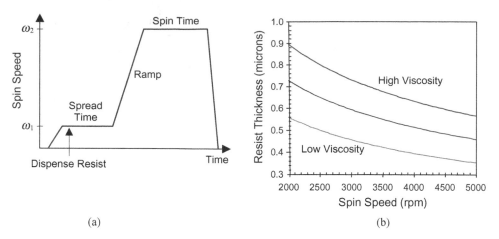

(a)   (b)

Figure 1.14 *Photoresist spin coat cycle: (a) pictorial representation (if $\omega_1 > 0$, the dispense is said to be dynamic), and (b) photoresist spins speed curves for different resist viscosities showing how resist thickness after post-apply bake varies as (spin speed)$^{-1/2}$*

of the post-apply bake, the film can have a thickness controlled to within 1–2 nm across the wafer and wafer-to-wafer.

The rheology of resist spin coating is complex and yet results in some simple, and seemingly unexpected, properties of the final resist film. Spinning results in centrifugal forces pushing the liquid photoresist toward the edge of the wafer where excess resist is flung off. The frictional force of viscosity opposes this centrifugal force. As the film thins, the centrifugal force (which is proportional to the mass of the resist on the wafer) decreases. Also, evaporation of solvent leads to dramatic increases in the viscosity of the resist as the film dries (as the resist transitions from a liquid to a solid, the viscosity can increase by 7–10 orders of magnitude). Eventually the increasing viscous force exceeds the decreasing centrifugal force and the resist stops flowing. This generally occurs within the first second of the spin cycle (often before the wafer has fully ramped to its final spin speed). The remaining portion of the spin cycle causes solvent evaporation without mass flow of the resist solids. The separation of the spin cycle into a very quick radial mass flow (coating stage) followed by a long evaporation of solvent (drying stage) provides for some of the basic and important properties of spin coating. Since the overall spin time is much longer than the coating stage time, the final thickness of resist is virtually independent of the initial volume of resist dispensed onto the wafer above a certain threshold. For laminar flow of air above the spinning wafer, the amount of drying (mass transfer of solvent) will be proportional to the square root of the spin speed. And since most of the thinning of the resist comes from the drying stage, the final thickness of the resist will vary inversely with the square root of the spin speed. Finally and most importantly, both the coating stage and drying stage produce a film whose thickness is not dependent on the radial position on the wafer.

Although theory exists to describe the spin coat process rheologically,[12,13] in practical terms the variation of photoresist thickness and uniformity with the process parameters are determined experimentally. The photoresist spin speed curve (Figure 1.14b) is an essential tool for setting the spin speed to obtain the desired resist thickness. As mentioned above, the final resist thickness varies as one over the square root of the spin speed (ω) and is roughly proportional to the liquid photoresist viscosity (v) to the 0.4–0.6 power:

$$thickness \propto \frac{v^{0.4}}{\omega^{0.5}} \tag{1.4}$$

For a given desired resist thickness, the appropriate spin speed is chosen according to the spin curve. However, there is a limited range of acceptable spin speeds. Speeds less than 1000 rpm are harder to control and do not produce uniform films. If the spin speed is too high, turbulent airflow at the edge of the wafer will limit uniformity. The onset of turbulence depends on the Reynolds number Re, which for a rotating disk is

$$Re = \frac{\omega r^2}{v_{air}} \tag{1.5}$$

where r is the wafer radius, v_{air} is the kinematic viscosity of air (about $1.56 \times 10^{-5}\,m^2/s$ at standard conditions). The onset of turbulence begins for Reynolds numbers of about 300 000.[14,15] Instabilities in the flow, in the form of spiral vortices, can occur at Reynolds numbers as low as 100 000 without careful design of the spin coat chamber. [Note that sometimes the square root of the expression (1.5) is used as the Reynolds number, so that the threshold for turbulence is 550.] For a 300-mm wafer, this means the maximum spin speed is on the order of 2000 rpm. If the desired resist thickness cannot be obtained over the acceptable range of spin speeds, a different viscosity resist formulation can be chosen. Typical resist viscosities range from 5 to 35 cSt (1 Stoke = 1 cm²/s). As a point of reference, water has a viscosity of about 1 cSt at room temperature.

Unfortunately, the forces that give rise to uniform resist coatings also cause an unwanted side effect: edge beads. The fluid flow discussion above described a balance of the centrifugal and viscous forces acting on the resist over the full surface of the wafer. However, at the edge of the wafer, a third force becomes significant. Surface tension at the resist–air interface results in a force pointing inward perpendicular to the resist surface. Over most of the wafer, this force is pointing downward and thus does not impact the force balance of spinning plus friction. However, at the edge of the wafer, this force must point inward toward the center of the wafer (Figure 1.15). The extra force adding to the viscous force will stop the flow of resist sooner at the edge than over the central portion of the wafer,

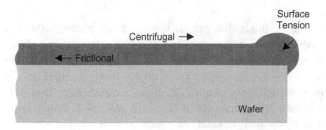

Figure 1.15 *A balance of spin-coat forces at the wafer edge leads to the formation of a resist edge bead*

resulting in an accumulation of resist at the edge. This accumulation is called an *edge bead*, which usually exists within the outer 1–2 mm of the wafer and can be 10–30 times thicker than the rest of the resist film.

The existence of an edge bead is detrimental to the cleanliness of subsequent wafer processing. Tools which grab the wafer by the edge will flake off the dried edge bead, resulting in very significant particulate contamination. Consequently, removal of the edge bead is required. Within the spin-coat chamber and immediately after the resist spin coating is complete, a stream of solvent (called an EBR, edge bead remover) is directed at the edge of the wafer while it slowly spins. Resist is dissolved off the edge and over the outer 1.5–2 mm of the wafer surface.

1.3.3 Post-Apply Bake

After coating, the resulting resist film will contain between 20 and 40% solvent by weight. The post-apply bake (PAB) process, also called a softbake or a prebake, involves drying the photoresist after spin coat by removing most of this excess solvent. The main reason for reducing the solvent content is to stabilize the resist film. At room temperature, an unbaked photoresist film will lose solvent by evaporation, thus changing the properties of the film with time. By baking the resist, the majority of the solvent is removed and the film becomes stable at room temperature. There are four major effects of removing solvent from a photoresist film: (1) film thickness is reduced; (2) post-exposure bake and development properties are changed; (3) adhesion is improved; and (4) the film becomes less tacky and thus less susceptible to particulate contamination. Typical post-apply bake processes leave between 3 and 10% residual solvent in the resist film (depending on resist and solvent type, as well as bake conditions), sufficiently small to keep the film stable during subsequent lithographic processing.

Unfortunately, there can be other consequences of baking photoresists. At temperatures greater than about 70 °C, the photosensitive component of a typical resist mixture, called the photoactive compound (PAC), may begin to decompose. Also, the resin, another component of the resist, can cross-link and/or oxidize at elevated temperatures. Both of these effects are undesirable. Thus, one must search for the optimum post-apply bake conditions that will maximize the benefits of solvent evaporation and minimize the detriments of resist decomposition. For chemically amplified resists, residual solvent can significantly influence diffusion and reaction properties during the post-exposure bake, necessitating careful control over the post-apply bake process. Fortunately, these modern resists do not suffer from significant decomposition of the photosensitive components during post-apply bake.

There are several methods that can be used to bake photoresists. The most obvious method is an oven bake. Convection oven baking of conventional photoresists at 90 °C for 30 minutes was typical during the 1970s and early 1980s, but currently the most popular bake method is the hot plate. The wafer is brought either into intimate vacuum contact with or close proximity to a hot, high-mass metal plate. Due to the high thermal conductivity of silicon, the photoresist is heated to near the hot plate temperature quickly (in about 5 seconds for hard contact, or about 20 seconds for proximity baking). The greatest advantage of this method is an order of magnitude decrease in the required bake time over convection ovens, to about 1 minute, and the improved uniformity of the bake.

In general, proximity baking is preferred to reduce the possibility of particle generation caused by contact with the backside of the wafer.

When the wafer is removed from the hot plate, baking continues as long as the wafer is hot. The total bake process cannot be well controlled unless the cooling of the wafer is also well controlled. In other words, the bake process should be thought of in terms of an integrated thermal history, from the start of the bake till the wafer has sufficiently cooled. As a result, hot plate baking is always followed immediately by a chill plate operation, where the wafer is brought in contact or close proximity to a cool plate (kept at a temperature slightly below room temperature). After cooling, the wafer is ready for its lithographic exposure.

1.3.4 Alignment and Exposure

The basic principle behind the operation of a photoresist is the change in solubility of the resist in a developer upon exposure to light. In the case of the standard diazonaphthoquinone positive photoresist, the PAC, which is not soluble in the aqueous base developer, is converted to a carboxylic acid on exposure to UV light in the range of 350–450 nm. The carboxylic acid product is very soluble in the basic developer. Thus, a spatial variation in light energy incident on the photoresist will cause a spatial variation in solubility of the resist in developer.

Contact and proximity lithography are the simplest methods of exposing a photoresist through a master pattern called a photomask (Figure 1.16). Contact lithography offers reasonably high resolution (down to about the wavelength of the radiation), but practical problems such as mask damage (or equivalently, the formation of mask defects) and resulting low yield make this process unusable in most production environments. Proximity printing reduces mask damage by keeping the mask a set distance above the wafer (e.g. 20 μm). Unfortunately, the resolution limit is increased significantly. For a mask–wafer gap of g and an exposure wavelength of λ,

$$Resolution \sim \sqrt{g\lambda} \qquad (1.6)$$

Because of the high defect densities of contact printing and the poor resolution of proximity printing, by far the most common method of exposure is *projection printing*.

Projection lithography derives its name from the fact that an image of the mask is projected onto the wafer. Projection lithography became a viable alternative to contact/proximity printing in the mid-1970s when the advent of computer-aided lens design and

Contact   Proximity   Projection

Figure 1.16 *Lithographic printing in semiconductor manufacturing has evolved from contact printing (in the early 1960s) to projection printing (from the mid-1970s to today)*

improved optical materials and manufacturing methods allowed the production of lens elements of sufficient quality to meet the requirements of the semiconductor industry. In fact, these lenses have become so perfect that lens defects, called aberrations, play only a small role in determining the quality of the image. Such an optical system is said to be *diffraction-limited*, since it is diffraction effects and not lens aberrations which, for the most part, determine the shape of the image.

There are two major classes of projection lithography tools – *scanning* and *step-and-repeat* systems. Scanning projection printing, as pioneered by the Perkin-Elmer company,[16] employs reflective optics (i.e. mirrors rather than lenses) to project a slit of light from the mask onto the wafer as the mask and wafer are moved simultaneously past the slit. Exposure dose is determined by the intensity of the light, the slit width and the speed at which the wafer is scanned. These early scanning systems, which use polychromatic light from a mercury arc lamp, are 1 : 1, i.e. the mask and image sizes are equal. Step-and-repeat cameras (called steppers for short), first developed by GCA Corp., expose the wafer one rectangular section (called the image field) at a time and can be 1 : 1 or reduction. These systems employ refractive optics (i.e. lenses) and are usually quasi-monochromatic. Both types of systems (Figure 1.17) are capable of high-resolution imaging, although reduction imaging is best for the highest resolutions in order to simplify the manufacture of the photomasks.

Scanners replaced proximity printing by the mid-70s for device geometries below 4 to 5 μm. By the early 1980s, steppers began to dominate as device designs pushed to 2 μm and below. Steppers have continued to dominate lithographic patterning throughout the 1990s as minimum feature sizes reached the 250-nm levels. However, by the early 1990s a hybrid *step-and-scan* approach was introduced by SVG Lithography, the successor to Perkin-Elmer. The step-and-scan approach uses a fraction of a normal stepper field (for example, 26 × 8 mm), then scans this field in one direction to expose the entire 4× reduction mask (Figure 1.18). The wafer is then stepped to a new location and the scan is repeated. The smaller imaging field simplifies the design and manufacture of the lens, but at the expense of a more complicated reticle and wafer stage. Step-and-scan technology is the technology of choice today for below 250-nm manufacturing.

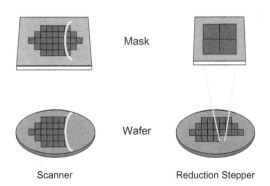

Figure 1.17 *Scanners and steppers use different techniques for exposing a large wafer with a small image field*

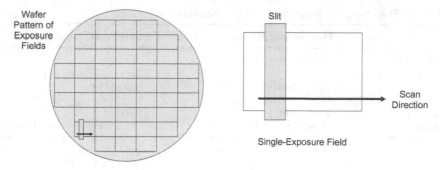

Figure 1.18 *In step-and-scan imaging, the field is exposed by scanning a slit that is about 25 × 8 mm across the exposure field*

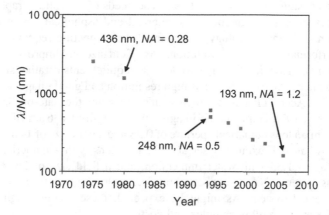

Figure 1.19 *The progression of λ/NA of lithographic tools over time (year of first commercial tool shipment)*

Resolution, the smallest feature that can be printed with adequate control, has two basic limits: the smallest image that can be projected onto the wafer, and the resolving capability of the photoresist to make use of that image. From the projection imaging side, resolution is determined by the wavelength of the imaging light (λ) and the numerical aperture (NA) of the projection lens according to the Rayleigh resolution criterion:

$$Resolution \propto \frac{\lambda}{NA} \qquad (1.7)$$

Lithography systems have progressed from blue wavelengths (436 nm) to UV (365 nm) to deep-UV (248 nm) to today's mainstream high-resolution wavelength of 193 nm (see Figure 1.19 and Table 1.2). In the meantime, projection tool numerical apertures have risen from 0.16 for the first scanners to amazingly high 0.93 NA systems producing features well under 100 nm in size. In addition, immersion lithography, where the bottom of

Table 1.2   *The change in projection tool specifications over time*

	First Stepper (1978)	Immersion Scanner (2006)
Wavelength	436 nm	193 nm
Numerical Aperture	0.28	1.2
Field Size	10 × 10 mm	26 × 33 mm
Reduction Ratio	10	4
Wafer Size	4″ (100 mm)	300 mm
Throughput	20 wafers per hour (0.44 cm²/s)	120 wafers per hour (24 cm²/s)

the lens is immersed in a high refractive index fluid such as water, enables numerical apertures greater than one, with the first such 'hyper NA' tools available in 2006.

The main imaging lens of a stepper or scanner is the most demanding application of commercial lens design and fabrication today. The needs of microlithographic lenses are driving advances in lens design software, spherical and aspherical lens manufacturing, glass production and lens metrology. There are three competing requirements of lithographic lens performance – higher resolution, large field size and improved image quality (lower aberrations). Providing for any two of these requirements is rather straightforward (for example, a microscope objective has high resolution and good image quality but over a very small field). Accomplishing all three means advancing the state-of-the-art in optics. The first stepper in 1978 employed an imaging wavelength of 436 nm (the g-line of the mercury spectrum), a lens numerical aperture of 0.28 and a field size of 14 mm in diameter. Today's tools use an ArF excimer laser at 193 nm, a lens with a numerical aperture of 0.93 dry, and up to 1.35 with water immersion, and a field size of 26 × 33 mm. The 'hyper-NA' lens systems (NA > 1) are catadioptric, employing both mirrors and refractive lenses in the optical system. As might be expected, these modern high-performance imaging systems are incredibly complex and costly.

Before the exposure of the photoresist with an image of the mask can begin, this image must be aligned with the previously defined patterns on the wafer. This alignment process, and the resulting overlay of the two or more lithographic patterns, is critical since tighter overlay control means circuit features can be packed closer together. Closer packing of devices through better overlay is nearly as critical as smaller devices through higher resolution in the drive toward more functionality per chip. Along with alignment, wafer focus is measured at several points so that each exposure field is leveled and brought into proper focus.

Another important aspect of photoresist exposure is the *standing wave* effect. Monochromatic light, when projected onto a wafer, strikes the photoresist surface over a range of angles, approximating plane waves. This light travels down through the photoresist and, if the substrate is reflective, is reflected back up through the resist. The incoming and reflected light waves interfere to form a standing wave pattern of high and low light intensity at different depths in the photoresist. This pattern is replicated in the photoresist, causing ridges in the sidewalls of the resist feature as seen in Figure 1.20. As pattern dimensions become smaller, these ridges can significantly affect the quality of the feature. The interference that causes standing waves also results in a phenomenon called *swing curves*, the sinusoidal variation in linewidth with changing resist thickness. These detri-

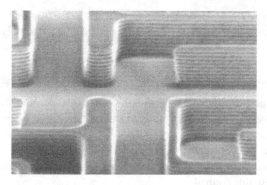

Figure 1.20 *Photoresist pattern on a silicon substrate (i-line exposure pictured here) showing prominent standing waves*

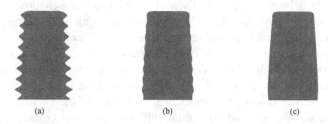

Figure 1.21 *Diffusion during a post-exposure bake is often used to reduce standing waves. Photoresist profile simulations as a function of the PEB diffusion length: (a) 20 nm, (b) 40 nm and (c) 60 nm*

mental effects are best cured by coating the substrate with a thin absorbing layer called a *bottom antireflective coating* (BARC) that can reduce the reflectivity of the substrate as seen by the photoresist to much less than 1 %.

1.3.5 Post-exposure bake

One method of reducing the standing wave effect is called the post-exposure bake (PEB).[17] The high temperatures used (100–130 °C) cause diffusion of the photoactive compound, thus smoothing out the standing wave ridges (Figure 1.21). It is important to note that the detrimental effects of high temperatures on photoresist, as discussed above concerning PAB, also apply to the PEB. Thus, it becomes very important to optimize the bake conditions. Also, the rate of diffusion of the exposure products is dependent on the PAB conditions – the presence of solvent enhances diffusion during a PEB. Thus, a low-temperature post-apply bake results in greater diffusion for a given PEB temperature.

For a conventional resist, the main importance of the PEB is diffusion to remove standing waves. For another class of photoresists, called chemically amplified resists, the PEB is an essential part of the chemical reactions that create a solubility differential between exposed and unexposed parts of the resist. For these resists, exposure generates a small amount of a strong acid that does not itself change the solubility of the resist. During the post-exposure bake, this photogenerated acid catalyzes a reaction that changes the solubility of the polymer resin in the resist. Control of the PEB is extremely critical for chemically amplified resists.

1.3.6   Development

Once exposed, the photoresist must be developed. Most commonly used photoresists employ aqueous bases as developers. In particular, tetramethyl ammonium hydroxide (TMAH) is used in concentrations of 0.2–0.26 N. Development is undoubtedly one of the most critical steps in the photoresist process. The characteristics of the resist–developer interactions determine to a large extent the shape of the photoresist profile and, more importantly, the linewidth control.

The method of applying developer to the photoresist is important in controlling the development uniformity and process latitude. In the past, batch development was the predominant development technique. A boat of some 10–20 wafers or more is developed simultaneously in a large beaker, usually with some form of agitation. With the push toward in-line processing in the late 1970s, however, other methods have become prevalent. During *spin development* wafers are spun, using equipment similar to that used for spin coating, and developer is poured onto the rotating wafer. The wafer is also rinsed and dried while still spinning. *Spray development* uses a process identical to spin development except the developer is sprayed, rather than poured, on the wafer by a nozzle that produces a fine mist of developer over the wafer (Figure 1.22). This technique reduces developer usage significantly and gives more uniform developer coverage.

Another in-line development strategy is called *puddle development*. Again using developers specifically formulated for this process, the developer is poured onto a slowly spinning wafer that is then stopped and allowed to sit motionless for the duration of the development time. The wafer is then spin rinsed and dried. Note that all three in-line processes can be performed in the same piece of equipment with only minor modifications, and combinations of spray and puddle techniques are frequently used. Puddle development has the advantage of minimizing developer usage but can suffer from developer depletion – clear regions (where most of the resist is being dissolved) result in excessive dissolved resist in the developer, which depletes the developer and slows down development in these clear regions relative to dark regions (where most of the resist

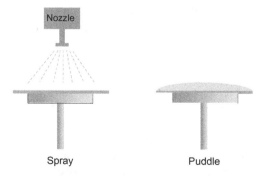

Figure 1.22   *Different developer application techniques are commonly used*

remains on the wafer). When this happens, the development cycle is often broken up into two shorter applications of the puddle in what is called a double-puddle process.

1.3.7 Postbake

The postbake (not to be confused with the post-exposure bake that comes before development) is used to harden the final resist image so that it will withstand the harsh environments of implantation or etching. The high temperatures used (120–150 °C) will cross-link the resin polymer in the photoresist, thus making the image more thermally stable. If the temperature used is too high, the resist will flow causing degradation of the image. The temperature at which flow begins is essentially equal to the glass transition temperature of the resist and is a measure of its thermal stability. In addition to cross-linking, the postbake can remove residual solvent, water, and gasses, and will usually improve adhesion of the resist to the substrate. Removal of these volatile components makes the resist more vacuum compatible, an important consideration for ion implantation.

Other methods are also used to harden a photoresist image. Exposure to high intensity deep-UV light cross-links the resin at the surface of the resist forming a tough skin around the pattern. Deep-UV hardened photoresist can withstand temperatures in excess of 200 °C without dimensional deformation. Plasma treatments and electron beam bombardment have also been shown to effectively harden photoresist. Commercial deep-UV hardening systems are available and widely used. Most of these hardening techniques are used simultaneously with high-temperature baking.

1.3.8 Measure and Inspect

Either before or after the postbake step, some sample of the resist patterns are inspected and measured for quality control purposes. Critical features and test patterns are measured to determine their dimensions (called a *critical dimension*, CD) and the overlay of the patterns with respect to previous lithographically defined layers. Wafers can also be inspected for the presence of random defects (such as particles) that may interfere with the subsequent pattern transfer step. This step of measurement and inspection is called ADI, *after develop inspect*, as opposed to measurements taken after pattern transfer, which are called *final inspect* (FI).

Inspection and measurement of the wafers before pattern transfer offer a unique opportunity: wafers (or entire lots) that do not meet CD or overlay specifications can be *reworked*. When a wafer is reworked, the patterned resist is stripped off and the wafers are sent back to the beginning of the lithography process. Wafers that fail to meet specifications at FI (after pattern transfer is complete) cannot be reworked and must be scrapped instead. Since reworking a wafer is far more cost-beneficial than scrapping a wafer, significant effort is put into verifying the quality of wafers at ADI and reworking any wafers that have potential lithography-limited yield problems.

1.3.9 Pattern Transfer

After the small patterns have been lithographically printed in photoresist, these patterns must be transferred into the substrate. As discussed in section 1.1.1, there are three basic pattern transfer approaches: subtractive transfer (etching), additive transfer (selective deposition) and impurity doping (ion implantation). Etching is the most common pattern transfer approach. A uniform layer of the material to be patterned is deposited on the substrate. Lithography is then performed such that the areas to be etched are left unprotected (uncovered) by the photoresist. Etching is performed either using wet chemicals such as acids, or more commonly in a dry plasma environment. The photoresist 'resists' the etchant and protects the material covered by the resist. When the etching is complete, the resist is stripped leaving the desired pattern etched into the deposited layer. Additive processes are used whenever workable etching processes are not available, for example, for copper interconnects (copper does not form volatile etching by-products, and so is very difficult to etch in a plasma). Here, the lithographic pattern is used to open areas where the new layer is to be grown (by electroplating, in the case of copper). Stripping of the resist then leaves the new material in a negative version of the patterned photoresist. Finally, doping involves the addition of controlled amounts of contaminants that change the conductive properties of a semiconductor. Ion implantation uses a beam of dopant ions accelerated at the photoresist-patterned substrate. The resist blocks the ions, but the areas uncovered by resists are embedded with ions, creating the selectively doped regions that make up the electrical heart of the transistors. For this application, the 'stopping power' of the resist (the minimum thickness of resist required to prevent ions from passing through) is the parameter of interest.

1.3.10   Strip

After the imaged wafer has been pattern transferred (e.g. etched, ion implanted, etc.), the remaining photoresist must be removed. There are two classes of resist stripping techniques: wet stripping using organic or inorganic solutions, and dry (plasma) stripping. A simple example of an organic stripper is acetone. Although commonly used in laboratory environments, acetone tends to leave residues on the wafer (scumming) and is thus unacceptable for semiconductor processing. Most commercial organic strippers are phenol-based and are somewhat better at avoiding scum formation. However, the most common wet strippers for positive photoresists are inorganic acid-based systems used at elevated temperatures.

Wet stripping has several inherent problems. Although the proper choice of strippers for various applications can usually eliminate gross scumming, it is almost impossible to remove the final monolayer of photoresist from the wafer by wet chemical means. It is often necessary to follow a wet strip by a plasma 'descum' step to completely clean the wafer of resist residues.[18] Also, photoresist which has undergone extensive hardening (e.g. deep-UV hardening) and been subjected to harsh processing conditions (e.g. high-energy ion implantation) can be almost impossible to strip chemically. For these reasons, plasma stripping has become the standard in semiconductor processing. An oxygen plasma is highly reactive toward organic polymers but leaves most inorganic materials (such as are mostly found under the photoresist) untouched.

Problems

1.1. When etching an oxide contact hole with a given process, the etch selectivity compared to photoresist is found to be 2.5. If the oxide thickness to be etched is 140 nm and a 50% overetch is used (that is, the etch time is set to be 50% longer than that required to just etch through the nominal oxide thickness), what is the minimum possible photoresist thickness (that is, how much resist will be etched away)? For what reasons would you want the resist to be thicker than this minimum?

1.2. For a certain process, a 300-keV phosphorous implant is masked well by a 1.0-μm-thick photoresist film.

    (a) For this implant process, how many multiples of the straggle does this resist thickness represent?

    (b) If the implant energy is increased to 450 keV, how much should the photoresist thickness be increased?

    (c) If the dopant is also changed to arsenic (with the energy at 450 keV), what resist thickness will be needed?

1.3. A photoresist gives a final resist thickness of 320 nm when spun at 2800 rpm.

    (a) What spin speed should be used if a 290-nm-thick coating of this same resist is desired?

    (b) If the maximum practical spin speed for 200-mm wafers is 4000 rpm, at what thickness would a lower viscosity formulation of the resist be required?

1.4. Resolution in optical lithography scales with wavelength and numerical aperture according to a modified Rayleigh criterion:

$$R = k_1 \frac{\lambda}{NA}$$

where k_1 can be thought of as a constant for a given lithographic approach and process. Assuming $k_1 = 0.35$, plot resolution versus numerical aperture over a range of NAs from 0.5 to 1.0 for the common lithographic wavelengths of 436, 365, 248, 193 and 157 nm. From this list, what options (NA and wavelength) are available for printing 90-nm features?

References

1   Glawischnig, H. and Parks, C.C., 1996, SIMS and modeling of ion implants into photoresist, *Proceedings of the 11th International Conference on Ion Implantation Technology*, 579–582.

2   Wong, A.K., Ferguson, R., Mansfield, S., Molless, A., Samuels, D., Schuster, R. and Thomas, A., 2000, Level-specific lithography optimization for 1-Gb DRAM, *IEEE Transactions on Semiconductor Manufacturing*, **13**, 76–87.

3   Moore, G.E., 1965, Cramming more components onto integrated circuits, *Electronics*, **38**, 114–117.

4   Moore, G.E., 1975, Progress in digital integrated electronics, *IEDM Technical Digest*, **21**, 11–13.

5   Moore, G.E., 1995, Lithography and the future of Moore's law, *Proceedings of SPIE: Optical/Laser Microlithography VIII*, **2440**, 2–17.

6   *The National Technology Roadmap for Semiconductors*, 1994, Semiconductor Industry Association (San Jose, CA).

7   Noyce, R., 1977, Microelectronics, *Scientific American*, **237**, 63–69.

8   Tobey, A.C., 1979, Wafer stepper steps up yield and resolution in IC lithography, *Electronics*, **52**, 109–112.

9   Lyman, J., 1985, Optical lithography refuses to die, *Electronics*, **58**, 36–38.

10   Collins, R.H. and Deverse, F.T., 1970, U.S. Patent No. 3,549,368.
11   Levinson, H.J., 2005, *Principles of Lithography*, second edition, SPIE Press (Bellingham, WA), p. 59.
12   Meyerhofer, D., 1978, Characteristics of resist films produced by spinning, *Journal of Applied Physics*, **49**, 3993–3997.
13   Bornside, D.E., Macosko, C.W. and Scriven, L.E., 1989, Spin coating: one-dimensional model, *Journal of Applied Physics*, **66**, 5185–5193.
14   Kobayashi, R., 1994, 1994 review: laminar-to-turbulent transition of three-dimensional boundary layers on rotating bodies, *Journal of Fluids Engineering*, **116**, 200–211.
15   Gregory, N., Stuart, J.T. and Walker, W.S., 1955, On the stability of three-dimensional boundary layers with application to the flow due to a rotating disk, *Philosophical Transactions of the Royal Society of London Series A*, **248**, 155–199.
16   Markle, D.A., 1974, A new projection printer, *Solid State Technology*, **17**, 50–53.
17   Walker, E.J., 1975, Reduction of photoresist standing-wave effects by post-exposure bake, *IEEE Transactions on Electron Devices*, **ED-22**, 464–466.
18   Kaplan, L.H. and Bergin, B.K., 1980, Residues from wet processing of positive resists, *Journal of The Electrochemical Society*, **127**, 386–395.

2

Aerial Image Formation – The Basics

Projection printing means projecting the image of a photomask (also called a reticle) onto a resist-coated wafer. Projection imaging tools are sophisticated reduction cameras with stages that allow, through a combination of stepping or stepping and scanning motions, exposure of many (reduced) copies of a mask pattern onto a large wafer. The image of the mask that is projected into the photoresist defines the information content used by the photoresist to form the final resist image. Understanding the limits and capabilities of projection imaging is the first step in understanding the limits and capabilities of lithography.

The imaging process is a well-studied optical phenomenon. We will begin at the most basic level with a mathematical description of light, followed by the theory of image formation. The impact of partial coherence will then be added. In the following chapter, nonideal and vector descriptions of image formation will be added to the basic description provided here.

2.1 Mathematical Description of Light

A mathematical description of light must, of course, begin with Maxwell's equations. However, these fundamental equations can be quite cumbersome and general solutions under diverse boundary conditions almost always entail detailed numerical calculations. However, under some circumstances a simplified form of Maxwell's equations, in particular what is called the wave equation, can be used and often allows reasonably compact solutions. In the case of imaging theory, decomposing an arbitrary wave into a summation of plane waves provides the simplest approach for calculating the images projected into a photoresist-coated wafer.

Fundamental Principles of Optical Lithography: The Science of Microfabrication, Chris Mack.
© 2007 John Wiley & Sons, Ltd.

2.1.1   Maxwell's Equations and the Wave Equation

Light is an electromagnetic wave with coupled electric and magnetic fields traveling through space and the materials that occupy that space. Maxwell's equations describe two fundamental, coupled fields, the electric field E (units of volts per meter) and the magnetic field H (units of amperes per meter), both vector quantities. Further, the effect of these fields on a material can give rise to four other quantities: three vectors, the electric displacement D (units of coulomb per square meter), the magnetic induction B (units of tesla, or webers per square meter) and the electric current density J (units of amps per square meter), and the scalar electric charge density ρ (units of coulomb per cubic meter). These six quantities are related by the four Maxwell's equations (shown here in MKS or SI units):

$$\nabla \times H - \frac{\partial D}{\partial t} = J \quad \text{(Maxwell's extension of Ampere's law)}$$

$$\nabla \times E + \frac{\partial B}{\partial t} = 0 \quad \text{(Faraday's law of induction)} \tag{2.1}$$

$$\nabla \cdot D = \rho \qquad \text{(Gauss's law)}$$

$$\nabla \cdot B = 0$$

where the standard curl and divergence operators are used (see Appendix B). The first two equations above show how a time-varying electric field (or displacement) causes a change in the magnetic field (or induction), and vice versa. The second two equations show how static charge affects the electric and magnetic fields (the last equation set to zero is a statement that there are no magnetic monopoles and therefore no net magnetic charge). A combination of Ampere's law and Gauss's law leads to the continuity equation, an expression of the conservation of electrical charge.

Additionally, the properties of the materials involved provide relationships between the current density and the electric field, the electric displacement and the electric field, and the magnetic induction and the magnetic field. In general, these relationships can be quite complicated, but in many circumstances they become simple. If the material that the electromagnetic radiation is propagating through is moving slowly relative to the speed of light, and the fields involved are time-harmonic (a concept that will be defined below), then the three material equations become

$$J = \sigma E$$

$$D = \varepsilon E$$

$$B = \mu H \tag{2.2}$$

where σ is the conductivity of the material, ε is its dielectric constant (sometimes called the electrical permittivity) and μ is its magnetic permeability, all of which are frequency dependent. Over the range of electric and magnetic field strengths of interest here, these three material properties can be considered constants. The first of the three material equations in Equation (2.2) is the differential form of Ohm's law. Note that for perfectly transparent materials (also called dielectrics), $\sigma = 0$. Thus, absorption in a material is related to the generation of current density under the influence of an electric field (i.e. conduction). Also, in general, the materials of interest to the lithographer are nonmagnetic, and thus $\mu = \mu_0$, the magnetic permeability of free space.

Consider a further simplification where the materials involved are homogeneous (the three material properties are not a function of position). In that case, the three material properties σ, ε and μ, become simple numbers (scalars), rather than tensors. By using these material relationships, the first two Maxwell's equations can be rewritten in terms of electric and magnetic fields only.

$$\nabla \times H - \varepsilon \frac{\partial E}{\partial t} = \sigma E$$

$$\nabla \times E + \mu \frac{\partial H}{\partial t} = 0$$

(2.3)

The goal now will be to take these two coupled first-order differential equations and create two noncoupled second-order equations. When these materials are free of charge ($\rho = 0$), $\nabla \cdot E = 0$ so that

$$\nabla^2 E = \nabla(\nabla \cdot E) - \nabla \times (\nabla \times E) = -\nabla \times (\nabla \times E)$$ (2.4)

Taking the curl of one of Equation (2.3) and substituting in the other,

$$\nabla^2 E = \varepsilon \mu \frac{\partial^2 E}{\partial t^2} + \mu \sigma \frac{\partial E}{\partial t}$$

$$\nabla^2 H = \varepsilon \mu \frac{\partial^2 H}{\partial t^2} + \mu \sigma \frac{\partial H}{\partial t}$$

(2.5)

As a final simplification, for a nonabsorbing medium ($\sigma = 0$), these equations become classic wave equations:

$$\nabla^2 E = \frac{1}{v^2} \frac{\partial^2 E}{\partial t^2}$$

$$\nabla^2 H = \frac{1}{v^2} \frac{\partial^2 H}{\partial t^2}$$

(2.6)

where $v = 1/\sqrt{\varepsilon\mu}$, the velocity of the wave. Letting ε_0 and μ_0 be the values of these material properties in vacuum, the speed of light in vacuum is then $c = 1/\sqrt{\varepsilon_0\mu_0}$. These constants have values of

$c = 2.99792458 \times 10^8$ m/s

$\varepsilon_0 = 8.8541878 \times 10^{-12}$ farads/m (1 farad = 1 coulomb/volt) = $1/(4\pi c^2) \times 10^7$ farads/m

$\mu_0 = 4\pi \times 10^{-7}$ henries/m (1 henry = 1 weber/amp = 1 volt-second/amp)

From this, the refractive index of a material can be defined as the ratio of the speed of light in vacuum to that in the material:

$$n \equiv \frac{c}{v} = \sqrt{\frac{\varepsilon\mu}{\varepsilon_0\mu_0}}$$ (2.7)

It is sometimes useful to define the dielectric constant and magnetic permeability relative to the free-space values:

$$\varepsilon_r \equiv \frac{\varepsilon}{\varepsilon_0}, \quad \mu_r \equiv \frac{\mu}{\mu_0} \tag{2.8}$$

so that

$$n = \sqrt{\varepsilon_r \mu_r} \tag{2.9}$$

And since we are mostly concerned with nonmagnetic materials with $\mu_r = 1$, the refractive index will generally be just the square root of the relative dielectric constant of the material (remembering that we have so far only looked at materials where $\sigma = 0$).

2.1.2   General Harmonic Fields and the Plane Wave in a Nonabsorbing Medium

Since the electric and magnetic fields are related to each other (the magnetic field is always perpendicular to the electric field and both fields are always perpendicular to the direction of propagation, as will be proven in section 2.1.3) and since photoresist reacts chemically to the magnitude of the electric field only, we can usually describe light by considering just the electric field. A general harmonic electric field vector E (due to monochromatic light of frequency ω) at any point P and time t can be described by a deceptively simple-looking sinusoidal equation:

$$E(P,t) = A(P)\cos(\omega t + \Phi(P)) \tag{2.10}$$

where A is the amplitude (like the electric field, a vector) and Φ is the phase, both of which are position dependent, in general, and will depend on initial and boundary conditions. As an example of one form of Equation (2.10), consider a 'plane wave' of light traveling in the $+z$ direction. The term plane wave refers to the shape of the *wavefront*, i.e. the shape of the function $\Phi(P)$ = constant. Thus, a plane wave traveling in the $+z$ direction would require a constant phase (and, consequently, a constant amplitude) in the x–y plane. Such a plane wave would be described by the equation

$$E(P,t) = A\cos(\omega t - kz + \delta) \tag{2.11}$$

where k is a constant called the *propagation constant* or the *wave number* and δ is a phase offset.

How does the wave represented by Equation (2.11) propagate? We can think of the wave as having a certain shape, and this shape travels through time. Thus, this wave will have the given shape at all points in space and time such that $\omega t - kz$ = constant, giving the same electric field. In other words, the wavefront (a plane in this case) travels through space and time according to

$$z - z_0 = \frac{\omega}{k}t \tag{2.12}$$

where z_0 is a constant corresponding to the position of the plane wave at $t = 0$. This is simply a plane of light traveling in the $+z$ direction at speed ω/k. Thus, since

$$v = \frac{c}{n} = \frac{\omega}{k} \tag{2.13}$$

and since $c = \omega\lambda/2\pi$ (a general property of waves), we obtain

$$k = \omega\sqrt{\varepsilon\mu} = 2\pi n/\lambda \tag{2.14}$$

where λ remains the vacuum wavelength of the light. This relationship of the propagation constant to refractive index assumes a nonabsorbing medium ($\sigma = 0$). In the section below, the propagation of a plane wave will be expanded to include absorption.

2.1.3 Phasors and Wave Propagation in an Absorbing Medium

Although Equation (2.10) completely describes an arbitrary time-harmonic electro-magnetic field, a more compact and convenient representation is possible based on the assumption that the frequency of the light does not change (quite a good assumption under normal optical conditions). A sinusoid can be related to a complex exponential by

$$E(P,t) = A(P)\cos(\omega t + \Phi(P)) = \text{Re}\{U(P)e^{-i\omega t}\} \tag{2.15}$$

where

$$U(P) = A(P)e^{-i\Phi(P)}$$

and $U(P)$ is called the *phasor* representation of the sinusoidal electric field $E(P,t)$. Notice that this phasor representation shows no time dependence. Study of the basic behavior of light has shown that the time dependence of the electric field typically does not change as light travels, interferes and interacts with matter. Thus, suppressing the time depen-dence and expressing the electric field as a phasor have become quite common in the mathematical analysis of optical systems. Fields that satisfy this assumption are called *time-harmonic fields*.

Separating out the time dependence for a time-harmonic wave makes evaluation of the time derivatives in Maxwell's equations straightforward. Using Equation (2.15) in Equa-tion (2.5), and showing only the electric field for simplicity,

$$\nabla^2 E = \varepsilon\mu\frac{\partial^2 E}{\partial t^2} + \mu\sigma\frac{\partial E}{\partial t} = -\omega^2\varepsilon\mu E - i\omega\mu\sigma E \tag{2.16}$$

Since every term in this equation contains the oscillatory $e^{-i\omega t}$, the time dependence can be suppressed by dividing this term out.

$$\nabla^2 U = -\omega^2\varepsilon\mu U - i\omega\mu\sigma U \tag{2.17}$$

Finally, this equation can be put into a standard form, known as the *Helmholtz equation*:

$$\nabla^2 U(P) + k^2 U(P) = 0 \tag{2.18}$$

where

$$k^2 = \omega^2\varepsilon\mu + i\omega\mu\sigma$$

The use of the symbol k in the Helmholtz equation is not coincidental. If $\sigma = 0$ (no absorption), k becomes the propagation constant defined in Equation (2.14). Thus, the

Helmholtz equation defines a complex propagation constant that accounts for absorption. Likewise, a complex refractive index can be defined.

$$\mathbf{n}^2 = \frac{k^2\lambda^2}{4\pi^2} = \frac{\omega^2\varepsilon\mu\lambda^2}{4\pi^2} + i\frac{\omega\mu\sigma\lambda^2}{4\pi^2} = \mu c^2\left(\varepsilon + i\frac{\sigma}{\omega}\right) \tag{2.19}$$

Defining the complex refractive index in terms of real and imaginary parts,

$$\mathbf{n} = n + i\kappa \tag{2.20}$$

allows for the real and imaginary parts of the refractive index to be expressed in terms of the electrical and magnetic properties of the material as well as the frequency of oscillation:

$$n^2 = \frac{1}{2}\mu c^2\varepsilon\left(\sqrt{1+\left(\frac{\sigma}{\varepsilon\omega}\right)^2}+1\right)$$

$$\kappa^2 = \frac{1}{2}\mu c^2\varepsilon\left(\sqrt{1+\left(\frac{\sigma}{\varepsilon\omega}\right)^2}-1\right) \tag{2.21}$$

Note that when $\sigma = 0$, $\kappa = 0$ and $\mathbf{n} = n$. For weakly absorbing materials, $\sigma/\varepsilon\omega \ll 1$ and Equations (2.21) can be approximated as

$$n \approx c\sqrt{\mu\varepsilon} = \sqrt{\mu_r\varepsilon_r}$$

$$\kappa \approx \frac{1}{2}\left(\frac{c\sigma}{\omega}\right)\sqrt{\frac{\mu}{\varepsilon}} = \frac{\lambda\sigma}{4\pi}\sqrt{\frac{\mu}{\varepsilon}} \tag{2.22}$$

Again assuming a nonmagnetic material, this weakly absorbing material will have an imaginary part that will be well approximated as

$$\frac{4\pi\kappa}{\lambda} \approx \frac{\sigma Z_0}{n} \quad \text{where} \quad Z_0 = \sqrt{\frac{\mu_0}{\varepsilon_0}} = 376.73\,\Omega \tag{2.23}$$

and Z_0 is called the characteristic impedance of free space.

Since converting back to the time domain involves taking the real part of the phasor, the sign of the phase in $U(P)$ and in the time-dependent term could just as easily have been chosen to be positive rather than negative. This (arbitrary) sign convention is not consistent among authors and there is no absolute standard. The sign convention represented by Equation (2.15) is used by Goodman[1] and in other standard optics textbooks dealing with imaging. Using this convention, a plane wave traveling in the $+z$ direction [i.e. Equation (2.11)] would be written as

$$U(P) = Ae^{ikz} \tag{2.24}$$

(where the phase offset δ has been ignored). This negative phase sign convention of Equation (2.15) is the most common for imaging applications. Unfortunately, many publications and textbooks in the area of thin film interference effects and coatings use the positive phase sign convention. For this thin film sign convention, Equation (2.24) would represent a plane wave traveling in the −z direction. Lithography requires both imaging calculations and thin film interference calculations. As such, both of these competing 'standard' sign conventions are often seen in lithography literature, though the negative sign convention of Equation (2.15) will be used throughout this book.

Equation (2.24) describes a plane wave traveling in the +z direction. Generalizing to an arbitrary direction, let us first define the propagation vector *k* as the propagation constant multiplied by a unit vector pointing in the direction of propagation. Any point in space (x,y,z) is described by its position vector *r*, a vector pointing from the origin to the point (x,y,z). The phasor representation of the electric field of the plane wave is then given by

$$U(x,y,z) = Ae^{ik \cdot r} = Ae^{ik(\alpha x + \beta y + \gamma z)} \tag{2.25}$$

where α, β and γ are called the *direction cosines* of the propagation. As we shall see later in this chapter, the z-axis will coincide with the optical axis of an imaging system, so that the z-direction will be the main direction of light propagation. Thus, it will be convenient to define the direction of propagation using spherical coordinates about the z-axis. Letting θ be the direction *k* makes with the z-axis and ϕ be the angle that the *k*–*z* plane makes with the x-axis, the direction cosines become

$$\alpha = \sin\theta\cos\phi$$
$$\beta = \sin\theta\sin\phi \tag{2.26}$$
$$\gamma = \sqrt{1 - \alpha^2 - \beta^2} = \cos\theta$$

Another interesting and common solution to the wave equation is the spherical wave. For this wave, the wavefront is a sphere and the amplitude of the electric field is inversely proportional to the distance from the center of the sphere. In the equation below,

$$U(r) = \frac{Ae^{ikr}}{r} \tag{2.27}$$

the spherical waves are propagating outward from r = 0, so that the spherical wave can be thought of as emanating from a point source of light located at r = 0.

It is useful to derive some important properties of the plane wave. The time derivative of the field of a plane wave can be related to its spatial derivative with respect to the direction of propagation. Considering a plane wave traveling in the +z direction, as given by Equation (2.11), but applied to the magnetic field:

$$\frac{\partial \mathbf{H}}{\partial t} = -\frac{\omega}{k}\frac{\partial \mathbf{H}}{\partial z} = -\frac{\omega}{k}\left(\frac{\partial H_x}{\partial z}\hat{x} + \frac{\partial H_y}{\partial z}\hat{y}\right) \tag{2.28}$$

Of course, an analogous equation applies to the electric field as well. The curl of the electric field for this plane wave becomes

$$\nabla \times \mathbf{E} = -\frac{\partial E_y}{\partial z}\hat{x} + \frac{\partial E_x}{\partial z}\hat{y} \tag{2.29}$$

These two equations can now be inserted into the second Maxwell equation, using the material property to relate B to H. The result, separated for each vector component, becomes

$$x\text{-component:} \quad -\frac{\partial E_y}{\partial z} - \frac{\omega\mu}{k}\frac{\partial H_x}{\partial z} = 0$$

$$y\text{-component:} \quad +\frac{\partial E_x}{\partial z} - \frac{\omega\mu}{k}\frac{\partial H_y}{\partial z} = 0 \tag{2.30}$$

Equations (2.30) can be integrated directly. Ignoring the constant of integration (any constant background electric or magnetic field will not concern us here),

$$E_y = -\frac{\omega\mu}{k}H_x = -\sqrt{\frac{\mu}{\varepsilon}}H_x = -ZH_x, \quad E_x = \frac{\omega\mu}{k}H_y = \sqrt{\frac{\mu}{\varepsilon}}H_y = ZH_y \tag{2.31}$$

where Z is called the characteristic impedance of the material. This gives a very important result. For a plane harmonic wave traveling through an isotropic material such that the material equations (2.2) apply, the electric and magnetic fields are related to each other by a multiplicative constant. Thus, knowing either one of these quantities is sufficient to determine the other. Moreover, it is the x-component of the magnetic field that is related to the y-component of the electric field, and vice versa. In other words, the electric field is always perpendicular to the magnetic field, and both are perpendicular to the direction of propagation. If Z is real, meaning that there is no absorption, the electric and magnetic fields will be in phase with each other.

2.1.4   Intensity and the Poynting Vector

When determining the intensity (also called irradiance†) transmitted into a material, it is very important to understand the exact definition of intensity. By definition, the *intensity* of light I is the magnitude of the (time averaged) Poynting vector, the energy per second crossing a unit area perpendicular to the direction of propagation of the light. The Poynting vector S is given by

$$S = E \times H \tag{2.32}$$

where E is the electric field, and H is the magnetic field. (The derivation of this definition of intensity, not presented here, calculates the work performed by the electric and magnetic fields on the electrons surrounding the atoms of the medium.[2])

For quasi-monochromatic time-harmonic fields, as represented by Equation (2.15), it will be convenient to express them as

$$E(P,t) = \text{Re}\{U(P)e^{-i\omega t}\} = \frac{1}{2}[U(P)e^{-i\omega t} + U*(P)e^{+i\omega t}]$$

$$H(P,t) = \text{Re}\{V(P)e^{-i\omega t}\} = \frac{1}{2}[V(P)e^{-i\omega t} + V*(P)e^{+i\omega t}] \tag{2.33}$$

---

†By international standard, intensity is defined as power per unit solid angle while irradiance is power per unit area. However, the older nomenclature of intensity as power per unit area is still common (used in the classic textbook by Born and Wolf, for example), and will be used throughout this book.

where U^* is the complex conjugate of U. Using these equations, the cross product of the electric and magnetic fields becomes

$$S = E \times H = \frac{1}{4}[U \times V e^{-2i\omega t} + U^* \times V + U \times V^* + U^* \times V^* e^{+2i\omega t}] \qquad (2.34)$$

Taking the time average of the Poynting vector is straightforward. For times much longer than the period of oscillation of the wave, the time harmonic terms will average to zero. Thus,

$$\langle S \rangle = \frac{1}{4}[U^* \times V + U \times V^*] = \frac{1}{2}\mathrm{Re}\{U \times V^*\} \qquad (2.35)$$

To apply Equation (2.35) to a specific case, consider the monochromatic plane wave traveling in the z-direction. Remembering that electric and magnetic fields are perpendicular to each other and the direction of propagation, and using the results of Equation (2.31),

$$U \times V^* = (U_x V_y^* - U_y V_x^*)\hat{z} = \frac{k}{\omega\mu}(U_x U_x^* + U_y U_y^*)\hat{z} \qquad (2.36)$$

Using the phasor expression for the plane wave, Equation (2.24),

$$U \times V^* = \frac{k}{\omega\mu}|U|^2\,\hat{z} = \frac{k}{\omega\mu}|A|^2\,\hat{z} \qquad (2.37)$$

Thus, replacing $\mathrm{Re}\{\omega/k\}$ with the speed of the wave from Equations (2.7) and (2.13), the intensity of the plane wave becomes

$$I = |\langle S \rangle| = \frac{n}{2c\mu}|A|^2 = \frac{1}{2}\sqrt{\frac{\varepsilon}{\mu}}|A|^2 \qquad (2.38)$$

Since in general our media will be nonmagnetic (so that $\mu = \mu_0$), this final expression for intensity says that the intensity, within a multiplicative factor, is equal to the magnitude of the phasor electric field squared times the real part of the refractive index of the medium. Throughout this book, we will often have occasion to describe the intensity relative to some incoming wave, and so the multiplicative constant, $1/(2c\mu_0)$, will most often be ignored. And since throughout the rest of this book electric fields will always be expressed in phasor form, our general form for intensity will become

$$I = n|E|^2 \qquad (2.39)$$

2.1.5   Intensity and Absorbed Electromagnetic Energy

The Poynting vector describes the flow of electromagnetic energy density. But the true purpose of knowing the intensity of light is to understand its impact on materials, which in our case means the impact of light on a photoresist. While the photochemistry of resists will be discussed in Chapters 5 and 6, suffice it to say here that the quantity of interest will be the electromagnetic energy absorbed by a material. Consider a monochromatic plane wave traveling through a uniform material of complex refractive index n. Separating the real and imaginary parts of the refractive index,

$$E = Ae^{ikz} = Ae^{i2\pi nz/\lambda} = A(e^{-2\pi\kappa z/\lambda})e^{i2\pi nz/\lambda} \tag{2.40}$$

The intensity can then be written, from Equation (2.39),

$$I = nA^2(e^{-4\pi\kappa z/\lambda}) = I_0 e^{-\alpha z} \tag{2.41}$$

where $I_0 = I(z = 0)$ and α is the absorption coefficient of the material, given by

$$\alpha = \frac{4\pi\kappa}{\lambda} \tag{2.42}$$

Equation (2.41) is the familiar *Lambert law of absorption*, empirically verified for homogeneous materials.

The imaginary part of the refractive index was related previously to the electrical and magnetic properties of the material in Equation (2.21). For a weakly absorbing material, such as a photoresist, the absorption coefficient becomes, from Equation (2.23)

$$\alpha \approx \frac{\sigma Z_0}{n} \quad \text{where} \quad Z_0 = \sqrt{\frac{\mu_0}{\varepsilon_0}} = 376.73\Omega \tag{2.43}$$

Thus, absorption of electromagnetic energy results from the conductivity of the material at the frequency of the electromagnetic radiation. The electromagnetic wave gives energy to the electrons of the atoms of the material in proportion to its conductivity, in the form of absorbed photons of light.

2.2   Basic Imaging Theory

Using the basic understanding of light from the previous section, and especially the phasor representation of a plane wave, we are now ready to examine the behavior of an optical imaging system. Consider the generic projection system shown in Figure 2.1. It consists of a *light source*, a *condenser lens*, the *mask*, the *objective lens* and finally the resist-coated wafer. The combination of the light source and the condenser lens is called the *illumination system*. In optical design terms, a lens is a system of (possibly many) lens elements. Each lens element is an individual piece of glass (refractive element) or a mirror

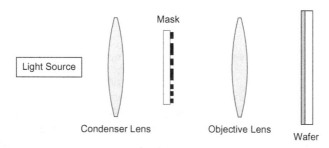

Figure 2.1   *Block diagram of a generic projection imaging system*

(reflective element) or other optical element. The purpose of the illumination system is to deliver light to the mask (and eventually into the objective lens) with sufficient intensity, the proper directionality and spectral characteristics, and adequate uniformity across the field (i.e. across the mask). The mask consists of a transparent substrate on which a pattern has been formed. This pattern changes the transmittance of the light and in its simplest form is just an opaque film. The light then passes through the clear areas of the mask and diffracts on its way to the objective lens. The purpose of the objective lens is to pick up a portion of the diffraction pattern and project an image onto the wafer which, one hopes, will resemble the mask pattern (or, more correctly, the desired pattern as expressed in the original design data).

2.2.1   Diffraction

The first and most basic phenomenon occurring in projection imaging is the diffraction of light. Diffraction is usually thought of as the bending of light as it passes through an aperture, which is certainly an appropriate description for diffraction by a lithographic mask. More correctly, diffraction theory simply describes how light propagates, including the effects of the surroundings (boundaries). Maxwell's equations describe how electromagnetic waves propagate, but result in partial differential equations of vector quantities which, for general boundary conditions, are extremely difficult to solve without the aid of a powerful computer and sophisticated numerical algorithms. A simpler approach is applicable to the propagation of light within a homogenous medium (for example, describing how light travels from the photomask to the objective lens of our imaging system). This simple approach leads to the development of *diffraction integrals* derived from the Helmholtz equation (2.18). While the derivation of these diffraction integrals will not be presented here, some historical background will help to put their usefulness in perspective.

A simple interpretation of the physical principle behind diffraction is best captured by *Huygens' principle*. First described by the famous Dutch scientist Christian Huygens in his 1690 *Treatise on Light*, any wavefront can be thought of as a collection of radiating point sources (Figure 2.2). The new wavefront at some later time can be constructed by summing up the wavefronts from all of the radiated spherical waves. As Figure 2.2 shows, one result of these self-luminous wavefront points is the apparent bending of light around an opaque edge – an example of diffraction. In 1818, Augustin-Jean Fresnel formed a

Plane Wave Propagation   Diffraction by a Slit

Figure 2.2 *Pictorial representation of Huygens' principle, where any wavefront can be thought of as a collection of point sources radiating spherical waves*

mathematical theory of diffraction by turning this summation into an integral and accounting for the phase of the light when adding together the propagating spherical waves. He
also included an empirical obliquity factor to keep these radiating spherical waves from
propagating backwards as readily as forward. This scalar diffraction theory was put on a
more rigorous footing by the Prussian mathematician and physicist Gustav Kirchhoff in
1882, who required the diffracting waves to satisfy the Helmholtz equation and conservation of energy, thus deriving Fresnel's previously empirical obliquity term. Fresnel's diffraction integral is in fact a simplification of Kirchhoff's formulation for the case when
the distance away from the diffracting plane (that is, the distance from the mask to the
objective lens) is much greater than the wavelength of light. Finally, if the distance to the
objective lens is very large, or if the mask is illuminated by a spherical wave which
converges to a point at the entrance to the objective lens, Fresnel diffraction simplifies
to *Fraunhofer diffraction* (named for Joseph von Fraunhofer, who used a lens to create
the far-field diffraction pattern). Comparison of these different diffraction regions is given
in Figure 2.3.

In order to establish a mathematical description of diffraction by a mask, we must first
describe the electric field transmittance of a mask pattern $t_m(x,y)$, where the mask is in
the x–y plane and $t_m(x,y)$ has in general a magnitude, phase and vector direction. An exact
calculation of the mask transmittance requires a solution to Maxwell's equations for the
given materials and physical structure of the mask and for a specific incident light wave.
As we shall see in the sections below, any arbitrary incident illumination can be conveniently broken up into a summation of plane waves of different polarizations and propagation directions. Thus, the transmittance of different polarizations of incident light as a
function of the incident angle of light is required.

In some circumstances, however, a simpler approach will prove adequate. For cases
where the feature sizes on the mask are larger than about 2λ and where the topography
on the mask (for example, the thickness of a chrome absorber) is less than about $\lambda/2$, it
is possible to ignore diffraction by the topography of the mask and simply assume that
the mask transmittance is a perfect shadow of the mask. In other words, the transmittance
directly below a specific feature on the mask is assumed to be the same as the transmittance under an infinitely large area of that material. At an edge between two materials,
the transmittance of the boundary is assumed to be a step function between the transmittances of the two materials. This assumed transmittance function is called the *Kirchhoff*

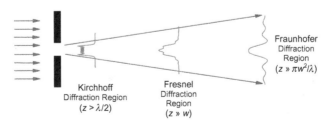

Figure 2.3 *Comparison of the diffraction 'regions' where various approximations become
accurate. Diffraction is for a slit of width* w *illuminated by light of wavelength* λ, *and* z *is the
distance away from the mask*

boundary condition. Note that under this assumption, the transmittance for each polarization is assumed to be the same, so that incident polarization can be ignored when employing the Kirchhoff boundary condition. Also, the effect of incident angle can be treated very simply in the case of the Kirchhoff boundary condition, as will be discussed in section 2.3.

For a simple chrome-glass mask, the mask E-field transmittance for a normally incident plane wave of illumination under the Kirchhoff boundary condition assumption is binary: $t_m(x,y)$ is 1 under the glass and 0 under the chrome. Let the $x'–y'$ plane be the diffraction plane, that is, the entrance to the objective lens, and let z be the distance from the mask to this diffraction plane. We will also assume monochromatic light of wavelength λ and that the entire system is in a medium of refractive index n. Defining the *spatial frequencies* of the diffraction pattern (which are simply scaled coordinates in the $x'–y'$ plane) as $f_x = nx'/(z\lambda)$ and $f_y = ny'/(z\lambda)$, the electric field of our diffraction pattern, $T_m(f_x,f_y)$, is given by the *Fraunhofer diffraction integral*:

$$T_m(f_x,f_y) = \int_{-\infty}^{\infty} \int_{-\infty}^{\infty} E_i(x,y)t_m(x, y)e^{-2\pi i(f_x x + f_y y)}dxdy \qquad (2.44)$$

where E_i is the electric field incident on the mask (and is just 1 for our unit amplitude, normally incident plane wave). For many scientists and engineers, this equation should be quite familiar – it is simply a *Fourier transform*. Thus, the diffraction pattern (i.e. the electric field distribution as it enters the objective lens) is just the Fourier transform of the mask pattern transmittance (the electric field directly under the mask). This is the principle behind an extremely useful approach to imaging called Fourier optics (for more information, consult Goodman's classic textbook[1]).

Figure 2.4 shows two mask patterns – one an isolated space, the other a series of equal lines and spaces – both infinitely long in the y-direction (the direction out of the page). The resulting one-dimensional mask field transmittance functions, $t_m(x)$, look like a square pulse and a square wave, respectively. The Fourier transforms for normally incident plane wave illumination ($E_i = 1$) are easily computed directly from Equation (2.44) or found in tables or textbooks (see Problem 2.4) and are also shown in Figure 2.4. The isolated space

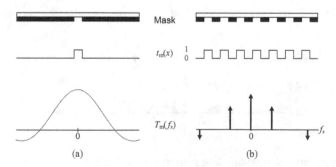

Figure 2.4   *Two typical mask patterns, (a) an isolated space and (b) an array of equal lines and spaces, and the resulting Fraunhofer diffraction patterns assuming normally incident plane wave illumination. Both t$_m$ and T$_m$ represent electric fields*

gives rise to a *sinc* function (sin(x)/x) diffraction pattern, and the equal lines and spaces yield discrete *diffraction orders*:

$$\text{Isolated space:} \quad T_{\mathrm{m}}(f_x) = \frac{\sin(\pi w f_x)}{\pi f_x}$$

$$\text{Dense space:} \quad T_{\mathrm{m}}(f_x) = \frac{1}{p} \sum_{j=-\infty}^{\infty} \frac{\sin(\pi w f_x)}{\pi f_x} \delta\left(f_x - \frac{j}{p} \right)$$

(2.45)

where δ is the Dirac delta function, w is the spacewidth and p is the pitch (the linewidth plus the spacewidth). The delta function is the mathematical representation of a point of light. (As a review, the properties of the delta function are discussed in Appendix C.)

It is interesting to consider how different these two diffraction patterns are. The isolated space results in a continuous distribution of energy in the form of a sinc function. The dense array of spaces produces discrete points of light, whose brightness follows the sinc function envelop (that is, the diffraction pattern of just one of its spaces). To visualize how this happens, consider the intermediate cases of two or three spaces of equal width, separated by lines of the same width. Figure 2.5 shows the intensity of the diffraction patterns (electric field amplitude squared) for these cases. For the multiple space cases, diffraction patterns from adjacent spaces interfere with each other, producing narrow interference fringes. The greater the number of spaces, the greater the interference is and thus the narrower the peaks. As the number of spaces in the array approaches infinity, the fringes become infinitely narrow points of light.

Let's take a closer look at the diffraction pattern for equal lines and spaces. At the center of the objective lens entrance ($f_x = 0$), the diffraction pattern has a bright spot called the *zero order*. The zero order is the light that passes through the mask and is not bent. To either side of the zero order are two peaks called the *first diffraction orders*. These peaks occur at spatial frequencies of $\pm 1/p$, where p is the pitch of the mask pattern (linewidth plus spacewidth). Since the position of these diffraction orders relative to the zero

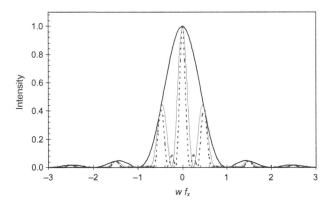

Figure 2.5 *Magnitude of the diffraction pattern squared (intensity) for a single space (thick solid line), two spaces (thin solid line) and three spaces (dashed lines) of width w. For the multiple-feature cases, the linewidth is also equal to w*

order depends on the mask pitch, their relative positions contain information about the pitch. It is this information that the objective lens will use to reproduce the image of the mask. In fact, in order for the objective lens to form a true image of a pattern of lines and spaces, at least two diffraction orders must be captured by the objective lens and used to form the image. In addition to the first order, there can be many higher orders, with the jth order occurring at a spatial frequency of j/p.

While in principle the number of diffraction orders (the range of values of j) is infinite, only some of these orders are capable of propagating from the mask to the lens. If a wave has a spatial frequency such that

$$\sqrt{f_x^2 + f_y^2} > \frac{1}{\lambda} \tag{2.46}$$

these waves are said to *evanescent*. This condition is equivalent to a wave whose direction cosines have $\alpha^2 + \beta^2 > 1$ so that the direction cosine γ is imaginary. The amplitudes of these evanescent waves decay exponentially as

$$e^{-\varsigma z} \quad \text{where} \quad \varsigma = 2\pi n \sqrt{f_x^2 + f_y^2 - \frac{1}{\lambda^2}} \tag{2.47}$$

Thus, for practical purposes, diffraction patterns need only be considered out to spatial frequencies less than the onset of evanescent waves.

Summarizing, given a mask in the x–y plane described by its electric field transmission $t_m(x,y)$ multiplied by the incident electric field $E_i(x,y)$, the electric field T_m as it enters the objective lens (the x'–y' plane) is given by

$$T_m(f_x, f_y) = \mathcal{F}\{E_i(x,y)t_m(x,y)\} \tag{2.48}$$

where the symbol $\mathcal{F}$ represents the Fourier transform and f_x and f_y are the spatial frequencies and are simply scaled coordinates in the x'–y' diffraction plane. Often, $t_m(x,y)$ is approximated as the shadow transmittance of the mask using the Kirchhoff boundary condition. For normally incident plane wave illumination, $E_i = 1$.

2.2.2   Fourier Transform Pairs

Some important diffraction patterns can now be calculated. Table 2.1 shows several common 1D functions of some interest to lithographers, as well as their Fourier transforms. These transform pairs can be applied to different problems with the addition of two important theorems. If $\mathcal{F}\{g(x,y)\} = G(f_x, f_y)$, the *shift theorem* of the Fourier transform says

$$\mathcal{F}\{g(x - a, y - b)\} = G(f_x, f_y)e^{-i2\pi(f_x a + f_y b)} \tag{2.49}$$

where a and b are constants. In other words, a shift in position of the mask produces a linear phase variation across the diffraction pattern. The *similarity theorem* of the Fourier transform is

$$\mathcal{F}\{g(ax, by)\} = \frac{1}{|ab|}G\left(\frac{f_x}{a}, \frac{f_y}{b}\right) \tag{2.50}$$

Table 2.1   *1D Fourier transform pairs useful in lithography*

$g(x)$	Graph of $g(x)$	$G(f_x)$
$rect(x) = \begin{cases} 1, & \|x\| < 0.5 \\ 0, & \|x\| > 0.5 \end{cases}$		$\dfrac{\sin(\pi f_x)}{\pi f_x}$
$step(x) = \begin{cases} 1, & x > 0 \\ 0, & x < 0 \end{cases}$		$\dfrac{1}{2}\delta(f_x) - \dfrac{i}{2\pi f_x}$
Delta function $\delta(x)$		1
$comb(x) = \displaystyle\sum_{j=-\infty}^{\infty} \delta(x-j)$		$\displaystyle\sum_{j=-\infty}^{\infty} \delta(f_x - j)$
$\cos(\pi x)$		$\dfrac{1}{2}\delta\left(f_x + \dfrac{1}{2}\right) + \dfrac{1}{2}\delta\left(f_x - \dfrac{1}{2}\right)$
$\sin(\pi x)$		$\dfrac{i}{2}\delta\left(f_x + \dfrac{1}{2}\right) - \dfrac{i}{2}\delta\left(f_x - \dfrac{1}{2}\right)$
Gaussian $e^{-\pi x^2}$		$e^{-\pi f_x^2}$
$circ(r) = \begin{cases} 1, & \|r\| < 1 \\ 0, & \|r\| > 1 \end{cases}$ $r = \sqrt{x^2 + y^2}$		$\dfrac{J_1(2\pi\rho)}{\pi\rho}$ $\rho = \sqrt{f_x^2 + f_y^2}$

This theorem shows that making a pattern smaller ($a, b < 1$) spreads the diffraction pattern to higher spatial frequencies. Finally, by recognizing that the Fourier transform is a linear operation (see Problem 2.7), superposition of known Fourier transform pairs can be used to build more complicated functions.

Another important general transform that is extremely useful in lithography is the generalized repeating pattern. Consider a two-dimensional pattern that repeats in x and y with pitches p_x and p_y, respectively. Let $t_r(x,y)$ be the transmittance of a single repeating unit such that the transmittance is nonzero only in the range $-p_x/2 < x < p_x/2$ and $-p_y/2 < y < p_y/2$. The total transmittance of the mask is then

$$t_m(x,y) = \sum_{j=-\infty}^{\infty} \sum_{k=-\infty}^{\infty} t_r(x+jp_x, y+kp_y) = t_r(x,y) \otimes comb\left(\frac{x}{p_x}, \frac{y}{p_y}\right) \qquad (2.51)$$

where the '$\otimes$' symbol denotes convolution. Letting $T_r(f_x, f_y)$ be the Fourier transform of the isolated repeating unit transmittance, the Fourier transform of Equation (2.51) gives

$$T_m(f_x, f_y) = \frac{1}{p_x p_y} \sum_{j=-\infty}^{\infty} \sum_{k=-\infty}^{\infty} T_r(f_x, f_y) \delta\left(f_x - \frac{j}{p_x}, f_y - \frac{k}{p_y}\right) \qquad (2.52)$$

Thus, any repeating pattern produces a diffraction pattern made up of delta functions (diffraction orders). The amplitudes of the diffraction orders are given by the Fourier transform of the isolated repeating unit mask pattern.

2.2.3   Imaging Lens

We are now ready to describe what happens next and follow the diffracted light as it enters the objective lens. In general, the diffraction pattern extends throughout the $x'-y'$ plane. However, the objective lens, being only of finite size, cannot collect all of the light in the diffraction pattern. Typically, lenses used in microlithography are circularly symmetric and the entrance to the objective lens can be thought of as a circular aperture. Only those portions of the mask diffraction pattern that fall inside the aperture of the objective lens go on to form the image. Of course, we can describe the size of the lens aperture by its radius, but a more common and useful description is to define the maximum angle of diffracted light that can enter the lens. Consider the geometry shown in Figure 2.6. Light passing through the mask is diffracted at various angles. Given a lens of a certain size placed at a certain distance from the mask, there is some maximum angle of diffraction, θ_{max}, for which the diffracted light just makes it into the lens. Light emerging from the mask at larger angles misses the lens and is not used in forming the image. The most convenient way to describe the size of the lens aperture is by its *numerical aperture*,

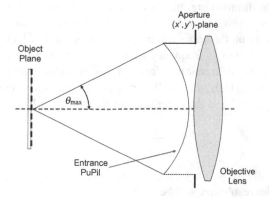

Figure 2.6   *The numerical aperture is defined as* NA = n sin θ_{max} *where* θ_{max} *is the maximum half-angle of the diffracted light that can enter the objective lens, and* n *is the refractive index of the medium between the mask and the lens*

defined as the sine of the maximum half-angle of diffracted light that can enter the lens times the index of refraction of the surrounding medium, n.

$$NA = n \sin \theta_{\max} \qquad (2.53)$$

Besides the diffraction plane, it is important to define the *entrance pupil* of the objective lens. Despite the simplicity of the single-lens diagram in Figure 2.6, the diffraction plane (also called the *pupil plane* or *aperture plane*) is actually buried inside the lens in most cases. The entrance pupil (or opening) is spherical in shape and is the image of the diffraction plane as seen from the entrance of the lens. The diffraction plane can be constructed simply as the geometric projection of the spherical entrance pupil onto a plane. (Likewise, the *exit pupil* is the image of the aperture plane as viewed from the exit side of the lens.) Thus, one can easily express the spatial frequency in terms of the angle of diffraction:

$$f_x = \frac{nx'}{\lambda z} = \frac{n \sin \theta_x}{\lambda} \qquad (2.54)$$

with, of course, an equivalent expression for the y-direction. Alternately, the spatial frequencies can be related to the direction cosines as defined in Equation (2.25):

$$f_x = \frac{n\alpha}{\lambda}, \quad f_y = \frac{n\beta}{\lambda} \qquad (2.55)$$

Using this description, the maximum spatial frequency that can enter the objective lens is given by NA/λ.

Clearly, the numerical aperture is going to be quite important. A large numerical aperture means that a larger portion of the diffraction pattern is captured by the objective lens. For a small numerical aperture, much more of the diffracted light is lost. In fact, we can use this viewpoint to define *resolution*, at least from the limited perspective of image formation. Consider the simple case of a mask pattern of equal lines and spaces. As we have seen, the resulting diffraction pattern is a series of discrete diffraction orders. In order to produce an image that even remotely resembles the original mask pattern, it is necessary for the objective lens to capture the zero order and at least one higher diffraction order. If the light illuminating the mask is a normally incident plane wave (the only type of illumination we have discussed so far), the diffraction pattern will be centered in the objective lens (more on this topic will be covered in section 2.3.1). Since the positions of the ±1st diffraction orders in frequency space are given by $\pm 1/p$, the requirement that a lens must capture these diffraction orders to form an image puts a lower limit on the pitch that can be imaged. Thus, the smallest pitch ($p_{\min}$) that still produces an image will put the first diffraction order at the outer edge of the objective lens:

$$\frac{1}{p_{\min}} = \frac{NA}{\lambda} \qquad (2.56)$$

If we let R represent the resolution element (the linewidth or the spacewidth) of our equal line/space pattern, the resolution will be given by

$$R = 0.5 \frac{\lambda}{NA} \qquad (2.57)$$

This classic equation is often called the *theoretical resolution* of an imaging system. Note that several very specific assumptions were made in deriving this resolution equation: a mask pattern of equal lines and spaces was used, and the illumination was a single wavelength normally incident plane wave (called coherent illumination). In reality, this equation defines the *pitch resolution*, the smallest pitch that can be imaged, for normally incident plane waves. For other features, for example an isolated line, there is no clear resolution cutoff. For other types of illumination, this cutoff will change. As a result, this approximate resolution expression is often generalized as

$$R = k_1 \frac{\lambda}{NA} \tag{2.58}$$

Such resolution equations are best interpreted as scaling equations, with k_1 as the scaled resolution. See Chapter 10 for a more thorough discussion of resolution.

2.2.4   Forming an Image

To proceed further, we must now describe how the lens affects the light entering it. Obviously, we would like the image to resemble the mask pattern. Since diffraction gives the Fourier transform of the mask, if the lens could give the *inverse* Fourier transform of the diffraction pattern, the resulting image would resemble the mask pattern. In fact, lenses are designed to behave precisely in this way. We can define an ideal imaging lens as one that produces an image at the focal plane that is identically equal to the Fourier transform of the light distribution entering the lens. A property of the Fourier transform is

$$\mathcal{F}\{\mathcal{F}\{g(x,y)\}\} = g(-x,-y) \tag{2.59}$$

Thus, one Fourier transform (diffraction) followed by a second Fourier transform (focusing by the imaging lens) produces an image that is a 180° rotated version of the original. It is common to define an image coordinate system that is rotated by 180° so that this rotation can be ignored. In the rotated image coordinate system, the focusing behavior of the imaging lens is described as a simple inverse Fourier transform.

It is the goal of lens designers and manufacturers to create lenses as close as possible to this ideal. Does an ideal lens produce a perfect image? No. Because of the finite size of the lens aperture, only a portion of the diffraction pattern enters the lens. Thus, even an ideal lens cannot produce a perfect image unless the lens is infinitely big. Since in the case of an ideal lens the image is limited only by the diffracted light that does not make it through the lens, we call such an ideal system *diffraction limited*.

In order to write our final equation for the formation of an image, let us define the objective lens *pupil function P* (a pupil is just another name for an aperture) as the transmittance of the lens from the entrance pupil to the exit pupil. The pupil function of an ideal lens simply describes what portion of light makes it through the lens: it is one inside the aperture and zero outside.

$$P(f_x,f_y) = \begin{cases} 1, & \sqrt{f_x^2 + f_y^2} < NA/\lambda \\ 0, & \sqrt{f_x^2 + f_y^2} > NA/\lambda \end{cases} \tag{2.60}$$

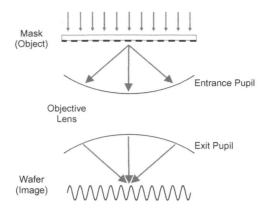

Figure 2.7   *Formation of an aerial image: a pattern of lines and spaces produces a diffraction pattern of discrete diffraction orders (in this case, three orders are captured by the lens). The lens turns these diffraction orders into plane waves projected onto the wafer, which interfere to form the image*

Thus, the product of the pupil function and the diffraction pattern describes the light transmitted through the objective lens, that is, the light at the exit pupil. Combining this with our description of how a lens behaves gives us our final expression for the electric field at the image plane (that is, at the wafer):

$$E(x,y) = \mathcal{F}^{-1}\{T_{\mathrm{m}}(f_x, f_y)P(f_x, f_y)\} \tag{2.61}$$

where the symbol $\mathcal{F}^{-1}$ represents the inverse Fourier transform. The *aerial image* is defined as the intensity distribution in air at the wafer plane and is simply the square of the magnitude of the electric field (see section 2.1.4).

Consider the full imaging process (Figure 2.7). First, light passing through the mask is diffracted. The diffraction pattern can be described as the Fourier transform of the electric field at the bottom of the mask. Since the objective lens is of finite size, only a portion of the diffraction pattern actually enters the lens. The numerical aperture defines the maximum angle of diffracted light that enters the lens and the pupil function is used to mathematically describe this behavior. Finally, the effect of the lens is to take the inverse Fourier transform of the light exiting the lens to give an image which (one hopes) resembles the mask pattern. If the lens is ideal, the quality of the resulting image is only limited by the amount of the diffraction pattern collected. This type of imaging system is called diffraction limited.

2.2.5   Imaging Example: Dense Array of Lines and Spaces

Again referring to the simplified case of an infinite array of lines and spaces, the diffraction pattern is made up of discrete diffraction orders, points of light represented mathematically as delta functions. The image of such a pattern is calculated as the inverse Fourier transform of the diffraction orders that make it through the lens (those with spatial frequencies less than NA/λ). The inverse Fourier transform of a delta function is just 1, so that a point source in the pupil plane of the lens produces a plane wave at the wafer plane. Letting N be the largest diffraction order that passes through the lens, and using the dense space diffraction pattern of Equation (2.45),

$$E(x) = \int_{-\infty}^{\infty} \frac{1}{p} \sum_{j=-N}^{N} \frac{\sin(\pi w f_x)}{\pi f_x} \delta\left(f_x - \frac{j}{p}\right) e^{i2\pi f_x x} df_x$$

$$E(x) = \sum_{j=-N}^{N} a_j e^{i2\pi jx/p} \quad \text{where} \quad a_j = \frac{\sin(j\pi w/p)}{j\pi}$$

(2.62)

The term $e^{i2\pi jx/p}$ represents the electric field of a unit-amplitude plane wave spread out over the focal plane, whose incident angle when striking the focal plane is given by the angle of the jth diffraction order ($\sin\theta = j\lambda/p$). The amplitude of this plane wave is a_j, the amplitude of the jth diffraction order. The electric field image at the wafer plane is the sum (the interference) of the various plane waves reaching the wafer. This interpretation of the line/space image result is extremely useful. As Equation (2.52) shows, any repeating pattern on the mask will produce a diffraction pattern made up of delta functions (point sources) called diffraction orders. Each diffraction order in turn produces one plane wave striking the wafer. The summation (interference) of all of the plane waves produces the image. This is sometimes called *plane wave decomposition*, since the image is decomposed into plane wave basis functions.

Since the diffraction pattern for this case is centered in the entrance pupil, Equation (2.62) can be simplified using Euler's theorem.

$$E(x) = a_0 + 2\sum_{j=1}^{N} a_j \cos(2\pi jx/p)$$

(2.63)

The zero order, given by a_0, provides a DC offset to the electric field. Each pair of diffraction orders, ±1, ±2, etc., provides a cosine at harmonics of the pitch p. For the case of equal lines and spaces, the diffraction order amplitudes become

$$a_j = \frac{\sin(j\pi/2)}{j\pi} = \begin{cases} \dfrac{1}{2}, & j=0 \\[2mm] 0, & j = even \\[2mm] \dfrac{1}{|j|\pi}, & j = \pm1, \pm5, \pm9, \ldots \\[2mm] -\dfrac{1}{|j|\pi}, & j = \pm3, \pm7, \pm11, \ldots \end{cases}$$

(2.64)

A graph of this diffraction pattern for ±3 orders is given in Figure 2.8.

Consider the case of lines and spaces where $N = 1$ (that is, the lens captures the zero order and the ±1 orders). The electric field image is then

$$E(x) = a_0 + 2a_1 \cos(2\pi x/p) = \frac{1}{2} + \frac{2}{\pi} \cos(2\pi x/p)$$

(2.65)

and the intensity image is

$$I(x) = |E(x)|^2 = a_0^2 + 4a_0 a_1 \cos(2\pi x/p) + 4a_1^2 \cos^2(2\pi x/p)$$

(2.66)

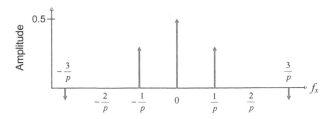

Figure 2.8 *Graph of the diffraction pattern for equal lines and spaces plotted out to ±3rd diffraction orders. For graphing purposes, the delta function is plotted as an arrow, the height equal to the amplitude multiplier of the delta function*

For equal lines and spaces, the image becomes

$$I(x) = \frac{1}{4} + \frac{2}{\pi}\cos(2\pi x/p) + \frac{4}{\pi^2}\cos^2(2\pi x/p) \tag{2.67}$$

or

$$I(x) = \frac{1}{4} + \frac{2}{\pi^2} + \frac{2}{\pi}\cos(2\pi x/p) + \frac{2}{\pi^2}\cos(4\pi x/p) \tag{2.68}$$

It is useful to interpret each term in Equation (2.67) physically. The zero order contributes a DC bias to the final image, seen as the first term on the right-hand side of Equation (2.67). The second term represents the interference of the zero order with the first orders and contains the main cosine function that determines the overall shape of the aerial image. The final cosine-squared term represents the interference of the +1 order with the −1 order. Since the distance between these orders is $2/p$ in spatial frequency space, this term contributes a frequency-doubled sinusoid to the image.

As a note, the image calculations above perform a *scalar* addition of the electric fields of the zero and first orders. That is, the vector directions of the electric fields of each plane wave interfering at the wafer plane are assumed to be parallel. For the case where this is not true, a full vector treatment of the image calculation, as given in Chapter 3, is required.

The number of diffraction orders captured by the objective lens determines the quality and characteristics of the resulting aerial image. If the numerical aperture of the lens is increased (or if the wavelength is decreased), increasing numbers of diffracted orders can be captured and used to form the image. Each added diffraction order contains information that improves the quality of the aerial image (Figure 2.9).

2.2.6   Imaging Example: Isolated Space

Consider now the image of an isolated space under coherent illumination. Taking the inverse Fourier transform of the portion of the diffraction pattern [from Equation (2.45)] that makes it through the lens,

$$E(x) = \int_{-NA/\lambda}^{NA/\lambda} \frac{\sin(\pi w f_x)}{\pi f_x} e^{i2\pi f_x x} df_x \tag{2.69}$$

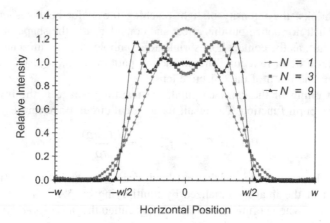

Figure 2.9 *Aerial images for a pattern of equal lines and spaces of width w as a function of the number of diffraction orders captured by the objective lens (coherent illumination). N is the maximum diffraction order number captured by the lens*

Breaking up the complex exponential into a sine and cosine using Euler's identity, any odd part of the function will integrate to zero. Thus, keeping only the even part,

$$E(x) = 2 \int_{0}^{NA/\lambda} \frac{\sin(\pi w f_x)\cos(2\pi x f_x)}{\pi f_x}\, df_x \qquad (2.70)$$

Using a trigonometric identity,

$$E(x) = \int_{0}^{NA/\lambda} \frac{\sin(2\pi f_x[x+w/2]) - \sin(2\pi f_x[x-w/2])}{\pi f_x}\, df_x \qquad (2.71)$$

This integral can be expressed in terms of the commonly tabulated Sine integral (Si) to give the final image intensity:

$$I(x) = \frac{1}{\pi^2}\left[\mathrm{Si}\left(\frac{2\pi NA}{\lambda}(x+w/2)\right) - \mathrm{Si}\left(\frac{2\pi NA}{\lambda}(x-w/2)\right)\right]^2 \qquad (2.72)$$

where

$$\mathrm{Si}(\theta) = \int_{0}^{\theta} \frac{\sin z}{z}\, dz$$

2.2.7 The Point Spread Function

A common method of characterizing the resolving capability of an imaging system is to consider the smallest possible contact hole that can be printed. (The term 'contact' hole refers to the use of this small hole in ICs for making electrical contact between two metal layers separated by an insulator.) Consider a mask pattern of an isolated rectangular hole in a chrome (totally dark) background. Now let that hole shrink to an infinitesimal pinhole.

As the hole shrinks, the intensity of light reaching the wafer becomes infinitesimally small, but we shall normalize out this effect and consider only the shape of the resulting aerial image. Thus, as the contact hole shrinks to a pinhole, the transmittance of the photomask becomes a delta function. The diffraction pattern, then, is just 1, a plane wave uniformly filling the pupil of the objective lens.

For uniform filling of the entrance pupil, the resulting image is the inverse Fourier transform of the pupil function. The result for an ideal circular aperture is

$$E(x,y) = \mathcal{F}^{-1}\{P(f_x, f_y)\} = \frac{J_1(2\pi\rho)}{\pi\rho} \tag{2.73}$$

where J_1 is the Bessel function of the first kind, order one, and ρ is the radial distance from the center of the image normalized by multiplying by NA/λ. The intensity aerial image of a pinhole, when normalized in this way, is called the *point spread function* (PSF) of the optical system.

$$PSF_{ideal} = \left| \frac{J_1(2\pi\rho)}{\pi\rho} \right|^2 \tag{2.74}$$

The PSF (also called the *Airy disk*, after George Biddle Airy) is a widely used metric of imaging quality for optical system design and manufacture and is commonly calculated for lens designs and measured on fabricated lenses using special bench-top equipment. Figure 2.10 shows a graph of the ideal PSF.

How wide is the PSF? For large contact holes, the normalized intensity at a position corresponding to the mask edge (that is, at the desired contact hole width) is about 0.25–0.3. If we use this intensity range to measure the width of the PSF, the result is a contact hole between 0.66 and 0.70 λ/NA wide. Thus, this width represents the smallest possible contact hole that could be imaged with a conventional chrome on glass (i.e. not phase-shifted) mask. If the contact hole size on the mask approaches or is made smaller than

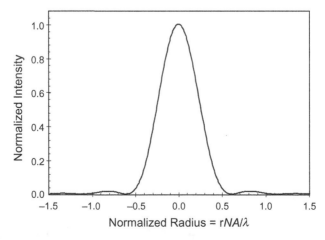

Figure 2.10 *Ideal point spread function (PSF), the normalized image of an infinitely small contact hole. The radial position is normalized by multiplying by* NA/λ

this value, the printed image is controlled by the PSF, not by the dimensions of the mask! Making the contact size on the mask smaller only reduces the intensity of the image peak. Thus, this width of the PSF, the ultimate resolution of a chrome on glass (COG) contact hole, is a *natural resolution* of the imaging system.

The PSF can also be interpreted using linear systems theory. For the coherent illumination discussed so far, the Fourier transform of the electric field image is equal to the pupil function multiplied by the Fourier transform of the mask transmittance.

$$\mathcal{F}\{E(x,y)\} = T_m(f_x,f_y)P(f_x,f_y) \tag{2.75}$$

Thus, the electric field image is linear in Fourier space, with the pupil function serving the role of the system transfer function. In real space, applying the convolution theorem of the Fourier transform to Equation (2.75) gives

$$E(x,y) = t_m(x,y) \otimes \mathcal{F}^{-1}\{P(f_x,f_y)\} \tag{2.76}$$

The inverse Fourier transform of the pupil function, as we saw above, is just the electric field PSF. Thus, for coherent illumination, the electric field image is the electric field mask transmittance convolved with the electric field PSF.

In linear systems theory, the electric field PSF would be called the *impulse response* of the imaging system. An infinitely small pinhole on the mask produces a delta function or 'impulse' of light as the object, so that the PSF, being the image of that impulse object, is called the impulse response of the imaging system. For fields, the Green's function plays the equivalent role of the impulse response and the PSF will sometimes be referred to as a Green's function.

2.2.8   Reduction Imaging

The above discussion makes a useful simplifying assumption – that the imaging system is 1×, or unit magnification, so that the dimensions of the mask patterns are nominally the same as those of the image being produced. In reality, most imaging tools used in lithography are 4× to 10× reduction systems (with 4× the most common), so that the mask is made 4× to 10× bigger than the desired pattern sizes on the wafer. Fortunately, this reduction can be accounted for by a simple scaling relationship so that imaging with reduction is treated mathematically the same as for a 1× system. For a reduction ratio R, all lateral (x and y) dimensions at the wafer are equal to the mask dimensions divided by R. Additionally, the sine of the angles of diffraction are R times greater on the wafer side (exit side) of the lens compared to the mask side (entrance side). Thus, the common numerical aperture used to describe the size of the lens is more properly defined as the exit pupil numerical aperture, the entrance pupil NA being R times smaller (Figure 2.11). Likewise, spatial frequencies on the wafer side are R times larger than spatial frequencies on the mask side (a property of any good imaging system known as the *Abbe sine condition*):

$$n_w \sin\theta_w = Rn_m \sin\theta_m \tag{2.77}$$

One can readily see that diffraction calculations using the mask dimensions and the mask side numerical aperture produce the same results as calculations using the mask scaled to wafer dimensions when using the wafer side NA. Additionally, all longitudinal

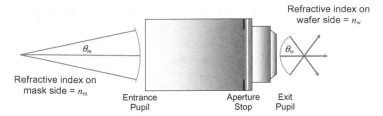

Refractive index on wafer side = n_w

θ_w

Refractive index on mask side = n_m

θ_m

Entrance Pupil   Aperture Stop   Exit Pupil

Figure 2.11 *An imaging lens with reduction R scales both the lateral dimensions and the sine of the angles by R*

(z, the focus direction) dimensions at the wafer are reduced by approximately R^2 compared to focus variations at the mask (see Chapter 3 for a more exact description).

Reduction (or magnification) in an imaging system adds an interesting complication. Light entering the objective lens can be assumed to leave the lens with no loss in energy (the lossless lens assumption). However, if there is reduction or magnification in the lens, the *intensity* distribution of the light entering will be different from that leaving since the intensity is the energy spread over a changing area. The result is a radiometric correction well known in optics.[3] Conservation of energy requires that the amplitude of the electric field passing through the pupil at a wafer-side diffraction angle θ_w (corresponding to an angle at the mask side of θ_m) be modified by the radiometric correction

$$Radiometric\ Correction = \sqrt{\frac{\cos\theta_m}{\cos\theta_w}} = \left(\frac{1-\left(\frac{n_w}{n_m}\right)^2\frac{\sin^2\theta_w}{R^2}}{1-\sin^2\theta_w}\right)^{0.25} \tag{2.78}$$

where R is the reduction factor (for example, 4.0 for a 4× reduction imaging tool, see Figure 2.12) and Equation (2.77) was used to covert mask-side angles to the wafer side. This radiometric correction can be incorporated directly into the lens pupil function in terms of the wafer-side spatial frequencies:

$$P(f_x, f_y) = P_{ideal}(f_x, f_y)\left(\frac{1-\frac{\lambda^2(f_x^2+f_y^2)}{n_m^2 R^2}}{1-\frac{\lambda^2(f_x^2+f_y^2)}{n_w^2}}\right)^{0.25} \tag{2.79}$$

Note that for R of about four and higher, the radiometric correction can be well approximated as

$$P(f_x, f_y) \approx P_{ideal}(f_x, f_y)\left(1-\frac{\lambda^2(f_x^2+f_y^2)}{n_w^2}\right)^{-0.25} \tag{2.80}$$

The existence of this radiometric correction phenomenon gives rise to an interesting question about image intensity normalization. In essence, for aerial image normalization,

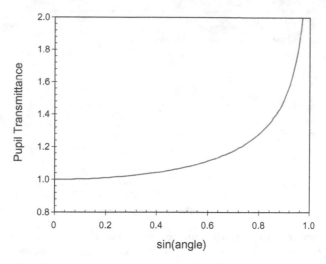

Figure 2.12 *A plot of the radiometric correction as a function of the wafer-side diffraction angle (R = 4, $n_w/n_m = 1$)*

there are two choices: wafer level and mask level normalization. For example, one could define a unit relative intensity as the intensity obtained at the wafer for a large open-field exposure (blank glass mask in the reticle plane). Alternatively, one could define the unit intensity as the intensity transmitted by a large open area on a mask (and suitably multiplied by the reduction ratio squared to take the area change between mask and wafer into account). If the radiometric correction is ignored, then these normalization approaches are identical by invocation of the 'lossless lens' approximation: the energy leaving the mask is 100% transmitted to the wafer and thus the intensity just below the mask is identical to the intensity striking the wafer for a blank quartz mask. However, if the radiometric correction is taken into account, this simple, intuitive relationship no longer holds (see Figure 2.13). Conservation of energy requires that, for a system with magnification or reduction, the intensity of light will change from one side of the lens to the other if that light is traveling at a non-normal direction. As a result, when the radiometric correction is taken into account, the calculated aerial images will be different for a system with a nonunit reduction ratio depending on whether mask-side or wafer-side normalization is used. The location of the energy sensor on the imaging tool, whether at the mask plane or the wafer plane, dictates which normalization scheme is most appropriate (most lithographic imaging tools have their dose sensors at the wafer plane).

More importantly than intensity normalization, the radiometric correction impacts the shape and quality of an aerial image. By increasing the pupil transmittance for high spatial frequencies, the radiometric correction can actually improve image quality for some masks and illuminators by making the higher-frequency diffraction orders (which carry the information about the smallest features on the mask) brighter.

Of course, the purpose of reduction in imaging has nothing to do with the radiometric correction. Masks are made larger by some multiple in order to simplify the making of the mask. Making masks with features that are 4× bigger than on the wafer is far easier

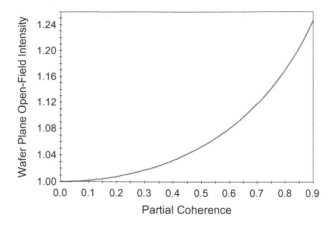

Figure 2.13   *One impact of the radiometric correction: the intensity at the wafer plane for a unit amplitude mask illumination as a function of the partial coherence factor* (R = 4, NA = 0.9)

than making 1× features on the mask. However, there are two important trade-offs with increasing reduction ratio. For a given image field size, the physical dimensions of the mask grow as the reduction ratio. For a 25 × 25 mm image field and a 4× mask, the resulting mask size would need to accommodate a 100 × 100 mm region. If the reduction ratio were made larger, the mask size could become too large to handle effectively, especially with respect to maintaining proper registration of the features over the larger mask area. Additionally, higher reduction ratios lead to more complex lens designs (the symmetry of 1× imaging leads to the simplest lens designs).

2.3   Partial Coherence

Although we have completely described the behavior of one simple, ideal imaging system, we must add an important complication before we describe the operation of a projection system for lithography. So far, we have assumed that the mask is illuminated by *spatially coherent* light. Coherent illumination means simply that the light striking the mask arrives from only one direction. We have further assumed that the coherent illumination on the mask is normally incident. But real lithographic tools use a more complicated form of illumination, where light strikes the mask from a range of angles.

(As a note, the coherence described here is a *spatial* coherence, as opposed to the related but different concept of *temporal* coherence. Spatial coherence can be thought of as the phase relationship between light at two points in space at one instance in time. Temporal coherence looks at light at one point in space but at different times. While temporal coherence often entails spatial coherence, the existence of spatial coherence usually does not imply anything about the temporal coherence of the light. Unfortunately, most technical literature will refer to one or the other of these properties as simply 'coherence'. Thus, in this book, coherence means spatial coherence unless specifically described otherwise.)

2.3.1 Oblique Illumination

For spatially coherent illumination (a plane wave of illumination normally incident on the mask), diffraction resulted in a pattern that was centered in the entrance to the objective lens. What would happen if we changed the direction of the illumination so that the plane wave struck the mask at some angle θ'? While the incident electric field will still have a magnitude of 1, the phase of this plane wave will vary linearly across the mask. For a plane wave tilted relative to the x-axis,

$$E_i(x,y) = e^{i2\pi\sin\theta' x/\lambda} \tag{2.81}$$

Under the Kirchhoff boundary condition assumption, the effect is simply to shift the position of the diffraction pattern with respect to the lens aperture (in terms of spatial frequency, the amount shifted is $f'_x = \sin\theta'/\lambda$), as seen in Figure 2.14 (see Problem 2.16). Recalling that only the portion of the diffraction pattern passing through the lens aperture is used to form the image, it is quite apparent that this shift in the position of the diffraction pattern can have a profound effect on the resulting image. Letting f'_x and f'_y be the shift in the spatial frequency due to the tilted illumination, Equation (2.61) becomes

$$E(x,y,f'_x,f'_y) = \mathcal{F}^{-1}\{T_m(f_x - f'_x, f_y - f'_y)P(f_x,f_y)\}$$
$$I(x,y,f'_x,f'_y) = |E(x,y,f'_x,f'_y)|^2 \tag{2.82}$$

The impact of shifting the diffraction pattern within the objective lens pupil on the quality of the resulting image is hard to generalize. In some cases, the result might be loss of an important diffraction order that otherwise would have made it through the lens. In another case, the shift might enable an otherwise lost diffraction order to be captured. The impact will be dependent on the incident angle, the mask pattern (for example, the pitch) and on the spatial frequency cutoff, NA/λ. Although the lithographic impact is complicated, some general rules will be developed in Chapter 10.

One of the most important lithographic consequences of tilting the illumination is a change in the resolution. Equation (2.57) showed that for normally incident illumination, the smallest half-pitch that could be imaged was $0.5\lambda/NA$. By tilting the illumination so

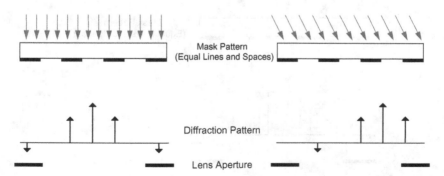

Figure 2.14 *The effect of changing the angle of incidence of plane wave illumination on the diffraction pattern is simply to shift its position in the lens aperture. A positive illumination tilt angle is depicted here*

that the zero order of light just passes through the right edge of the lens, the pitch can be reduced until the −1st order just passes through the left edge of the lens. Thus, the distance between diffraction orders (1/p) is equal to the diameter of the lens (2NA/λ) and the smallest half-pitch becomes

$$R = 0.25 \frac{\lambda}{NA} \tag{2.83}$$

Tilting the illumination has the potential for improving resolution by a factor of 2.

2.3.2   Partially Coherent Illumination

If the illumination of the mask is composed of light coming from a range of angles rather than just one angle, the illumination is called *partially coherent*. If one angle of illumination causes a shift in the diffraction pattern, a range of angles will cause a range of shifts, resulting in broadened diffraction orders (Figure 2.15). The most common type of partially coherent source (called a 'conventional' source) provides a uniform range of angles illuminating the mask, with each angle having equal intensity. One can characterize the range of angles used for the illumination in several ways, but the most common is the *partial coherence factor*, σ (also called the degree of partial coherence, the pupil filling function, or just the partial coherence). The partial coherence is defined as the sine of the half-angle of the illumination cone divided by the objective lens numerical aperture:

$$\sigma = \frac{n \sin(\theta_{max})}{NA} = \frac{source\ diameter}{lens\ diameter} \tag{2.84}$$

(where a wafer-side equivalent illumination angle is used, calculated from the Abbe sine condition). It is thus a measure of the angular range of the illumination relative to the angular acceptance of the lens. As we shall see in section 2.3.7, the illumination system creates an image of the light source at the objective lens pupil. Thus, a conventional illumination source will appear as a circle of uniform intensity when imaged into the pupil

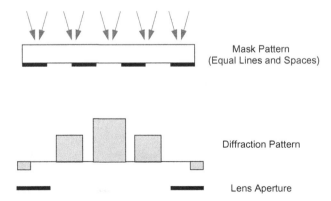

Figure 2.15   *The diffraction pattern is broadened by the use of partially coherent illumination (plane waves over a range of angles striking the mask)*

(Figure 2.16). Another form of the definition of the partial coherence factor of a conventional source is the diameter of the source disk divided by the diameter of the objective lens aperture. Finally, if the range of angles striking the mask extends from -90 to $90°$ (that is, all possible angles), the illumination is said to be *incoherent*. Table 2.2 helps define the terminology of spatial coherence. Note that the maximum partial coherence factor (assuming the mask remains in air) is R/NA, where R is the reduction ratio. Thus, perfectly incoherent illumination is possible only in the limit of very small numerical apertures. However, from an imaging perspective any partial coherence factor greater than 2 is essentially incoherent and is usually treated as such.

An arbitrarily shaped illumination source can be described as an *extended source*, where the source shape is divided into a collection of individual point sources. Each point is a source of spatially coherent illumination, producing a plane wave striking the mask at one angle, and resulting in an aerial image given by Equation (2.82). Two point sources from the extended source, however, do not interact coherently with each other. Thus, the contributions of these two sources must be added to each other incoherently (that is, the intensities are added together). Illumination systems for lithographic tools are carefully designed to have this property. If two spatially separated points on the source were phase-related, light from those points would interfere at the mask plane, creating nonuniform illumination (the phenomenon of speckle is an example of such coherent interaction of source points).

The full aerial image is determined by calculating the coherent aerial image from each point on the source, and then integrating the intensity over the source (an approach known as the Abbe method of partially coherent image calculation). The source can be defined

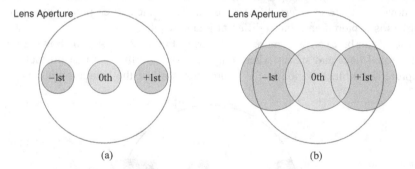

(a)   (b)

Figure 2.16 *Top-down view of the lens aperture showing the diffraction pattern of a line/ space pattern when partially coherent illumination is used: (a) $\sigma = 0.25$; and (b) $\sigma = 0.5$*

Table 2.2 *Partial coherence types (conventional illumination)*

Illumination Type	Partial Coherence Factor	Source Shape
Coherent	$\sigma = 0$	Point source
Incoherent	$\sigma = \infty$	Infinite size source
Partially Coherent	$0 < \sigma < \infty$ (but generally $0 < \sigma < 1$)	Circular disk-shaped source

by a source function, $S(f_x', f_y')$, which is just the intensity of the source as a function of position (or angle). The total intensity of the image is then

$$I_{\text{total}}(x,y) = \frac{\iint I(x,y,f_x',f_y')S(f_x',f_y')\mathrm{d}f_x'\mathrm{d}f_y'}{\iint S(f_x',f_y')\mathrm{d}f_x'\mathrm{d}f_y'} \qquad (2.85)$$

Often, the pupil coordinates (f_x, f_y) are normalized in the same way as the partial coherence factor, creating pupil coordinates in 'sigma' space. These coordinates are

$$\sigma_x = \frac{f_x\lambda}{NA}, \quad \sigma_y = \frac{f_y\lambda}{NA} \qquad (2.86)$$

In sigma coordinates, the lens pupil is always the unit circle. Also, the source shape function S can be conveniently normalized as

$$\tilde{S}(\sigma_x',\sigma_y') = \frac{S(\sigma_x',\sigma_y')}{\iint S(\sigma_x',\sigma_y')\mathrm{d}\sigma_x'\mathrm{d}\sigma_y'} \qquad (2.87)$$

so that

$$I_{\text{total}}(x,y) = \iint I(x,y,\sigma_x',\sigma_y')\tilde{S}(\sigma_x',\sigma_y')\mathrm{d}\sigma_x'\mathrm{d}\sigma_y' \qquad (2.88)$$

The examples of Figure 2.16 illustrate an important difficulty in calculating images with partially coherent illumination. In Figure 2.16a, all of the zero and first diffraction orders are captured. When Equation (2.85) is evaluated, the aerial image for each source point, $I(x,y,f_x',f_y')$, will be identical, being made up of three diffraction orders. Thus, the partially coherent image is equal to the coherent image. For the case shown in Figure 2.16b, however, there is a very different result. For some source points, only one of the first orders is captured and thus a different image results.

This idea can be expressed more clearly using Figure 2.17. Again showing a case of imaging equal lines and spaces where only the zero and the two first diffraction orders are captured, the black region in the figure shows all of the source points where three

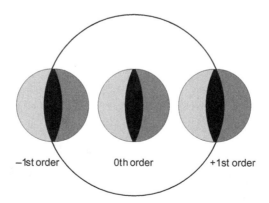

Figure 2.17 *Example of dense line/space imaging where only the zero and first diffraction orders are used. Black represents three-beam imaging, lighter and darker grays show the area of two-beam imaging*

diffraction orders are captured. This region of the source produces 'three beam' imaging, equal to the coherent aerial image of Equation (2.66). The gray areas, however, result in only the zero order and one first order captured by the lens (the light gray area forms an image from the 0 and +1st orders, the dark gray area forms an image from the 0 and −1st orders). These areas form a 'two-beam' image, given by

$$I(x) = a_0^2 + a_1^2 + 2a_0 a_1 \cos(2\pi x/p) \tag{2.89}$$

The final image will be a weighted average of the two-beam and three-beam images, where the weights are their respective source intensity fractions. Thus, for the case of a uniform intensity source shape and a repeating pattern of lines and spaces, the solution to the partially coherent imaging integrals of Equation (2.85) becomes a geometry problem of calculating the overlapping areas of the various circles for the source and the lens.

This approach, often called the Kintner method,[4] is readily solvable for the case of a circularly shaped source. Every image takes the form of

$$I(x) = s_1 I_{1\text{-beam}} + s_2 I_{2\text{-beam}} + s_3 I_{3\text{-beam}} \tag{2.90}$$

where

$$I_{1\text{-beam}} = a_0^2$$
$$I_{2\text{-beam}} = a_0^2 + a_1^2 + 2a_0 a_1 \cos(2\pi x/p)$$
$$I_{3\text{-beam}} = a_0^2 + 4a_0 a_1 \cos(2\pi x/p) + 4a_1^2 \cos^2(2\pi x/p)$$

The weights s_i take on different values for different regimes of pitch and partial coherence, as given in Table 2.3, and always add up to 1. This table makes use of the following values (making sure that the inverse tangent is defined over the range from 0 to π):

$$\chi = \frac{1 + \left(\dfrac{\lambda}{pNA}\right)^2 - \sigma^2}{\dfrac{2\lambda}{pNA}} \tag{2.91}$$

Table 2.3 *Kintner weights for conventional source imaging of small lines and spaces*

Regime	Description	s_1	s_2	s_3
$\dfrac{\lambda}{pNA} \le 1 - \sigma$	All three-beam imaging	0	0	1
$1 - \sigma \le \dfrac{\lambda}{pNA} \le \sqrt{1 - \sigma^2}$	Combination of two- and three-beam imaging	0	2γ	$1 - 2\gamma$
$\sqrt{1 - \sigma^2} \le \dfrac{\lambda}{pNA} \le 1$	Combination of one-, two- and three-beam imaging	$\eta + 2\gamma - 1$	$2(1 - \gamma - \eta)$	η
$1 \le \dfrac{\lambda}{pNA} \le 1 + \sigma$	Combination of one- and two-beam imaging	$2\gamma - 1$	$2(1 - \gamma)$	0
$1 + \sigma \le \dfrac{\lambda}{pNA}$	Only the zero order is captured, all one-beam	1	0	0

$$\gamma = \frac{\sigma^2 \tan^{-1}\left(\dfrac{\sqrt{1-\chi^2}}{\chi - \dfrac{\lambda}{pNA}}\right) - \cos^{-1}(\chi) + \dfrac{\lambda}{pNA}\sqrt{1-\chi^2}}{\pi\sigma^2} \qquad (2.92)$$

$$\eta = \frac{2\left[\cos^{-1}\left(\dfrac{\lambda}{pNA}\right) - \dfrac{\lambda}{pNA}\sqrt{1-\left(\dfrac{\lambda}{pNA}\right)^2}\right]}{\pi\sigma^2} \qquad (2.93)$$

2.3.3   Hopkins Approach to Partial Coherence

H.H. Hopkins reformulated the Abbe extended source method for partially coherent image calculations in an interesting and occasionally useful way.[5,6] Equation (2.82) gives the electric field for a single source point by shifting the diffraction pattern relative to a fixed pupil. A change of variables will show that this is exactly equivalent to fixing the diffraction pattern and shifting the pupil by the same amount:

$$E(x, f_x') = \int_{-\infty}^{\infty} P(f_x + f_x')T_m(f_x)e^{2\pi i(f_x + f_x')x}df_x \qquad (2.94)$$

where only one dimension is shown for simplicity. Calculating the image intensity for this single source point as the product of the electric field and its complex conjugate,

$$I(x, f_x') = E(x, f_x')E^*(x, f_x')$$

$$= \left[\int_{-\infty}^{\infty} P(f_x + f_x')T_m(f_x)e^{2\pi i(f_x + f_x')x}df_x\right]\left[\int_{-\infty}^{\infty} P^*(\tilde{f}_x + f_x')T_m^*(\tilde{f}_x)e^{-2\pi i(\tilde{f}_x + f_x')x}d\tilde{f}_x\right] \qquad (2.95)$$

$$= \int_{-\infty}^{\infty}\int_{-\infty}^{\infty} P(f_x + f_x')P^*(\tilde{f}_x + f_x')T_m(f_x)T_m^*(\tilde{f}_x)e^{2\pi i(f_x - \tilde{f}_x)x}df_x d\tilde{f}_x$$

The next step in the Abbe extended source method is to integrate over all of the source points as in Equation (2.85) or (2.88). The critical step in the Hopkins formulation is to change the order of integration and to integrate over the source first, before carrying out the inverse Fourier transform integrals of Equation (2.95). Since only the pupil functions in Equation (2.95) depend on the source point, The integration over the source gives an intermediate variable

$$TCC(f_x, \tilde{f}_x) = \int_{-\infty}^{\infty} P(f_x + f_x')P^*(\tilde{f}_x + f_x')\tilde{S}(f_x')df_x' \qquad (2.96)$$

so that

$$I_{total}(x) = \int_{-\infty}^{\infty}\int_{-\infty}^{\infty} TCC(f_x, \tilde{f}_x)T_m(f_x)T_m^*(\tilde{f}_x)e^{2\pi i(f_x - \tilde{f}_x)x}df_x d\tilde{f}_x \qquad (2.97)$$

The intermediate variable *TCC* in Equation (2.96) is called the *transmission cross-coefficient*. It has the useful property of being independent of the mask transmittance function – it depends only on the pupil function and the source shape. For applications where the pupil and source functions are fixed, but images must be calculated for a wide range for different masks, precalculation of the *TCC* function can greatly speed up image calculations. Optical proximity correction (OPC), discussed in Chapter 10, is one example where this approach becomes useful. Physically, the *TCC* can be thought of as an intensity-based description of how two separate pupil points interact when forming an image. Of course, the one-dimensional formulations above can be easily extended to two dimensions.

2.3.4 Sum of Coherent Sources Approach

An approximate solution to the Hopkins imaging equations called *Sum of Coherent Sources* (SOCS) provides extremely fast computation times and is thus very useful for some imaging calculation like those used for optical proximity correction.[7] To begin, we will convert the final Hopkins equation (2.97) from the spatial frequency domain to the spatial domain. From the definition of the Fourier transform of the mask transmittance,

$$T_m(f_x) = \int_{-\infty}^{\infty} t_m(x) e^{-i2\pi f_x x} dx, \quad T_m * (\tilde{f}_x) = \int_{-\infty}^{\infty} t_m^*(x) e^{i2\pi \tilde{f}_x x} dx \qquad (2.98)$$

Also, noting that the electric field point spread function is the Fourier transform of the pupil function,

$$h(x) = \int_{-\infty}^{\infty} P(f_x) e^{-i2\pi f_x x} df_x \qquad (2.99)$$

and defining the *mutual coherence function* as the Fourier transform of the normalized source shape,

$$J_0(x) = \int_{-\infty}^{\infty} \tilde{S}(f_x) e^{-i2\pi f_x x} df_x \qquad (2.100)$$

the Hopkins imaging integral can be converted to an integration over x (that is, an integration over the object plane) rather than an integration over the pupil plane:

$$I(x) = \int_{-\infty}^{\infty} \int_{-\infty}^{\infty} J_0(x' - x'') h(x - x') h^*(x - x'') t_m(x') t_m^*(x'') dx' dx'' \qquad (2.101)$$

Defining a function W that depends only on the source and the pupil,

$$W(a,b) = J_0(a - b) h(a) h^*(b) \qquad (2.102)$$

this new Hopkins integral becomes

$$I(x) = \int_{-\infty}^{\infty} \int_{-\infty}^{\infty} W(x - x', x - x'') t_m(x') t_m^*(x'') dx' dx'' \qquad (2.103)$$

So far, there is no real advantage to this formulation of the Hopkins imaging equation. Let us now propose an interesting decomposition for W as the weighted sum of its eigenvectors ϕ_n:

$$W(a,b) = \sum_{n=1}^{\infty} \lambda_n \phi_n(a) \phi_n^*(b) \tag{2.104}$$

where λ_n is the eigenvalue for the eigenvector ϕ_n. Expressed in this way, the Hopkins integral becomes

$$
\begin{aligned}
I(x) &= \sum_{n=1}^{\infty} \lambda_n \int_{-\infty}^{\infty} \int_{-\infty}^{\infty} \phi_n(x-x')\phi_n^*(x-x'')t_m(x')t_m^*(x'')\mathrm{d}x'\mathrm{d}x'' \\
&= \sum_{n=1}^{\infty} \lambda_n \int_{-\infty}^{\infty} \phi_n(x-x')t_m(x')\mathrm{d}x' \int_{-\infty}^{\infty} \phi_n^*(x-x'')t_m^*(x'')\mathrm{d}x'' \\
&= \sum_{n=1}^{\infty} \lambda_n \left| \phi_n(x) \otimes t_m(x) \right|^2
\end{aligned}
\tag{2.105}
$$

Thus, for a given source shape and pupil function, a set of eigenvectors and eigenvalues can be determined, each eigenvector playing the role of a coherent PSF-like function. These eigenvectors are then convolved with the mask transmittance function to give an electric field at the wafer plane. Converting to intensity, the weighted sum of these image components becomes the total image. The eigen functions also have the convenient property of being orthogonal to each other. Note that for the case of coherent illumination, $J_0(x) = 1$, $\lambda_1 = 1$, $\phi_1 = h(x)$ and $\lambda_n = 0$ for $n > 1$. In fact, one simple decomposition is to let each n represent a coherent point on the source, so that λ_n is the intensity of the source point and ϕ_n is just the coherent PSF with a phase tilt corresponding to the position of that source point. Thought of in this way, Equation (2.105) is just an expression of the Abbe extended source formulation expressed as convolutions in the object plane rather than products in the pupil plane.

Of course, for any real calculation of an image using this approach, the infinite summation must be cut off at some number N:

$$I(x) \approx \sum_{n=1}^{N} \lambda_n \left| \phi_n(x) \otimes t_m(x) \right|^2 \tag{2.106}$$

where higher values of N lead to greater accuracy in the calculation. The key insight into the use of SOCS is to find a set of eigen functions that allows N to be minimized. Letting each term in the summation represent one source point often requires hundreds of terms to produce the desired accuracy. By cleverly choosing a different set of eigen functions, the required number of terms in the summation can be reduced by an order of magnitude or so. However, since a new set of eigen functions and eigenvalues must be determined every time the source or the pupil function is changed, this approach tends to be used only for cases where the source and pupil are fixed and images for a large number of mask patterns must be determined.

2.3.5   Off-Axis Illumination

Off-axis illumination refers to any illumination shape that significantly reduces or eliminates the 'on-axis' component of the illumination, that is, the light striking the mask at
near normal incidence (Figure 2.14). By tilting the illumination away from normal incidence, the diffraction pattern of the mask is shifted within the objective lens. While all
the reasons why this is desirable are not yet obvious (resolution is one benefit, but another
involves depth of focus, a concept not yet discussed), there are many lithographic applications where off-axis illumination is preferred. The most popular illumination shapes,
besides the conventional disk shape, are annular illumination, some form of quadrupole
illumination, and dipole illumination, although many other shapes are used for special
applications (see Chapter 10).

The idealized source shapes pictured in Figure 2.18 can be thought of as 'top-hat' illuminators: the intensity is one inside the source shape and zero outside of it. However,
since these source shapes are created using a variety of optical components, the real source
distributions of a real projection tool can be significantly more complicated. Figure 2.19
shows some actual measured source shapes from two different production lithographic

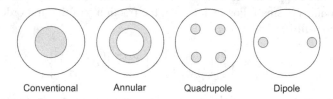

Conventional   Annular   Quadrupole   Dipole

Figure 2.18 *Conventional illumination source shape, as well as the most popular off-axis
illumination schemes, plotted in spatial frequency coordinates. The outer circle in each
diagram shows the cutoff frequency of the lens*

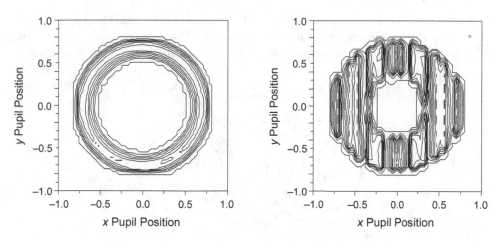

Figure 2.19 *Examples of measured annular source shapes (contours of constant intensity)
from two different lithographic projection tools showing a more complicated source distribution than the idealized 'top-hat' distributions*

exposure tools. While both of these source shapes are annular, they exhibit a gradual falloff of intensity with pupil radius rather than a sharp cutoff, and a fine structure that is specific to the optical method of generating the shape.

In many cases, the real source shape can be approximated by the 'designed' source shape (the idealized top-hat illumination) convolved with an illumination point spread function, to account for the finite resolution of the illumination system. Often, the illumination point spread function can reasonably be approximated as a Gaussian.

2.3.6   Imaging Example: Dense Array of Lines and Spaces Under Annular Illumination

Off-axis illumination, with annular illumination being a common example, is most often used to push the resolution limits of small, dense patterns. Thus, for the case of a dense array of lines and spaces, these small features will generally only admit the zero and ±first diffraction orders through the lens (Figure 2.20). As before, the image will depend on the fraction of the first order that is captured by the lens, and whether this fraction results in only two-beam imaging, or a combination of two-beam and three-beam imaging.

While a full Kintner-like solution is certainly possible (see section 2.3.2), here we will develop a simplified approximate solution that will be reasonably accurate whenever the width of the annulus is small. For such a case, the fraction γ of the first order that passes through the lens will be[8]

$$\gamma = \frac{1}{\pi}\cos^{-1}\left\{\frac{1}{2}\left[\left(\frac{\lambda}{pNA}\right)\frac{1}{\sigma} + \left(\frac{pNA}{\lambda}\right)\left(\sigma - \frac{1}{\sigma}\right)\right]\right\} \tag{2.107}$$

where σ is the radial position of the center of the annulus in σ-coordinates. Using the notation of Equation (2.90), the final image will be the weighted sum of one-beam, two-beam and three-beam images, with weights given in Table 2.4.

2.3.7   Köhler Illumination

In the descriptions of imaging above, the mask was always illuminated by a plane wave, either normally or obliquely incident. Also, in the drawings and analyses of the diffraction patterns given above, the assumption was made that the mask pattern was positioned

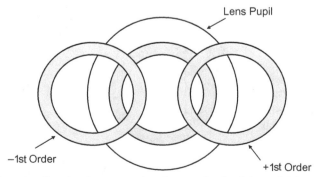

Figure 2.20   *Annular illumination example where only the 0th and ±1st diffracted orders pass through the imaging lens*

Table 2.4 *Kintner weights for thin annular source imaging of small lines and spaces*

Regime	Description	s_1	s_2	s_3
$\dfrac{\lambda}{pNA} \leq 1-\sigma$	All three-beam imaging	0	0	1
$1-\sigma \leq \dfrac{\lambda}{pNA} \leq \sqrt{1-\sigma^2}$	Combination of two- and three-beam imaging	0	$2(1-\gamma)$	$2\gamma - 1$
$\sqrt{1-\sigma^2} \leq \dfrac{\lambda}{pNA} \leq 1+\sigma$	Combination of one- and two-beam imaging	$1-2\gamma$	2γ	0
$1+\sigma \leq \dfrac{\lambda}{pNA}$	Only the zero order is captured, all one-beam	1	0	0

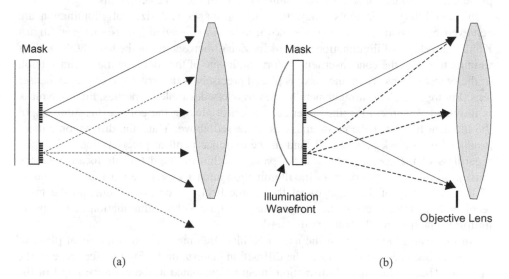

Figure 2.21 *The impact of illumination: (a) when illuminated by a plane wave, patterns at the edge of the mask will not produce diffraction patterns centered in the objective lens pupil, but (b) the proper converging spherical wave produces diffraction patterns that are independent of field position*

directly on the optical axis (the center line running through the middle of the lens). The mask and the resulting image field, however, are reasonably large so that imaging takes place for mask features that are a long way from the optical axis. This poses a significant problem: diffraction patterns produced by features toward the edge of the mask will produce diffraction patterns that are not centered in the lens (see Figure 2.21a).

The shifting behavior of oblique illumination offers a solution to the problem. It is very desirable that patterns at all positions of the mask be printed with about equal imaging characteristics. Thus, identical mask patterns at different field positions should produce diffraction patterns that are positioned identically within the objective lens pupil. The diffraction pattern for the line/space pattern at the mask edge of Figure 2.21a can be made to behave identically with the mask pattern on the optical axis by shifting the diffraction pattern up so that it is centered in the lens. This shift, as in section 2.3.1, can be accomplished by tilting the illumination. In fact, at every point on the mask the diffraction patterns can be made to coincide if the illumination is properly tilted for that mask point. The ideal tilt is achieved across the entire mask if the illumination is made to be a spherical wave converging to a point in the entrance pupil of the objective lens (Figure 2.21b).

Note that all of the previous calculations of diffraction patterns assumed a plane wave of illumination striking the mask. If a converging spherical wave is used instead as the illumination, the previous results are still valid so long as the radius of curvature of the spherical wave (that is, the distance from the mask to the objective lens entrance pupil) is sufficiently large compared to the feature dimensions on the mask, so that the spherical wave appears like a plane wave over the region of interest on the mask. This is most certainly true in the Fraunhofer diffraction region. Thus, even though spherical wave illumination will be used, plane wave illumination can still be assumed in the diffraction calculations.

In most lithographic tools today, the various shapes and sizes of illumination are achieved by an optical arrangement known as *Köhler illumination*. Named for August Köhler, developer of illumination systems for Zeiss microscopes in the late 1800s, Köhler illumination uses the condenser lens to form an image of the source at the entrance pupil of the objective lens. Then, the mask is placed precisely at the exit pupil of the condenser lens (Figure 2.22). This arrangement has several very desirable properties. First, the mask is illuminated with converging spherical waves, producing the correct directionality of the incident light over the entire mask as discussed above. Thus, the diffraction pattern produced by a mask pattern will land in the entrance pupil at precisely the same spot, regardless of the position of the mask pattern within the field (i.e. its location on the mask). Second, the uniformity of the illumination intensity over the mask is independent of the uniformity of the intensity of the source itself (since every point on the mask receives light from every point on the source). Thus, Köhler illumination helps ensure uniform imaging for all points on the mask.

An interesting side effect of the use of Köhler illumination is that the focal plane of the source becomes the far field of the diffraction pattern, in the Fraunhofer sense of the far field. That is, the Fresnel diffraction integrals evaluated at the entrance pupil of the objective lens become equal to the Fraunhofer diffraction integrals without further

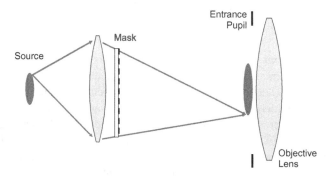

Figure 2.22 *Köhler illumination where the source is imaged at the entrance pupil of the objective lens and the mask is placed at the exit pupil of the condenser lens*

approximation. By making only the Fresnel diffraction approximations, the diffraction pattern can still be calculated as the Fourier transform of the mask transmittance. In essence, the condenser lens brings the 'infinite' diffraction plane in closer, to the image plane of the condenser lens.

2.3.8   Incoherent Illumination

As mentioned above, as the size of the illumination source becomes large relative to the size of the pupil (that is, as σ for an ideal conventional source goes to infinity), the illumination is said to be spatially incoherent. For this case, integration over the source can be greatly simplified. Using the Hopkins formulation, the TCC becomes

$$TCC(f_x, \tilde{f}_x) \propto \int_{-\infty}^{\infty} P(f_x + f_x')P^*(\tilde{f}_x + f_x')\mathrm{d}f_x' = \int_{-\infty}^{\infty} P(f_x')P^*(\tilde{f}_x - f_x + f_x')\mathrm{d}f_x' \quad (2.108)$$

As the final form of this equation indicates, the TCC will only be a function of the difference in spatial frequencies $\tilde{f}_x - f_x$. Thus,

$$TCC(\tilde{f}_x - f_x) \propto P(f_x) \otimes P^*(f_x) \quad (2.109)$$

where the proportionality is made into an equality through proper normalization. For incoherent illumination, the *TCC* is called the *Optical Transfer Function* (OTF). The magnitude the OTF is called the *Modulation Transfer Function* (MTF). It is usually normalized by dividing by the area of the pupil. For an ideal circular pupil, this convolution can be easily carried out (being interpreted geometrically as the area overlap between shifted circles, Figure 2.23) giving

$$MTF(f) = \begin{cases} \dfrac{2}{\pi}\left[\cos^{-1}\left(\dfrac{f}{2NA/\lambda}\right) - \dfrac{f}{2NA/\lambda}\sqrt{1 - \left(\dfrac{f}{2NA/\lambda}\right)^2}\right], & f \le 2\dfrac{NA}{\lambda} \\ 0, & \text{otherwise} \end{cases} \quad (2.110)$$

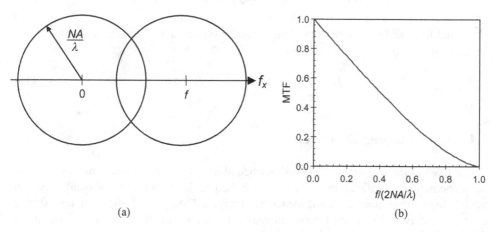

(a)   (b)

Figure 2.23   *The MTF(f) is the overlapping area of two circles of diameter NA/λ separated by an amount f, as illustrated in (a) and graphed in (b)*

where $f = \tilde{f}_x - f_x$ for the case of our 1D features. One can see that the cutoff frequency for the incoherent MTF is $2NA/\lambda$, corresponding to the zero order at one edge of the lens and the first order at the opposite edge.

Using the MTF in Equation (2.97),

$$I_{\text{total}}(x) = \int\limits_{-\infty}^{\infty} \int\limits_{-\infty}^{\infty} MTF(f)T_{\text{m}}(f_x)T_{\text{m}}^*(f_x + f)e^{-2\pi i f x}\,\mathrm{d}f_x\mathrm{d}f$$

$$(2.111)$$

$$I_{\text{total}}(x) = \int\limits_{-\infty}^{\infty} MTF(f)e^{-2\pi i f x}\mathrm{d}f \int\limits_{-\infty}^{\infty} T_{\text{m}}(f_x)T_{\text{m}}^*(f_x + f)\mathrm{d}f_x$$

But, the convolution of the diffraction pattern with its complex conjugate can be easily computed using the convolution theorem of the Fourier transform:

$$\int\limits_{-\infty}^{\infty} T_{\text{m}}(f_x)T_{\text{m}}^*(f_x + f)\mathrm{d}f_x = \mathcal{F}\{t_{\text{m}}(x)t_{\text{m}}^*(x)\}$$

$$(2.112)$$

The product of the electric field transmittance of the mask with its complex conjugate is just the intensity transmittance of the mask. Thus, Equation (2.111) becomes

$$I_{\text{total}}(x) = \mathcal{F}^{-1}\{MTF(f_x)IT_{\text{m}}(f_x)\}$$

$$(2.113)$$

where $IT_{\text{m}}(f_x)$ is the Fourier transform of the intensity transmittance of the mask.

As we saw at the beginning of this chapter, for coherent illumination the electric field image is a linear function of the electric field mask transmittance in Fourier space, with the pupil function as the linear transfer function. Equation (2.113) shows that for incoherent illumination, the intensity image is a linear function of the intensity mask transmittance in Fourier space, with the MTF as the linear transfer function. For partially coherent illumination, the relationship between the image and the object is nonlinear.

As an example of an incoherent image, consider a binary pattern of lines and spaces. The resulting aerial image is

$$I(x) = a_0 + 2\sum_{j=1}^{N} MTF(j/p)a_j \cos(2\pi j x/p)$$

$$(2.114)$$

For equal lines and spaces where only the 0 and ±1st diffraction orders are captured, the aerial image becomes

$$I(x) = \frac{1}{2} + \frac{2}{\pi} MTF(1/p)\cos(2\pi x/p)$$

$$(2.115)$$

2.4   Some Imaging Examples

While most of the imaging examples described above were one-dimensional for the sake of mathematical simplicity, the approach described in this chapter can of course be used to calculate more general two-dimensional images. Figure 2.24 shows a few simple examples of two-dimensional aerial images calculated with the lithography simulator PROLITH. Figure 2.25 shows images for an isolated binary edge mask feature as a function of the partial coherence factor.

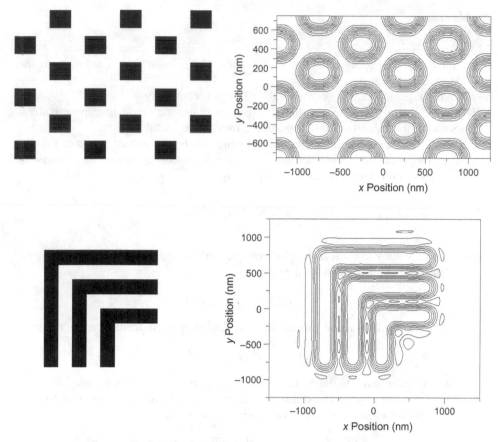

Figure 2.24 *Some examples of two-dimensional aerial image calculations (shown as contours of constant intensity), with the mask pattern on the left (dark representing chrome)*

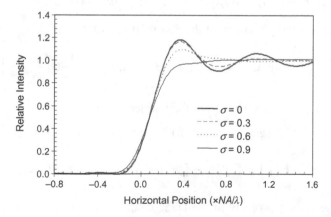

Figure 2.25 *Aerial images of an isolated edge as a function of the partial coherence factor*

Problems

2.1.   Derive Equations (2.5).

2.2.   Show that the spherical wave of Equation (2.27) is in fact a solution to the Helmholtz equation.

2.3.   Using Equation (2.38), determine the SI units for intensity.

2.4.   Derive Equations (2.45) for the isolated space and the dense line/space pattern. Some of the theorems from Appendix C may prove useful.

2.5.   Show that the diffraction pattern for an dense line/space pattern becomes the diffraction pattern for an isolated space as the pitch becomes infinite (for a constant spacewidth w). This is equivalent to proving that

$$\lim_{p \to \infty} \frac{1}{p} \sum_{j=-\infty}^{\infty} \delta\left(f_x - \frac{j}{p}\right) = 1$$

2.6.   Complimentary mask features (for example, an isolated line and an isolated space of the same width) are defined by

$$t_m^c(x,y) = 1 - t_m(x,y)$$

Prove that the diffraction patterns of complimentary mask features are given by

$$t_m^c(f_x,f_y) = \delta(f_x,f_y) - T_m(f_x,f_y)$$

Use this expression to derive the diffraction pattern of an isolated line.

2.7.   Show that the Fourier transform is a linear operation, that is, show that for two functions $f(x,y)$ and $g(x,y)$, and two constants a and b,

$$\mathcal{F}\{af(x,y) + bg(x,y)\} = aF(f_x,f_y) + bG(f_x,f_y)$$

2.8.   Prove the shift theorem of the Fourier transform:

$$\text{If } \mathcal{F}\{g(x,y)\} = G(f_x,f_y), \ \mathcal{F}\{g(x-a,y-b)\} = G(f_x,f_y)e^{-i2\pi(f_x a + f_y b)}$$

2.9.   Prove the similarity theorem of the Fourier transform:

$$\text{If } \mathcal{F}\{g(x,y)\} = G(f_x,f_y), \quad \mathcal{F}\{g(ax,by)\} = \frac{1}{|ab|}G\left(\frac{f_x}{a},\frac{f_y}{b}\right)$$

2.10. Prove the convolution theorem of the Fourier transform:

$$\text{If } \mathcal{F}\{g(x,y)\} = G(f_x,f_y) \quad \text{and} \quad \mathcal{F}\{h(x,y)\} = H(f_x,f_y),$$

$$\mathcal{F}\left\{\int\int_{-\infty}^{\infty} g(\xi,\eta)h(x-\xi,y-\eta)d\xi d\eta\right\} = G(f_x,f_y)H(f_x,f_y)$$

2.11. Prove the scaling property of the Dirac delta function [in Appendix C, Equation (C.10)].

2.12. Derive the diffraction pattern of an isolated edge:

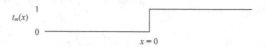

2.13. Derive the diffraction pattern of an isolated double space pattern:

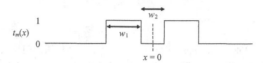

(Hint: Superposition plus the shift theorem may prove useful.)

2.14. Derive the diffraction pattern of an alternating phase shift mask (a pattern of lines and spaces where every other space is shifted in phase by 180°, resulting in a transmittance of −1):

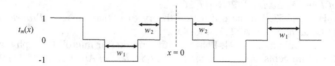

Graph the resulting diffraction pattern out to the ±5th diffraction orders for the case of $w_1 = w_2$.

2.15. Prove Equation (2.59).

2.16. In section 2.3, it was claimed that illuminating a mask with a plane wave at an angle θ' resulted, under the Kirchhoff boundary condition assumption, in a diffraction pattern equal to the normally incident diffraction pattern shifted by an amount $\sin\theta/\lambda$. Prove this statement using the definition of the Fraunhofer diffraction pattern, Equation (2.44).

2.17. Consider a binary pattern of lines and spaces. Which diffraction orders pass through the lens under these circumstances:
  (a) $\lambda = 248\,\text{nm}$, $NA = 0.8$, pitch = 300 nm, on-axis coherent illumination
  (b) Same as (a), but pitch = 400 nm
  (c) Same as (b), but illumination tilted by an angle $\sin\theta = 0.5$

2.18. Equations (2.65) and (2.67) give the electric field and intensity images for three-beam images, where the 0 and ±1st diffraction orders are used to form the image. Derive similar expressions for two-beam images, where only the 0 and +1st diffraction orders are captured by the lens.

2.19. For a repeating line/space pattern and coherent illumination, derive expressions for the aerial image intensity at the center of the line and the center of the space as a function of the number of diffraction orders captured.

2.20. For $\lambda = 248\,\text{nm}$, $NA = 0.8$ and pitch = 400 nm, below what value of σ is the image entirely made up of three-beam interference? At what σ value does one-beam imaging first appear?

2.21. For coherent illumination and a lens of a given *NA* and λ, derive a formula for the electric field image of an isolated edge. The result will be in terms of the sine integral, defined as

$$\mathrm{Si}(z) = \int_0^z \frac{\sin y}{y} \, dy$$

References

1   Goodman, J.W., 1968, *Introduction to Fourier Optics*, McGraw-Hill (New York, NY).
2   Wong, A.K., 2005, *Optical Imaging in Projection Microlithography*, SPIE Press (Bellingham, WA), pp. 3–8.
3   Born, M. and Wolf, E., 1980, *Principles of Optics*, sixth edition, Pergamon Press, (Oxford, UK), pp. 113–117.
4   Kintner, E.C., 1978, Method for the calculation of partially coherent imagery, *Applied Optics*, **17**, 2747–2753.
5   Hopkins, H.H., 1951, The concept of partial coherence in optics, *Proceedings of the Royal Society of London, Series A*, **208**, pp. 263–277.
6   Hopkins, H.H., 1953, On the diffraction theory of optical images, *Proceedings of the Royal Society of London, Series A*, **217**, 408–432.
7   Pati, Y.C. and Kailath, T., 1994, Phase-shifting masks for microlithography: automated design and mask requirements, *Journal of the Optical Society of America A*, **11**, 2438–2452.
8   Straaijer, A., 2005, Formulas for lithographic parameters when printing isolated and dense lines, *Journal of Microlithography, Microfabrication, and Microsystems*, **4**, 043005–1.

3

Aerial Image Formation – The Details

The impact of aberrations and defocus must now be added to the description of image formation provided in the previous chapter. The unique aspects of imaging while scanning the mask and wafer past the stationary imaging lens will also be included. Next, a discussion of the vector nature of light and the impact of polarization on imaging will be added and immersion lithography will be described. Finally, a preliminary discussion of image quality will conclude this chapter.

3.1 Aberrations

According to Webster, an aberration is '. . . a departure from what is right, true or correct'. In optical imaging, 'right, true or correct' can be thought of as the ideal, 'diffraction-limited' imaging performance of a lens (which was rigorously defined in the previous chapter using Fourier optics). Thus, a lens aberration is any deviation of the real performance of that lens from its ideal performance. As one might imagine, aberrations are undesirable intrusions of reality into our attempts to achieve imaging perfection.

3.1.1 The Causes of Aberrations

In practice, aberrations come from three sources – aberrations of design, aberrations of construction and aberrations of use. Aberrations of construction are probably the most tangible sources of errors and include incorrect shapes and thickness of the glass elements that make up the lens, inhomogeneous glass used in their construction, improper mounting, spacings or tilts of the various lens elements, or other imperfections in the manufacture of the lens (Figure 3.1). Aberrations of use include all ways in which improper use of the lens degrades its performance: using the wrong wavelength or wavelength

---

Fundamental Principles of Optical Lithography: The Science of Microfabrication, Chris Mack.
© 2007 John Wiley & Sons, Ltd.

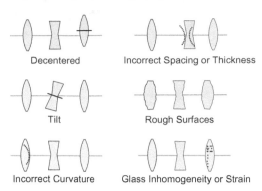

Figure 3.1   *Examples of typical aberrations of construction*

spectrum, tilt of the mask or wafer plane, or incorrect environmental conditions (e.g. changes in the refractive index of air due to changes in temperature, humidity or barometric pressure). But in a sense, it is the aberrations of design which are the fundamental problem for lithographic lenses.

An 'aberration of design' does not mean mistakes or problems caused by the designer of the lens. In fact, it means quite the opposite. Aberrations are fundamental to the nature of imaging and it is the goal of the lens designer to 'design out' as many of these aberrations as possible. The aberrations of design are those aberrations, inserted into the imaging system by nature, that the designer was not able to extract. The root cause of these aberrations for refractive optical elements is the nonlinear nature of Snell's law.

Snell's law describes the refraction of light as it passes from one material to another. Light incident at the boundary of two materials at an angle θ_1 with respect to the normal will refract to a new angle θ_2 in material 2 given by Snell's law:

$$n_1 \sin \theta_1 = n_2 \sin \theta_2 \qquad (3.1)$$

where n_1 and n_2 are the refractive indices of the two materials. It is this refraction between air and glass that allows a curved glass surface to focus light. In terms of refraction, the goal of imaging can be stated very simply: light emanating in all directions from some point on the object should be collected by the lens and focused to its ideal image point. Thus, a ray of light coming from the object point should pass through the lens and arrive at the image point, regardless of its angle.

Does Snell's law permit this ideal behavior? For a lens made of a single element (i.e. a single piece of glass) with spherical front and back surfaces, it is not difficult to show that ideal imaging behavior would be obtained *only* if Snell's law were linear: $n_1 \theta_1 = n_2 \theta_2$. Thus, it is the nonlinear nature of the refraction of light that causes nonideal imaging behavior. In the limit of small angles, Snell's law does behave about linearly ($\sin \theta \approx \theta$ for small θ). When all the angles of light passing through the imaging system are small enough so that Snell's law is approximately linear, we find that the lens behaves ideally in this *paraxial* (near the axis) approximation. Although a spherically shaped lens was used as an example, there is no single shape for a lens surface that provides ideal imaging for many different object points. Thus, fundamentally, the nonlinear nature of Snell's law results in nonideal imaging behavior (i.e. aberrations) for any single lens element.

If aberrations are fundamental to the nature of imaging, how can we make a lens that approaches the ideal imaging behavior? Since each lens element has aberrations by its very nature, these aberrations are corrected by combining different lens elements with differing aberrations. If two lens elements are brought together that have aberrations of similar magnitude but in opposite directions, the combination will have smaller aberrations than either individual lens element. A simple example is a pair of eyeglasses. One lens, your eye, has aberrations of a certain magnitude and direction. A second lens, in the glasses, is designed to have aberrations of about the same magnitude but in the opposite direction. The combination yields residual aberrations that are small enough to be acceptable. One of the simplest examples of aberration correction through lens design is a cemented doublet of positive and negative elements made from two different types of glass, which has the added advantage of reducing chromatic aberrations – the variation of imaging behavior with wavelength (Figure 3.2).

The combination of two lens elements can reduce the aberrations of the overall lens system, but not eliminate them. If a third element is added, the total amount of aberrations can be reduced further. Each lens element, with its two curved surfaces, thickness, position and type of glass, adds more degrees of freedom to the design to allow the overall level of aberrations to be reduced. Thus, the lens designer uses these degrees of freedom (in addition to the position of an aperture stop) to 'design out' the aberrations that nature has embedded into the behavior of a single element (see Figure 3.3 for an example of a modern lithographic lens). An aberration free lens system, however, is an ideal that can never be achieved in practice. It would take an infinite number of lens elements to make the designed behavior perfect. Before this point is reached, however, aberrations of construction would begin to dominate. The optimum number of elements is reached when the addition of one more lens element adds more aberrations of construction than it reduces the aberrations of design.

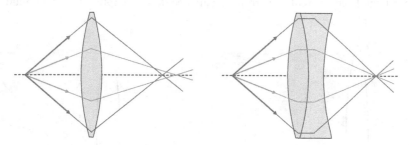

Figure 3.2   *The reduction of spherical aberration by the use of a cemented doublet*

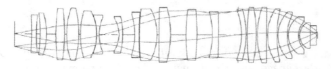

Figure 3.3   *Example of a simple* NA = 0.8, 248-nm *lens design*[1]

Since the nonlinear nature of Snell's law leads to aberrations, larger angles passing through the lens produce greater nonideal behavior. Thus, for a given lens design, increases in numerical aperture (NA) or field size result in large increases in aberrations. In fact, however, new generations of lithographic lenses with higher NAs always seem to exhibit *lower* levels of aberrations compared to previous generations due to improvements in lens design and manufacturing. The increasing use of aspherical lens elements, with their greater degrees of freedom for the lens design, has helped to keep the number of lens elements lower than would be otherwise required (though at a cost of increased manufacturing complexity). Still, lens size and weight grow considerably for each generation of lithographic tool, with 500-kg lenses more than 2 m tall now common.

Unfortunately, even the low levels of aberrations found in a state-of-the-art lens are not small enough to be ignored. Invariably, the high NA lens is used to print ever smaller features that are increasingly sensitive to even the slightest deviation from perfection. The impact of these aberrations is a reduction of image quality, and ultimately a loss of critical dimension (CD) control and overlay performance (see Chapter 8). A thorough understanding of the aberration behavior of a lens is a necessity for those seeking to push the limits of their lithography tools.

3.1.2 Describing Aberrations: the Zernike Polynomial

How are aberrations characterized? In the simple geometrical optics approach, optical *rays* represent the propagation of light through the optical system. By tracing the paths of the light through a lens (using a lens design software program, for example), the deviations in the actual paths of the rays from the ideal paths tell us the magnitude of the aberration (Figure 3.4). The magnitude of the aberration can be expressed as an optical path difference (OPD) between the ideal and actual rays. Rays point in the direction that light is traveling and are thus perpendicular to the *wavefront* of the light, a surface of constant phase. By propagating these wavefronts through the lens, the deviation of the ideal wavefront from the real wavefront can also be used to express the magnitude of the

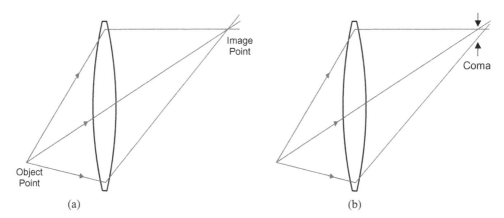

(a)   (b)

Figure 3.4 *Ray tracing shows that (a) for an ideal lens, light coming from the object point will converge to the ideal image point for all angles, while (b) for a real lens, the rays do not converge to the ideal image point*

aberration (Figure 3.5). The two descriptions are completely equivalent since light undergoes a phase change of 2π every time it travels a distance of one wavelength, providing a simple conversion between OPD and phase error.

The aberration behavior of a lens can be predicted using lens design software, or measured using interferometry. In both cases, the result is a map of the phase error (or OPD) of the light exiting from the lens for a given point in the field. It is important to note that the aberrations will vary as a function of field position (center of the field versus the upper left corner, for example). It typically takes 25 to 50 measurements across the field to fully characterize how these aberrations vary for a typical lithographic lens.

A map of the phase error across the exit pupil of the lens is a useful, but potentially large, set of data. One of the most common ways to better understand such data is to curve fit the data to a function, typically a high-order polynomial. In that way, this large set of experimental data can be represented by relatively few polynomial coefficients. Further, by carefully choosing the form of the polynomial fitting function, the coefficients themselves can take on physically meaningful interpretations. By far the most common such polynomial is the *Zernike polynomial*, named for Frits Zernike, the 1953 Nobel prize winning physicist. This orthogonal polynomial series has an infinite number of terms, but is typically cut off after 36 terms, with powers of the relative radial pupil position R and trigonometric functions of the polar angle ϕ. Thus, knowing the numerical values of the 36 Zernike coefficients allows one to describe to a reasonable level of accuracy the aberration behavior of a lens at a single field point.

The Zernike polynomial can be arranged in many ways, but most lens design software and lens measuring equipment employ the fringe or circle Zernike polynomial, defined as

$$W(R,\phi) = \frac{OPD}{\lambda} = \sum_{i=0}^{\infty} Z_i F_i(R,\phi) \qquad (3.2)$$

where $W(R,\phi)$ is the optical path difference relative to the wavelength, Z_i is called the *i*th Zernike coefficient, and $F_i(R,\phi)$ are the polynomial terms, defined in Table 3.1. The

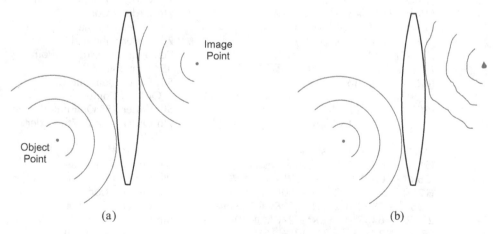

(a)   (b)

Figure 3.5 *Wavefronts showing the propagation of light (a) for an ideal lens and (b) for a lens with aberrations*

phase error due to aberrations will be simply $2\pi W(R,\phi)$. The polar coordinates on the unit circle (R,ϕ) are related to the Cartesian spatial frequency coordinates by

$$f_x = \frac{NA}{\lambda}R\cos\phi, \quad f_y = \frac{NA}{\lambda}R\sin\phi \tag{3.3}$$

Table 3.1 *Polynomial functions and common names for the first 36 terms in the Fringe Zernike polynomial*

Term	Fringe Zernike Formula, F_i	Common Name
Z_0	1	Piston
Z_1	$R\cos\phi$	x-Tilt
Z_2	$R\sin\phi$	y-Tilt
Z_3	$2R^2 - 1$	Power (paraxial focus)
Z_4	$R^2\cos2\phi$	3rd Order Astigmatism
Z_5	$R^2\sin2\phi$	3rd Order 45° Astigmatism
Z_6	$(3R^2 - 2)R\cos\phi$	3rd Order x-Coma
Z_7	$(3R^2 - 2)R\sin\phi$	3rd Order y-Coma
Z_8	$6R^4 - 6R^2 + 1$	3rd Order Spherical
Z_9	$R^3\cos3\phi$	Trefoil (3rd Order 3-Point)
Z_{10}	$R^3\sin3\phi$	45° Trefoil
Z_{11}	$(4R^2 - 3)R^2\cos2\phi$	5th Order Astigmatism
Z_{12}	$(4R^2 - 3)R^2\sin2\phi$	5th Order 45° Astigmatism
Z_{13}	$(10R^4 - 12R^2 + 3)R\cos\phi$	5th Order x-Coma
Z_{14}	$(10R^4 - 12R^2 + 3)R\sin\phi$	5th Order y-Coma
Z_{15}	$20R^6 - 30R^4 + 12R^2 - 1$	5th Order Spherical
Z_{16}	$R^4\cos4\phi$	Quadrafoil (3rd Order 4-Point)
Z_{17}	$R^4\sin4\phi$	45° Quadrafoil
Z_{18}	$(5R^2 - 4)R^3\cos3\phi$	5th Order Trefoil (5th Order 3-Point)
Z_{19}	$(5R^2 - 4)R^3\sin3\phi$	5th Order 45° Trefoil
Z_{20}	$(15R^4 - 20R^2 + 6)R^2\cos2\phi$	7th Order Astigmatism
Z_{21}	$(15R^4 - 20R^2 + 6)R^2\sin2\phi$	7th Order 45° Astigmatism
Z_{22}	$(35R^6 - 60R^4 + 30R^2 - 4)R\cos\phi$	7th Order x-Coma
Z_{23}	$(35R^6 - 60R^4 + 30R^2 - 4)R\sin\phi$	7th Order y-Coma
Z_{24}	$70R^8 - 140R^6 + 90R^4 - 20R^2 + 1$	7th Order Spherical
Z_{25}	$R^5\cos5\phi$	Pentafoil (3rd Order 5-Point)
Z_{26}	$R^5\sin5\phi$	45° Pentafoil
Z_{27}	$(6R^2 - 5)R^4\cos4\phi$	5th Order Quadrafoil (5th Order 4-Point)
Z_{28}	$(6R^2 - 5)R^4\sin4\phi$	5th Order 45° Quadrafoil
Z_{29}	$(21R^4 - 30R^2 + 10)R^3\cos3\phi$	7th Order Trefoil (7th Order 3-Point)
Z_{30}	$(21R^4 - 30R^2 + 10)R^3\sin3\phi$	7th Order 45° Trefoil
Z_{31}	$(56R^6 - 105R^4 + 60R^2 - 10)R^2\cos2\phi$	9th Order Astigmatism
Z_{32}	$(56R^6 - 105R^4 + 60R^2 - 10)R^2\sin2\phi$	9th Order 45° Astigmatism
Z_{33}	$(126R^8 - 280R^6 + 210R^4 - 60R^2 + 5)R\cos\phi$	9th Order x-Coma
Z_{34}	$(126R^8 - 280R^6 + 210R^4 - 60R^2 + 5)R\sin\phi$	9th Order y-Coma
Z_{35}	$252R^{10} - 630R^8 + 560R^6 - 210R^4 + 30R^2 - 1$	9th Order Spherical
Z_{36}	$924R^{12} - 2772R^{10} + 3150R^8 - 1680R^6 + 420R^4 - 42R^2 + 1$	11th Order Spherical

It is the magnitude of the Zernike coefficients that determines the aberration behavior of a lens. They have units of optical path length relative to the wavelength. (Note that different notational schemes are in use for the Zernike coefficients. Another popular scheme labels the piston term Z_1 rather than Z_0, so that all the indices differ by 1. Unfortunately, there is no universal standard and naming conventions can vary.)

Zernike polynomials have some important properties that make their use extremely attractive. First, they form a complete set over the unit circle, meaning that any piecewise-continuous function can be fitted exactly within this unit circle with an infinite-order Zernike polynomial. Second, they are *orthogonal* over the unit circle. Mathematically, this means that the product of two different polynomial terms, when integrated over the pupil, will always be zero, whereas the integral of one polynomial term squared will be nonzero. Practically, this means that each polynomial term behaves independently of the others, so that the addition or removal of one polynomial term does not affect the best-fit coefficients of the other polynomial terms. Third, the Zernike polynomial terms are separable into functions of R multiplied by functions of ϕ, providing rotational symmetry. The ϕ functions are trigonometric, so that a rotation of the *x*-axis (defining ϕ) does not change the form of the equation.

The impact of aberrations on the aerial image can be calculated by modifying the pupil function of the lens given in Chapter 2 to include the phase error due to aberrations.

$$P(f_x, f_y) = P_{\text{ideal}}(f_x, f_y)e^{i2\pi W(R,\phi)} \tag{3.4}$$

Figure 3.6 shows several examples of plots of $W(R,\phi)$ for different simple combinations of third-order Zernike terms. Implicit in the use of Equation (3.4) is that this lens pupil function is also a function of field position and wavelength. The wavelength dependence is discussed in section 3.1.6. While the field dependence of the pupil function is sometimes modeled mathematically (the Seidel aberration equations being the most common), in lithography applications the most common approach is to characterize each field position with its own pupil function.

Although the phase variation across the pupil of a typical lithographic lens can be quite complicated, there are a few simple examples that can be used to describe the majority of the lithographic effects due to aberrations. It is also important to note that the impact of a given aberrated lens on the printing of a mask feature is a strong function of the feature itself. The reason is quite simple: different mask features diffract light differently, sending light through different parts of the lens. As an example, a pattern of equal lines and spaces illuminated with coherent (single direction) light will result in a diffraction pattern of discrete diffraction 'orders', rays of light passing through the lens at certain distinct points. Changing the pitch of the line/space pattern changes the position within the lens that the diffraction orders will pass (Figure 3.7). We can think of the diffraction patterning as 'sampling' the lens differently depending on the mask pattern. In general, the phase error seen by these different mask patterns will be different, giving rise to feature type, size and orientation dependencies to the effects of aberrations.

3.1.3   Aberration Example – Tilt

One of the simplest forms of an aberration is *tilt*, a linear variation of phase from one side of the lens to the other as given by the Zernike terms Z_1 and Z_2 (Figure 3.8a). As the

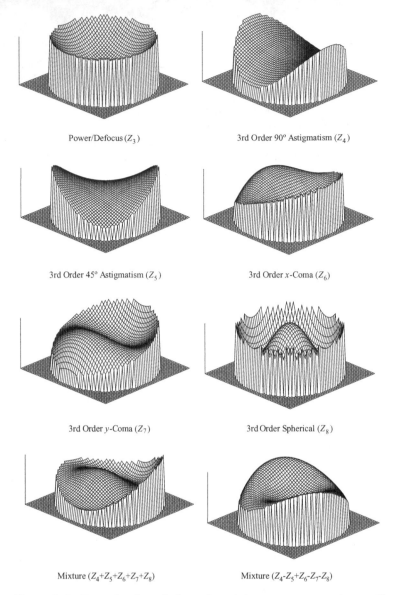

Power/Defocus (Z_3)

3rd Order 90° Astigmatism (Z_4)

3rd Order 45° Astigmatism (Z_5)

3rd Order x-Coma (Z_6)

3rd Order y-Coma (Z_7)

3rd Order Spherical (Z_8)

Mixture $(Z_4+Z_5+Z_6+Z_7+Z_8)$

Mixture $(Z_4-Z_5+Z_6-Z_7-Z_8)$

Figure 3.6 *Example plots of aberrations (phase error across the pupil)*

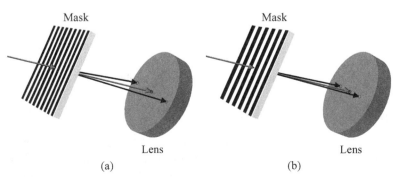

Mask

Mask

Lens

Lens

(a)

(b)

Figure 3.7 *Diffraction patterns from (a) a small pitch, and (b) a larger pitch pattern of lines and spaces will result in light passing through a lens at different points in the pupil. Note also that y-oriented line/space features result in a diffraction pattern that samples the lens pupil only along the x-direction*

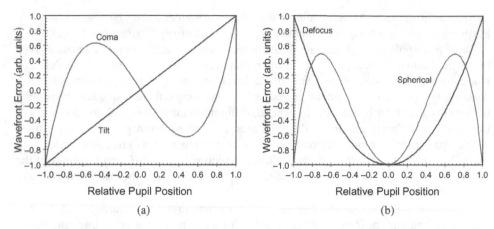

Figure 3.8   *Phase error across the diameter of a lens for several simple forms of aberrations: (a) the odd aberrations of tilt and coma, and (b) the even aberrations of defocus and spherical*

name implies, the aberration of tilt will 'tilt' the wavefront exiting the lens, changing the direction of the light. The result is a simple shift in the position of the final printed feature (as can be easily shown using the shift property of the Fourier transform). For pure tilt, the resulting positional error of the pattern is independent of feature size and type, but is very dependent on the orientation of the pattern. As an example, a pattern of lines and spaces oriented in the y-direction will give rise to a diffraction pattern with diffraction orders spread out in the x-direction (Figure 3.7). The resulting placement error (the amount that the image has shifted in x with respect to the unaberrated image) will be

$$placement\ error = -\frac{Z_1 \lambda}{NA} \tag{3.5}$$

In this case, an x-tilted phase variation will cause a pattern placement error for the features, whereas a y-tilt aberration (given by the coefficient Z_2) will have no effect.

The variation of the tilt Zernike coefficients with field position is called *distortion*. Since tilt causes the same image placement error for all features, distortion is independent of feature size and type. As will be seen below in the section on coma, higher-order odd aberrations will cause pattern placement errors as well, but with varying placement error as a function of feature size and type.

3.1.4   Aberration Example – Defocus, Spherical and Astigmatism

The next simplest form of an aberration is called *paraxial defocus* (also called power), a parabolic variation in phase about the center of the lens (Figure 3.8b). Since this aberration is radially symmetric, all feature orientations will behave the same way. However, smaller features, with the diffraction orders farther out toward the edge of the lens, will have a greater amount of phase error. Thus, small features are more sensitive to the aberration of defocus. More details of the effects of defocus are given in section 3.4. Further explanation of the lithographic impact of focus errors is given in Chapter 8.

There is a whole class of aberrations that behave somewhat like defocus across a given radial cut through the lens pupil. These aberrations, called *even* aberrations, have the same phase error at either side of the center of the lens along any given diagonal (that is, for a fixed ϕ). For even aberrations, the primary effect on a single feature size and orientation is basically a shift in best focus. In general, different feature sizes and orientations will experience a different amount of the even aberration phase error, giving different focal shifts. Spherical aberration can be thought of as causing a focus error that varies with pitch. As Figure 3.8b shows, different pitches, which will give different positions of the first diffraction order, will experience different phase errors and thus different effective focus errors (see Problem 3.5).

A variation in best focus as a function of the orientation of the features is called *astigmatism*. Examining the Zernike terms 4 and 5, for a given orientation of line (and thus a given ϕ), the variation of phase with R is quadratic. Thus, for a given orientation, astigmatism adds a phase error identical to paraxial defocus. But the amount of effective defocus error due to astigmatism will depend on the orientation of the line (that is, on ϕ). One of the major consequences of the presence of astigmatism is a focus-dependent horizontal–vertical (H–V) bias, a difference in linewidth between horizontally oriented and vertically oriented patterns. More on H–V bias will be given in Chapter 8.

3.1.5   Aberration Example – Coma

Finally, *odd* aberrations are those that do not have radial symmetry about the center of the lens. Coma is a common example of an odd aberration (see Figure 3.8). Consider, for example, the sixth term in the Zernike polynomial, representing x-oriented 3rd order coma:

$$Z_6(3R^2 - 2)R\cos\phi \tag{3.6}$$

y-oriented coma (Zernike term number 7) has the same form as the above expression, but with a $\sin\phi$ instead of the cosine term. The combination of these two terms can define any amount of 3rd order coma directed in any orientation. For simplicity, then, we can consider coma in the direction of maximum magnitude as given by the above equation with the angle equal to zero.

What are the main lithographic effects of 3rd order coma? One important effect is a feature-size dependent pattern placement error. Since the x-tilt Zernike term takes the form $Z_1 R\cos\phi$, comparison with Equation (3.6) shows that coma acts as an effective tilt with coefficient $Z_6(3R^2 - 2)$. Consider a pattern of lines and spaces of pitch p. Diffraction will produce discrete orders (directions) of light traveling through the lens, the zero order going straight through the lens and the plus and minus first orders entering the lens at relative radial positions of $R = \pm\lambda/pNA$. For coherent illumination, the pattern placement error can be derived directly from the above expressions for x-tilt and 3rd order x-coma.

$$placement\ error = -\frac{Z_6\lambda}{NA}\left(3\left(\frac{\lambda}{pNA}\right)^2 - 2\right) \tag{3.7}$$

It is important to note that the above expression only applies to line/space patterns when $\lambda/NA < p < 3\lambda/NA$ and for coherent illumination. The effect of partially coherent illumination is to spread out the orders in a circle centered about each coherent diffraction order, thus averaging out some of the tilt caused by the coma. Figure 3.9 shows how pattern placement error varies with pitch for a pattern of equal lines and spaces in the presence of coma for different amounts of partial coherence. One can observe a general, though not universal, trend: lower partial coherence increases the sensitivity to aberrations.

A second important effect of coma is asymmetric linewidth errors. A single isolated line, printed in the presence of coma, will show an asymmetry of the right to left side of the photoresist profile. An even greater effect, however, is the difference in linewidth between the rightmost and leftmost lines of a five-bar pattern. Figure 3.10 shows how the difference between right and left linewidths varies almost linearly with the amount of coma, with smaller features more sensitive to a given amount of coma. In fact, this lithographic measurement can be used as a test for the amount of coma in a lens.

The third important impact of coma is a change in the shape of the final resist profile, especially through focus. Coma causes an asymmetric degradation of the resist profile, affecting one side of the feature more than the other, as seen in Figure 3.11. At best focus, this degradation in profile shape is usually not noticeable. But out of focus, there is a characteristic loss of resist off one side of the profile at the part of the profile that is farthest away from the plane of best focus. The reason for this behavior can be seen from aerial image *isophotes* (contours of constant intensity through focus and horizontal position) in the presence of coma, as shown in Figure 3.12.

3.1.6 Chromatic Aberrations

An assumption of the Zernike description of aberrations is that the light being used is monochromatic. The phase error of light transmitted through a lens applies to a specific

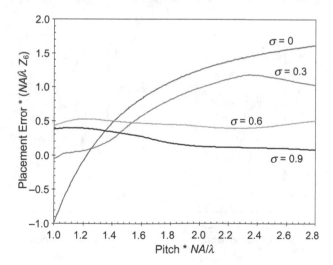

Figure 3.9 *The effect of coma on the pattern placement error of a pattern of equal lines and spaces (relative to the magnitude of the 3rd order x-coma Zernike coefficient Z_6) is reduced by the averaging effect of partial coherence*

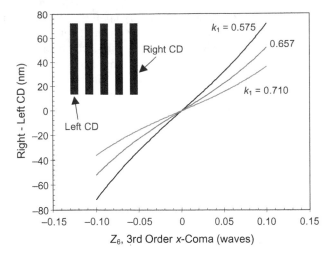

Figure 3.10   *The impact of coma on the difference in linewidth between the rightmost and leftmost lines of a five bar pattern (simulated for i-line, NA = 0.6, sigma = 0.5). Note that the y-oriented lines used here are most affected by x-coma. Feature sizes (350, 400 and 450 nm) are expressed as* $k_1 = linewidth *NA/\lambda$

Figure 3.11   *Variation of the resist profile shape through focus in the presence of coma*

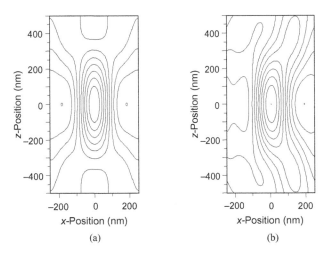

Figure 3.12   *Examples of isophotes (contours of constant intensity through focus and horizontal position) for (a) no aberrations, and (b) 100 mλ of 3rd order coma. (NA = 0.85, λ = 248 nm, σ = 0.5, 150-nm space on a 500-nm pitch)*

wavelength. One expects this phase error, and thus the coefficients of the resulting Zernike polynomial, to vary with the wavelength of the light used. The property of a lens element that allows light to bend is the index of refraction of the lens material. Since the index of refraction of all materials varies with wavelength (a property called *dispersion*), lens elements will focus different wavelengths differently. This fundamental problem, called *chromatic aberration*, can be alleviated by using two different glass materials with different dispersions such that the chromatic aberrations of one lens element cancel the chromatic aberrations of the other. The cemented doublet of Figure 3.2 is the simplest example of a chromatic-corrected lens (called an *achromat*) where each element is made from a glass with different dispersion characteristics. As with all aberrations, this cancellation is not perfect, meaning that all lenses will have some level of residual chromatic aberrations.

The effects of chromatic aberrations depend on two things: the degree to which the Zernike polynomial coefficients vary with wavelength (the magnitude of the chromatic aberrations), and the range of wavelengths used by the imaging tool. For example, a typical i-line stepper might use a range of wavelengths on the order of 5–10 nm, whereas a KrF excimer laser-based deep-UV stepper may illuminate the mask using light with a wavelength range of less than 1 pm. The obvious difference in the magnitude of the wavelength ranges between these two light sources does not mean, however, that chromatic aberrations are a problem for i-line but not for deep-UV. i-line lenses are designed to use the (relatively) wide range of wavelengths by chromatically correcting the lens with several different types of glass. A typical deep-UV lens, on the other hand, makes no attempt at chromatic correction since only fused silica is used for all the elements in the lens (due to a lack of suitable alternative materials). As a result, chromatic aberrations are a concern in deep-UV lithographic lenses even when extremely narrow bandwidth light sources are used.

In principle, every Zernike coefficient is a function of wavelength. In practice, for lenses with no chromatic corrections (i.e. where every lens element is made from the same material), one Zernike term dominates: the term that describes defocus. The plane of best focus shifts with changes in wavelength in a nearly linear fashion, a phenomenon called *longitudinal chromatic aberration*. This can be seen quite readily by considering an ideal thin lens in air. Such a lens is made of glass of index n and has front and back radii of curvature R_1 and R_2. The focal length f, reduction ratio R, and distances from the lens to the object and image, d_o and d_i, respectively, are all related by the thin-lens equations:

$$\frac{1}{f} = (n-1)\left(\frac{1}{R_1} - \frac{1}{R_2}\right) = \frac{1}{d_o} + \frac{1}{d_i}, \quad R = \frac{d_o}{d_i} \tag{3.8}$$

A change in refractive index of the lens (with wavelength, for example) results in a change in focal length. For a fixed object distance and reduction ratio, a change in focal length changes the image distance, that is, focus. Taking the derivative with respect to wavelength,

$$\frac{\Delta focus}{\Delta \lambda} = \frac{d(d_i)}{d\lambda} = -\frac{d_i}{(n-1)}\frac{dn}{d\lambda} \tag{3.9}$$

The term $dn/d\lambda$ is called the dispersion of the lens material. For KrF-grade fused silica at 248.3 nm, $n = 1.508$ and the dispersion is about -5.6×10^{-7}/pm. As an example, a thin lens of fused silica with an image distance of 0.2 m would result in a change in focus per change in wavelength of 0.22 μm/pm. Of course, more complicated lens systems would have an effective image distance different from the physical image distance, but the basic trend and magnitude for a lens with no chromatic correction is similar. For immersion lithography, the slope of the focus versus wavelength dependence will increase by the fluid refractive index.

Since the center wavelength of most excimer lasers is easily adjustable over a reasonably large range, this effect can be readily measured for any given stepper. Figure 3.13a shows a typical example.[2] To account for this behavior, the Zernike term Z_3 (the paraxial defocus term) can be used. The coefficient of the third Zernike term Z_3 can be related to a focus shift $\Delta\delta$ by

$$Z_3 = \Delta\delta \frac{NA^2}{4\lambda_0} \tag{3.10}$$

where λ_0 is the center wavelength of the illumination spectrum. Each wavelength produces a different focus shift $\Delta\delta$, which leads to a different value of Z_3. Higher-order focus terms could be added as well.

A typical defocus/wavelength behavior for a KrF (248 nm) lens is shown in Figure 3.13a. An excimer laser light source will emit light over a range of wavelengths with a characteristic spectrum (output intensity versus wavelength) that can be roughly approximated as somewhere between a Gaussian shape and a Lorentzian shape. The full width half maximum (FWHM) of the laser output spectrum is called the bandwidth of the laser, with typical line-narrowed KrF and ArF lasers having less than 1-pm bandwidths. Each wavelength in the laser spectrum will be projected through the imaging lens, forming an aerial image shifted in focus according to the wavelength response characteristic for that

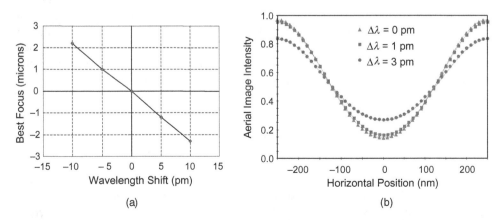

(a)   (b)

Figure 3.13 *Chromatic aberrations: (a) measurement of best focus as a function of center wavelength shows a linear relationship with slope 0.255 μm/pm for this 0.6 NA lens; (b) degradation of the aerial image of a 180-nm line (500-nm pitch) with increasing illumination bandwidth for a chromatic aberration response of 0.255 μm/pm*

lens. The total aerial image will be the sum of all the images from all the wavelengths in the source, resulting in a final aerial image that is somewhat smeared through focus.

Figure 3.13b shows a typical example. For this 180-nm image, the 1-pm bandwidth shows a slight degradation of the aerial image, but the 3-pm bandwidth is clearly unacceptable. As one might expect, smaller features will be more sensitive to the effects of chromatic aberrations than larger features. Thus, as features are pushed to smaller sizes, even these seemingly miniscule sub-picometer excimer bandwidths will require careful consideration as to their chromatic aberration effects. Also, since dense line/space features tend to respond very differently to focus errors compared to isolated lines, changes in laser bandwidth can result in noticeable changes in *iso-dense bias*, the difference in resist linewidth between an isolated and a dense feature of the same mask width.

For rigorous treatment of chromatic aberrations, a mathematical description of the light source spectral output is required. A modified Lorentzian distribution has been shown to match measured excimer laser spectrums quite well.[3] Letting I_{rso} be the relative spectral output intensity,

$$I_{rso}(\lambda) = \frac{\Gamma^n}{\Gamma^n + [2(\lambda - \lambda_0)]^n} \tag{3.11}$$

where λ_0 is the center wavelength, Γ is the bandwidth, and n is an empirically determined parameter that is typically between 2 and 3. Figure 3.14 shows a typical example from a KrF laser.

One very simple way to avoid the problem of chromatic aberrations is to use mirrors for the imaging elements. Since mirrors reflect light in about the same way for a wide range of wavelengths, near-perfect chromatic correction is possible. However, mirror optical systems pose many other challenges, not the least of which is the extreme difficulty of using multiple focusing mirrors without those mirrors getting in the way of each other.

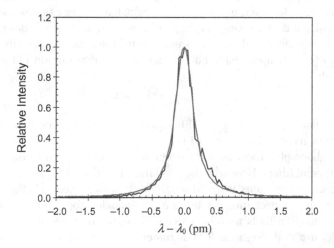

Figure 3.14 *Measured KrF laser spectral output[1] and best-fit modified Lorentzian (Γ = 0.34 pm, n = 2.17, λ_0 = 248.3271 nm)*

3.1.7 Strehl Ratio

The overall level of the aberrations of a lens can be characterized in several ways. The *peak wavefront deviation* is the maximum phase difference (or OPD) between any two points in the pupil (sometimes called the peak-to-valley wavefront error). The *RMS (root mean square) wavefront deviation* provides an average level of phase error or OPD across the pupil. Another approach is to look at the degradation of the point spread function (PSF) caused by the aberrations. Since the PSF samples the imaging lens pupil evenly, its degradation provides an overall measure of how aberrations might affect image quality. Since the ideal PSF is normalized to have a peak value of 1, the *Strehl ratio* is defined as the peak intensity of the PSF of the real lens at best focus, a value that is always less than 1. For anything but the highest levels of aberrations, the Strehl ratio can be related directly to the RMS wavefront deviation:

$$Strehl\ Ratio = \mathrm{e}^{-(2\pi W_{\mathrm{RMS}})^2}$$

(3.12)

where W_{RMS} is the RMS wavefront error in waves. For small levels of aberrations (certainly an applicable constraint for lithographic lenses), the Strehl ratio becomes

$$Strehl\ Ratio \approx 1 - (2\pi W_{\mathrm{RMS}})^2$$

(3.13)

As an example, a Strehl ratio of 0.9 corresponds to an RMS wavefront error of 50 milliwaves. For a Strehl ratio of 0.95, the RMS wavefront error is 35 milliwaves. State-of-the-art lithographic lenses will have Strehl ratios greater than 0.95, with the best lenses having RMS wavefront errors of less than 10 milliwaves.

3.2 Pupil Filters and Lens Apodization

Pupil filters are special filters placed inside the objective lens in order to purposely modify the pupil function $P(f_x, f_y)$. In general, the ideal pupil function will provide the best overall imaging capabilities. However, under special circumstances (for example, when imaging one specific mask pattern), a change in the pupil function may result in desirable imaging properties such as enhanced depth of focus. A pupil filter can be described by a pupil filter function $F(f_x, f_y)$ that can have both a variation in transmission (T) and phase (θ) across the pupil:

$$F(f_x, f_y) = T(f_x, f_y)\mathrm{e}^{i\theta(f_x, f_y)}$$

(3.14)

The final pupil function is then the pupil function given by Equation (3.4) multiplied by the filter function given in Equation (3.14).

In general, lithographic tools used for integrated circuit (IC) production do not allow the insertion of pupil filters. However, the pupil filter function of Equation (3.14) can also be used to describe variations in pupil amplitude transmittance, whether intentionally induced or not, that impact imaging but are not intended for 'filtering' purposes. For example, some imaging tools introduce a central obscuration into the pupil in order to extract light for use in through-the-lens alignment systems. Such an obscuration blocks all light transmitted through the central portion of the lens, for example, when the sine of the diffracted angle is less than ±0.05. 1× steppers made by Ultratech Stepper and early steppers made by ASML include such central obscurations for through-the-lens alignment purposes.

Additionally, nonideal lens antireflection coatings can lower the transmittance of the lens at the highest spatial frequencies, sometimes by as much as 5–10 %. Such nonideal transmittances can be empirically described by a high-order Gaussian:

$$T(f_x, f_y) = e^{-[(f_x^2 + f_y^2)/NA^2]^q} \tag{3.15}$$

where the power q might be on the order of 5–10. In general, a lowering of the transmittance of a lens at high spatial frequencies is called *apodization*. Such transmission errors can also be described using the Jones pupil, as defined below in section 3.6.4.

3.3 Flare

The goal of an imaging lens is to collect diffracted light, spreading out away from an object, and focus that light down to an imaging plane, creating an image that resembles the original object. A typical lens performs this task through the use of curved surfaces of materials with indices of refraction different than air, relying on the principle of refraction. Implicit in this description is the idea that light travels in only one general direction: from the object to the image. The astute student of optics, however, might detect a problem: the difference in index of refraction between two media (such as air and glass) that gives rise to refraction will also give rise to an unwanted phenomenon – reflection.

The design of a lens, including the ray tracing algorithms used to optimize the individual shapes and sizes of each glass or fused silica element in the lens, makes use of the assumption that all light traveling through the lens continues to travel from object to image *without any reflections*. How is this achieved in practice? One of the hidden technologies of lens manufacturing is the use of antireflection coatings on nearly every glass surface in a lens. These lens coatings are designed to maximize the transmittance of light at the interface between materials through interference (a subject that will be treated extensively in Chapter 4). These coatings, usually made of two layers, have specifically designed thicknesses and refractive indices that reduce reflections through interference among several reflected beams. The coatings are quite effective at reducing reflections at a specific wavelength and over a range of incident angles. They had better be, since a typical microlithographic lens will have over 50 surfaces requiring coatings.

Although the lens coatings used in lithography tools are quite good, they are not perfect. As a result, unwanted reflections, though small in magnitude, are inevitably causing light to bounce around within a lens. Eventually some of this light will make its way to the wafer. For the most part, these spurious reflections are reasonably random, resulting in a nearly uniform background light level exposing the wafer called *flare*. Flare is defined as the fraction of the total light energy reaching the wafer that comes from unwanted reflections and scatterings within the lens.

Lens coating nonideality is not the only source of flare. Flare is caused by anything that causes the light to travel in a 'nonray trace' direction. In other words, reflections at an interface, scattering caused by particles or surface roughness, or scattering caused by glass inhomogeneity all result in stray light called flare (Figure 3.15). Defects such as

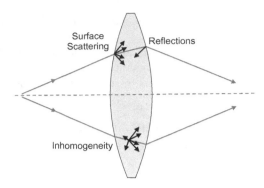

Figure 3.15   *Flare is the result of unwanted scattering and reflections as light travels through an optical system*

these can be built into the lens during manufacturing, or can arise due to lens degradation (aging, contamination, etc.). External sources of flare, such as light that reflects off the wafer, travels up through the lens and reflects off the backside of the chrome mask, can also contribute. The lithographic consequence of this stray light is probably obvious: degradation of image quality. Since flare is a nearly uniform background light level exposing the wafer, it provides exposure to nominally dark features (such as a line), reducing their quality.

While mirrors do not suffer from the problems of nonideal antireflection coatings inherent in refractive optical elements, they are far more sensitive to surface roughness scattering. For a given RMS surface roughness, a mirror will scatter greater than one order of magnitude more light than an equally rough refractive lens surface.

3.3.1   Measuring Flare

Of course, it is the goal of lens manufacturing and maintenance to produce lenses with very low levels of flare. How can one tell if flare poses a problem for a particular stepper or scanner? Fortunately, measuring flare is reasonably straightforward. Consider the imaging of an island feature whose dimension is extremely large compared to the resolution limits of the imaging tool (say, a 100-μm square island in positive resist). In the absence of flare, the imaging of such a large feature will result in very nearly zero light energy at the center of the image of the island. The presence of flare, on the other hand, will provide light to this otherwise dark region on the wafer (Figure 3.16). A positive photoresist can be used as a very sensitive detector for low levels of flare.

The dose-to-clear (E_0) is defined as the minimum dose required to completely remove the photoresist during development for a large open-frame exposure. A related concept is the island dose-to-clear ($E_{0\text{-island}}$), the minimum dose required to completely wash away a large island structure during a normal development process. In the absence of flare, a large island would take nearly an infinite dose to produce enough exposure at its center to make the resist soluble in developer. With flare present, however, this dose is reduced considerably. In fact, by measuring the normal dose-to-clear and the dose-to-clear for the large island, the amount of flare can be determined as[4]

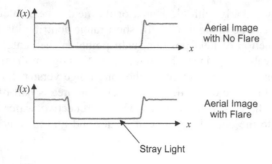

Figure 3.16 *Plots of the aerial image intensity I(x) for a large island mask pattern with and without flare*

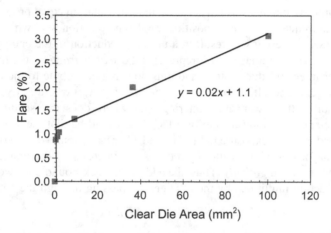

Figure 3.17 *Using framing blades to change the field size (and thus total clear area of the reticle), flare was measured at the center of the field (from Mack and Kaufman[4])*

$$Flare = \frac{E_0}{E_{0-island}} \tag{3.16}$$

For example, if the dose-to-clear of a resist is $35\,mJ/cm^2$, then an imaging tool with 5% flare would mean that a large island will clear with a dose of $700\,mJ/cm^2$.

Although flare is a characteristic of an imaging tool, it is also a function of how that tool is used. For example, the amount of flare experienced by any given feature is a function of both the local environment around that feature (short range flare) and the total amount of energy going through the lens (long range flare).[5] A dark field reticle produces images with almost no flare, whereas a reticle which is almost 100% clear will result in the maximum possible flare. The data in Figure 3.17 shows the two distinct regions clearly. Here, a clear field reticle was used with one flare target placed in the middle of the field. Framing blades were used to change the field size and thus the total clear area of the exposure field. As can be seen, flare quickly rises to a level of about 1%, then grows approximately linearly with clear area of the field. The *y*-intercept of the linear portion of the curve can be thought of as the short range contribution to the total flare.

Flare is also a function of field position, with points in the center of the field often experiencing flare levels 50% higher than points near the edge of the field. This phenomenon may be thought of as a side effect of the long-range versus short-range scattering discussed above. At the edge of the field, the local clear area is half of that at the center of the field. Finally, flare is also increased when a reflective wafer is exposed. Light reflected back up into the optical system can scatter and find its way back to the wafer as stray light.

3.3.2   Modeling Flare

Consider a simple mechanism for the generation of DC (long range) flare. Light passing through the lens is scattered by one or more of the mechanisms shown in Figure 3.15. In its simplest manifestation, flare would cause all light to scatter in equal proportion regardless of its spatial frequency and field position (angle and position with which it enters the lens). In this case, scattering will result in a uniform reduction in the energy that coherently interacts to from the image. If *SF* represents the *scatter fraction*, the fraction of the light energy (or intensity) that scatters and thus does not contribute to the coherent creation of the image, the resulting image would be $(1-SF)I_0(x,y)$ where $I_0(x,y)$ is the image that would result if there were no scattering. But where does this scattered light go? Another simple approximation for uniform DC flare would be that the scattered light is uniformly spread over the exposure field. Thus, a DC or background dose would be added to the aerial image. But while scattering reduces the intensity of the image *locally*, the background dose is added *globally*. Thus, the additional background dose would be equal to the scatter fraction multiplied by the total energy passing through the lens:

$$I(x,y) = SF \cdot EF + (1-SF)I_0(x,y) \tag{3.17}$$

where *EF* is the energy fraction of the reticle, the ratio of the total energy reaching the wafer for this reticle compared to the energy that would reach the wafer for a perfectly clear reticle. As a reasonable approximation, the total energy passing through the lens is proportional to the clear area of the reticle. (Because of diffraction, only a portion of the energy passing through the reticle actually passes through the lens. For example, for a mask pattern of equal lines and spaces where only the zero and ±1 first orders pass through the lens, the energy going through the lens is about 90% of the energy that passes though the mask.) Thus, a simple but still quite accurate DC flare model would be

$$I(x,y) = SF \cdot CF + (1-SF)I_0(x,y) \tag{3.18}$$

where *CF* is the clear area fraction for the field.

For a 100% clear area full-field mask, the standard measurement of flare for a clear field reticle would yield the scatter fraction used in Equation (3.18). Since the clear area fraction is generally known (at least approximately) for any mask, the DC flare can be calculated for any mask once the scatter fraction for the lens has been measured. For a dark field mask, as the clear area fraction goes to zero, the effect of flare is just a loss of dose for the image. It is doubtful that this dose loss is observable in normal lithographic practice since dark field exposure layers (such as contacts and vias) generally operate at different numerical aperture/partial coherence combinations than clear field reticles, and dose calibration is not consistent across different stepper lens and illumination settings.

A slight correction to Equation (3.18) would occur if some fraction of the scattered light was assumed to be lost (absorbed into the lens housing, for example). However, for the low levels of flare expected for normal lithographic imaging tools (much less than 10%), the effect of some of the scatter fraction being lost and not reaching the wafer would be a simple dose recalibration and would not be observable.

Equation (3.18) does not take into account the impact of short-range flare effects, as observed in Figure 3.17. The scattering PSF approach[6] can account for these short-range flare effects. The local scattered light is described as the convolution of the local light intensity (the aerial image) with a scattering point spread function, PSF_{scat}.

$$I_{scat}(x,y) = PSF_{scat}(x,y) * I_0(x,y) \qquad (3.19)$$

Note that if the scattering PSF is a constant, then the convolution becomes a calculation of the total energy of the aerial image and the scattered intensity is equivalent to the scatter fraction times the energy fraction. Generally this global scattering is removed from the PSF_{scat} and treated separately. A typical scattering PSF is a Gaussian, with standard deviations in the range of 10–20 μm, though other functional forms are used as well.

3.4 Defocus

The inverse Fourier transform expression for calculating the aerial image (see Chapter 2) applies only to the image at the focal plane. What happens when the imaging system is out of focus, i.e. what is the image intensity distribution some small distance away from the plane of best focus?

3.4.1 Defocus as an Aberration

The impact of focus errors on the resulting aerial image can be described as an aberration of a sort. Consider a perfect spherical wave converging (i.e. focusing) down to a point. An ideal projection system will create such a wave coming out of the lens aperture (called the *exit pupil*), as shown in Figure 3.18a. If the wafer to be printed were placed in the same plane as the focal point of this wave, we would say that the wafer was in focus. What happens if the wafer were removed from this plane by some distance δ, called the defocus distance? Figure 3.18b shows such a situation. The spherical wave with the solid line represents the actual wave focused to a point a distance δ away from the wafer (this defocus distance is arbitrarily defined to be positive when the wafer is moved away from the lens). If, however, the wave had a different shape, as given by the dotted curve, then the wafer would be in focus. Note that the only difference between these two different waves is the radius of curvature of the wavefront. Since the dotted curve is the wavefront we want for the given wafer position, we can say that the actual wavefront is in error because it does not focus where the wafer is located. (This is just a variation of 'the

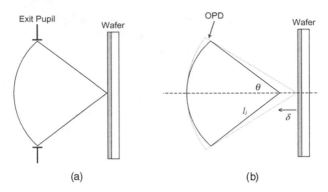

Figure 3.18   *Focusing of light can be thought of as a converging spherical wave: (a) in focus, and (b) out of focus by a distance δ. The optical path difference (OPD) can be related to the defocus distance δ, the angle θ and the radius of curvature of the converging wave (also called the image distance)* l_i

customer is always right' attitude – the wafer is always right; it is the optical wavefront that is out of focus.)

By viewing the actual wavefront as having an error in curvature relative to the desired wavefront (i.e. the one that focuses on the wafer), we can quantify the effect of defocus. Looking at Figure 3.18b, it is apparent that the distance from the desired to the 'defocused' wavefront goes from zero at the center of the exit pupil and increases as we approach the edge of the pupil. This distance between wavefronts is called the *optical path difference* (OPD). The OPD is a function of the defocus distance and the position within the pupil and can be obtained from the geometry shown in Figure 3.18b. Describing the position within the exit pupil by the light propagation angle θ, the OPD (assuming $l_i \gg \delta$) is given by

$$OPD = n\delta(1 - \cos\theta) \tag{3.20}$$

where n is the refractive index of the medium between the lens and the wafer (generally air, though possibly an immersion fluid, as discussed in section 3.7).

As we have seen before, the spatial frequency and the NA define positions within the pupil as the sine of an angle. Thus, the above expression for OPD will be more useful if expressed as a function of sin θ:

$$OPD = n\delta(1 - \cos\theta) = n\delta\left(1 - \sqrt{1 - \sin^2\theta}\right) \tag{3.21}$$

Expanding the square root as a Taylor series,

$$OPD = \frac{1}{2}n\delta\left(\sin^2\theta + \frac{\sin^4\theta}{4} + \frac{\sin^6\theta}{8} + \ldots\right) \approx \frac{1}{2}n\delta\sin^2\theta \tag{3.22}$$

where the final approximation is accurate only for relatively small angles. Figure 3.19 shows the accuracy of this approximation as a function of the angle θ.

So how does this OPD affect the formation of an image? The OPD acts just like an aberration, modifying the pupil function of the lens. For light, this path length traveled

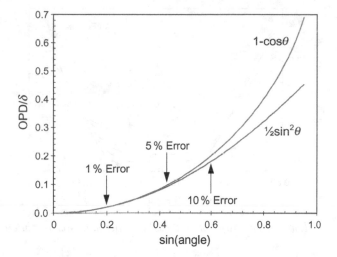

Figure 3.19 *Comparison of the exact and approximate expressions for the defocus optical path difference (OPD) shows an increasing error as the angle increases. An angle of 37° (corresponding to the edge of an NA = 0.6 lens) shows an error of 10 % for the approximate expression. At an NA of 0.93, the error in the approximate expression is 32 %*

(the OPD) is equivalent to a change in phase. Thus, the OPD can be expressed as a phase error, $\Delta\Phi$, due to defocus:

$$P_{\text{defocus}}(f_x, f_y) = P_{\text{ideal}}(f_x, f_y)e^{i2\pi OPD/\lambda} = P_{\text{ideal}}(f_x, f_y)e^{i\Delta\Phi}$$

$$\Delta\Phi = 2\pi OPD/\lambda = 2\pi n\delta(1 - \cos\theta)/\lambda \approx \pi n\delta\sin^2\theta/\lambda \tag{3.23}$$

where again, the final approximation is only valid for small angles. We are now ready to see how defocus affects the diffraction pattern and the resulting image. Our interpretation of defocus is that it causes a phase error as a function of radial position within the aperture. Light in the center of the aperture has no error; light at the edge of the aperture has the greatest phase error. This is very important when we remember what a diffraction pattern looks like as it enters the lens aperture. Recall that diffraction by periodic patterns results in discrete diffraction orders – the zero order is the undiffracted light passing through the center of the lens; higher orders contain information necessary to reconstruct the image. Thus, the effect of defocus is to add a phase error to the higher-order diffracted light relative to the zero order. When the lens recombines these orders to form an image, this phase error will result in a degraded image (Figure 3.20).

Considering defocus as an aberration, the pupil error due to defocus [Equation (3.21)] can be expressed in the terms of a Zernike polynomial. In the paraxial approximation (small angles), there is a simple correspondence to the third term in the polynomial expression (3.2).

$$Z_3 \approx \frac{\delta NA^2}{4\lambda} \tag{3.24}$$

If the exact defocus expression is used, higher-order Zernike defocus terms are required (Z_8, Z_{15}, Z_{24}, etc.).

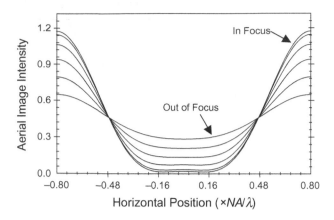

Figure 3.20 *Aerial image intensity of a 0.8λ/NA line and space pattern as focus is changed*

3.4.2   Defocus Example: Dense Lines and Spaces and Three-Beam Imaging

As an example of using Equation (3.23) when calculating an aerial image, consider a pattern of small lines and spaces so that only the 0 and ±1 diffraction orders are used to form the image (that is, three-beam imaging). For coherent illumination, the zero order will pass through the center of the lens and undergo no phase error due to defocus. The first diffracted orders will pass through the pupil at spatial frequencies of ±1/p and experience a phase error of

$$\Delta\Phi = 2\pi n\delta\left(1 - \sqrt{1 - (\lambda/np)^2}\right)/\lambda \tag{3.25}$$

Calculating the electric field image including the pupil function defocus,

$$E(x) = \int_{-\infty}^{\infty} \sum_{j=-1}^{1} a_j \delta\left(f_x - \frac{j}{p}\right) e^{i\Delta\Phi} e^{i2\pi f_x x} df_x \tag{3.26}$$

$$E(x) = a_0 + 2a_1 e^{i\Delta\Phi} \cos(2\pi x/p)$$

The intensity image becomes

$$I(x) = a_0^2 + 4a_0 a_1 \cos(\Delta\Phi)\cos(2\pi x/p) + 2a_1^2[1 + \cos(4\pi x/p)] \tag{3.27}$$

The impact of defocus is to reduce the magnitude of the main cosine term in the image. In other words, as the first orders go out of phase with respect to the zero order, the interference between the zero and first orders decreases. When $\Delta\Phi = 90°$, there is no interference and the $\cos(2\pi x/p)$ term disappears.

Partial coherence can significantly change the focus response of an aerial image. The effect of partial coherence is to spread the diffraction order points into larger spots, the shape of each order's spot being determined by the shape of the source. Each point on the source is independent, i.e. incoherent, with no fixed phase relationship to any other source point. Each source point produces a coherent aerial image, the total image being the (incoherent) sum of all the intensity images from each source point. Since only the exact center source point produces an image affected by focus in the way described by Equation (3.27), a more thorough treatment is required for partially coherent imaging.

Consider the case, as we did above, where only the first diffraction orders make it through the lens and that these orders are completely inside the lens (not clipped by the aperture). This occurs when

$$\frac{\lambda}{NA(1-\sigma)} < p < \frac{3\lambda}{NA(1+\sigma)} \tag{3.28}$$

A quick look at this constraint shows that it can possibly be true only when $\sigma < 0.5$. Thus, while our example here will not be applicable to many situations, it will still be informative. Second, we shall use the paraxial approximation for the effects of focus, so that the OPD due to a defocus error δ will be a quadratic function of the sine of the incident angle or the spatial frequency.

Now the aerial image in the presence of defocus can be calculated analytically by integrating over the source. Considering only equal lines and spaces for the sake of simplicity,

$$I(x) = a_0^2 + 4a_0 a_1 \cos(\Delta\Phi)\frac{2J_1(a)}{a}\cos(2\pi x/p) + 2a_1^2\left[1 + \frac{J_1(2a)}{a}\cos(4\pi x/p)\right] \tag{3.29}$$

where $a = 2\pi\delta\sigma NA/p$ and J_1 is the Bessel function of the first kind, order one. This defocus function $2J_1(a)/a$ is plotted in Figure 3.21 and has the familiar Airy disk form. Comparing Equations (3.27) and (3.29), the effect of partial coherence is to cause a faster decline in the main cosine interference term [the Airy disk falloff is on top of the $\cos(\Delta\Phi)$ falloff

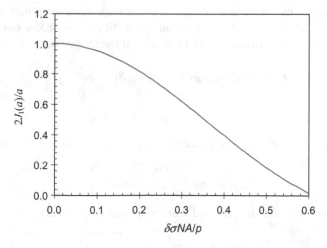

Figure 3.21 *The Airy disk function as it falls off with defocus*

that occurs for coherent illumination] and to add a degradation to the higher-order interference of the + first and − first orders.

It can be useful to simplify the above aerial image Equation (3.29) for the case of small amounts of defocus. Both the cosine term and the Airy disk terms can be expanded as series approximations, keeping only the low-order terms:

$$
\cos(\Delta\Phi) \approx 1 - \frac{\Delta\Phi^2}{2}
$$
$$
\frac{2J_1(a)}{a} \approx 1 - \frac{a^2}{8}
$$

(3.30)

Using these approximations, and keeping only terms out to second order in defocus,

$$
I(x,\delta) \approx I(x,0) - 2a_0 a_1 \cos(2\pi x/p)\left(\Delta\Phi^2 + \frac{a^2}{4}\right) - a_1^2 \cos(4\pi x/p)a^2
$$
$$
= I(x,0) - \left(\frac{\pi\delta\lambda}{p^2}\right)^2 \left[2a_0 a_1 \cos(2\pi x/p)\left(1 + \left(\frac{\sigma NAp}{\lambda}\right)^2\right) -
$$

(3.31)

$$
4a_1^2 \cos(4\pi x/p)\left(\frac{\sigma NAp}{\lambda}\right)^2\right]
$$

Thus, the final result is an example where the effects of defocus can be put into a separable form as

$$
I(x,\delta) \approx I(x,0) - \delta^2 f(x)
$$

(3.32)

Most images can be put into such a form for small amounts of defocus.

3.4.3   Defocus Example: Dense Lines and Spaces and Two-Beam Imaging

Consider the case of coherent off-axis illumination of a dense line/space pattern. In particular, let's assume that the illumination is a plane wave tilted such that the diffraction pattern is shifted by f_x' and only the zero and −1st diffraction orders pass through the lens. The portion of the diffraction pattern passing through the lens will be

$$
T_m(f_x)P(f_x) = a_0\delta(f_x - f_x') + a_1\delta\left(f_x + \frac{1}{p} - f_x'\right)
$$

(3.33)

With defocus included in the pupil function, the electric field image becomes

$$
E(x) = a_0 e^{i\Delta\Phi_0} e^{i2\pi f_x' x} + a_1 e^{i\Delta\Phi_1} e^{-i2\pi x/p} e^{i2\pi f_x' x}
$$
$$
= e^{i\Delta\Phi_0} e^{i2\pi f_x' x}\left[a_0 + a_1 e^{i(\Delta\Phi_1 - \Delta\Phi_0)} e^{-i2\pi x/p}\right]
$$

(3.34)

where the phase errors for the zero and first orders are, respectively,

$$
\Delta\Phi_0 = 2\pi n\delta\left(1 - \sqrt{1 - (\lambda f_x')^2}\right)\Big/\lambda, \quad \Delta\Phi_1 = 2\pi n\delta\left(1 - \sqrt{1 - \left(\lambda\left[\frac{1}{p} - f_x'\right]\right)^2}\right)\Big/\lambda
$$

(3.35)

The intensity image is

$$I(x) = \left|a_0 + a_1 e^{i(\Delta\Phi_1 - \Delta\Phi_0)} e^{-i2\pi x/p}\right|^2 = a_0^2 + a_1^2 + 2a_0 a_1 \cos(2\pi x/p - \Delta\Phi_1 + \Delta\Phi_0) \quad (3.36)$$

Note that the final image depends only on the difference in phase between the zero and first orders.

The impact of defocus on this tilted-illumination image is clear from Equation (3.36): focus errors cause a shift in the position of the sinusoidal image. Such a focus-dependent pattern placement error is called *nontelecentricity*. *Telecentricity*, on the other hand, means that the resulting image will not shift in position as the wafer goes out of focus. Wafer-side telecentricity is a requirement for lithographic imaging. Telecentricity can be recovered by adding a second illumination angle, causing the same shift in the diffraction pattern but in the opposite direction (that is, a shift of $-f_x'$). The resulting image will be the same as that given by Equation (3.36), except with the opposite sign of the phase error. The average of the two aerial images coming from these two symmetrically tilted plane waves will be

$$I(x) = \frac{I(x, f_x') + I(x, -f_x')}{2} = a_0^2 + a_1^2 + 2a_0 a_1 \cos(\Delta\Phi_1 - \Delta\Phi_0)\cos(2\pi x/p) \quad (3.37)$$

Consider now a very special case, where the tilt of the illumination is chosen to be $f_x' = \pm 1/2p$. For this case, the zero order and the first order will be evenly spaced about the center of the lens and the phase error due to defocus will be the same for both orders. Thus, the resulting intensity image is

$$I(x) = a_0^2 + a_1^2 + 2a_0 a_1 \cos(2\pi x/p) \quad (3.38)$$

Remarkably, this is the same image that is obtained when at best focus. That is, defocus has no impact on the final aerial image for this special case of coherent two-beam imaging where the zero and first orders are evenly spaced about the center of the lens. The possibility of essentially infinite depth of focus when the tilt is adjusted properly for a specific pitch is the reason why off-axis illumination is so popular for high-resolution imaging.

In reality, the infinite depth of focus promised by Equation (3.38) is never realized because real illumination sources are not coherent – they are partially coherent. Thus, instead of two point sources separated by $\pm f_x'$, a real source would be two disks centered at these spatial frequencies. The resulting illumination is called dipole illumination. The image from such a dipole source, assuming that all of the zero order and one first order go through the lens, would be the integral of Equation (3.37) over the range of spatial frequency shifts corresponding to the full shape of the source. Using the paraxial approximation for the effects of focus, so that the OPD due to a defocus error δ will be a quadratic function of the sine of the incident angle or the spatial frequency, the phase difference from zero to first orders for any given source point is

$$\Delta\Phi_1 - \Delta\Phi_0 \approx \pi n \delta\lambda \left[\left(\frac{1}{p} - f_x'\right)^2 - f_x'^2\right] = \frac{\pi n \delta\lambda}{p}\left[\frac{1}{p} - 2f_x'\right] \quad (3.39)$$

Since each dipole disk is centered at $f'_x = \pm 1/2p$, let a point on this source be expressed as $f'_x = 1/2p - \Delta f'_x$:

$$\Delta\Phi_1 - \Delta\Phi_0 = \frac{2\pi n \delta \lambda}{p} \Delta f'_x = \frac{2\pi n \delta NA}{p} \Delta \sigma_x \tag{3.40}$$

where the final expression comes from converting spatial frequencies into sigma coordinates.

Using Equation (3.40) in Equation (3.37) and integrating over the source, where each disk has a radius of σ, the resulting two-beam image through focus becomes

$$I(x) = a_0^2 + a_1^2 + 2a_0 a_1 \frac{2J_1(a)}{a} \cos(2\pi x/p) \tag{3.41}$$

where $a = 2\pi\delta\sigma NA/p$ and J_1 is the Bessel function of the first kind, order one. Referring again to Figure 3.21, one can see that when the size of the dipole disk is bigger than zero, defocus will cause a degradation in the quality of the aerial image. Note from the definition of the term a that the sensitivity to defocus is in direct proportion to σ, the radius of the dipole disk.

3.4.4   Image Isofocal Point

Figure 3.20 shows an interesting phenomenon where at a particular horizontal position, the aerial image curves cross. In other words, at some x-position for this 1D image, the intensity is not a function of focus (although in general this is only approximately true). The point at which intensity does not depend of focus is called the *image isofocal point*. Mathematically, it is the value of x at which

$$\frac{\partial I}{\partial \delta} = 0 \tag{3.42}$$

Consider the coherent three-beam image case of Equation (3.27). The isofocal point is given by

$$\frac{\partial I}{\partial \delta} = -4a_0 a_1 \cos(2\pi x/p)\sin(\Delta\Phi)\left(2\pi n\left(1 - \sqrt{1 - (\lambda/p)^2}\right)\right)/\lambda = 0 \tag{3.43}$$

Solving for x, the isofocal point is $\pm jp/4$. where j is any odd number. Examining the partially coherent dipole two-beam image of Equation (3.41), the same isofocal point is obtained. In fact, this same isofocal point always results whenever the image is made up of 2-beam or a combination of 1-beam and 2-beam imaging.

For the partially coherent three-beam image case, the result is far more complicated. However, a reasonably simple answer comes from looking for the isofocal point for small amounts of defocus, so that Equation (3.31) can be used. Setting the term in this equation that multiplies the defocus to zero,

$$2a_0 a_1 \cos(2\pi x/p)\left(1 + \left(\frac{\sigma NAp}{\lambda}\right)^2\right) = -4a_1^2 \cos(4\pi x/p)\left(\frac{\sigma NAp}{\lambda}\right)^2 \tag{3.44}$$

$$\frac{\cos(4\pi x/p)}{\cos(2\pi x/p)} = -\frac{a_0}{2a_1}\left(1 + \left(\frac{\lambda}{\sigma NAp}\right)^2\right) \tag{3.45}$$

Consider the case of equal lines and spaces, so that the isofocal value for x can be defined as $p/4 + \Delta x$. Assuming that Δx will be small,

$$\frac{\cos(4\pi x/p)}{\cos(2\pi x/p)} = \frac{\cos(4\pi\Delta x/p)}{\sin(2\pi\Delta x/p)} \approx \frac{p}{2\pi\Delta x} \tag{3.46}$$

Thus, the isofocal point will be approximately at

$$\Delta x \approx -\frac{2p}{\pi^2}\left(1+\left(\frac{\lambda}{\sigma NAp}\right)^2\right)^{-1} \tag{3.47}$$

Recall that this approximate position of the image isofocal point applies to three-beam imaging of equal lines and spaces under conventional illumination such that the pupil does not clip any of the first orders. The minus sign shows that the image isofocal point will occur in the space region of the image (since $x = 0$ is the center of the space).

3.4.5   Focus Averaging

Frequently, the final aerial image is an average of the aerial image through some range of focus, a phenomenon called *focus averaging* (also called focus drilling, and sometimes known by the confusing term 'focus blur'). Chromatic aberrations, discussed in section 3.1.6 produce just such a result, as do stage scanning errors, to be discussed in section 3.5 below. Occasionally, such focus averaging is intentionally induced in order to extend the tolerable range of focus.[7]

The impact of focus averaging can be seen in an approximate form using the separable defocus image expression of Equation (3.32), which is appropriate for small levels of defocus. If $P(\delta)$ is the probability distribution of a defocus error δ, the focus-averaged image will be

$$I(x,\delta) \approx I(x,0) - f(x)\int_{-\infty}^{\infty} \delta^2 P(\delta)\mathrm{d}\delta \tag{3.48}$$

Consider a uniform probability distribution over a range of focus from δ_{min} to δ_{max}. The resulting image will be

$$I(x) \approx I(x,0) - f(x)\left[\frac{\delta_{max}^3 - \delta_{min}^3}{3(\delta_{max} - \delta_{min})}\right] \tag{3.49}$$

If the focus averaging range is symmetric about best focus (so that $\delta_{min} = -\delta_{max}$),

$$I(x) \approx I(x,0) - f(x)\left[\frac{\delta_{max}^2}{3}\right] \tag{3.50}$$

If now some unintentional focus error δ is added,

$$I(x,\delta) \approx I(x,0) - f(x)\left[\frac{\delta_{max}^2}{3} + \delta^2\right] \tag{3.51}$$

Another common probability distribution for focus averaging is a Gaussian with standard deviation σ_F. The resulting image, including unintentional focus error δ, is

$$I(x,\delta) \approx I(x,0) - f(x)[\sigma_F^2 + \delta^2] \tag{3.52}$$

For an arbitrary focus averaging function, an effective defocus error (assuming small focus errors) can be defined by

$$\delta_{eff}^2 \approx \sigma_F^2 + \delta^2 \tag{3.53}$$

where σ_F^2 is the variance of the focus averaging probability distribution.

3.4.6 Reticle Defocus

While the main focus problem in optical lithography may be controlling the focal position of the wafer, it is certainly possible for the reticle to be out of focus as well. Like the wafer, reticle focus errors can be described as an OPD with essentially the same form as for wafer focus errors:

$$OPD = n_m \delta_m (1 - \cos \theta_m) \approx \frac{1}{2} n_m \delta_m \sin^2 \theta_m \tag{3.54}$$

where the subscript 'm' refers to the mask side. Because of the reduction ratio, angles on the mask side are much lower than on the wafer side, and so the paraxial approximation for the OPD on the right-hand side of Equation (3.54) will be quite accurate. As discussed in Chapter 2, angles on the mask side can be related to angles on the wafer side by

$$R n_m \sin \theta_m = n_w \sin \theta_w \tag{3.55}$$

Thus, the OPD caused by the mask defocus can be described in terms of the wafer-side angles:

$$OPD \approx \frac{1}{2} \left(\frac{\delta_m}{R^2} \right) \left(\frac{n_w}{n_m} \right) n_w \sin^2 \theta_m \tag{3.56}$$

Comparing this expression to the paraxial effect of wafer defocus, the equivalent wafer defocus to cause this same amount of OPD would be

$$\delta_w \approx \left(\frac{\delta_m}{R^2} \right) \left(\frac{n_w}{n_m} \right) \tag{3.57}$$

For imaging in air, where both the wafer-side and mask-side refractive indices are 1, focus errors of the reticle are like focus errors on the wafer, but reduced by the reduction ratio squared. For a typical reduction ratio of 4, this means that the image is 16 times less sensitive to reticle focus errors than to wafer focus errors. (We often say that the longitudinal magnification is equal to the lateral magnification squared.) Of course, this relationship is only approximate since wafer focus errors are not accurately described by the paraxial form of the OPD for reasonably large numerical apertures. However, reticle focus errors can be readily accounted for using the Zernike third-order defocus term:

$$Z_3 = \left(\frac{\delta_m}{R^2} \right) \frac{NA^2}{4\lambda} \tag{3.58}$$

3.4.7   Rayleigh Depth of Focus

Depth of Focus (DOF) is defined as the range of focus that can be tolerated, before the image quality is degraded beyond usefulness. Lord Rayleigh, more than 100 years ago, gave us a simple approach to estimate depth of focus in an imaging system. Here we'll express his method and results in modern lithographic terms. We'll consider the coherent imaging of an array of small lines and spaces so that only the zero and the plus and minus first diffraction orders pass through the lens to form the image.

As discussed above in section 3.4.1, a common way of thinking about the effect of defocus on an image is to consider the defocusing of a wafer as equivalent to causing an aberration – an error in curvature of the actual wavefront relative to the desired wavefront (i.e. the one that focuses on the wafer). Equations (3.20)–(3.22) show how the OPD varies across the pupil for a given amount of defocus. How much OPD can our line/space pattern tolerate? Consider the extreme case. If the OPD of the first diffraction orders were set to a quarter of the wavelength, the zero and first diffracted orders would be exactly 90° out of phase with each other. At this much OPD, the zero order would not interfere with the first orders at all and no pattern would be formed. The true amount of tolerable OPD must be less than this amount:

$$OPD_{max} = k_2 \frac{\lambda}{4}, \text{ where } k_2 < 1 \tag{3.59}$$

Substituting this maximum permissible OPD into Equation (3.20), we can find the DOF.

$$DOF = 2\delta_{max} = \frac{k_2}{2} \frac{\lambda}{n(1-\cos\theta)} \tag{3.60}$$

At the time of Lord Rayleigh, lens numerical apertures were relatively small and the medium of imaging was air, so that $n = 1$. Thus, the largest angles going through the lens were also quite small and the cosine term could be approximated in terms of $\sin\theta$:

$$DOF \approx k_2 \frac{\lambda}{\sin^2\theta} = k_2 \frac{p^2}{\lambda} \tag{3.61}$$

This paraxial form is reasonably accurate for numerical apertures below 0.5. At this point, Lord Rayleigh made a crucial application of this formula that is often forgotten. While Equation (3.61) would apply to any small pattern of lines and spaces, Lord Rayleigh essentially looked at the extreme case of the smallest pitch that could be imaged – the resolution limit. The smallest pitch that can be printed would put the first diffracted order at the largest angle that could pass through the lens, defined by the numerical aperture, *NA*. For this one pattern, the general expression [Equation (3.61)] becomes the more familiar and specific Rayleigh DOF criterion:

$$DOF = k_2 \frac{\lambda}{NA^2} \tag{3.62}$$

The Rayleigh depth of focus criterion, as usually expressed in Equation (3.62), is rife with caveats. It was derived for a pattern of equal lines and spaces imaged under coherent illumination, assumed that the feature was at the resolution limit, and made the paraxial approximation, which is not very accurate at numerical apertures in common lithographic use. In general, it does a very poor job of predicting the DOF observed in lithographic tools.

3.5   Imaging with Scanners Versus Steppers

In general, increasing the numerical aperture of a lens while keeping the field size constant requires significant improvements in lens design and manufacturing in order to keep aberration levels low (or better yet, to reduce them). A lens with a smaller field size will always be easier to design and manufacture. However, the field size must be at least the size of the largest chip to be manufactured, and larger field sizes provide for higher imaging productivity since throughput is improved if multiple chips can be imaged at once. This tension between smaller field sizes to improve aberration performance and larger field sizes to improve manufacturing productivity has led to a transition from steppers to *step-and-scan* tools.

A step-and-repeat camera (stepper) images a single field (for example, 20 × 20 mm in size on the wafer) while the mask and wafer are stationary, then steps the wafer to a new location and repeats the exposure. Eventually the whole wafer is covered with exposed fields. A scanner, on the other hand, will scan the mask and wafer simultaneously past an illuminated exposure field (in the shape of a slit), allowing for an exposure area limited only by the length of the slit and the travel of the mask and wafer. Step-and-scan tools are a hybrid between these two approaches. The mask and wafer are scanned past a small slit (for example, 26 × 8 mm) until the full mask is exposed (up to 26 × 33 mm at wafer dimensions). The wafer is then stepped to a new location and the scanning is repeated (Figure 3.22). The smaller image field (the slit) makes the imaging system simpler, though at the expense of more complicated mechanical mask and wafer motion. Thus, scanning can be considered a compromise between throughput, stage complexity and lens complexity for large-field imaging.

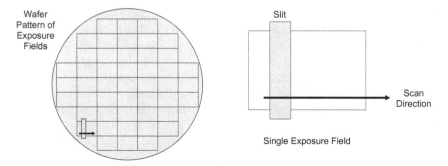

Figure 3.22   *A wafer is made up of many exposure fields (with a maximum size that is typically 26 × 33 mm), each with one or more die. The field is exposed by scanning a slit that is about 26 × 8 mm across the exposure field*

For scanners, exposure dose is controlled by the scan speed. Since step-and-scan tools generally use excimer laser light sources, which emit light in short pulses, the scan speed V needed to achieve a specific dose is given by

$$V = W_s \frac{f}{n} \qquad (3.63)$$

where W_s is the slit width, f is the repetition rate of the laser and n is the number of pulses needed to achieve the proper dose. The pulse-to-pulse dose repeatability of a good excimer laser is a few percent. Since the overall dose repeatability is improved by a factor of $\sqrt{n}$ (assuming each pulse is independent), to achieve a dose control of a few tenths of a percent the number of pulses must be on the order of 100. For a slit width of 8 mm, a laser repetition rate of 2 kHz, and 50 pulses required, the wafer stage velocity would need to be 0.32 m/s (the reticle stage must move at exactly 4× this speed for a 4× reduction system). Improved laser repetition rates and reduced pulse-to-pulse variability allow for faster scan speeds and thus better throughput.

Besides all of the practical differences between the operation of steppers and step-and-scan tools, there are differences in imaging performance as well. Lenses are fabricated to be circularly symmetric, while chips are necessarily rectangular in shape. Thus, different-sized rectangular fields are possible for a given diameter lens. If the field size has a diameter of 28 mm, the maximum square field (for a stepper, for example), is just under 20 mm square. The same lens, however, can accommodate a 26 × 8 mm slit in a step-and-scan configuration, allowing for larger die sizes and/or more productive printing of multiple die in the larger field. In addition, the lens can be rotated during testing and assembly to find the 26 × 8 mm slice of the circular field that provides the best imaging quality.

As was discussed in sections 3.1 and 3.3, the nonideal imaging behavior of a lens (aberrations, flare) varies as a function of field position. The result is an unwanted variation in the dimensions and placements of the printed patterns as a function of field position. Scanning can significantly change the nature of this spatial variation across the field. When scanning the mask and wafer across the stationary slit, each point on the wafer is exposed by an average of aerial images – one for each field position along the scan direction of the slit. In other words, any variations in aberrations or flare across the field in the scan direction will be averaged out. Thus, the variations in feature size and overlay are significantly reduced in the scan direction. Of course, in the slit direction there is no such averaging and so the full spatial variation of lithographic results will remain in this direction (see Chapter 8 for a more detailed discussion of spatial CD variations).

The beneficial effects of aberration averaging mean that step-and-scan tools generally exhibit improved CD control compared to steppers. However, scanners also introduce new sources of process variation that, if not properly controlled, could erase these CD uniformity benefits. Scanning the mask and wafer past the imaging slit requires extremely precise synchronization of the mask and wafer position during the scan. Scan synchronization errors act exactly like wafer vibrations during a stepper exposure. The result is a blurring of the projected image, sometimes described as 'image fading'. While systematic synchronization errors are possible, random errors are more common and are characterized by a *moving standard deviation* (MSD) of the stage position (see Figure 3.23). The

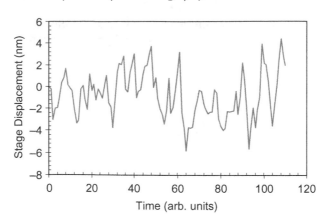

Figure 3.23 *Example stage synchronization error (only one dimension is shown), with an MSD of 2.1 nm*

resulting 'averaged' image is the convolution of the static image with a Gaussian of this standard deviation. Additionally, focus variations during the scan produce an averaging of the aerial image through focus, which will degrade image quality at best focus.

3.6   Vector Nature of Light

As James Clerk Maxwell has shown us, light is an electromagnetic wave. Electric and magnetic fields oscillate at some characteristic frequency while traveling at the speed of light. These fields have a magnitude, phase and direction. It is the direction (also called the polarization) of the electric field that defines its vector nature, and a key property of an electromagnetic wave is that its electric and magnetic field vectors are at right angles to each other, and at right angles to the direction of propagation. Figure 3.24 shows a standard textbook picture of a traveling ray of electromagnetic energy.

The direction of the electric field plays a critical role in the phenomenon of interference. Two plane waves, approaching a wafer at different angles, will interfere to form fringe patterns of light and dark making the simplest of lithographic patterns – an array of lines and spaces. This phenomenon of interference is the key to pattern formation. Without it, light hitting the wafer from different directions would simply add up to give a uniform intensity. We can see mathematically the effect of interference by examining how two electric fields combine to form a resultant electric field and intensity. Ignoring the refractive index of the medium, the intensity I is the square of the magnitude of the electric field E. If two electric fields are combined, what is the intensity of the combination? If the two electric fields do not interfere, the total intensity is the sum of the individual intensities.

$$I = |E_1|^2 + |E_2|^2 \tag{3.64}$$

If, however, the two electric fields interfere completely, the total intensity will be

$$I = |E_1 + E_2|^2 \tag{3.65}$$

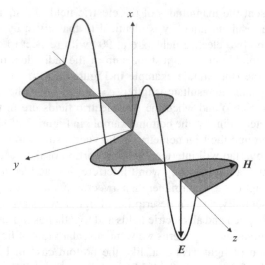

Figure 3.24   *A monochromatic plane wave traveling in the z-direction. The electric field vector is shown as **E** and the magnetic field vector as **H***

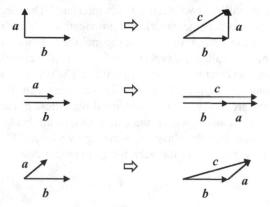

Figure 3.25   *Examples of the sum of two vectors **a** and **b** to give a result vector **c**, using the geometric 'head-to-tail' method*

Two electric fields interfere with each other only to the extent that their electric fields oscillate in the same direction. If the electric fields are at right angles to each other, there will be no interference. Thus, to determine the amount of interference between two electric fields, one must first determine the amount of directional overlap between them. Vector mathematics gives us some simple tools to calculate directional overlap and thus the amount of interference. Figure 3.25 shows the standard head-to-tail method of geometrically adding two vectors. In the top case, the two vectors are at right angles to each other, so the head-to-tail construction forms a simple right triangle. Thus, the length of the resultant vector is given by the familiar Pythagorean theorem: $c^2 = a^2 + b^2$. If the vector lengths represent the magnitudes of the electric field, the square of the length is the intensity and the resultant intensity is identical to that given by Equation (3.64). In other words, when the two electric fields are at 90° with respect to each other, they do not interfere and the total intensity is just the sum of the individual intensities.

At the other extreme (the middle example in Figure 3.25), two vectors in the same direction add directly and the resultant length is $c = a + b$. Thus, the intensity of the sum is given by Equation (3.65) and when the two electric fields are in the same direction, they interfere completely. Finally, the bottom example in Figure 3.25 shows an intermediate case. Working through the trigonometry of the vector sum shows that the portion of the two vectors that overlap will interfere, and the portion of the vectors that are at right angles will add in quadrature (i.e. they won't interfere, just add as intensities).

When thinking of light as a scalar rather than a vector quantity, we ignore the subtleties discussed above. Essentially, a scalar description of light always assumes that the electric fields are 100% overlapped and all electric fields add together as in Equation (3.65). Since interference is what gives us the patterns we want, a scalar view of light is too optimistic. The vector description of light says that, like the bottom case in Figure 3.25, there is usually some fraction of the electric fields of our two vectors that don't overlap and thus don't contribute to interference. The noninterfering light is still there, but it adds as a uniform intensity that degrades the quality of the image.

Consider the interference of two plane waves approaching the wafer at moderate angles, as in Figure 3.26. How will the two electric fields interfere? The electric field can point in any direction perpendicular to the direction of propagation. This arbitrary direction can in turn be expressed as the sum of any two orthogonal (basis) directions. The most convenient basis directions are called transverse electric or TE (the electric field pointing out of the page of the drawing) and transverse magnetic or TM (the electric field pointing in the page of the drawing). As Figure 3.26 shows, the TE case means that the electric fields of the two planes wave are always 100% overlapped regardless of the angle between the plane waves. For the TM case, however, the extent of overlap between the two vectors grows smaller as the angle between the plane waves grows larger. TM is the 'bad' polarization in lithography, especially at the very high numerical apertures that allow large angles between light rays.

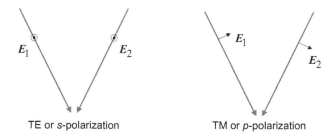

Figure 3.26 *Two plane waves with different polarizations will interfere very differently. For transverse electric (TE) polarization (electric field vectors pointing out of the page), the electric fields of the two vectors overlap completely regardless of the angle between the interfering beams*

We can calculate how much the two electric fields for the TM case will interfere with each other. Suppose that the two rays in Figure 3.26 are traveling at an angle θ with respect to the vertical direction (i.e. the direction normal to the wafer). The electric fields E_1 and E_2 will have an angle between them of 2θ. The amount of the electric field vector E_2 that points in the same direction as E_1 is just the geometric projection, $E_2\cos(2\theta)$. Thus, the intensity will be given by the coherent (electric field) sum of the parts that overlap plus the incoherent (intensity) sum of the parts that don't overlap.

$$I = |E_1 + E_2 \cos(2\theta)|^2 + |E_2 \sin(2\theta)|^2 \qquad (3.66)$$

Note that for $\theta = 0$, this equation reverts to the perfectly coherent (interfering) sum of Equation (3.65). For $\theta = 45°$, the two electric fields are perpendicular to each other and Equation (3.66) becomes the perfectly incoherent (noninterfering) sum of Equation (3.64). For typical photoresists (with refractive indices between 1.6 and 1.7), this 45° angle inside the resist will be obtained for numerical apertures in the 1.15–1.2 range.

3.6.1 Describing Polarization

As we saw above, the behavior of a wave of light depends not only on the amplitude and phase of the wave (its scalar properties), but on the direction that the electric field is pointing (its vector property). In particular, the amount of interference between two electromagnetic waves depends on the relative directions of the two electric fields. If the electric fields point in the same direction, they will interfere completely. If they are orthogonal to one another, there will be no interference. The direction of the electric field of an electromagnetic wave, and how that direction varies in time or space, is called its *polarization*. Recall that Maxwell's equations require that the electric and magnetic fields are always perpendicular to each other and both always point perpendicular to the direction of propagation. As a result, it is sufficient to consider only the direction of the electric field.

Since an electromagnetic wave travels through time and space, there are three important ways of looking at (or thinking about) the direction of the electric field. First, imagine 'riding the wave': a given point on the wave moves through time and space at the speed of light. On this point on the wave, the electric field stays pointing in a constant direction, regardless of the type of polarization that the wave has. While an important conceptual idea, this way of thinking about the wave doesn't give us any information about its polarization. The second approach is to look at an instant of time, and see how the electric field direction is changing through space (this works best for simple harmonic waves, such as a plane wave). Finally, the third viewpoint is a fixed location in space, monitoring the change in the direction of the electric field through time. These last two perspectives allow for equivalent descriptions of polarization: the variation of the electric field direction in space at an instant in time, or the variation of the electric field direction in time at a fixed point in space.

With the above perspectives in mind, how can light be polarized? The simplest type of polarization is called *linear* polarization, shown in Figure 3.27. At an instant in time, the electric field E is always pointing in one direction (in this case, the y-direction, though in both positive and negative directions) for all points in space (Figure 3.27a). At one point in space, the electric field changes magnitude sinusoidally through time, but

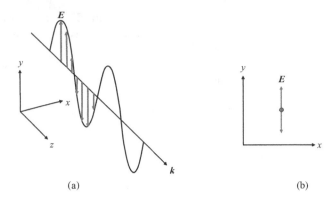

Figure 3.27   *Linear polarization of a plane wave showing (a) the electric field direction through space at an instant in time, and (b) the electric field direction through time at a point in space. The* **k** *vector points in the direction of propagation of the wave*

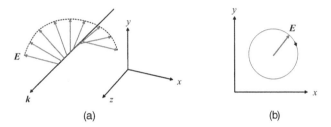

Figure 3.28   *Right circular polarization of a plane wave showing (a) the electric field direction through space at an instant in time, and (b) the electric field direction through time at a point in space*

always points in the same direction (Figure 3.27b). A second type of polarization is called *circular* polarization, as shown in Figure 3.28. Here, the electric field direction takes on a spiral form through space at a given instant in time (Figure 3.28a). At a given point in space, the electric field direction rotates through a circle, keeping a constant magnitude (Figure 3.28b). Both right-circular and left-circular polarization are possible, depending on the direction of rotation of the polarization. Note that the name given to the polarization comes from the shape traced out by the electric field through time at a given point in space. It is this viewpoint that is usually the easiest one to picture and understand.

Figure 3.29 compares some common types of polarization. Besides linear and circular polarization, *elliptical* polarization is somewhere in between the two. *Random* polarization is by far the most common type of polarization available for light sources, and until recently the only polarization available to lithographers. (Note that the common name for this type of polarization, unpolarized light, is actually a misnomer since light always has a polarization, even if it is random.) Here, the electric field direction changes randomly

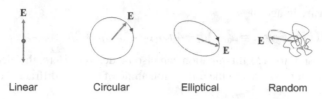

Linear   Circular   Elliptical   Random

Figure 3.29 *Examples of several types of polarizations (plotting the electric field direction through time at a point in space)*

(and quickly) through time. When two waves of random polarization interfere with each other, some statistical average of amounts of electric field overlap must be used. Fortunately, a rigorous analysis provides a simple result: the intensity resulting from the interference of two randomly polarized waves can be calculated by breaking each wave into two linearly polarized orthogonal waves (for example, TE and TM polarizations), interfering each polarization separately, and summing the two resulting intensities.

In reality, none of the polarizations described above are achieved perfectly in practice. Practical light sources almost always produce some combination of polarized light and unpolarized light. To quantify the degree to which a light source supplies properly polarized illumination, a metric called *degree of polarization* (DOP) is used:

$$DOP = \frac{I_{\text{good pol}} - I_{\text{bad pol}}}{I_{\text{good pol}} + I_{\text{bad pol}}} \tag{3.67}$$

where $I_{\text{good pol}}$ is the desired polarization and $I_{\text{bad pol}}$ is the undesired polarization. If one is trying to make linear TE polarized light, for example, the good polarization is TE and the bad polarization is TM. Perfectly polarized light would have a DOP of 1.0. Randomly polarized light, being made up of equal parts TE and TM on average, will have a DOP of zero. Polarized light sources for use in lithography are expected to have DOPs ≥ 0.9.

Mathematically, the polarization of a plane wave can be characterized by the phase difference between orthogonal components of the electric field. For example, consider a plane wave traveling in the z-direction with electric field components in the x- and y-directions. If the difference in phase between the x- and y-components of the electric field ($\Delta\phi$) is a multiple of π, the plane wave will be linearly polarized. If $\Delta\phi$ is an odd multiple of $\pi/2$, the wave will be circularly polarized. For other fixed phase differences, the wave will be elliptically polarized.

3.6.2 Polarization Example: TE Versus TM Image of Lines and Spaces

For TE illumination of line/space patterns, the electric fields of each of the diffracted orders overlap completely so that the vector sum of the electric fields equals the scalar sum. An example of a coherent illumination aerial image where only the zero and first diffracted orders are used was given in Chapter 2. If the lines are oriented in the y-direction, the only electric field component for this polarization will be in the y-direction as well:

$$E_y(x) = a_0 + 2a_1 \cos(2\pi x/p) \tag{3.68}$$

giving an intensity image of

$$I(x) = |E_y(x)|^2 = a_0^2 + 2a_1^2 + 4a_0a_1\cos(2\pi x/p) + 2a_1^2\cos(4\pi x/p) \tag{3.69}$$

A similar expression for TM illumination can also be derived. Here, the electric field will have both x and z components. If the diffraction angle of the first diffracted orders is given by $\sin\theta = \lambda/p$,

$$\begin{aligned} E_x &= a_0 + 2a_1\cos\theta\cos(2\pi x/p) \\ E_z &= 2a_1\sin\theta\sin(2\pi x/p) \end{aligned} \tag{3.70}$$

giving

$$\begin{aligned} I(x) &= |E_x(x)|^2 + |E_z(x)|^2 \\ &= a_0^2 + 2a_1^2 + 4a_0a_1\cos\theta\cos(2\pi x/p) + 2a_1^2\cos(2\theta)\cos(4\pi x/p) \end{aligned} \tag{3.71}$$

The effect of using the 'bad' polarization is to decrease the amount of interference by the cosine of the angle between the interfering orders. Thus, the interference between the zero and first orders is lessened by the factor $\cos\theta$ and the interference between the plus and minus first orders is decreased by $\cos2\theta$. For randomly polarized light, the final aerial image would be the average of the two images given above for TE and TM polarization.

The difference between TE and TM polarization, as shown in the above example, can be quite large for small-pitch patterns. However, when considering only the aerial image (that is, the image in air) this difference is greatly exaggerated. When the image propagates from the air to the resist, refraction will decrease the angle of the first orders inside the resist according to Snell's law. The lower angles in the resist will have greater overlap of the TM electric fields and thus a better quality image. For example, if the first order diffraction angle in air is $\sin\theta = 0.9$, then $\cos\theta = 0.44$ and the interference with the zero order is less than 50% effective. However, in resist the first-order angle is reduced by the refractive index (typically about 1.7), so that $\sin\theta_{resist} = 0.53$ and $\cos\theta_{resist} = 0.85$. The image in resist will have far higher quality than the aerial image. And of course, it is the image in resist that actually exposes the resist, not the aerial image. This topic will be discussed in greater detail in Chapter 4.

3.6.3   Polarization Example: The Vector PSF

An interesting (and somewhat nonintuitive) consequence of the vector nature of light is that linearly polarized illumination will produce an asymmetric PSF.[8] A cross section of the PSF in the direction of polarization is wider than in the perpendicular direction (30% greater for the case of NA = 0.866, see Figure 3.30).

3.6.4   Polarization Aberrations and the Jones Pupil

As discussed at the beginning of this chapter, aberrations are characterized by means of the pupil function, defined as the ratio of the electric field at the exit pupil to the electric field at the entrance pupil. When thinking of electric fields as decomposed into orthogonal polarization components (TE and TM being the most convenient decomposition), an interesting question arises: will the pupil function for TE light passing through the lens

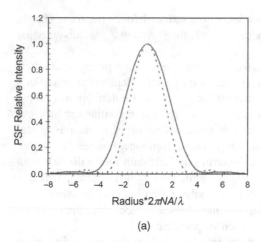
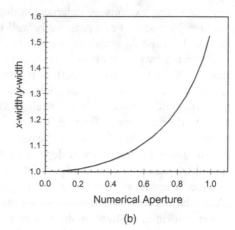

Figure 3.30 *The point spread function (PSF) for linearly x-polarized illumination: (a) cross sections of the PSF for NA = 0.866 (solid line is the PSF along the x-axis, dashed line is the PSF along the y-axis); (b) ratio of the x-width to the y-width of the PSF as a function of numerical aperture*

be the same as for TM light? Further, if TE light only enters the lens, will any of the light exiting the lens be TM? To include the variation of pupil transmittance with polarization, a more complete description of the pupil is required.

If the electric field entering the lens is broken down into TE and TM components, given by E_{iTE} and E_{iTM}, respectively, and the electric field at the exit pupil is given by E_{oTE} and E_{oTM}, these electric fields can be expressed as *Jones vectors*:[9]

$$\begin{bmatrix} E_{iTE} \\ E_{iTM} \end{bmatrix} \quad \text{and} \quad \begin{bmatrix} E_{oTE} \\ E_{oTM} \end{bmatrix} \tag{3.72}$$

Describing the transmittance of the lens by the *Jones pupil* matrix J, the input and output Jones vectors are related by[10,11]

$$\begin{bmatrix} E_{oTE} \\ E_{oTM} \end{bmatrix} = \begin{bmatrix} J_{11} & J_{12} \\ J_{21} & J_{22} \end{bmatrix} \begin{bmatrix} E_{iTE} \\ E_{iTM} \end{bmatrix} \tag{3.73}$$

Note that if $J_{12} = J_{21} = 0$ and $J_{11} = J_{12}$, then the Jones pupil description of the lens transmittance reverts to the scalar pupil description where $J_{11} = J_{12} = P$, the scalar pupil function. It is thus convenient to define the scalar pupil function as the average of the diagonal elements of the Jones pupil:

$$P = \frac{J_{11} + J_{22}}{2} \tag{3.74}$$

Using this definition of the scalar pupil, a reduced Jones pupil $\tilde{J}$ can be defined by

$$J = P\tilde{J} = P \begin{bmatrix} \tilde{J}_{11} & \tilde{J}_{12} \\ \tilde{J}_{21} & \tilde{J}_{22} \end{bmatrix} \tag{3.75}$$

For a lens that exhibits no polarization dependence, the reduced Jones matrix is simply the identity matrix. For a reasonably well-behaved lens, then, $\tilde{J}_{11}$ and $\tilde{J}_{22}$ are about equal to 1, and $\tilde{J}_{12}$ and $\tilde{J}_{21}$ are about equal to 0.

There are two main physical causes for what are sometimes called polarization aberrations. When light hits an interface (such as the surface of a lens element) at an oblique angle, the transmittance of that interface is different for TE and TM polarizations (a topic that will be discussed in some detail in Chapter 4). Antireflection coatings reduce but do not eliminate this difference. Thus, any lens that is a collection of refractive optical elements can be thought of as a weakly polarizing optical component, since the overall transmittance of the lens will depend on the incoming polarization and will vary with spatial frequency (the outer edges of the lens showing the biggest differences). This difference in amplitude transmittance as a function of polarization is called *diattenuation*. (A similar but less commonly observed phenomenon in lenses is *dichroism*, the variation in the absorption of light by the lens as a function of polarization.)

The second major cause of polarization effects in a lens is *birefringence*, a difference in the refractive index of a material as a function of the direction of the electric field vector. Birefringence leads to the general observation of *retardance*, a difference in the phase of the transmitted light as a function of its polarization. Some materials, most notably certain crystals, exhibit intrinsic birefringence. Most optical materials become at least somewhat birefringent when stressed. Common optical materials used in lithographic lenses exhibit small amounts of both intrinsic and stress-induced birefringence at 193-nm wavelength, and relatively large amounts of birefringence at a wavelength of 157 nm. Lens antireflection coatings can also contribute to retardance. Birefringence is usually expressed as the OPD experienced by two orthogonal polarizations per unit length traveled through the material, with units of nm/cm.

A useful mathematic decomposition of the Jones pupil expresses this matrix as linear combination of Pauli spin matrices:

$$\tilde{J} = \begin{bmatrix} 1 & 0 \\ 0 & 1 \end{bmatrix} + a_1 \begin{bmatrix} 1 & 0 \\ 0 & -1 \end{bmatrix} + a_2 \begin{bmatrix} 0 & 1 \\ 1 & 0 \end{bmatrix} + a_3 \begin{bmatrix} 0 & -i \\ i & 0 \end{bmatrix} \tag{3.76}$$

The magnitudes of a_1, a_2 and a_3 then indicate the deviation of the reduced Jones pupil from an identity matrix, and can be easily related to the elements of the reduced Jones pupil matrix:

$$a_1 = \frac{\tilde{J}_{11} - \tilde{J}_{22}}{2}, \quad a_2 = \frac{\tilde{J}_{12} + \tilde{J}_{21}}{2}, \quad a_3 = \frac{\tilde{J}_{12} - \tilde{J}_{21}}{-2i} \tag{3.77}$$

These (complex number) coefficients have important physical interpretations. The real parts of a_1 and a_2 describe the diattenuation of the lens along the axes and at 45° to the axes, respectively. The imaginary parts of a_1 and a_2 describe the retardance of the lens along the axes and at 45° to the axes, respectively. The coefficient a_3 describes the circularly directed diattenuation and retardance.

All of these various forms for describing polarization-dependent pupil transmission suffer from a common problem – the excess complication that comes from expanding a scalar pupil (with its two pupil maps) into a vector pupil (with eight pupil maps). The reduced Jones pupil form of Equation (3.75) preserves the intuition and experience of the scalar pupil, but does not reduce the overall level of complexity. Recently, Geh formulated an alternate decomposition of the Jones pupil into a scalar pupil plus a partial polarizer (producing diattenuation), a rotator (that simply rotates the polarization state), and a retarder (giving birefringence).[12] The resulting decomposition is significantly more intuitive.

One should note that the Jones pupil cannot describe the phenomenon of *depolarization*, the decrease in the degree of polarization as light travels through a lens. Depolarization is generally due to scattering mechanisms – the same mechanisms that give rise to flare. However, in general the amount of scattering will be approximately the same for each polarization. Thus, the mechanism used to account for flare described in section 3.3 can be coupled with the Jones pupil description of the nonscattered light transmittance to give a very reasonable and accurate description of lithographic imaging. The flare (scattered light) reaching the wafer is then assumed to be randomly polarized, thus accounting for depolarization.

3.7 Immersion Lithography

Although the scientific principles underlying immersion lithography have been known for well over 100 years (immersion microscopes were in common use by 1880), only recently has this technology attracted widespread attention in the semiconductor industry. Despite this rather late start, the potential of immersion lithography for improved resolution and depth of focus has changed the industry's roadmap and is destined to extend the life of optical lithography to new, smaller limits.

From a practical perspective, immersion lithography at 193 nm uses ultrapure degassed water, with refractive index of 1.437, as the immersion liquid. A small puddle of water is kept between the lens and the wafer during wafer exposure (Figure 3.31). New water is constantly pumped under the lens to keep the optical properties of the fluid consistent, and to prevent the buildup of contaminants. A special wafer chuck keeps the puddle intact as the edge of the wafer is scanned under the puddle.

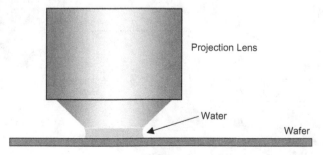

Projection Lens

Water

Wafer

Figure 3.31 *Immersion lithography uses a small puddle of water between the stationary lens and the moving wafer. Not shown is the water source and intake plumbing that keeps a constantly fresh supply of immersion fluid below the lens*

3.7.1   The Optical Invariant and Hyper-NA Lithography

The story of immersion lithography begins with Snell's Law. Light traveling through material 1 with refractive index n_1 strikes a surface with angle θ_1 relative to the normal to that surface. The light transmitted into material 2 (with index n_2) will have an angle θ_2 relative to that same normal as given by Snell's law.

$$n_1 \sin \theta_1 = n_2 \sin \theta_2 \qquad (3.78)$$

Now picture this simple law applied to a film stack made up of any number of thin parallel layers (Figure 3.32a). As light travels through each layer, Snell's law can be repeatedly applied:

$$n_1 \sin \theta_1 = n_2 \sin \theta_2 = n_3 \sin \theta_3 = n_4 \sin \theta_4 = \ldots = n_k \sin \theta_k \qquad (3.79)$$

Thanks to Snell's law, the quantity $n\sin\theta$ is *invariant* as a ray of light travels through this stack of parallel films. Interestingly, the presence or absence of any film in the film stack in no way affects the angle of the light in other films of the stack. If films 2 and 3 were removed from the stack in Figure 3.32a, for example, the angle of the light in film 4 would be exactly the same.

We find another, related invariant when looking at how an imaging lens works. A well-made imaging lens (with low levels of aberrations) will have a *Lagrange invariant* (often just called the optical invariant) that relates the angles entering and exiting the lens to reduction ratio of that lens, as described in Chapter 2:

$$R = \frac{n_w \sin \theta_w}{n_m \sin \theta_m} \qquad (3.80)$$

where n_m is the refractive index of the media on the mask side of the lens, θ_m is the angle of a ray of light entering the lens relative to the optical axis, n_w is the refractive index of the media on the wafer side of the lens and θ_w is the angle of a ray of light exiting the lens relative to the optical axis (Figure 3.32b). Note that, other than a scale factor given by the magnification of the imaging lens and a change in the sign of the angle to account for the focusing property of the lens, the Lagrange invariant makes a lens seem like a

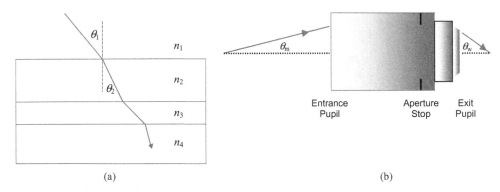

(a)   (b)

Figure 3.32   *Two examples of an 'optical invariant', (a) Snell's law of refraction through a film stack, and (b) the Lagrange invariant of angles propagating through an imaging lens*

thin film obeying Snell's law. (It is often convenient to imagine the imaging lens as 1×, scaling all the object dimensions by the magnification, thus allowing $R = 1$ and making the Lagrange invariant look just like Snell's law).

These two invariants can be combined when thinking about how a photolithographic imaging system works. Light diffracts from the mask at a particular angle. This diffracted order propagates through the lens and emerges at an angle given by the Lagrange invariant. This light then propagates through the media between the lens and the wafer and strikes the photoresist. Snell's law dictates the angle of that ray in the resist, or any other layers that might be coated on the wafer. Taking into account the magnification scale factor, the quantity $n\sin\theta$ for a diffracted order is constant from the time it leaves the mask to the time it combines inside the resist with other diffraction orders to form an image of the mask.

If we replace the air between the lens and the wafer with water, the optical invariant says that the angles of light inside the resist will be the same, presumably creating the exact same image. The impact of immersion lithography comes from two sources: the maximum possible angle of light that can reach the resist, and the phase of that light.

Consider again the chain of angles through multiple materials as given by Equation (3.79). Trigonometry will never allow the sine of an angle to be greater than 1. Thus, the maximum value of the invariant will be limited by the material in the stack with the smallest refractive index. If one of the layers is air (with a refractive index of 1.0), this will become the material with the smallest refractive index and the maximum possible value of the invariant will be 1.0. If we look then at the angles possible inside of the photoresist, the maximum angle possible would be $\sin\theta_{max,resist} = 1 / n_{resist}$. Now suppose that the air is replaced with a fluid of a higher refractive index, but still smaller than the index of the photoresist. In this case, the maximum possible angle of light inside the resist will be greater: $\sin\theta_{max,resist} = n_{fluid} / n_{resist}$. At a wavelength of 193 nm, resists have refractive indices of about 1.7 and water has a refractive index of about 1.44. The fluid does not make the angles of light larger, but it *enables* those angles to be larger. If one were to design a lens to emit larger angles, immersion lithography will allow those angles to propagate into the resist. The numerical aperture of the lens (defined as the maximum value of the invariant $n\sin\theta$ that can pass through the lens) can be made to be much larger using immersion lithography, with the resulting improvements in resolution one would expect.

Thus, by using an immersion fluid between the lens and the wafer, numerical apertures greater than 1 have been achieved. These so-called hyper-NA immersion lithography tools are certainly not the first NA > 1 imaging tools, since immersion microscopes have been common for over 100 years. The challenges in making immersion practical for high-volume semiconductor manufacturing, however, are quite great, especially considering the need for high-speed scanning of the wafer past the lens while immersed in the fluid. Practical numerical apertures for immersion systems are probably in the range of 93 % of the refractive index of the fluid. For water, with $n = 1.44$ at 193 nm, this means a numerical aperture of up to 1.35. High-index fluids ($n = 1.65$, for example) could enable numerical apertures as high as 1.5. Beyond this level, the maximum angle is limited by the refractive index of the photoresist, which would have to be raised to allow for further increases in numerical aperture. From the lens perspective, the refractive index of the various lens materials for 193-nm lithography is on the order of 1.5, so that concave lens elements would be required in order to couple light into a higher-index fluid. Alternatively, one could attempt to develop higher index glasses for use as lens elements. Both of these possibilities pose quite challenging problems.

The second way that an immersion fluid changes the results of imaging comes from manner in which the fluid affects the phase of the light as it reaches the wafer. Light, being a wave, undergoes a phase change as it travels. If light of (vacuum) wavelength λ travels some distance Δz through some material of refractive index n, it will undergo a phase change $\Delta\phi$ given by

$$\Delta\varphi = 2\pi n \Delta z / \lambda \qquad (3.81)$$

A phase change of 360° will result whenever the optical path length (the refractive index times the distance traveled) reaches one wavelength. This is important in imaging when light from many different angles combine to form one image. All of these rays of light will be in phase only at one point – the plane of best focus. When out of focus, rays traveling at larger angles will undergo a larger phase change than rays traveling at smaller angles. As a result, the phase difference between these rays will result in a blurred image.

For a given diffraction order (and thus a given angle of the light inside the resist), the angle of the light inside an immersion fluid will be less than if air were used. These smaller angles will result in smaller OPDs between the various diffracted orders when out of focus, and thus a smaller degradation of the image for a given amount of defocus. In other words, for a given feature being printed and a given numerical aperture, immersion lithography will provide a greater depth of focus (DOF). A more thorough description of the impact of immersion on DOF will be given in the following section.

3.7.2   Immersion Lithography and the Depth of Focus

From the derivation of the Rayleigh DOF criterion given above in section 3.4.7, we can state the restrictions on this conventional expression of the Rayleigh DOF: relatively low numerical apertures imaging a binary mask pattern of lines and spaces at the resolution limit. Equation (3.60) can be thought of as the high-NA version of the Rayleigh DOF criterion, which still assumes we are imaging a small binary pattern of lines and spaces, but is appropriate at any numerical aperture. It also accounts for immersion lithography through the refractive index n. The angle θ can be related to the pitch by letting this angle be that of the first diffraction order.

$$n \sin\theta = \frac{\lambda}{p} \qquad (3.82)$$

Combining Equations (3.60) and (3.82), one can see how immersion will improve the DOF of a given feature:

$$\frac{DOF(immersion)}{DOF(dry)} = \frac{1 - \sqrt{1 - (\lambda/p)^2}}{n - \sqrt{n^2 - (\lambda/p)^2}} \qquad (3.83)$$

As Figure 3.33 shows, the improvement in DOF is at least the refractive index of the fluid, and grows larger from there for the smallest pitches.

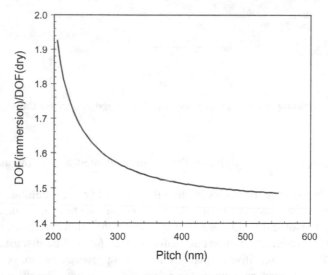

Figure 3.33 *For a given pattern of small lines and spaces, using immersion improves the depth of focus (DOF) by at least the refractive index of the fluid (in this example, λ = 193 nm, n_{fluid} = 1.46)*

3.8 Image Quality

While much of this chapter has focused on the nature of imaging and those factors that affect the resulting aerial image, along the way we have had occasion to mention the 'quality' of the aerial image without actually defining what is meant by this term. While it is intuitively easy to see how defocus and aberrations degrade the image, we will need to define a numerical metric of image quality in order to quantify these effects. While much more will be said on this topic in Chapter 9, here we will simply define a metric related to the ability of the image to define and control the position of the edge to be printed in photoresist.

3.8.1 Image CD

The aerial image, when propagated into a photoresist, interacts with the resist to create a final feature of a certain width. It is often convenient to estimate the width of the final resist feature by the width of the aerial image. Consider an ideal, infinite contrast positive resist. For such a resist, any exposure above some threshold dose E_{th} will cause all of the resist to be removed. Thus, for an aerial image $I(x)$ and an exposure dose E, the critical dimension (CD) will equal $2x$ when

$$E_{th} = EI(x) \qquad (3.84)$$

For a given dose, an image intensity threshold value can be defined as

$$I_{th} = \frac{E_{th}}{E} \qquad (3.85)$$

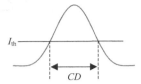

Figure 3.34   *Defining image CD: the width of the image at a given threshold value I_{th}*

The image CD is simply the width of the image evaluated at this intensity threshold (Figure 3.34).

This simple threshold model for estimating the resist CD can be significantly improved by including a 'resist bias'. As Figure 3.20 shows and section 3.4.4 discusses, aerial images exhibit an isofocal point, an x-position where the intensity varies the least with defocus. Likewise, resist CD exhibits an isofocal value (as will be discussed in Chapter 8) where the CD is least sensitive to changes in focus. In general, the image isofocal point will not occur at the same position as the resist isofocal point. The difference between these points is called the *resist bias*.

Our simple threshold model can be modified to account for this empirical observation by adding a bias to Equation (3.84):

$$I_{th} = I(x+b) \tag{3.86}$$

where b is called the edge bias and the CD is still $2x$ for a symmetrical image. It should be evident that the threshold intensity and the resist bias are not independent parameters, but are related by the slope of the aerial image.

$$\frac{dI_{th}}{db} = \frac{dI}{dx}\bigg|_{x+b} \tag{3.87}$$

Integrating this relationship,

$$I_{th} = I_{th0} + \frac{dI}{dx}b \tag{3.88}$$

where I_{th0} is the threshold one would obtain if the bias were zero.

There are still only two independent parameters in Equation (3.88). If I_{th0} and b are used as the parameters, then the threshold intensity used to extract the edge position from the image will vary as a function of the image slope. This is a simple form of a *variable threshold resist model*. Adding a dose dependence, as in Equation (3.85), we obtain

$$I(x = CD/2) = I_{th} = \frac{E_{th0}}{E} + \frac{dI}{dx}b \tag{3.89}$$

where the slope is evaluated at $I = I_{th}$. Since this requires knowing I_{th} before evaluating the slope, either an iterative procedure must be used or one must assume that the slope is constant over the range of values of b commonly encountered. A second form of the model avoids this complication by using I_{th0} (or E_{th0}) and I_{th} as the parameters. In this case, the bias will vary as a function of the image slope and the result will be a *variable bias resist model*. For the variable bias resist model,

$$I_{th} = I(x+b), \quad b = \frac{I_{th} - I_{th0}}{\mathrm{d}I/\mathrm{d}x} \tag{3.90}$$

where the slope is now evaluated at a constant $I = I_{th}$.

3.8.2 Image Placement Error (Distortion)

The threshold model used above to define the width of an aerial image can also be used to define its center position, called the image placement. Since this center position is the average of the two edge positions, a constant bias does not impact image placement. However, a variable threshold or variable bias resist model, as discussed above, will influence the center position whenever the image slope is different on the right and left sides of the image.

Any deviation in the image placement from the ideal (desired) position is called *image placement error* (as described above in section 3.1.3). When image placement error does not depend on the size or surroundings of the feature in question, this error is called *distortion*. The most common cause of distortion is the aberration of tilt. Variation of image placement error with feature size or type can be caused by any odd Zernike term other than tilt.

3.8.3 Normalized Image Log-Slope (NILS)

The slope of the image intensity I as of function of position x ($\mathrm{d}I/\mathrm{d}x$) measures the steepness of the image in the transition from bright to dark. However, for the slope to be useful as a metric of image quality, the intensity must be normalized by dividing the slope by the intensity at the point where the slope is being measured. The resulting metric is called the image log-slope (ILS):

$$Image\ Log - Slope = \frac{1}{I}\frac{\mathrm{d}I}{\mathrm{d}x} = \frac{\mathrm{d}\ln(I)}{\mathrm{d}x} \tag{3.91}$$

where this log-slope is measured at the nominal (desired) line edge (Figure 3.35). Since variations in the photoresist edge positions (linewidths) are typically expressed as a percentage of the nominal linewidth, the position coordinate x can also be normalized by multiplying the log-slope by the nominal linewidth w, to give the normalized image log-slope (NILS).

$$NILS = w\frac{\mathrm{d}\ln(I)}{\mathrm{d}x} \tag{3.92}$$

The normalized image log-slope can now be used to calculate the image quality of various aerial images that we have discussed in this chapter. Consider first a generic 1D image of a line-space pattern of pitch p. This image can be represented as a Fourier series

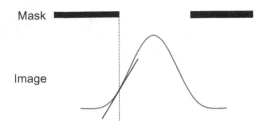

Mask

Image

Figure 3.35 *Image Log-Slope (or the Normalized Image Log-Slope, NILS) is the best single metric of image quality for lithographic applications*

of some number of terms N (typically equal to twice the highest diffraction order number that passes through the lens).

$$I(x) = \sum_{j=0}^{N} \beta_j \cos(2\pi jx/p) \tag{3.93}$$

The NILS calculated at the edge position $x = w/2$ will be

$$NILS = -\left(\frac{2\pi w}{p}\right)\frac{\displaystyle\sum_{j=0}^{N} j\beta_j \sin(\pi jw/p)}{\displaystyle\sum_{j=0}^{N} \beta_j \cos(\pi jw/p)} \tag{3.94}$$

For the special case of equal lines and spaces, so that $w = p/2$,

$$NILS = -\pi \frac{\displaystyle\sum_{j=0}^{N} j\beta_j \sin(\pi j/2)}{\displaystyle\sum_{j=0}^{N} \beta_j \cos(\pi j/2)} = -\pi \frac{\beta_1 - \beta_3 + \beta_5 - \ldots}{\beta_0 - \beta_2 + \beta_4 - \ldots} \tag{3.95}$$

For small-pitch patterns, only a few diffraction orders make it through the lens. Near the resolution limit, where the first diffraction orders are the highest ones transmitted through the lens, $N = 2$ and

$$NILS = -\pi \frac{\beta_1}{\beta_0 - \beta_2} \tag{3.96}$$

One can see from this simple derivation that a high-magnitude NILS results from larger amounts of high-order components (larger β_1 and β_2) and less zero-order component (smaller β_0).

Note that NILS can be positive or negative, depending on whether the pattern is a line or a space, and whether the right or the left edge of the pattern is being considered. As a convention, the sign of the NILS expression is usually adjusted to produce a positive value of NILS. Table 3.2 shows the calculated value of NILS for each of the equal line/space aerial images discussed in this chapter. Of the images shown, the coherent, in-focus

Table 3.2 *Calculation of NILS for various equal line/space images presented in this chapter*

Image Type	Equation	β_0	β_1	β_2	NILS
Coherent, in focus, TE	(3.69)	$\dfrac{1}{4}+\dfrac{2}{\pi^2}$	$\dfrac{2}{\pi}$	$\dfrac{2}{\pi^2}$	8
Coherent, in focus, TM	(3.71)	$\dfrac{1}{4}+\dfrac{2}{\pi^2}$	$\dfrac{2}{\pi}\cos\theta$	$\dfrac{2}{\pi^2}\cos(2\theta)$	$\dfrac{8\cos\theta}{1+\dfrac{8}{\pi^2}\sin^2\theta}$
Coherent, out of focus, TE	(3.27)	$\dfrac{1}{4}+\dfrac{2}{\pi^2}$	$\dfrac{2}{\pi}\cos(\Delta\Phi)$	$\dfrac{2}{\pi^2}$	$8\cos(\Delta\Phi)$
Partially coherent, out of focus, TE	(3.29)	$\dfrac{1}{4}+\dfrac{2}{\pi^2}$	$\dfrac{2}{\pi}\cos(\Delta\Phi)\dfrac{2J_1(a)}{a}$	$\dfrac{2}{\pi^2}\dfrac{J_1(2a)}{a}$	$\dfrac{8\cos(\Delta\Phi)\dfrac{2J_1(a)}{a}}{1+\dfrac{8}{\pi^2}\left(1-\dfrac{J_1(2a)}{a}\right)}$
Incoherent, in focus, TE	See Chapter 2	$\dfrac{1}{2}$	$\dfrac{2}{\pi}MTF(1/p)$	0	$4MTF(1/p)$

image from TE illumination produces the highest NILS (equal to 8). Defocus or the use of TM illumination causes a decrease in the NILS, and thus a degradation of image quality.

In Chapter 2, the Kintner method was used to calculate the partially coherent in-focus image of small lines and spaces as the weighted some of one-beam, two-beam and three-beam images. This Kintner formulation can also be put into the Fourier series form of Equation (3.93). For TE illumination where only the zero and ±first diffraction orders are used,

$$\beta_0 = a_0^2 + a_1^2(s_2 + 2s_3)$$
$$\beta_1 = 2a_0a_1(s_2 + 2s_3) \tag{3.97}$$
$$\beta_2 = 2a_1^2(s_3)$$

where s_2 and s_3 are the fractions of two-beam and three-beam imaging, respectively, as given in Tables 2.3 or 2.4.

3.8.4   Focus Dependence of Image Quality

All of the image quality metrics described above – image CD, image placement and image log-slope – will in general vary with focus. The variation of image CD with focus is sometimes called isofocal bias, a topic that will be discussed at length in Chapter 8. The variation of image log-slope with focus is a common focus characterization technique, and is discussed extensively in Chapter 9. The variation of image placement with focus is called telecentricity error, and is commonly due to asymmetric source shapes (discussed briefly in Chapter 8).

Problems

3.1. Derive Equation (3.5).

3.2. Equation (3.7), which was derived for coherent illumination, predicts that coma will result in a positive placement error for lines and spaces when the pitch is less than $\sqrt{1.5}\lambda/NA$, a negative placement error for pitches greater than this amount, and no placement error when the pitch equals $\sqrt{1.5}\lambda/NA$. Show why this is true using pictures of the wavefront error and the diffraction pattern.

3.3. Using the Zernike polynomial term F_4, explain the effect of 3rd-order astigmatism on x-oriented lines and spaces versus y-oriented lines and spaces.

3.4. Calculate the RMS wavefront error, W_{RMS}, and the resulting Strehl ratio for the case when

   (a) Only 3rd-order spherical aberration is present
   (b) Only 3rd-order x-coma is present

   Note that
   $$W_{RMS}^2 = \frac{1}{\pi} \int_0^1 \int_0^{2\pi} W^2(R,\phi) R dR d\phi$$

3.5. Consider coherent imaging of equal lines and spaces where only the zero and ±first diffraction orders enter the lens. Derive an expression, similar to Equation (3.24), that shows how spherical aberration results in a pitch-dependent paraxial focus error.

3.6. Calculate the impact of long-range flare on the image log-slope.

3.7. Consider the case of dense equal lines and spaces (only the 0 and ±1st orders are used) imaged with coherent illumination. Show that the peak intensity of the image in the middle of the space falls off approximately quadratically with defocus for small amounts of defocus.

3.8. Using the geometry of Figure 3.18b, derive Equation (3.20).

3.9. Compare the depth of focus predictions of the high-NA version of the Rayleigh equation [Equation (3.60)] to the paraxial version of Equation (3.61) by plotting predicted DOF versus pitch (use $k_2 = 0.6$, $\lambda = 248\,nm$, pitch in the range from 250 to 500 nm, and assume imaging in air).

3.10. Consider two unit-amplitude plane waves traveling in air at angles $\pm\theta$ with respect to the z-axis (as in Figure 3.26). Show that the resulting interference patterns, depending on the polarizations of the two beams, will be

   TE-polarization: $I(x) = 2 + 2\cos(4\pi x \sin(\theta) / \lambda)$
   TM-polarization: $I(x) = 2 + 2\cos(2\theta)\cos(4\pi x \sin(\theta) / \lambda)$

   What is the pitch of the resulting line/space image?

3.11. Derive Equation (3.83). What is the limit of this equation for $p \gg \lambda$?

3.12. Consider a coherent, three-beam image of a line/space pattern under TE illumination ($NA = 0.93$, $\lambda = 193\,nm$, linewidth = 100 nm, pitch = 250 nm). Plot the resulting aerial image at best focus and at defocus values of 50 and 100 nm for:
   (a) imaging in air
   (b) immersion imaging ($n_{fluid} = 1.44$)

3.13. Consider an in-focus, TE image where only the zero order and one of the first orders pass through the lens. For this image, what are the values of β_0, β_1 and β_2 in Equation (3.93)? What is the value of the NILS for the case of equal lines and spaces?

3.14. For the line/space image of Equation (3.93):

(a) derive an equation for the image contrast, defined as

$$Image\ Contrast = \frac{I_{max} - I_{min}}{I_{max} + I_{min}}$$

For simplicity, assume the maximum intensity occurs in the middle of the space, and the minimum intensity occurs in the middle of the line.

(b) For the case where the second harmonic is the largest harmonic in the image (i.e. when $N = 2$), compare the resulting image contrast to the NILS of Equation (3.96).

3.15. Consider an in-focus, coherent, three-beam image of an equal line/space pattern under TE illumination ($NA = 0.93$, $\lambda = 193\,nm$, linewidth $= 130\,nm$, pitch $= 260\,nm$). Use a threshold model ($E_{th} = 20\,mJ/cm^2$) to plot the resulting line CD versus dose. What dose produces the nominal feature size?

References

1 Ulrich, W., Rostalski, H.-J. and Hudyma, R., 2004, Development of dioptric projection lenses for deep ultraviolet lithography at Carl Zeiss, *Journal of Microlithography, Microfabrication, and Microsystems*, **3**, 87–96.

2 Kroyan, A., Bendik, J., Semprez, O., Farrar, N., Rowan, C. and Mack, C.A., 2000, Modeling the effects of excimer laser bandwidths on lithographic performance, *Proceedings of SPIE: Optical Microlithography XIII*, **4000**, 658–664.

3 Lai, K., Lalovic, I., Fair, B., Kroyan, A., Progler, C., Farrar, N., Ames, D. and Ahmed, K., 2003, Understanding chromatic aberration impacts on lithographic imaging, *Journal of Microlithography, Microfabrication, and Microsystems (JM3)*, **2**, 105–111.

4 Flagello, D.G. and Pomerene, A.T.S., 1987, Practical characterization of 0.5 μm optical lithography, *Proceedings of SPIE: Optical Microlithography VI*, **772**, 6–20.

5 Mack, C.A. and Kaufman, P.M., 1988, Mask bias in submicron optical lithography, *Journal of Vacuum Science & Technology B*, **B6**, 2213–2220.

6 Lai, K., Wu, C. and Progler, C., 2001, Scattered light: the increasing problem for 193 nm exposure tools and beyond, *Proceedings of SPIE: Optical Microlithography XIV*, **772**, 1424–1435.

7 Fukuda, H., Hasegawa, N., Tanaka, T. and Hayashida, T., 1987, A new method for enhancing focus latitude in optical lithography: FLEX, *IEEE Electron Device Letters*, **EDL-8**, 179–180.

8 Richards, B. and Wolf, E., 1959, Electromagnetic diffraction in optical systems. II. Structure of the image field in an aplanatic system, *Proceedings of the Royal Society of London, Series A*, **253**, 358–379.

9 Jones, R.C., 1941, New calculus for the treatment of optical systems, *Journal of the Optical Society of America*, **31**, 488–493.

10 Totzeck, M., Gräupner, P., Heil, T., Göhnermeier, A., Dittmann, O., Krähmer, D., Kamenov, V., Ruoff, J. and Flagello, D., 2005, Polarization influence on imaging, *Journal of Microlithography, Microfabrication, and Microsystems*, **4**, p. 031108.

11 McIntyre, G., Kye, J., Levinson, H. and Neureuther, A., 2006, Polarization aberrations in hyper-numerical aperture projection printing: a comparison of various representations, *Journal of Microlithography, Microfabrication, and Microsystems*, **5**, p. 033001.

12   Geh, B., Ruoff, J., Zimmermann, J., Gräupner P., Totzeck, M., Mengel, M., Hempelmann, U. and Schmitt-Weaver, E., 2007, The impact of projection lens polarization properties on lithographic process at hyper-NA, *Proceedings of SPIE: Optical Microlithography* **6520**, 65200F.

4

Imaging in Resist: Standing Waves and Swing Curves

The previous chapter developed a comprehensive theory for calculating aerial images, the image of a photomask as projected by the imaging tool into air. But of course the real goal is to project that image into photoresist, where chemical reactions responding to the local exposure dose will transform the intensity image into a latent image of exposed and unexposed material. The coupling of the aerial image into the photoresist involves the behavior of light as it encounters interfaces between materials – reflection and transmission – and the propagation of light through an absorbing medium.

One of the unique aspects of the image in resist (as opposed to the aerial image) comes from reflections off the substrate (or films coated on the substrate). The total energy in the resist is the sum of the downward-propagating image and the upward-propagating reflected image. Since these two images will interfere, the results are highly dependent on the thickness of the resist, and possibly other films above or below the resist. Two consequences of this interference are (1) standing waves – a sinusoidal variation of dose through the thickness of the resist – and (2) swing curves – a sinusoidal dependence of the energy coupled into the resist as a function of resist thickness.

A thorough understanding of these phenomena, plus a recognition of their detrimental impact on linewidth control, leads to a search for methods to reduce standing waves and swing curves. A description of antireflection coatings, both top and bottom, will show how linewidth control can be reclaimed through effective reflection control.

Finally, examining how light propagates through the photoresist, including all of the various thin film interference effects, allows for a rigorous understanding of the difference between an aerial image and an image in resist.

Fundamental Principles of Optical Lithography: The Science of Microfabrication, Chris Mack.
© 2007 John Wiley & Sons, Ltd.

4.1   Standing Waves

Standing waves occur whenever two waves, traveling in opposite directions and with a fixed phase relationship to each other, combine. Examples of standing waves abound, from vibrational waves traveling down a guitar string and reflecting off the fixed end of that string, to acoustic waves bouncing within the metal tube of a pipe organ, to the design of impedance matching connections for coaxial television cables. In lithography, light passing through the photoresist is reflected off the substrate. This reflected light wave interferes with the light wave traveling down to produce a standing wave pattern.[1,2]

4.1.1   The Nature of Standing Waves

Before looking at the lithographic case in great detail, we'll examine the basic nature of standing waves by considering an exceedingly simplified version of light standing waves. Suppose a plane wave of monochromatic light traveling through air (which we'll assume to be nonabsorbing with an index of refraction of 1.0) is normally incident on a mirror. What is the total light intensity in the vicinity of the mirror? The total electric field will be the sum of the wave traveling toward the mirror and the reflected wave. Letting z be the direction the incident light is traveling in, the electric field E_I of a plane wave traveling in this direction has the mathematical form

$$E_I(z) = Ae^{ikz} \tag{4.1}$$

where A is the amplitude of the plane wave and k is the propagation constant (equal to $2\pi n/\lambda$ where n is the refractive index of the medium and λ is the vacuum wavelength of the light). (This equation requires an assumption about the sign convention for the phasor representation of a sinusoidal wave, as discussed in Chapter 2. Unfortunately, there is no standard and the convention commonly used for thin film calculations is the opposite to that normally used for imaging. In this book, a plane wave traveling in the +z-direction is represented by $\exp(+ikz)$ and the imaginary part of the index of refraction must be positive to represent an absorbing media.) For simplicity, let $z = 0$ define the position of the mirror. Letting ρ be the electric field reflectivity of the mirror, the reflected light electric field will be

$$E_R(z) = \rho E_I(0)e^{-ikz} = \rho Ae^{-ikz} \tag{4.2}$$

The total electric field is just the sum of the incident and reflected fields.

$$E_T(z) = E_R(z) + E_I(z) = A(e^{+ikz} + \rho e^{-ikz}) \tag{4.3}$$

The special case of a perfectly reflective mirror ($\rho = -1$) provides an especially simple result. The two complex exponentials in Equation (4.3) become a sinusoid by using the Euler identity.

$$E_T(z) = A(e^{+ikz} - e^{-ikz}) = i2A\sin(kz) \tag{4.4}$$

The intensity of the light in this region is the 'interference pattern' between the incident and reflected beams:

$$I_T(z) = |E_T(z)|^2 = 4A^2 \sin^2(kz) = 2A^2(1 - \cos(2kz)) \tag{4.5}$$

The most striking aspect of this simple result is that the traveling wave has been converted to a standing wave. The plane wave in Equation (4.1) is 'traveling', meaning that the phase of the electric field varies linearly with distance traveled. Literally, the wave moves through its ups and downs of the sinusoidal phase cycle as it travels, while the amplitude remains constant. In contrast, the wave in Equation (4.4) is stationary; it is 'standing' in place. The amplitude of the electric field varies sinusoidally with position z, but the phase remains fixed and independent of position.

(As an interesting historical footnote, the mathematical experiment described above is almost identical to a famous and seminal experiment performed in 1888 by Heinrich Hertz. Attempting to verify Maxwell's new theory of electromagnetic radiation, Hertz covered one of the walls of his laboratory with copper and reflected radio waves off this 'mirror'. Using a wire loop detector sensitive to the magnitude of the electric field, he measured the resulting sinusoidal standing wave, which agreed well with Maxwell's theory. Otto Wiener first demonstrated that light exhibits standing waves in 1890 by exposing a photographic film tilted at a small angle to a normally illuminated plane mirror.)

4.1.2   Standing Waves for Normally Incident Light in a Single Film

The very simple standing wave result derived above shows us the basic form of standing waves caused by plane waves traveling in opposite directions. For the case of standing waves in a photoresist, the simplest example is a thin layer of photoresist coated on a thick (semi-infinite) substrate. Let us begin with the simple geometry shown in Figure 4.1a. A thin photoresist (layer 2) rests on a thick substrate (layer 3) in air (layer 1). Each material has optical properties governed by its complex index of refraction, $n = n + i\kappa$, where n is the real index of refraction and κ is the imaginary part, sometimes called the extinction coefficient. This later name comes from the relationship between the imaginary part of the refractive index and α, the absorption coefficient of the material (as derived in Chapter 2).

$$\alpha = \frac{4\pi\kappa}{\lambda} \tag{4.6}$$

Consider now the propagation of light through this film stack. We will begin with illumination of the stack by a monochromatic plane wave normally incident on the resist. When

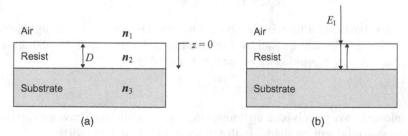

Figure 4.1   *Film stack showing (a) the geometry for the standing wave derivation, and (b) a normally incident electric field* E_I

this plane wave strikes the resist surface, some of the light will be transmitted and some will be reflected. The amount of each is determined by the transmission and reflection coefficients. Defined as the ratio of the transmitted to incident electric field, the *transmission coefficient* τ_{ij} for a normally incident plane wave transmitting from layer i to layer j is given by

$$\tau_{ij} = \frac{2n_i}{n_i + n_j} \tag{4.7}$$

In general, the transmission coefficient will be complex, indicating that when light is transmitted from one material to another, both the magnitude and the phase of the electric field will change. Similarly, the light reflected off layer j back into layer i is given by the *reflection coefficient* ρ_{ij}:

$$\rho_{ij} = \frac{n_i - n_j}{n_i + n_j} \tag{4.8}$$

Both of these equations are derived from the requirement that electric and magnetic field components that are within (or tangential to) the plane of the interface between materials i and j must be continuous across that boundary.

If an electric field E_I is incident on the photoresist (see Figure 4.1b), the transmitted electric field will be $\tau_{12}E_I$. The transmitted plane wave will now travel down through the photoresist. As it travels, the wave will change phase sinusoidally with distance and undergo absorption. Both of these effects are given by the standard description of a plane wave as a complex exponential:

$$E(z) = e^{ikz} = e^{i2\pi n z/\lambda} \tag{4.9}$$

where this $E(z)$ represents a plane wave traveling in the $+z$-direction. Using the coordinates defined in Figure 4.1a, the transmitted electric field propagating through the resist, E_0, will be given by

$$E_0(z) = \tau_{12}E_I e^{i2\pi n_2 z/\lambda} \tag{4.10}$$

Eventually, the wave will travel through the resist thickness D and strike the substrate, where it will be partially reflected. The reflected wave, E_1, just after reflection will be

$$E_1(z = D) = \tau_{12}E_I \rho_{23} e^{i2\pi n_2 D/\lambda} \tag{4.11}$$

The complex exponential term in the above equation has some physical significance: it represents the electric field transmitted from the top to the bottom of the photoresist and is called the *internal transmittance* of the resist, τ_D.

$$\tau_D = e^{i2\pi n_2 D/\lambda} \tag{4.12}$$

As the reflected wave travels back up through the resist, the distance traveled will be $(D-z)$ and the propagation will be similar to that described by Equation (4.9):

$$E_1(z) = \tau_{12}E_I \rho_{23} \tau_D e^{i2\pi n_2 (D-z)/\lambda} = \tau_{12}E_I \rho_{23} \tau_D^2 e^{-i2\pi n_2 z/\lambda} \tag{4.13}$$

So far, our incident wave (E_I) has been transmitted in the photoresist (E_0) and then reflected off the substrate (E_1), as pictured in Figure 4.1b. The resultant electric field in the photoresist will be the sum of E_0 and E_1.

$$E_0(z) + E_1(z) = \tau_{12}E_I(e^{i2\pi n_2 z/\lambda} + \rho_{23}\tau_D^2 e^{-i2\pi n_2 z/\lambda}) \tag{4.14}$$

This sum is not the total electric field in the photoresist. The wave E_1 will travel to the top of the resist where it will be reflected by the air–resist interface. The new wave, E_2, will travel down through the resist where it will be reflected by the substrate, giving another wave E_3. The next two waves are given by

$$E_2(z) = E_I\rho_{21}\rho_{23}\tau_{12}\tau_D^2 e^{ik_2 z}$$
$$E_3(z) = E_I\rho_{21}\rho_{23}^2\tau_{12}\tau_D^4 e^{-ik_2 z} \tag{4.15}$$

The total electric field within the thin film, $E_T(z)$, is the sum of each $E_j(z)$. Performing this summation gives

$$E_T(z) = E_I\tau_{12}(e^{ik_2 z} + \rho_{23}\tau_D^2 e^{-ik_2 z})S \tag{4.16}$$

where $S = 1 + \rho_{21}\rho_{23}\tau_D^2(1 + \rho_{21}\rho_{23}\tau_D^2(1 + \ldots$

The summation S is simply a geometric series and converges to

$$S = \frac{1}{1 - \rho_{21}\rho_{23}\tau_D^2} = \frac{1}{1 + \rho_{12}\rho_{23}\tau_D^2} \tag{4.17}$$

Thus, the total electric field in the resist will be[3]

$$\frac{E_T(z)}{E_I} = \frac{\tau_{12}(e^{i2\pi n_2 z/\lambda} + \rho_{23}\tau_D^2 e^{-i2\pi n_2 z/\lambda})}{1 + \rho_{12}\rho_{23}\tau_D^2} \tag{4.18}$$

Although we have determined an expression for the standing wave electric field, it is the intensity of the light that causes exposure. Calculation of the relative intensity from Equation (4.18) leads to a somewhat messy result, but a few algebraic simplifications allow for a reasonably useful form:

$$I(z) = \frac{n_2|E(z)|^2}{n_1|E_I|^2}$$
$$= T_{eff}\left[(e^{-\alpha z} + |\rho_{23}|^2 e^{-\alpha(2D-z)}) + 2|\rho_{23}|e^{-\alpha D}\cos(4\pi n_2(D-z)/\lambda + \phi_{23})\right] \tag{4.19}$$

where $\rho_{23} = |\rho_{23}|e^{i\phi_{23}}$, and

$$T_{eff} = \frac{\frac{n_2}{n_1}|\tau_{12}|^2}{|1 + \rho_{12}\rho_{23}\tau_D^2|^2} = \frac{T_{12}}{|1 + \rho_{12}\rho_{23}\tau_D^2|^2}, \text{ the effective transmittance into the photoresist.}$$

This equation is graphed in Figure 4.2 for an i-line photoresist with typical properties on a silicon substrate. By comparing the equation to the graph, many important aspects of the standing wave effect become apparent. The most striking feature of the standing wave

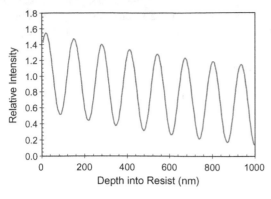
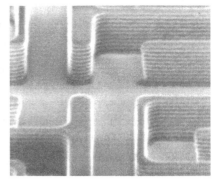

Figure 4.2 *Standing wave intensity in one micron of photoresist on a silicon substrate for an i-line exposure*

plot is its sinusoidal variation. The cosine term in Equation (4.19) shows that the period of the standing wave is given by

$$\text{Period} = \lambda / 2n_2 \qquad (4.20)$$

The amplitude of the standing waves is given by the multiplier of the cosine in Equation (4.19). It is quite apparent that there are two basic ways to reduce the amplitude of the standing wave intensity. The first is to reduce the reflectivity of the substrate (reduce ρ_{23}). The use of an antireflection coating on the substrate is one of the most common methods of reducing standing waves (see section 4.3). The second method for reducing the standing wave intensity that Equation (4.19) suggests is to increase absorption in the resist (reduce the $e^{-\alpha D}$ term). This is accomplished by adding a dye to the photoresist (increasing α). Finally, the 'bulk' intensity variation, the variation of the intensity averaged over the standing wave period, is given by the pair of exponential terms being added to the cosine in Equation (4.19). The bulk effect is a function of both absorption in the resist and the reflectivity of the substrate.

As an aside, it is worth mentioning one difficulty in using Equation (4.19): the conversion of the complex substrate reflectivity from rectangular to polar form. The calculation of the magnitude of the reflection coefficient is straightforward:

$$|\rho_{23}| = \sqrt{\frac{(n_2 - n_3)^2 + (\kappa_2 - \kappa_3)^2}{(n_2 + n_3)^2 + (\kappa_2 + \kappa_3)^2}} \qquad (4.21)$$

However, determining the angle of the reflectivity using an arctangent function (as one might commonly do in spreadsheet calculations) can introduce an ambiguity if the range over which the arctangent is defined is not properly taken into account. Defining two simplifying variables x and y,

$$\begin{aligned} x &= (n_2^2 + \kappa_2^2) - (n_3^2 + \kappa_3^2) \\ y &= 2(n_3\kappa_2 - n_2\kappa_3) \end{aligned} \qquad (4.22)$$

(1) If $\tan^{-1}$ is defined on the interval $[0,\pi)$, then

$$\theta_{23} = \tan^{-1}\left(\frac{y}{x}\right), \qquad y > 0$$

$$\theta_{23} = \tan^{-1}\left(\frac{y}{x}\right) + \pi, \quad y < 0 \tag{4.23}$$

(2) If $\tan^{-1}$ is defined on the interval $[-\pi/2, \pi/2)$, then

$$\theta_{23} = \tan^{-1}\left(\frac{y}{x}\right), \qquad x > 0$$

$$\theta_{23} = \tan^{-1}\left(\frac{y}{x}\right) + \pi, \quad x < 0 \tag{4.24}$$

The cases of $y = 0$ or $x = 0$ can be handled separately from the arctangent evaluation.

4.1.3 Standing Waves in a Multiple-Layer Film Stack

It is very common to have more than one film coated on a substrate. An analysis similar to that for one film yields the following result for the electric field in the top layer of an m-1 layer system:

$$\frac{E_2(z)}{E_I} = \frac{\tau_{12}\left(e^{i2\pi n_2 z/\lambda} + \rho'_{23}\tau_D^2 e^{-i2\pi n_2 z/\lambda}\right)}{1 + \rho_{12}\rho'_{23}\tau_D^2} \tag{4.25}$$

where

$$\rho'_{23} = \frac{n_2 - n_3 X_3}{n_2 + n_3 X_3}$$

$$X_3 = \frac{1 - \rho'_{34}\tau_{D3}^2}{1 + \rho'_{34}\tau_{D3}^2}$$

$$\rho'_{34} = \frac{n_3 - n_4 X_4}{n_3 + n_4 X_4}$$

$$.$$

$$.$$

$$.$$

$$X_m = \frac{1 - \rho_{m,m+1}\tau_{Dm}^2}{1 + \rho_{m,m+1}\tau_{Dm}^2}$$

$$\rho_{m,m+1} = \frac{n_m - n_{m+1}}{n_m + n_{m+1}}$$

$$\tau_{Dj} = e^{ik_j D_j}$$

and all other parameters are as previously defined. The parameter ρ'_{23} is the effective reflection coefficient between the thin resist film and what lies beneath it. Once the effective reflection coefficient of the substrate film stack is determined, Equation (4.25) becomes identical to Equation (4.18).

To better understand the nature and impact of the effective reflection coefficient of the substrate, let's consider two special cases where one layer (layer 3) lies between the resist and the substrate. For the first case, let this layer 3 be thick and absorbing enough so that no reflected light from the substrate makes it back to the top of layer 3. In other words, let $\tau_{D3}^2 = 0$. From the equations above, X_3 will equal 1 and $\rho'_{23} = \rho_{23}$. In other words, a sufficiently thick absorbing layer acts as a substrate. For the second case, assume layer 3 is optically matched to the photoresist so that $n_2 = n_3$. For this case, $\rho'_{23} = \rho_{34}\tau_{D3}^2$. Ignoring any small amount of absorption that may occur in layer 3 (the imaginary part of n_2 is usually very small), τ_{D3}^2 will have a magnitude of about 1 and will thus only impact the phase of ρ'_{23}. In other words, the magnitude of the effective substrate reflection coefficient will be determined by the reflection coefficient of layer 3 with the substrate and the phase of the effective reflection coefficient will be determined by the thickness of layer 3. (The resist on oxide on silicon case shown in Figure 4.3 is very similar to this idealized case.) This impact of the phase of the reflection coefficient on lithography is explored below in section 4.6.

If the thin film in question is not the top film (layer 2), the intensity can be calculated in layer j from

$$E_j(z) = E_{\text{I,eff}} \tau_{j-1,j}^* \frac{(e^{ik_j z_j} + \rho'_{j,j+1}\tau_{Dj}^2 e^{-ik_j z_j})}{1 + \rho_{j-1,j}^* \rho'_{j,j+1}\tau_{Dj}^2} \tag{4.26}$$

where $\tau_{j-1,j}^* = 1 + \rho_{j-1,j}^*$. The effective reflection coefficient ρ^* is analogous to the coefficient ρ', looking in the opposite direction. $E_{\text{I,eff}}$ is the effective intensity incident on layer

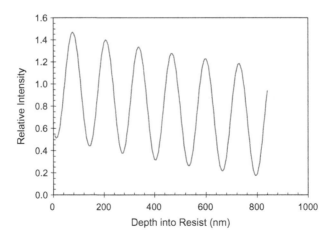

Figure 4.3 *Standing wave intensity within a photoresist film at the start of exposure (850 nm of resist on 100-nm SiO$_2$ on silicon, $\lambda = 436$ nm). Note the impact of the oxide film on the phase of the effective substrate reflectivity, which affects the intensity at the bottom of the resist*

j. Both $E_{1,\text{eff}}$ and ρ^* are defined in detail in Reference 3. Equation (4.26) is needed when using a top antireflection coating, as described in section 4.4.

If the resist film is not homogeneous, the equations above are, in general, not valid. Let us, however, examine one special case in which the inhomogeneity takes the form of small variations in the imaginary part of the index of refraction of the film in the *z*-direction, leaving the real part constant. Photoresist bleaching (described in detail in Chapter 5) will cause such a variation of the absorption coefficient with depth into the resist. In this case, the absorbance *Abs* is no longer simply αz, but becomes

$$Abs(z) = \int_{0}^{z} \alpha(z')\mathrm{d}z' \tag{4.27}$$

Equations (4.18), (4.25) and (4.26) are still approximately valid if this anisotropic expression for absorbance is used. Thus, $I(z)$ can be found if the absorption coefficient is known as a function of *z*.

Figure 4.3 shows a typical result of the standing wave intensity within a photoresist film coated on an oxide on silicon film stack.

4.1.4 Oblique Incidence and the Vector Nature of Light

For simplicity, first consider a plane wave traveling through a uniform material and striking a partially reflective substrate at $z = 0$ (see Figure 4.4). The plane wave, described by its propagation vector $\mathbf{k}$ (defined as the propagation constant multiplied by a unit vector pointing in the direction of propagation), makes an angle θ with respect to the *z*-axis (which is also normal to the reflecting surface). Taking the *x*–*z* plane as the plane of incidence, any point in space (x,z) is described by its position vector $\mathbf{r}$. The electric field of the plane wave (before it reflects off the substrate) is then given by

$$E(x,z) = A\mathrm{e}^{i\mathbf{k}\cdot\mathbf{r}} = A\mathrm{e}^{ik(x\sin\theta + z\cos\theta)} \tag{4.28}$$

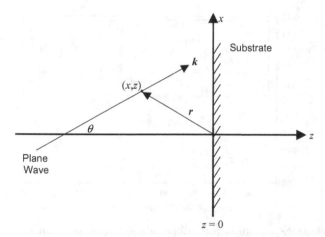

Figure 4.4 *Geometry used for describing plane waves and standing waves for oblique incidence*

where A is the amplitude of the plane wave and the (complex) propagation constant k is given by $2\pi n/\lambda$ where n is the complex refractive index of the medium. This plane wave will reflect off the substrate with an angle- and polarization-dependent reflectivity $\rho(\theta)$ as given by the Fresnel formulae.

$$\rho_{ij\perp}(\theta_j) = \frac{n_i \cos(\theta_i) - n_j \cos(\theta_j)}{n_i \cos(\theta_i) + n_j \cos(\theta_j)}$$

$$\tau_{ij\perp}(\theta_j) = \frac{2n_i \cos(\theta_i)}{n_i \cos(\theta_i) + n_j \cos(\theta_j)}$$

$$\rho_{ij\|}(\theta_j) = \frac{n_i \cos(\theta_j) - n_j \cos(\theta_i)}{n_i \cos(\theta_j) + n_j \cos(\theta_i)}$$

$$\tau_{ij\|}(\theta_j) = \frac{2n_i \cos(\theta_i)}{n_i \cos(\theta_j) + n_j \cos(\theta_i)}$$

(4.29)

and the two angles θ_i and θ_j are related by Snell's law,

$$n_i \sin(\theta_i) = n_j \sin(\theta_j)$$

(4.30)

Here, $\|$ represents an electric field vector that lies in a plane defined by the direction of the incident light and a normal to the resist surface (i.e. in the plane of the paper in Figure 4.4). Other names for $\|$ polarization include *p-polarization* and TM (transverse magnetic) polarization. The polarization denoted by $\perp$ represents an electric field vector that lies in a plane perpendicular to that defined by the direction of the incident light and a normal to the resist surface (i.e. perpendicular to the plane of the paper in Figure 4.4). Other names for $\perp$ polarization include *s-polarization* and TE (transverse electric) polarization. Note that for light normally incident on the resist surface, both s- and p-polarization result in electric fields that lie along the resist surface and the four Fresnel formulae revert to the two standard definitions of reflection and transmission coefficients used earlier in Equations (4.7) and (4.8). Figure 4.5 shows how the intensity reflectivity (the square of

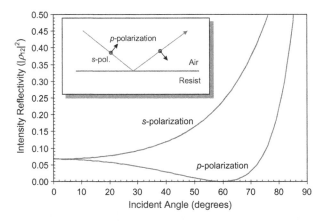

Figure 4.5 *Reflectivity (square of the reflection coefficient) as a function of the angle of incidence showing the difference between s- and p-polarization ($n_1 = 1.0$, $n_2 = 1.7$). Both air and resist layers are assumed to be infinitely thick*

the magnitude of the reflection coefficient) varies with incident angle for both *s*- and *p*-polarized illumination for a particular case.

The total electric field to the left of the substrate in Figure 4.4 is then the sum of the incident and reflected waves. This simple summation brings up the first interesting complication – the effect of polarization, the direction that the electric field is pointing. If the original electric field in Equation (4.28) is *s*-polarized (also called TE polarized) so that the electric field vector points directly out of the plane of Figure 4.4, then both the incident and reflected electric fields will point in the same direction and the vector sum will equal the scalar (algebraic) sum of the incident and reflected fields:

$$\text{s-polarized: } E(x,z) = Ae^{ikx\sin\theta}(e^{ikz\cos\theta} + \rho(\theta)e^{-ikz\cos\theta})$$

$$I(x,z) = |E(x,z)|^2 = |A|^2(1 + |\rho(\theta)|^2 + 2|\rho(\theta)|\cos(2kz\cos\theta + \phi_\rho)) \tag{4.31}$$

where ϕ_ρ is the phase angle of the complex reflectivity ρ and, for simplicity, the propagation medium is assumed to be nonabsorbing. The amplitude of the standing waves, given by the factor multiplying the cosine in the intensity expression above, is controlled by the magnitude of the electric field reflectivity. Note that this equation reverts to the simpler perfect mirror Equation (4.5) when $\rho = -1$.

Thus, for *s*-polarized light, the standing wave Equation (4.18) can be easily modified for the case of non-normally incident plane waves. Suppose a plane wave is incident on the resist film at some angle θ_1. The angle of the plane wave inside the resist will be θ_2 as determined from Snell's law. An analysis of the propagation of this plane wave within the resist will give an expression similar to Equation (4.18) but with the position z replaced with $z\cos\theta_2$. For this case, the electric field will point in the *y*-direction.

$$\frac{E_y(z,\theta_2)}{E_I} = \frac{\tau_{12}(\theta_2)(e^{i2\pi n_2 z\cos\theta_2/\lambda} + \rho_{23}(\theta_2)\tau_D^2(\theta_2)e^{-i2\pi n_2 z\cos\theta_2/\lambda})}{1 + \rho_{12}(\theta_2)\rho_{23}(\theta_2)\tau_D^2(\theta_2)} \tag{4.32}$$

The transmission and reflection coefficients are now functions of the angle of incidence (as well as the polarization of the incident light) and are given by the Fresnel formulae above. The internal transmittance becomes

$$\tau_D = e^{i2\pi n_2 D\cos\theta_2/\lambda} \tag{4.33}$$

The resulting relative intensity becomes

$$I(z) = T_{\text{eff}}(\theta_2)(e^{-\alpha z/\cos\theta_2} + |\rho_{23}(\theta_2)|^2 e^{-\alpha(2D-z)/\cos\theta_2})$$
$$+ T_{\text{eff}}(\theta_2)2|\rho_{23}(\theta_2)|e^{-\alpha D/\cos\theta_2}\cos(4\pi n_2\cos\theta_2(D-z)/\lambda + \phi_{23}) \tag{4.34}$$

(For an understanding of why the absorption terms vary as distance/$\cos\theta_2$, see section 4.8.2)

For *p*-polarized incident light, reduced overlap in electric fields causes the standing wave amplitude to be reduced by the factor $\cos(2\theta_2)$, as will be shown below.[4] The resulting electric field will have both *x*- and *z*-components (*x* being the dimension in the plane of incidence), whereas for the *s*-polarized case the electric field only points in the *y*-direction (out of the plane of incidence).

$$\frac{E_x}{E_I} = \frac{\cos(\theta_2)\tau_{12}(\theta_2)(e^{i2\pi n_2 z\cos\theta_2/\lambda} + \rho_{23}(\theta_2)\tau_D^2(\theta_2)e^{-i2\pi n_2 z\cos\theta_2/\lambda})}{1 + \rho_{12}(\theta_2)\rho_{23}(\theta_2)\tau_D^2(\theta_2)} \tag{4.35}$$

$$\frac{E_z}{E_I} = \frac{\sin(\theta_2)\tau_{12}(\theta_2)(e^{i2\pi n_2 z \cos\theta_2/\lambda} - \rho_{23}(\theta_2)\tau_D^2(\theta_2)e^{-i2\pi n_2 z \cos\theta_2/\lambda})}{1 + \rho_{12}(\theta_2)\rho_{23}(\theta_2)\tau_D^2(\theta_2)} \tag{4.36}$$

The total standing wave intensity, relative to the incident intensity, will be

$$
\begin{aligned}
I(z) &= \frac{n_2(|E_x|^2 + |E_z|^2)}{n_1|E_I|^2} \\
&= T_{\text{eff}}(\theta_2)(e^{-\alpha z/\cos\theta_2} + |\rho_{23}(\theta_2)|^2 \, e^{-\alpha(2D-z)/\cos\theta_2}) \\
&\quad + T_{\text{eff}}(\theta_2)2|\rho_{23}(\theta_2)|e^{-\alpha D/\cos\theta_2} \cos(2\theta_2)\cos(4\pi n_2 \cos\theta_2(D-z)/\lambda + \phi_{23})
\end{aligned}
\tag{4.37}
$$

If $\theta_2 = 45°$, the incident and reflected waves have no overlap of their electric fields and the resulting lack of interference means there will be no standing waves (only the bulk-effect term survives).

It is interesting to look at the impact of direction that the light is traveling on the definitions of the reflection and transmission coefficients. Completely reversing the direction of the light, if light approaches the interface through material j at an angle θ_j, the resulting reflection and transmission coefficients become

$$
\begin{aligned}
\rho_{ji} &= -\rho_{ij} \\
\tau_{ji} &= \frac{n_j \cos(\theta_j)}{n_i \cos(\theta_i)}\tau_{ij}
\end{aligned}
\tag{4.38}
$$

where these relationships hold for either polarization.

Figure 4.6 shows an example of the impact of incident angle on the resulting standing wave pattern in the resist. The change in incident angle causes a change in the period of the standing waves and, because of the angular dependence of T_{eff}, a difference in the amount of light transmitted into the photoresist film.

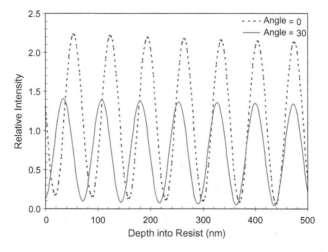

Figure 4.6 *Standing wave intensity within a photoresist film (500nm of resist on silicon, λ = 248nm) as a function of incident angle (s-polarization assumed)*

A second approximation can be made to simplify the evaluation of Equations (4.32), (4.35) or (4.36). For a complex resist refractive index, Snell's law produces the nonintuitive but mathematically correct result of a complex angle of refraction in the resist. For the case where the resist is only weakly absorbing (quite a normal situation), the imaginary part of the angle of refraction into the resist can be ignored with only a small loss of accuracy. The impact of making such an approximation is explored more fully below in section 4.8.

4.1.5 Broadband Illumination

The analysis presented above applies to monochromatic illumination. What if broadband (polychromatic) illumination were used? In general, the use of broadband illumination in imaging applications can be thought of as the incoherent superposition of individual monochromatic results. Thus, the standing wave intensity for a single wavelength is integrated over the wavelengths of the source, weighted by the source illumination spectrum. Figure 4.7 shows a typical mercury arc lamp output spectrum (before filtering in the illumination system of the projection tool). Table 4.1 describes the common wavelengths used in lithographic tools. For 248- and 193-nm lithography tools, where a line-narrowed excimer laser is used to generate the light, bandwidths are less than a picometer, so that these sources can be considered monochromatic from a standing wave perspective.

Although wavelength appears explicitly in the expressions for the standing wave electric field in the resist, an implicit dependence occurs in the indices of refraction of the various resist and substrate materials. The variation of the index of refraction of a material with wavelength, called dispersion, is extremely material dependent. Over a limited range of wavelengths, many photoresist materials' dispersion curves can be adequately described

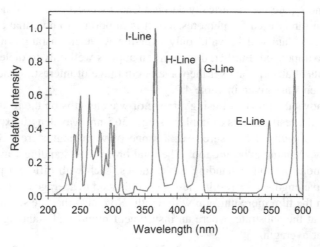

Figure 4.7 *Spectral output of a typical high-pressure mercury arc lamp. The illumination spectrum of an i-line or g-line lithographic exposure tool is usually a filtered portion of this lamp spectrum*

Table 4.1 *Wavelengths for various lithographic light sources (high pressure mercury arc lamps and free-running, unnarrowed excimer lasers)*

Source	Center Wavelength (nm)	Typical Unnarrowed Bandwidth (nm)
Mercury Arc Lamp g-line	435.8	5
Mercury Arc Lamp h-line	404.7	5
Mercury Arc Lamp i-line	365.0	6
KrF excimer laser	248.35	0.30
ArF excimer laser	193.3	0.45
F_2 excimer laser	157.63	0.002

Table 4.2 *Some values for Cauchy coefficients of photoresists (from http://www.microe. rit.edu/research/lithography/)*

Photoresist	C_1	C_2 (nm^2)	C_3 (nm^4)	Wavelength Range (nm)
SPR-500 i-line (unexposed)	1.6133	5481.6	1.4077×10^9	300–800
SPR-500 i-line (fully exposed)	1.5954	9291.9	4.2559×10^8	300–800
193-nm resist	1.5246	3484	1.49×10^8	190–800

by an empirical expression called the Cauchy equation (named for the 19th century French mathematician Augustin-Louis Cauchy):

$$n(\lambda) = C_1 + \frac{C_2}{\lambda^2} + \frac{C_3}{\lambda^4} \qquad (4.39)$$

where C_1, C_2 and C_3 are the empirically derived Cauchy coefficients (higher-order terms are possible, but rarely used for photoresists). It is important to note that any Cauchy fit to refractive index data will be valid only over the wavelength range of the data and should not be extrapolated. Further, this equation applies well only to dielectrics with no or small amounts of absorption over the wavelength range of interest. Some typical resist Cauchy coefficients are given in Table 4.2.

Figure 4.8 shows the impact of using a range of wavelengths on the resulting standing wave pattern in the resist. In this example, a single 365-nm-wavelength exposure is compared to a very broadband exposure over a range of wavelengths from 350 to 450 nm using a mercury arc lamp exposure tool. The total broadband standing wave result is the sum of many monochromatic standing wave curves, each with different periods due to the different wavelengths. As a result, these sinusoids tend to average out making the final intensity within the film more uniform with depth. Note also that the standing waves are at their largest at the substrate, where an insufficient number of standing wave periods are available for averaging.

An alternate viewpoint is to consider the *coherence length* of the light source. While commonly used to describe lasers, coherence length can in fact be used to describe any light source and is defined as the longest distance light can travel and still maintain a

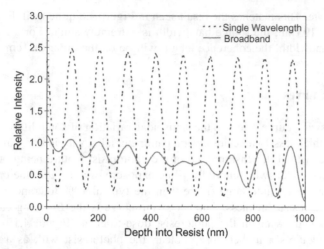

Figure 4.8 *Standing wave intensity within a photoresist film (1000 nm of resist on silicon), for monochromatic (λ = 365 nm) and broadband illumination (350–450 nm range of the mercury spectrum)*

fixed phase relationship all along that length (that is, still maintain the ability to interfere with itself). Suppose light of wavelength λ_1 travels a distance $L/2$, strikes a mirror at normal incidence, and travels back a distance $L/2$. Further suppose that the total distance traveled L is an integer number of wavelengths $N\lambda_1$. Consider now a second wavelength λ_2 such that this same distance is equal to $(N+1/2)\lambda_2$. For this case, when the first wavelength interferes constructively, the second wavelength will be interfering destructively. If both wavelengths are present, no interference pattern will be observed at the distance $L/2$ away from the mirror and L is called the coherence length.

Combining the two equations

$$L = N\lambda_1 = (N+1/2)\lambda_2 = \left(\frac{L}{\lambda_1}+1/2\right)\lambda_2 \qquad (4.40)$$

Solving for L,

$$L = \frac{\lambda_1\lambda_2}{2(\lambda_1-\lambda_2)} \qquad (4.41)$$

Letting $\lambda_1 - \lambda_2 = \Delta\lambda$, the bandwidth, and assuming this bandwidth is small compared to the wavelength so that $\lambda_1\lambda_2 = \lambda^2$, where the geometric mean λ is about equal to the arithmetic mean, the equation becomes

$$L = \frac{\lambda^2}{2\Delta\lambda} \qquad (4.42)$$

As an example, if both g-line and i-line light exposes a photoresist, the range of wavelengths is about 80 nm and the mean wavelength is about 400 nm. The coherence length will be 1000 nm, which means that there will be no standing waves at about 500 nm up from the substrate. This matches the result seen in Figure 4.8 quite well. For laser lithography at 248 or 193 nm, the source bandwidth is extremely small. For a 248-nm source with a 1-pm bandwidth, the coherence length will be on the order of 3 cm.

4.2   Swing Curves

Generically, a swing curve refers to the sinusoidal variation of some lithographic parameter with resist thickness.[5] There are several parameters that vary in this way, but the most important is the critical dimension (CD) of the photoresist feature being printed. Figure 4.9 shows a typical CD swing curve for i-line exposure of a 0.5-μm line on silicon. The change in linewidth is quite large (more than the typical 10% tolerance) for relatively small changes in resist thickness. Another swing curve is the E_0 swing curve, showing the same sinusoidal swing in the photoresist dose-to-clear (Figure 4.10). For a resist thickness that requires a higher dose-to-clear, the photoresist will, as a consequence, require a higher dose to achieve the desired line size. But if the exposure dose is fixed (as it was for the CD swing curve), the result will be an underexposed line that prints too large. Thus, it follows that the E_0 and CD swing curves result from the same effect, coupled by the exposure latitude of the feature. The final swing curve measures the reflectivity of the resist-coated wafer as a function of resist thickness (Figure 4.11). Although reflectivity is further removed from lithographic metrics such as E_0 or CD, it is the reflectivity swing curve that provides the most insight as to the cause of the phenomenon.

4.2.1   Reflectivity Swing Curve

The reflectivity swing curve shows that variations in resist thickness result in a sinusoidal variation in the reflectivity of the resist-coated wafer. Since the definition of reflectivity is the total reflected light intensity divided by the total incident intensity, an increase in

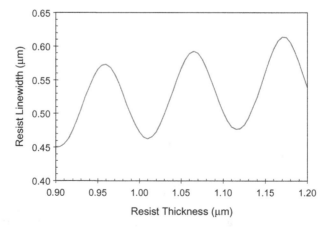

Figure 4.9   *CD swing curve showing a sinusoidal variation in the resist linewidth with resist thickness (i-line exposure of resist on silicon)*

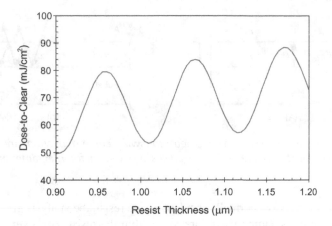

Figure 4.10 E_o *swing curve showing a sinusoidal variation in the resist dose-to-clear with resist thickness (i-line exposure of resist on silicon)*

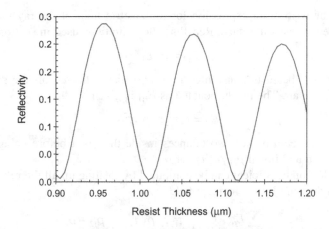

Figure 4.11   *Reflectivity swing curve showing a sinusoidal variation in the resist-coated wafer reflectivity with resist thickness (i-line exposure of resist on silicon)*

reflectivity results in more light that does not make it into the resist. Less light being coupled into the resist means that a higher dose is required to affect a certain chemical change in the resist, resulting in a larger E_0. Thus, the E_0 and CD swing curves can both be explained by the reflectivity swing curve. (The interested reader can convince him/ herself that the phases of the sinusoids of Figure 4.9 through Figure 4.11 make sense with respect to each other.)

What causes the reflectivity swing curve of Figure 4.11? Of course, the answer lies in the thin film interference effects that were discussed in the previous section on standing waves. Using the same simple geometry shown in Figure 4.1, a thin photoresist (layer 2) rests on a thick substrate (layer 3) in air (layer 1). If we illuminate this film stack with a

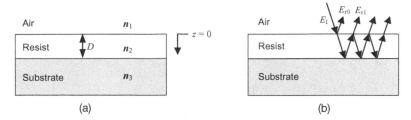

Figure 4.12 *Film stack showing (a) geometry for swing curve derivation, and (b) incident, transmitted and reflected waves [oblique angles are shown for diagrammatical purposes only]*

monochromatic plane wave normally incident on the resist, the analysis given before was used to determine the standing wave intensity within the resist. However, our goal here is to determine the total light reflected by the film stack. As shown in Figure 4.12, the total reflected light is made up of the incident beam reflecting off the air–resist interface and beams that have bounced off of the substrate and then were transmitted by the air–resist interface.

Let's begin by writing an expression for the electric field of the ray that is directly reflected by the air–resist interface. Recalling the definitions used in section 4.1.2,

$$E_{r0} = \rho_{12} E_I \tag{4.43}$$

The next 'reflected' beam is transmitted into the resist, reflected off the substrate, and transmitted into the air. The result, denoted as E_{r1}, is given by

$$E_{r1} = E_I \tau_{12} \tau_{21} \rho_{23} \tau_D^2 \tag{4.44}$$

The next reflected beam makes two bounces inside the resist before being transmitted out, resulting in an additional $\rho_{21} \rho_{23} \tau_D^2$ term.

The total reflection coefficient can be computed by totaling up all the reflected electric fields and then dividing by the incident field.

$$\rho_{\text{total}} = \frac{\sum E_{ri}}{E_I} = \rho_{12} + \frac{\tau_{12} \tau_{21} \rho_{23} \tau_D^2}{1 + \rho_{12} \rho_{23} \tau_D^2} = \frac{\rho_{12} + \rho_{23} \tau_D^2}{1 + \rho_{12} \rho_{23} \tau_D^2} \tag{4.45}$$

Note that since all of the reflected light will always be traveling in the same direction, Equation (4.45) applies equally well to s- or p-polarization. The intensity reflectivity of the film stack is the square of the magnitude of the electric field reflection coefficient. At first glance, the sinusoidal dependence of reflectivity with resist thickness is not obvious from Equation (4.45). The dependence is contained in the internal transmittance. Carrying out the calculation of reflectivity for the case of normal incidence,

$$R = \frac{|\rho_{12}|^2 + |\rho_{23}|^2 e^{-\alpha 2D} + 2|\rho_{12}\rho_{23}| e^{-\alpha D} \cos(4\pi n_2 D/\lambda - \phi_{12} + \phi_{23})}{1 + |\rho_{12}\rho_{23}|^2 e^{-\alpha 2D} + 2|\rho_{12}\rho_{23}| e^{-\alpha D} \cos(4\pi n_2 D/\lambda + \phi_{12} + \phi_{23})} \tag{4.46}$$

where ϕ_{12} and ϕ_{23} are the phase angles of the reflection coefficients ρ_{12} and ρ_{23}, respectively.

The discussion so far has been mostly mathematical. Equation (4.45) gives a rigorous result that, when expressed as Equation (4.46), leads to an understanding of the reflectivity swing curve. Physically, the reflectivity swing curve is the result of interference among the reflected rays. As pictured in Figure 4.12, the total reflected field is the sum of the various rays. How the initially reflected ray E_{r0} adds to the first transmitted and reflected ray E_{r1} depends on the phase of E_{r1}, which in turn depends on the resist thickness. At some thickness, E_{r1} will be in phase with E_{r0}, resulting in a maximum reflectivity. At another thickness E_{r1} will be out of phase with E_{r0}, resulting in a minimum reflectivity.

Equation (4.46) can also lead to a better understanding of swing curves. The period of all of the swing curves can be easily obtained from Equation (4.46) and is the same as the period of the standing waves in the photoresist. Likewise, the effects of increasing or reducing the reflectivities can be seen. Twice the multiplier of the cosine in Equation (4.46) is called the *swing amplitude*,[6] and is equal to $4|\rho_{12}\rho_{23}|e^{-\alpha D}$. If the substrate is non-reflective ($\rho_{23} = 0$), the film stack reflectivity becomes constant (the swing amplitude goes to zero). Thus, a bottom antireflection coating can reduce or eliminate the swing curve. Less obviously, if $\rho_{12} = 0$ the reflectivity will also become constant, eliminating the swing curve. This can be achieved by using a top antireflection coating. Physically, if the swing curve results from interference between E_{r0} and E_{r1}, eliminating either E_{r0} or E_{r1} will eliminate the interference and the swing. Finally, absorption in the resist will reduce the swing amplitude in Equation (4.46) by reducing the amplitude of E_{r1} and thus the amount of interference that can take place.

While not as obvious from the above discussion, the use of broadband illumination can reduce or eliminate swing curves as well. Recalling the discussion in section 4.1.5, swing curves will be reduced whenever the resist thickness approaches the coherence length of the light source.

Some refractive index values for common materials encountered in optical lithography are given in Table 4.3. For the case of a typical resist on silicon at 248 nm, $\alpha = 0.5\,\mu m^{-1}$, $|\rho_{12}| = 0.275$ and $|\rho_{23}| = 0.73$. For a nominal 300-nm resist thickness, that gives

$$R = \frac{0.47 + 0.345\cos(2\pi D/(70.5\,\text{nm}) - \phi_{12} + \phi_{23})}{1.03 + 0.345\cos(2\pi D/(70.5\,\text{nm}) + \phi_{12} + \phi_{23})} \qquad (4.47)$$

Table 4.3 *Some refractive index values of common materials at common lithographic wavelengths*[7]

Material	n at 436 nm	n at 365 nm	n at 248 nm	n at 193 nm
Photoresist	$1.65 + i0.022$	$1.69 + i0.027$	$1.76 + i0.010$	$1.71 + i0.018$
Silicon	$4.84 + i0.178$	$6.50 + i2.61$	$1.57 + i3.57$	$0.883 + i2.78$
Amorphous Silicon	$4.45 + i1.73$	$3.90 + i2.66$	$1.69 + i2.76$	$1.13 + i2.10$
Silicon Dioxide	$1.470 + i0.0$	$1.474 + i0.0$	$1.51 + i0.0$	$1.56 + i0.0$
Silicon Nitride	$2.06 + i0.0$	$2.09 + i0.0$	$2.28 + i0.0$	$2.66 + i0.240$
Aluminum	$0.580 + i5.30$	$0.408 + i4.43$	$0.190 + i2.94$	$0.113 + i2.20$
Copper	$1.17 + i2.33$	$1.27 + i1.95$	$1.47 + i1.78$	$0.970 + i1.40$
Chrome	$1.79 + i4.05$	$1.39 + i3.24$	$0.85 + i2.01$	$0.84 + i1.65$
Gallium Arsenide	$5.07 + i1.25$	$3.60 + i2.08$	$2.27 + i4.08$	$1.36 + i2.02$
Indium Phosphide	$4.18 + i0.856$	$3.19 + i1.95$	$2.13 + i3.50$	$1.49 + i2.01$
Germanium	$4.03 + i2.16$	$4.07 + i2.58$	$1.39 + i3.20$	$1.13 + i2.09$

where $\phi_{12} = -179.5°$ and $\phi_{23} = -134°$. The maximum reflectivity is about 0.59, and the minimum is about 0.18. For the case of a typical resist on silicon at 365 nm, $\alpha = 0.95 \, \mu m^{-1}$, $|\rho_{12}| = 0.257$ and $|\rho_{23}| = 0.634$ (before bleaching). For a nominal 600-nm resist thickness, that gives

$$R = \frac{0.195 + 0.184 \cos(2\pi D/(108 \, \text{nm}) - \phi_{12} + \phi_{23})}{1.008 + 0.184 \cos(2\pi D/(108 \, \text{nm}) + \phi_{12} + \phi_{23})} \tag{4.48}$$

The maximum reflectivity for this case is about 0.318, and the minimum is about 0.013.

4.2.2   Dose-to-Clear and CD Swing Curves

An approximate behavior of the E_0 swing curve can be obtained by assuming an ideal, threshold resist. For such a case, the resist will just clear away when the dose at the bottom of the resist (for an open-frame exposure) reaches some critical threshold. Using Equation (4.19) and assuming the standing waves in the resist are averaged out by some post-exposure bake mechanism (see Chapter 5), the average or bulk intensity will be

$$I_{\text{avg}}(z) = T_{\text{eff}}(e^{-\alpha z} + |\rho_{23}|^2 \, e^{-\alpha(2D-z)}) \tag{4.49}$$

where this equation was derived by averaging the standing wave intensity Equation (4.19) over one period. It is the effective transmittance T_{eff} that contains the swing curve effect of energy coupled into the resist as a function of resist thickness. Evaluating this expression at the bottom of the resist,

$$I_{\text{avg}}(z = D) = T_{\text{eff}}(1 + |\rho_{23}|^2)e^{-\alpha D} \tag{4.50}$$

The incident dose will equal the dose-to-clear, E_0, when the energy at the bottom of the resist (equal to $E_0 I_{\text{avg}}$) reaches some critical dose, E_{crit}. Thus,

$$E_0 = \frac{E_{\text{crit}}}{T_{\text{eff}}(1 + |\rho_{23}|^2)e^{-\alpha D}}$$

$$= E_{\text{crit}} \frac{1 + |\rho_{12}\rho_{23}|^2 \, e^{-\alpha 2D} + 2|\rho_{12}\rho_{23}|e^{-\alpha D}\cos(4\pi n_2 D/\lambda + \phi_{12} + \phi_{23})}{T_{12}(1 + |\rho_{23}|^2)e^{-\alpha D}} \tag{4.51}$$

Likewise, for a threshold resist, the dose to achieve a certain CD will follow this same equation though with a different critical dose. For a fixed dose, the CD obtained will depend on the exposure latitude of the feature. A simple expression that relates CD to exposure dose will be derived in Chapter 8, and takes the form

$$\frac{CD - CD_1}{CD_1} = \frac{d \ln CD}{d \ln E}\bigg|_{E=E_1} \left(1 - \frac{E_1}{E}\right) \tag{4.52}$$

where E_1 is a reference dose (such as the nominal dose) that produces the $CD = CD_1$, and the exposure latitude term $(d\ln CD/d\ln E)$ can be approximated as a constant.

In the context of swing curves, the best choice for E_1 will be the dose that produces the nominal CD when the resist thickness is at a minimum or maximum. For the case of operating at a swing curve minimum (resist thickness $= D_{min}$),

$$E_1 = E_{crit} \frac{1 + |\rho_{12}\rho_{23}|^2 e^{-\alpha 2D_{min}} - 2|\rho_{12}\rho_{23}|e^{-\alpha D_{min}}}{T_{12}(1 + |\rho_{23}|^2)e^{-\alpha D_{min}}} \tag{4.53}$$

Substituting Equation (4.51) for E and Equation (4.53) for E_1 into Equation (4.52) gives a CD swing curve expression.

4.2.3 Swing Curves for Partially Coherent Illumination

Just as for the standing wave expressions, the above swing curve equations will change as a function of the incident angle of illumination, so that actual swing curves are strongly dependent on the NA and illumination of the stepper. In fact, the range of angles coming from a partially coherent imaging source will produce a range of periods of standing waves and swing curves in the resist, reducing the amplitude in the same way that a range of wavelengths does.

Suppose that the incident light strikes the resist at an angle θ_1. The angle inside the resist will be θ_2, given by Snell's law. Then the swing curve expressions given above will be correct if the resist thickness D is replaced by $D\cos\theta_2$ and the reflection and transmission coefficients are calculated for the appropriate angle and polarization according to the Fresnel formulae [Equation (4.29)]. The swing curve period becomes

$$\text{Period} = \frac{\lambda}{2n_2 \cos\theta_2} \tag{4.54}$$

Consider a simple case. Light normally incident on a thin resist film coated on a reflective substrate is found to have a maximum of its swing curve at a certain thickness. This means that the path length traveled by the light through the resist is an integer multiple of the wavelength. If now the angle of incidence is increased, the path length through the resist will also increase. If the angle is large enough so that the path length increases by half a wavelength, this same resist thickness will correspond to a minimum of the swing curve. The angle inside the resist which causes this swing curve phase reversal is given by

$$\theta_2 = \cos^{-1}\left(\frac{m}{m+0.5}\right) \tag{4.55}$$

where m is the multiple of wavelengths that the swing curve maximum represents. For $m = 9$ (typical for a 1000-nm film used for i-line lithography), $\theta_2 \approx 18.7°$ inside the resist, corresponding to about a 32° angle incident on a typical photoresist and $\sin\theta_1 = 0.53$. For $m = 5$, (typical for a 250-nm resist film used for 193-nm lithography), $\theta_2 \approx 24.6°$ inside the resist, corresponding to about a 45° angle incident on a typical photoresist and $\sin\theta_1 = 0.71$. Both of these conditions are commonly met.

What does this mean for real swing curves in real lithographic situations? When measuring a dose-to-clear (E_0) swing curve, the light striking the resist is made up of zero-order light (that is, light which is not diffracted). If the illumination were coherent, the light would be normally incident on the resist. For conventional partially coherent illumination, the light is incident on the resist over a range of angles given by $\sin \theta_1 = \pm \sigma NA$ where NA is the numerical aperture of the objective lens and σ is the partial coherence factor. Each angle produces its own 'swing curve' and the total result is the superposition of all the individual responses to each angle. Since any change in σ or NA will change the range of angles striking the photoresist, the swing curve will change as well. Figure 4.13 shows how E_0 swing curves vary with σNA (affected by changing either the partial coherence or the numerical aperture).

Consider also a simple example of imaging small lines and spaces. For conventional illumination, the zero order will be centered around normal incidence at the resist surface with a range of angles determined by σNA. The $\pm$1st diffraction orders will strike the resist at an angle of $\sin^{-1}(\lambda/p)$ where p is the pitch of the line/space pattern. For 350-nm features imaged with i-line, the center of the first-order angular range will be about 31.4° in air – very close to the angle given above for swing curve phase reversal for a 1000-nm-thick resist film. For 130-nm lines and spaces imaged with 193-nm light, the first-order light also produces a swing that is nearly perfectly out of phase with the normally incident light. Thus, if the resist thickness were adjusted to give a maximum of the E_0 swing curve (i.e. the zero order is at a maximum of the swing curve), the first orders would effectively be at a minimum of the swing curve. The zero-order light would be maximally reflected out of the resist while the first-order light would be maximally coupled into the resist. When these orders combine to form the image in resist, the result will be significantly different than the case of imaging on a nonreflecting substrate. On the other hand, if the resist thickness were at an E_0 swing curve minimum, the first orders would be at a swing curve maximum. The lithographic response of these features (for example, the size of the focus-exposure process window) could be quite different when operating at an E_0 swing curve minimum versus a maximum.

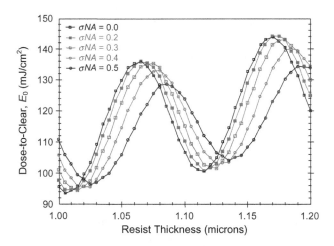

Figure 4.13   *The phase and amplitude of a dose-to-clear swing curve are affected by the range of angles striking the resist, which is controlled by the product of the partial coherence and the numerical aperture (σNA) for conventional illumination*

4.2.4 Swing Ratio

While detailed equations for swing curve effects are certainly possible, as the above discussions have shown, it is often convenient to define simple metrics to capture the essence of the effects. For swing curves, it is the amplitude of the swing that determines the degree of the problem caused by resist thickness variations. A simple metric for the magnitude of the swing is called the swing ratio, defined as the difference between adjacent min and max values of the swing curve divided by their average value. For example, for an E_0 swing curve, the swing ratio (SR) would be

$$ SR = \frac{E_{0\max} - E_{0\min}}{(E_{0\min} + E_{0\max})/2} \tag{4.56} $$

Since any two adjacent min/max pairs can be used, it will make a difference whether the max is at a greater or lesser thickness compared to the min, as will be seen below.

Using Equation (4.51),

$$ E_{0\min} = E_{\text{crit}} \frac{e^{\alpha D_{\min}} + |\rho_{12}\rho_{23}|^2 e^{-\alpha D_{\min}} - 2|\rho_{12}\rho_{23}|}{T_{12}(1+|\rho_{23}|^2)} $$

$$ E_{0\max} = E_{\text{crit}} \frac{e^{\alpha D_{\max}} + |\rho_{12}\rho_{23}|^2 e^{-\alpha D_{\max}} + 2|\rho_{12}\rho_{23}|}{T_{12}(1+|\rho_{23}|^2)} \tag{4.57} $$

which gives a swing ratio of

$$ SR = \frac{(e^{\alpha D_{\max}} - e^{\alpha D_{\min}}) + |\rho_{12}\rho_{23}|^2 (e^{-\alpha D_{\max}} - e^{-\alpha D_{\min}}) + 4|\rho_{12}\rho_{23}|}{\frac{1}{2}[(e^{\alpha D_{\max}} + e^{\alpha D_{\min}}) + |\rho_{12}\rho_{23}|^2 (e^{-\alpha D_{\max}} + e^{-\alpha D_{\min}})]} \tag{4.58} $$

Let $D_{\max} = D + \Delta D/2$ and $D_{\min} = D - \Delta D/2$, where a positive value of ΔD means that the max is at a resist thickness greater than the min. (Note that ΔD will be the swing period, that is $\lambda/2n_2$.) Equation (4.58) can be rewritten as

$$ SR = \frac{e^{\alpha D}(e^{\alpha \Delta D/2} - e^{-\alpha \Delta D/2}) + |\rho_{12}\rho_{23}|^2 e^{-\alpha D}(e^{-\alpha \Delta D/2} - e^{\alpha \Delta D/2}) + 4|\rho_{12}\rho_{23}|}{\frac{1}{2}[e^{\alpha D}(e^{\alpha \Delta D/2} + e^{-\alpha \Delta D/2}) + |\rho_{12}\rho_{23}|^2 e^{-\alpha D}(e^{-\alpha \Delta D/2} + e^{\alpha \Delta D/2})]} \tag{4.59} $$

If the amount of absorption is moderate, so that $\alpha \Delta D \ll 1$, this equation can be simplified to

$$ SR \approx \frac{e^{\alpha D}(\alpha \Delta D) + |\rho_{12}\rho_{23}|^2 e^{-\alpha D}(-\alpha \Delta D) + 4|\rho_{12}\rho_{23}|}{e^{\alpha D} + |\rho_{12}\rho_{23}|^2 e^{-\alpha D}} \tag{4.60} $$

In general,

$$ |\rho_{12}\rho_{23}|^2 e^{-2\alpha D} \ll 1 \tag{4.61} $$

(for example, for resist on silicon at 248 nm, the value of this term is about 0.01). Neglecting this term gives

$$SR \approx \alpha \Delta D + 4|\rho_{12}\rho_{23}|e^{-\alpha D} \qquad (4.62)$$

Thus, the swing ratio is equal to the swing amplitude plus an absorption term that describes the tilt of the swing curve over one swing period.

From Equation (4.62) it is easy to see how the standard approaches to reducing swing curves are captured by the swing ratio metric: (1) reduce substrate reflectivity $|\rho_{23}|$ by using a bottom antireflection coating; (2) reduce air–resist reflectivity $|\rho_{12}|$ by using a top antireflection coating; and (3) reduce $e^{-\alpha D}$ by increasing absorption. However, the sign of ΔD will also have an impact. If the maximum of the swing curve is at a greater thickness than the minimum, ΔD will be positive and the upward tilt of the swing curve due to absorption will make the swing ratio worse. If, however, the maximum of the swing curve is chosen to be at a lower resist thickness than the minimum, ΔD will be negative and bulk absorption will lower the swing ratio. In fact, it is possible to make the swing ratio go to zero:

$$SR \approx 0 \quad \text{when} \quad \alpha = \frac{4|\rho_{12}\rho_{23}|e^{-\alpha D}}{-\Delta D} \qquad (4.63)$$

Figure 4.14 shows an example of zero swing ratio for a specific case. The upward tilt of the swing curve due to absorption can make the swings look more like stair steps. Operating in the flat region (where $SR \approx 0$) can lead to greater CD control for these cases.

As mentioned in the previous section, light strikes the wafer over a range of angles, each angle setting up its own swing curve with its own period. If the range of angles is great enough, a very noticeable decrease in the total swing ratio can be observed (see Figure 4.13). For a simple case, we can calculate the impact of the range of angles on the swing ratio metric defined above. Let's consider the E_0 swing curve for conventional illumination. Further, let's assume that the effect of angle on the reflection and

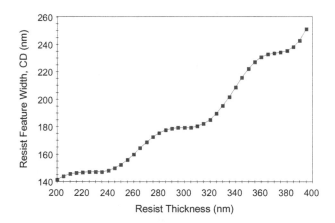

Figure 4.14 *Proper balancing of absorption and reflectivities can make the minimum of a swing curve (D = 310 nm) achieve the same CD as the previous swing curve maximum (D = 280 nm)*

transmission coefficients and on absorption is small compared to the effect on the period. Thus, we can put the E_0 swing curve formula [Equation (4.51)] into the basic approximate form

$$E_0 = a + b\cos(4\pi n_2 D\cos\theta_2/\lambda + \phi) \qquad (4.64)$$

where the swing ratio is $2b/a$.

Inside the resist, let $s = \sin\theta_2$ so that we must integrate over the range $\pm s_{max}$ where

$$s_{max} = \frac{\sigma NA}{n_2} \qquad (4.65)$$

Integrating over the source,

$$E_0 = a + b\frac{2}{s_{max}^2}\int_0^{s_{max}}\cos(4\pi n_2 D\sqrt{1-s^2}/\lambda + \phi)s\,ds \qquad (4.66)$$

This integral is not analytically solvable, but it can be approximated by letting

$$\cos\theta_2 = \sqrt{1-s^2} \approx 1 - \frac{s^2}{2} \qquad (4.67)$$

A typical value for s_{max} is about 0.5, so that this approximation is not too bad. Using the approximation and carrying out the integral gives

$$E_0 = a + b\frac{\sin(c)}{c}\cos\left(4\pi n_2 D\left(1 - \frac{s_{max}^2}{4}\right)\bigg/\lambda + \phi\right) \qquad (4.68)$$

where

$$c = \frac{\pi D(\sigma NA)^2}{\lambda n_2}$$

By using the same small angle approximation as was used in Equation (4.67),

$$E_0 = a + b\frac{\sin(c)}{c}\cos\left(4\pi n_2 D\cos\left(\frac{\theta_{2max}}{\sqrt{2}}\right)\bigg/\lambda + \phi\right) \qquad (4.69)$$

At first glance, it is not clear what the period of the new standing wave is, since there is a product of a sine and a cosine. However, the sine term has a period that is 10–20 times larger than the cosine term, so it is easiest to interpret the cosine as creating the swing curve and the sinc function as an amplitude modification. Thus, the period of the partially coherent swing curve is about

$$\text{Period} = \frac{\lambda}{2n_2\cos\theta_{eff}} \quad \text{where} \quad \theta_{eff} = \frac{\theta_{2max}}{\sqrt{2}} \qquad (4.70)$$

and the swing ratio is reduced by the factor $\sin(c)/c$. Consider a typical 193-nm imaging case where $NA = 0.93$ and $\sigma = 0.7$. The swing curve period for normally incident light would be 56 nm, but for the partially coherent light it grows slightly to 59 nm. The real impact is in the reduction of the swing ratio, which for a 300-nm-thick resist would produce $\sin(c)/c = 0.77$, meaning a 23% reduction in the swing ratio.

4.2.5   Effective Absorption

As we shall see, the optimum resist absorption is a strong function of the reflectivity of the substrate. Consider first, however, the simple case of a nonreflecting substrate so that light travels only downward through a resist film of thickness D. The absorption of light through the resist leads to an exposure dose error: a smaller dose at the bottom of the resist (E_{bottom}) compared to the top (E_{top}). The fraction of light making it to the bottom is given by

$$\frac{E_{\text{bottom}}}{E_{\text{top}}} = T_D = e^{-\alpha D} \tag{4.71}$$

As an example, for a resist with $\alpha = 0.5\,\mu\text{m}^{-1}$, a 500-nm-thick film will absorb 22% of the light so that $T_D = 0.78$.

   If the resist film is coated on a reflective substrate, reflected light traveling up through the film will also be absorbed. The reflected beam will be brighter at the bottom of the resist, so that the sum of the incident and reflected beams will have a smaller variation in dose from top to bottom than for the nonreflective substrate case. The amount can be quantified using one of the previously derived expressions for the standing wave intensity. The 'average' intensity (averaged over one standing wave period) was previously derived in Equation (4.50), and is given here normalized so that the intensity at the top of the resist is 1:

$$I_{\text{avg}}(z) = e^{-\alpha z} \left(\frac{1 + |\rho_{23}|^2\, e^{-\alpha 2D} e^{\alpha 2z}}{1 + |\rho_{23}|^2\, e^{-\alpha 2D}} \right) \tag{4.72}$$

The term $|\rho_{23}|^2 e^{-\alpha 2D}$ represents the fraction of the light intensity that makes it back to the top of the resist after traveling down through the resist, reflecting off the substrate, and traveling back up to the top. It can be thought of as a 'round-trip' transmittance and is an important factor in determining the difference between Equation (4.72) and simple bulk absorption.

   For small amounts of absorption ($\alpha 2D < 1$, for example), the z-dependent exponential term in the parentheses of Equation (4.72) can be expanded as Taylor series and an approximate expression for the bulk effect can be derived:

$$I_{\text{average}}(z) \approx e^{-\alpha_{\text{eff}} z} \tag{4.73}$$

where the effective absorption coefficient is given by

$$\alpha_{\text{eff}} = \alpha \left(\frac{1 - |\rho_{23}|^2\, e^{-\alpha 2D}}{1 + |\rho_{23}|^2\, e^{-\alpha 2D}} \right) \tag{4.74}$$

As discussed above, a more reflective substrate actually reduces the bulk intensity variation through the resist, which is expressed here as a lower effective absorption coefficient.

Because of the Taylor series expansion, the expression (4.74) used in Equation (4.73) provides a very good match for the actual intensity as a function of depth near the top of the resist. However, the approximate expression overestimates the amount of absorption, resulting in an intensity that is too low everywhere, with the greatest error at the bottom of the resist. A second approach toward defining an effective absorption is to match the intensity at the top and the bottom of the resist.[8] This results in an effective absorption coefficient of

$$\alpha_{\text{eff}} = \alpha - \frac{1}{D} \ln \left(\frac{1+|\rho_{23}|^2}{1+|\rho_{23}|^2 \, e^{-\alpha 2D}} \right) \tag{4.75}$$

For the case of small substrate reflectivity (when using a good bottom antireflection coating, for example), this expression simplifies to

$$\alpha_{\text{eff}} \approx \alpha \left[1 - 2|\rho_{23}|^2 \left(\frac{1-e^{-\alpha 2D}}{\alpha 2D} \right) \right] \tag{4.76}$$

For moderate to small amounts of absorption, this equation can be further simplified to

$$\alpha_{\text{eff}} \approx \alpha [1 - 2|\rho_{23}|^2 (1 - \alpha D)] \tag{4.77}$$

While the intensity using the effective absorption of Equation (4.75) matches the actual average intensity at the top and the bottom of the resist, it gives an intensity that is higher than the actual intensity everywhere else. To match the integrated intensity through the thickness of the resist, the effective absorption coefficient must satisfy the following equation:

$$\frac{\alpha_{\text{eff}}}{1-e^{-\alpha_{\text{eff}}D}} = \alpha \frac{1+|\rho_{23}|^2 \, e^{-\alpha 2D}}{1-|\rho_{23}|^2 \, e^{-\alpha 2D} - (1-|\rho_{23}|^2)e^{-\alpha D}} \tag{4.78}$$

The effective absorption coefficient that matches the integrated intensity will generally lie about halfway between the values given by Equations (4.74) and (4.75).

How can the effective absorption be used when designing resists for different reflectivity applications? One simple design criterion might be to fix the effective absorption coefficient. Suppose that a 0.7-μm-thick deep-UV resist with an absorption coefficient of $0.4\,\mu\text{m}^{-1}$ is currently providing acceptable resist profile results on a bare silicon wafer ($|\rho_{23}|^2 \approx 0.5$). In other words, for the parameters given, the effective absorption provides an acceptable dose variation from the top to the bottom of the resist. From Equation (4.74), the effective absorption coefficient is $0.22\,\mu\text{m}^{-1}$. Thus, for a resist to have approximately the same profile behavior on a perfectly nonreflecting substrate, its absorption coefficient would have to be lowered to this 0.22-μm^{-1} value. On the other hand, if one wanted to use an equivalent resist on an aluminum substrate with a reflectivity of 0.84 (not necessarily a good idea, given the swing curve effects), one could raise the absorption coefficient to $0.52\,\mu\text{m}^{-1}$ and still exhibit the same effective absorption.

Light traveling at an angle through the photoresist will result in an apparent difference in the absorption coefficient. As will be derived in section 4.8.2, a plane wave traveling through resist at some angle θ_2 will be absorbed according to

$$I(z) \propto \exp(-\alpha z / \cos \theta_2) \qquad (4.79)$$

Thus, an effective absorption coefficient can be defined as $\alpha/\cos \theta_2$. If a reflective substrate is present, Equations (4.74)–(4.78) can still be used if $\alpha/\cos \theta_2$ is substituted for α. If a range of angles is used, such as for partially coherent illumination, Equation (4.79) can be integrated over the angular range of the source. For conventional illumination, and for moderate amounts of absorption so that the exponential can be expanded as a Taylor series, the resulting effective absorption coefficient becomes

$$\alpha_{\text{eff}} \approx \alpha \left(\frac{2}{1 + \cos \theta_{2\max}} \right) \qquad (4.80)$$

where $\theta_{2\max}$ is the angle in resist corresponding to the maximum angle of the conventional partially coherent source.

4.3   Bottom Antireflection Coatings

As discussed above, reflections from the substrate can cause unwanted variations in the resist profile and swing curve effects. Reflections are caused by a difference in the complex index of refraction of two materials. Since lithography takes place on a variety of substrates with film stacks of many materials and thicknesses, each interface in the film stack can contribute to the overall reflectivity back into the photoresist. This complicated situation is made worse by the inevitable variations in the thicknesses, and sometimes the refractive indices, of the films. One possible solution to reflectivity problems is the bottom antireflection coating (BARC). Also, swing curves can be improved by the use of a top antireflection coating (TARC), described in the next section. These types of antireflection coatings have somewhat different goals, but their basic behavior is the same.

The goals of film stack optimization are to minimize standing waves in the resist, and to reduce the sensitivity of the process to film stack variations (including resist and BARC, but other layers as well). By far the most common way to accomplish all of these goals is by using an optimized BARC. When optimizing a lithography process for reflectivity, there are three basic tasks: (1) optimize the BARC, (2) optimize the resist thickness (from a swing curve perspective) and (3) understand the sensitivity to BARC, resist and film stack variations. For the first task, there are two classes of BARC problems:

- BARC on an absorbing substrate (such as metal) – the goal is to reduce the reflectivity (the thickness of the metal or what is underneath doesn't matter)
- BARC on a transparent substrate (such as silicon dioxide) – reduce the sensitivity to oxide thickness variations (while also keeping reflectivity low)

There is an unfortunate problem in that the substrate reflectivity experienced by the photoresist cannot be measured. Measuring the reflectivity of the substrate when not coated by photoresist is not useful to this task, and once coated the reflectivity becomes hidden from measurement. Thus, the approach that must be used is to calculate the

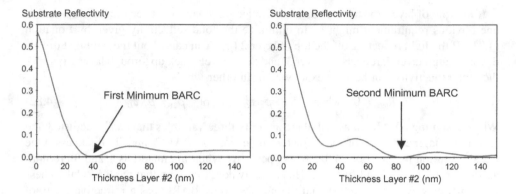

Figure 4.15 *Typical examples of substrate reflectivity versus BARC thickness for different resist/BARC/substrate stacks*

reflectivity from measured fundamental parameters, namely the thickness and complex refractive index of each layer in the film stack.

For a single-layer BARC, there are three parameters available for optimization: the thickness of the BARC, and the real and imaginary parts of its refractive index. For the simplest use case, a BARC is given and the goal is just to optimize its thickness. Figure 4.15 shows substrate reflectivity calculations for two different resist/BARC/substrate stacks. These reflectivity curves show a characteristic feature of a resist/BARC/substrate stack: the lowest reflectivity could be at the first minimum or at the second minimum depending on the parameters. Etch and process integration considerations determine the range of acceptable BARC thicknesses, and thus the preference for a first-minimum BARC or a second-minimum BARC.

4.3.1   BARC on an Absorbing Substrate

For the case where all of the BARC properties are available for optimization, a more detailed look at the problem is required. Recapping the basic thin film reflectivity theory, the electric field reflection coefficient (the ratio of reflected to incident electric fields) at the interface between two materials is a function of the complex indices of refraction for the two layers. For normal incidence, the reflection coefficient of light traveling through layer i and striking layer j is given by Equation (4.8). For the case of a bottom antireflection coefficient (BARC), assume the BARC (layer 2) is sandwiched between a resist (layer 1) and a very thick substrate (layer 3). The total reflectivity looking down on layer 2 includes reflections from both the top and bottom of the BARC film. The resulting reflectivity, taking into account all possible reflections, is

$$R_{\text{total}} = |\rho_{\text{total}}|^2 = \left| \frac{\rho_{12} + \rho_{23}\tau_D^2}{1 + \rho_{12}\rho_{23}\tau_D^2} \right|^2 \tag{4.81}$$

where the internal transmittance, τ_D, is the change in the electric field as it travels from the top to the bottom of the BARC, given by Equation (4.12).

If the role of layer 2 is to serve as an antireflection coating between materials 1 and 3, one obvious requirement might be to minimize the total reflectivity given by Equation (4.81). If the light reflecting off the top of layer 2 (ρ_{12}) can cancel out the light that travels down through layer 2, reflects off layer 3, and then travels back up through layer 2 ($\rho_{23}\tau_D^2$), then the reflectivity can become exactly zero. In other words,

$$R_{\text{total}} = 0 \quad \text{when} \quad \rho_{12} + \rho_{23}\tau_D^2 = 0 \quad \text{or} \quad \rho_{21} = \rho_{23}\tau_D^2 \qquad (4.82)$$

When designing a BARC material, there are only three variables that can be adjusted: the real and imaginary parts of the refractive index of the BARC, and its thickness. One classic solution to Equation (4.82) works perfectly when the materials 1 and 3 are non-absorbing: let $\tau_D^2 = -1$ and $\rho_{12} = \rho_{23}$. This is equivalent to saying that the BARC thickness is a 'quarter wave' ($D = \lambda/4n_2$), and the nonabsorbing BARC has a refractive index of $n_2 = \sqrt{n_1 n_3}$. While this BARC solution is ideal for applications like antireflective coatings on lens surfaces, it is not particularly useful for common lithography substrates, which are invariably absorbing.

Will a solution to Equation (4.82) always exist, even when the resist and substrate have complex refractive indices? Since all the terms in Equation (4.82) are complex, zero reflectivity occurs when both the real part and the imaginary part of Equation (4.82) are true. This requirement can be met by adjusting only two of the three BARC parameters (n, κ and D). In other words, there is not just one solution but a family of solutions to the optimum BARC problem. Expressing each reflection coefficient in terms of magnitude and phase,

$$\rho_{ij} = |\rho_{ij}|e^{i\theta_{ij}} \qquad (4.83)$$

Equation (4.82) can be expressed as two equalities:

$$D = \frac{\lambda}{4\pi\kappa_2}\ln\left|\frac{\rho_{23}}{\rho_{21}}\right| = \frac{\lambda}{4\pi n_2}(\theta_{21} - \theta_{23}) \qquad (4.84)$$

Unfortunately, the seemingly simple forms of Equations (4.82) and (4.84) are deceptive: solving for the unknown complex refractive index of the BARC is exceedingly messy. As a consequence, numerical solutions to Equation (4.82) or (4.84) are almost always used. Note that a second-minimum BARC can be optimized using Equation (4.84) by adding 2π to the angle difference on the right-hand side of the equation.

Consider a common case of a BARC for 193-nm exposure of resist on silicon. The ideal BARC is the family of solutions as shown in Figure (4.16), which gives the ideal BARC n and κ values as a function of BARC thickness. Each solution produces exactly zero reflectivity for normally incident monochromatic light. Within this family of solutions, available materials and the acceptable range of BARC thicknesses (usually constrained by coating and etch requirements) will dictate the final solution chosen.

Another criterion for choosing the optimum BARC is the sensitivity to BARC thickness variations. Often, BARC layers are coated over topography and are partially planarizing. This means that BARC thickness variations across the device are inevitable. Is one BARC solution from the family of solutions given in Figure 4.16 better from a BARC thickness

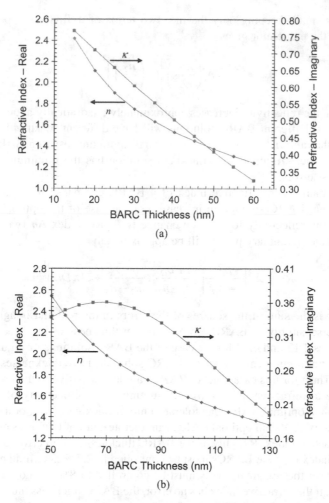

Figure 4.16 *Optimum BARC refractive index (real and imaginary parts, n and κ) as a function of BARC thickness for normal incidence illumination (resist index = 1.7 + i0.01536 and silicon substrate index = 0.8831 + i2.778) at 193 nm. (a) First minimum BARCs, and (b) second minimum BARCs*

sensitivity perspective? Consider only small errors in BARC thickness about the optimum value, so that $D = D_{opt} + \Delta$ where D_{opt} is the optimum thickness given by Equation (4.84). Since Equation (4.82) is true at the optimum thickness,

$$\rho_{12} + \rho_{23}\tau_D^2 = \rho_{12} + \rho_{23}\tau_{D_{opt}}^2 e^{i4\pi n_2 \Delta/\lambda} = \rho_{12}(1 - e^{i4\pi n_2 \Delta/\lambda}) \tag{4.85}$$

Using a similar operation on the denominator of the reflectivity expression (4.81) gives

$$R = \left| \frac{\rho_{12}(1 - e^{i4\pi n_2 \Delta/\lambda})}{1 - \rho_{12}^2 e^{i4\pi n_2 \Delta/\lambda}} \right|^2 \tag{4.86}$$

Assuming Δ is small, keeping only the first two terms of a Taylor's expansion of the exponential in the numerator gives

$$R \approx \Delta^2 \left(\frac{\pi n_1}{\lambda} \right)^2 \left| 1 - \left(\frac{n_2}{n_1} \right)^2 \right|^2 \tag{4.87}$$

As can be seen, the reflectivity increases approximately quadratically about the optimum BARC thickness. Different BARC solutions will have different sensitivities, depending on how close the ratio n_2/n_1 is to one. For the first minimum resist on BARC on silicon example above, the ~50-nm BARC thickness solution has the minimum sensitivity to BARC thickness errors.

Similarly, variations in refractive index of a BARC can cause the reflectivity of an otherwise optimal BARC solution to increase. For the case of the optimal normal-incidence BARC, the reflectivity for a given change is BARC index Δn (which can be an error in real and/or imaginary parts) will be approximately

$$R \approx |\Delta n|^2 \left| \frac{\rho_{12}}{1 - \rho_{12}^2} \right|^2 \left| \frac{1 + \rho_{12}}{n_1 - n_2} + \frac{1 - \rho_{23}}{n_2 - n_3} + i 4 \pi D / \lambda \right|^2 \tag{4.88}$$

Note that $|\Delta n|^2$ is the sum of the squares of the errors in the real and imaginary parts of the index. The sensitivity to BARC errors, either in thickness or in refractive index, is shown in Figures 4.17 and 4.18 for the case of the BARC solutions of Figure 4.16. Note that for the first minimum case, practical BARC solutions have thicknesses in the 20- to 50-nm range. Thus, for this case, thicker BARCs are less sensitive to thickness errors but more sensitive to refractive index errors. Assuming that a reflectivity of 0.1% can be tolerated, a 20-nm optimal BARC can tolerate 1 nm of thickness error, or a 0.053 change in refractive index. A 40-nm optimal BARC can tolerate 2 nm of thickness error, or a 0.04 change in refractive index. Roughly, if the BARC thickness has no error, the real part of the refractive index of these BARCs must be controlled to 2.5% assuming no error in the imaginary part, or the imaginary parts must be controlled to 8% for no error in the real part. Likewise, if the refractive index has no error, the BARC thickness must be controlled to roughly 5% for the given 0.1% reflectivity tolerance.

4.3.2   BARCs at High Numerical Apertures

Things become a bit more complicated for the more general case of light traveling at an angle with respect to the film stack normal. In lithographic terms, high numerical apertures allow large ranges of angles to pass through the lens and arrive at the wafer. Off-axis illumination of small pitch patterns produces images made of light concentrated at large angles. Small isolated features with large σ partially coherent illumination create light reaching the wafer over a wide range of angles. And of course, low numerical apertures result in limited ranges of angles reaching the wafer.

How does non-normal incidence affect Equation (4.81)? Each reflection coefficient ρ_{ij} is a function of the angle of light, and a function of polarization. Randomly polarized light (often called unpolarized light, the kind most commonly employed in lithographic tools) can be considered the incoherent sum of two linear and orthogonal polarizations. If θ_i is the incident (and reflected) angle inside the resist and θ_j is the transmitted angle

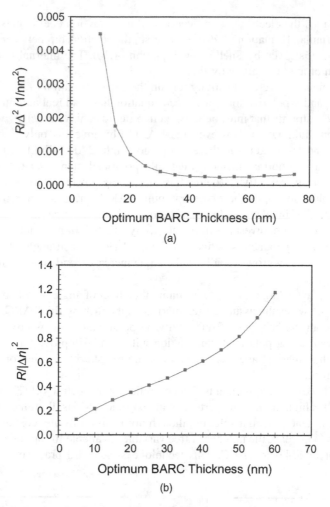

(a)

(b)

Figure 4.17 *Sensitivity of substrate reflectivity for the optimum first minimum BARCs of Figure 4.16a as a function of (a) BARC thickness errors, or (b) BARC refractive index errors*

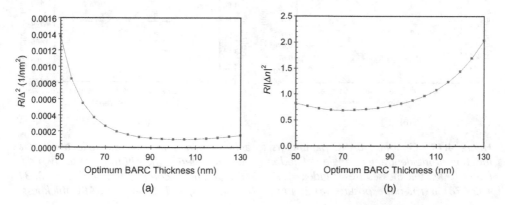

(a)

(b)

Figure 4.18 *Sensitivity of substrate reflectivity for the optimum second minimum BARCs of Figure 4.16b as a function of (a) BARC thickness errors, or (b) BARC refractive index errors*

in the BARC, then the electric field reflection and transmission coefficients are given by the Fresnel formulae [Equation (4.29)]. Of course, the relationship between incident and transmitted angle is given by Snell's law, Equation (4.30). The internal transmittance is also a function of angle as given by Equation (4.33).

The requirements for zero reflectivity remain the same. Equation (4.82) must be satisfied for both *s*- and *p*-polarization. Since each equation has both real and imaginary parts, there are four constraints that must be satisfied in order to achieve exactly zero reflectivity for unpolarized illumination. However, the BARC film gives us only three degrees of freedom (n, κ and D). In general, there can be no single BARC film that results in zero reflectivity for a non-normal incident randomly polarized plane wave. Calculating the unpolarized intensity reflectivity (defined as the average of the individual reflectivities for *s*- and *p*-polarization), the best case optimum BARC parameters are given in Figure 4.19 as a function of incident angle in air (before striking the resist).

Figure 4.20 shows the lowest possible reflectivity as a function of incident angle, using the optimum BARC parameters defined in Figure 4.19 for each angle. As can be seen, the best case reflectivity grows rapidly as the angle increases, and is worse for the thinner BARC film.

Polarization plays a large role in determining the effect of angle on reflectivity. Figures 4.19 and 4.20 show results assuming unpolarized light striking the BARC, while Figure 4.21 separates out the behavior of reflectivity by polarization for two example BARCs. The trend toward using polarized illumination will greatly improve the performance of BARCs at high numerical apertures since only the reflectivity for that one polarization need be optimized.

Stringent CD control requirements demand reflectivities below 0.1%, making BARC design difficult at high numerical apertures with extreme off-axis illumination (i.e. at large incident angles). One potential solution, though not pleasant from a cost and complexity perspective, is to increase the number of free variables available for optimization by using a two-layer or three-layer BARC. This technique, a standard practice in antireflective

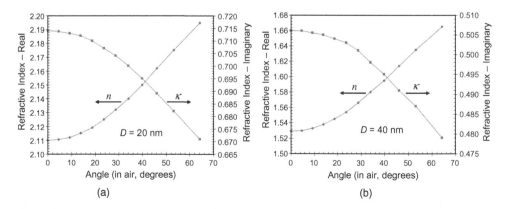

Figure 4.19 *Optimum BARC parameters to achieve minimum substrate reflectivity as a function of incident angle (angle defined in air, before entering the photoresist) for two different BARC thicknesses (resist index = 1.7 + i0.01536 and silicon substrate index = 0.8831 + i2.778) at 193-nm exposure: (a) 20-nm BARC thickness, and (b) 40-nm BARC thickness*

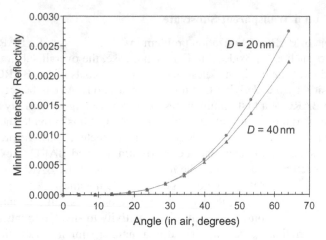

Figure 4.20 *The best case (minimum) reflectivity (using the BARC parameters shown in Figure 4.19) of the substrate as a function of incident angle for 20- and 40-nm-thick BARC films. Note that 60° corresponds to the maximum angle in air allowed for NA = 0.866*

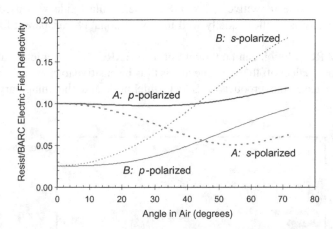

Figure 4.21 *An example of the variation of BARC reflectivity as a function of light angle and polarization for two different BARCs. The intensity reflectivity is the square of the electric field reflectivity plotted here, but interference makes the field reflectivity a better measure of the standing wave effects (resist index = 1.7 + i0.01536, silicon substrate index = 0.8831 + i2.778, BARC A index = 1.80 + i0.48, BARC A thickness = 30 nm, BARC B index = 1.53 + i0.54, BARC B thickness = 39 nm)*

coatings for lenses, would provide enough adjustable parameters to make the reflectivity go to zero at normal incidence and at one or more angles, thus providing low reflectivity over a wide range of angles. This approach has become necessary with the advent of immersion lithography and numerical apertures greater than 1.0. Alternately, a gradient index material, with low κ near the top of the BARC gradually increasing toward the bottom, can be used.

4.3.3   BARC on a Transparent Substrate

The second type of BARC optimization problem involves the use of a BARC on a transparent substrate, such as an oxide film. For such a case, the overall substrate reflectivity will be a function of the oxide thickness and one of the goals of the BARC design is to reduce the sensitivity to underlying film thickness variations. As can be seen from Figure 4.22, using the BARC at a first minimum results in a very large sensitivity to underlying oxide thickness variations. In fact, the thickest BARC films provide the most robust behavior when a wide range of oxide thicknesses are expected. Thus, the preferred design approach for this case is to determine the maximum allowed BARC thickness from an integration perspective, then optimize the n and κ values of the BARC to minimize the maximum reflectivity over the range of expected oxide thicknesses.

If the oxide thickness under the BARC varies by only a small amount, the use of a first-minimum BARC becomes practical. The sensitivity to small amounts of underlying oxide thickness variation Δ_{ox} can be derived in a manner similar to Equation (4.87):

$$R \approx \Delta_{ox}^2 \left(\frac{\pi n_{ox}}{\lambda} \right)^2 \left| \frac{\mathbf{n}_1}{\mathbf{n}_2} - \frac{\mathbf{n}_2}{\mathbf{n}_1} \right|^2 \qquad (4.89)$$

where n_{ox} is the oxide refractive index. Of course, while oxide was used here as an example, this analysis applies equally well to any reasonably transparent film under the BARC.

A 'thick' BARC solution can be thought of as a BARC that completely absorbs all of the light that may reflect off the substrate and is thus insensitive to changes in the substrate stack. For this case, the product of BARC thickness and the imaginary part of its

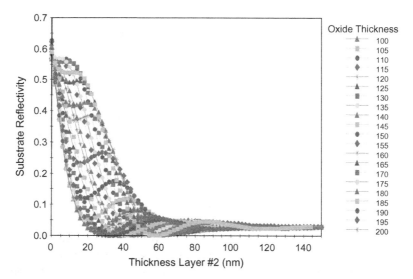

Figure 4.22 *Substrate reflectivity versus BARC thickness over a range of underlying oxide thicknesses (oxide on top of a silicon substrate)*

refractive index (κ_2) must be sufficiently high to make $\tau_D^2 \approx 0$. Maximum BARC thickness is usually determined by process considerations, thus fixing the minimum κ_2. Assuming the imaginary part of the resist refractive index is sufficiently small, the resulting reflectivity is

$$R_{\text{total}} = |\rho_{12}|^2 = \left| \frac{n_1 - n_2 - i\kappa_2}{n_1 + n_2 + i\kappa_2} \right|^2 = \frac{(n_1 - n_2)^2 + \kappa_2^2}{(n_1 + n_2)^2 + \kappa_2^2} \tag{4.90}$$

The optimum real part of the BARC refractive index to minimize reflectivity will then be

$$n_2 = \sqrt{n_1^2 + \kappa_2^2} \tag{4.91}$$

The resulting reflectivity is

$$R = \frac{n_2 - n_1}{n_2 + n_1} \approx \frac{\kappa_2^2}{4n_1^2 + \kappa_2^2} \tag{4.92}$$

Obviously, the thick BARC solution is insensitive to BARC thickness variations by design. The sensitivity to errors in BARC refractive index can be obtained from Equation (4.90). For a small change in the real part of the BARC refractive index,

$$\frac{\Delta R}{R} \approx \Delta n_2^2 \frac{n_1}{n_2(n_2^2 - n_1^2)} = \Delta n_2^2 \frac{n_1}{n_2 \kappa_2^2} \tag{4.93}$$

For a small change in the imaginary part of the BARC refractive index,

$$\frac{\Delta R}{R} \approx \frac{\Delta \kappa_2}{\kappa_2} \left(\frac{2n_1}{n_2} \right) \tag{4.94}$$

4.3.4   BARC Performance

How critical is BARC optimization? How low must the substrate reflectivity be before acceptable CD control can be expected? There is no single answer to these questions, since they are process and feature dependent. But consider the example shown in Figure 4.23. Here, 100-nm lines on a 280-nm pitch are simulated with a stepper using annular illumination, with a center sigma given by $\sigma NA = 0.54$. As can be seen, a substrate reflectivity of less than 0.1 % still leads to a noticeable swing behavior.

A common approach to characterizing the impact of substrate reflectance (that is, the impact of a nonideal BARC) is to relate reflectivity in the presence of film thickness variations to effective dose errors. Consider, for example, the case where topography on the wafer results in resist thickness variations that extend beyond one swing curve period. For such a case, the swing ratio is effectively the fractional dose error range resulting from resist thickness variations.

$$\frac{\Delta E}{E} \approx SR \approx \alpha \Delta D + 4|\rho_{12}\rho_{23}|e^{-\alpha D} \tag{4.95}$$

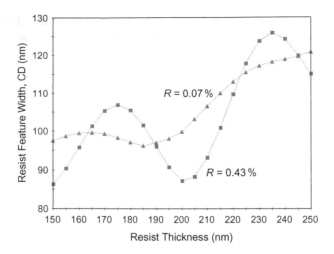

Figure 4.23 *CD swing curves (100-nm lines with a 280-nm pitch are printed with a stepper using annular illumination, with a center sigma given by σNA = 0.54) for two different BARCs with different levels of optimization, as given by the resulting substrate reflectivity R*

Consider a 193-nm resist with $\alpha = 1.2\,\mu m^{-1}$, $D = 200\,nm$, a resist thickness variation equal to the swing period of 56 nm, and $|\rho_{12}| = 0.26$. This results in a fractional effective dose error of

$$\frac{\Delta E}{E} \approx 0.067 + 0.83|\rho_{23}| \qquad (4.96)$$

For a BARC that allows 1% reflectance ($|\rho_{23}| = 0.1$), bulk absorbance adds 6.7% effective dose error while the swing effect adds another 8.3% dose error. Both of these numbers are clearly too high, so that a 193-nm lithography process must employ planarization to reduce the resist thickness variations caused by topography.

For a 248-nm resist with $\alpha = 0.5\,\mu m^{-1}$, $D = 400\,nm$, a resist thickness variation equal to the swing period of 70 nm, and $|\rho_{12}| = 0.275$, the fractional effective dose error is

$$\frac{\Delta E}{E} \approx 0.035 + 0.90|\rho_{23}| \qquad (4.97)$$

For a BARC that allows 1% reflectance ($|\rho_{23}| = 0.1$), bulk absorbance adds 3.5% effective dose error while the swing effect adds another 9% dose error.

A more thorough 'dose budget' analysis would consider variations in resist thickness, BARC thickness and underlying film stack simultaneously to determine the effective dose errors caused by changes in the energy coupled into the resist. Equation (4.95) assumed sufficient topography to cause one full swing period of resist thickness variations. When smaller amounts of resist thickness variation are present, the effective dose errors are also smaller. Consider a resist thickness $D = D_{min} \pm \Delta$ where D_{min} is the thickness at the swing curve minimum and Δ is small compared to the swing period.

Equation (4.51) can be used to derive the effective dose error for a small resist thickness error:

$$\frac{\Delta E}{E} \approx \left(\frac{[e^{\alpha\Delta}-1]+2|\rho_{12}\rho_{23}|e^{-\alpha D}[1-\cos(4\pi n_2\Delta/\lambda)]}{1-2|\rho_{12}\rho_{23}|e^{-\alpha D}} \right) \qquad (4.98)$$

which can be further approximated for small Δ as

$$\frac{\Delta E}{E} \approx \left(\frac{\alpha\Delta+4|\rho_{12}\rho_{23}|e^{-\alpha D}(2\pi n_2\Delta/\lambda)^2}{1-2|\rho_{12}\rho_{23}|e^{-\alpha D}} \right) \qquad (4.99)$$

and for a reasonably good BARC as

$$\frac{\Delta E}{E} \approx \alpha\Delta+4|\rho_{12}\rho_{23}|e^{-\alpha D}(2\pi n_2\Delta/\lambda)^2 \qquad (4.100)$$

The dose error comes from a bulk absorption term ($\alpha\Delta$) plus a fraction of the swing amplitude. Consider the standard 193-nm resist on BARC on silicon case described above, with $\alpha = 1.2\,\mu m^{-1}$, $D = 200\,nm$, $n_{resist} = 1.7$, and $|\rho_{air-resist}| = 0.26$. For Δ in nanometers,

$$\frac{\Delta E}{E} \approx 0.0012\Delta+0.0025|\rho_{23}|\Delta^2 \qquad (4.101)$$

For an 8.3-nm increase in resist thickness, the bulk effect causes about a 1% effective dose error. The swing effect will cause another 1% effective dose error for that 8.3-nm resist thickness error if the substrate reflectivity is 0.33%. The maximum allowed substrate reflectivity is a function of the range of resist thickness that must be tolerated and the amount of effective dose error one is willing to accept. Specifications of substrate reflectivity in the 0.1–0.5% range are not uncommon.

Swing curves are extremely important in lithography for one simple reason: topography on the wafer can lead to variations in resist thickness on the order of a period in the swing curve or more. In some processes, these effects are the leading cause of CD errors on some critical mask levels. Figure 4.24 shows an example of how a line printing over a topographical step on the wafer can lead to linewidth variations due to swing curve effects.

BARCs of course can dramatically reduce the linewidth variation caused by topography on the wafer. A second problem that BARCs can solve is *reflective notching*. As Figure 4.25 illustrates, light can reflect off the tilted edges of reflective topography. Patterns being printed near such a reflective edge may receive extra exposure from the reflection, causing a notch, usually near the top of the feature. BARCs can also significantly suppress reflective notching (Figure 4.26).

4.4 Top Antireflection Coatings

As we saw in Equation (4.46), the amplitude of the reflectivity swing curve is determined by the multiplier of the cosine term, $2|\rho_{12}\rho_{23}|e^{-\alpha D}$. It is clear that this term can be reduced in three ways: (1) increase absorption (increase the term αD) through the use of a dyed

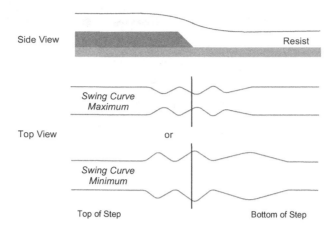

Figure 4.24 *Example of how resist thickness variations over topography produce linewidth variations due to swing curve effects when a BARC is not used*

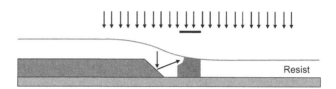

Figure 4.25 *Reflective notching occurs when nearby topography reflects light obliquely into an adjacent photoresist feature*

 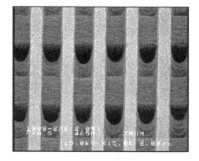

Figure 4.26 *Imaging of lines and spaces over reflective topography without BARC (left) showing reflective notching, and with BARC (right) showing the reflective notching effectively suppressed (photos courtesy of AZ Photoresist, used with permission)*

resist; (2) decrease substrate reflectivity (decrease ρ_{23}) by using a BARC; and (3) decrease resist reflectivity (decrease ρ_{12}) by using a top antireflection coating (TARC).[9,10] Looking again at Figure 4.12b, the reflectivity swing curve comes about through the interference of E_{r0} with the other reflected rays such as E_{r1}. By eliminating E_{r0} using a TARC, there can be no interference and thus no swing curve.

One can simplify the design of the TARC somewhat by imagining the photoresist as infinitely thick. Thus, Equation (4.81) applies where layer 2 is the TARC and layer 3 is the resist. If the role of layer 2 is to serve as an antireflection coating between materials 1 and 3, one obvious requirement might be to minimize the total reflectivity given by Equation (4.81). If the light reflecting off the top of layer 2 (ρ_{12}) can cancel out the light which travels down through layer 2, reflects off layer 3, and then travels back up through layer 2 ($\rho_{23}\tau_D^2$), then the reflectivity can become exactly zero [in other words, when Equation (4.82) is true for the TARC].

When designing an antireflection coating material, there are only three variables that can be adjusted: the real and imaginary parts of the refractive index of the coating, and its thickness. One classic solution to Equation (4.82) works very well when the materials 1 and 3 are not very absorbing. It is clear that Equation (4.82) is satisfied when $\tau_D^2 = -1$ and $\rho_{12} = \rho_{23}$. The requirement that $\tau_D^2 = -1$ means that two passes of the light through the TARC cause a 180° phase change with no absorption (since the magnitude is still one). From the definition of the internal transmittance, this means that the TARC thickness must be adjusted to a 'quarter wave':

$$D = \frac{\lambda}{4n_2} \qquad (4.102)$$

The requirement that $\rho_{12} = \rho_{23}$ will be satisfied when the index of refraction of the TARC is made to be

$$n_2 = \sqrt{n_1 n_3} \qquad (4.103)$$

Further, since the TARC does not absorb, the imaginary part of its index is zero. Thus, Equation (4.82) can only be true if both materials 1 and 3 have no imaginary parts to their indices of refraction. For a resist with a refractive index of 1.7 at a wavelength of 193 nm (in air), the optimum TARC will have a refractive index of about 1.30 and a thickness of 37 nm.

An antireflection layer defined by Equations (4.102) and (4.103) will have zero reflectivity. Also, due to the lack of absorption, 100% of the light striking the TARC will be transmitted into layer 3. Thus, this type of antireflection coating is commonly used for coating optical components (such as camera or stepper lenses) where the goal is not so much reducing the reflectivity as it is maximizing the transmittance. The 'maximum transmittance' type of antireflection coating is also used for top antireflection coatings. This perfect TARC solution, however, is only available for the special case when both layers 1 and 3 are transparent. Thus, since in reality resist will always be somewhat absorbing, a perfect TARC is not possible, with a residual reflectance off of the TARC proportional to the imaginary part of the resist refractive index squared. However, since resists have an imaginary part of the refractive index in the range of 0.01–0.02, the impact of resist absorption on TARC performance is quite negligible.

More problematically, it is very difficult to find practical materials with refractive indices low enough to make a near ideal TARC. Most available TARCs have refractive indices greater than 1.4 (versus an ideal index closer to 1.3). Letting n_{opt} be the optimum TARC refractive index given by Equation (4.103) and assuming a quarter wave thickness of TARC is always used, the reflectivity resulting from the use of a nonoptimum TARC refractive index is

$$R = \left[\frac{n_{TARC}^2 - n_{opt}^2}{n_{opt}^2 + n_{TARC}^2} \right]^2 \tag{4.104}$$

If the TARC refractive index is close to the optimum, this reflectivity is approximately

$$R \approx \left[\frac{n_{TARC} - n_{opt}}{n_{opt}} \right]^2 \tag{4.105}$$

Like BARCs, the performance of TARCs is degraded when a range in incident angles of light is used. Both the optimum TARC thickness and the optimum refractive index will be a function of incident angle and polarization.

4.5   Contrast Enhancement Layer

A *contrast enhancement layer* (CEL) is a highly bleachable coating placed on top of the photoresist that serves to enhance the contrast of an aerial image projected through it.[11] Unlike a TARC, where the real part of the refractive index is chosen to maximize transmittance and the imaginary part is kept as low as possible, the CEL works by having a high absorbance that 'bleaches', i.e. that becomes more transparent as it is exposed. The mechanism of CEL bleaching is similar to the bleaching that occurs in g-line and i-line photoresists, which will be covered in Chapter 5.

The CEL is initially opaque so that essentially no light is transmitted into the photoresist. Clear areas of the mask, and thus areas of high intensity of the aerial image, produce exposure of the CEL, which begins to bleach in these regions. As the CEL bleaches, light is transmitted into the photoresist where it can begin to expose the resist. In the regions of low intensity, however, the CEL remains more opaque and less transmitting. As a result, the image transmitted into the resist exhibits a steeper transition from bright to dark (Figure 4.27).

CELs are not commonly used for submicron semiconductor lithography. Their main drawback, besides the expense of the material and added processing, is the significant increase in exposure dose (in the range of 2–3 times higher) required to bleach the CEL.

4.6   Impact of the Phase of the Substrate Reflectance

The refractive index of a material is complex in general, the imaginary part being directly proportional to the absorption coefficient of the material. Thus, in general, the reflection coefficient between two materials will be a complex number. The magnitude of ρ determines the magnitude of the reflected light while its phase gives a phase change upon reflection. As we've seen, it is the magnitude of the reflection coefficient that impacts the amplitudes of standing waves and swing curves. But the phase will have an impact on

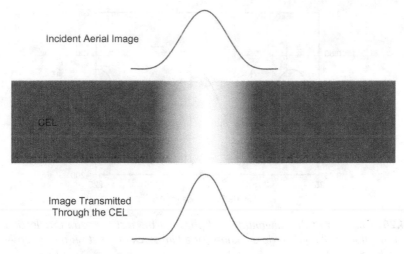

Figure 4.27   *Contrast Enhancement Layer (CEL) bleaching improves the quality of the aerial image transmitted into the photoresist*

lithography as well. Consider a typical resist on silicon at the i-line wavelength (365 nm). The magnitude of the reflection coefficient is about 0.63 with a phase of −169°. For resist on aluminum, $\rho = 0.93 \angle -138°$ (the notation meaning that the reflection coefficient has a magnitude of 0.93 and a phase angle of −138°). For more complicated layered substrates, both the magnitude and phase of the reflection coefficient will depend on the thicknesses of the various layers.

Consider a simple but common film stack: resist (1 μm thick) on silicon nitride (100 nm) on silicon dioxide (40 nm) on silicon. The reflection coefficient between the resist and the underlying film stack is a function of the optical properties of all of the materials, but also of the thickness of the nitride and the oxide films. For example, variation of the nitride thickness leads to a moderate variation in the magnitude and a large variation in the phase of the reflection coefficient, as shown in Figure 4.28. Oxide thickness variations produce similar effects.

What will be the lithographic effects of this nitride thickness variation? The ±7 % change in the magnitude of the reflection coefficient will have some subtle effects, but the large change in the phase of the reflection will cause two major problems. The resist swing curve is a function of the phase change of light that passes down and back up through the resist. Changes in resist thickness cause a change in this phase, giving rise to a sinusoidal variation in dose-to-clear (E_0) and linewidth. Any change in the phase upon reflection will produce the same effect [a variation in ϕ_{23} in Equation (4.51), for example]. Figure 4.29 shows two resist swing curves corresponding to two nitride thicknesses (minimum and maximum thicknesses from Figure 4.28). The nitride thickness variation produces its own swing curve, resulting in linewidth variations of the same magnitude as for resist thickness variations. Control of the nitride thickness (and oxide thickness, for that matter) is just as critical as resist thickness control.

The second effect of this phase change upon reflection is on the shape of the resulting resist profile. When the reflected light is 180° out of phase with the incident light,

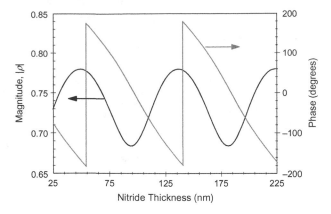

Figure 4.28   *Variation of the magnitude and phase of the resist/substrate reflection coefficient as a function of silicon nitride thickness for a film stack of resist on nitride on 40 nm of oxide on silicon*

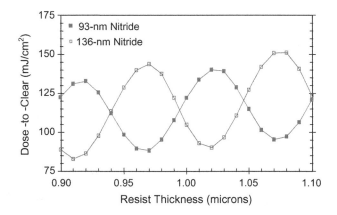

Figure 4.29   *Changes in nitride thickness cause a shift in the phase of the resist swing curve, making nitride thickness control as critical as resist thickness control*

destructive interference results in a minimum light intensity at the resist/substrate interface. When the reflected light is in phase with the incident light, constructive interference results in a maximum light intensity at the resist/substrate interface. Although standing waves are generally smoothed out by post-exposure bake diffusion, any asymmetry in the standing wave pattern inside the resist can lead to less than perfect reduction of the amplitude. When the phase change upon reflection is about $+90°$ or $-90°$, the local region near the interface has an average intensity that is higher or lower, respectively, than the average through the bulk. The result is resist undercutting and resist footing. Figure 4.30 shows typical resist profile shapes at different nitride thicknesses.

One can see that nitride thickness variations of just ±20 nm can have huge effects on both linewidth and resist profile shape. If oxide thickness can vary as well, the requirements for nitride thickness control become even tighter. In many cases, the

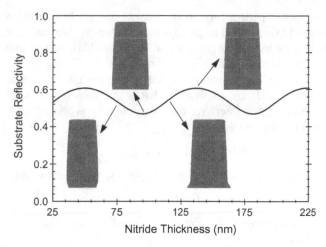

Nitride Thickness (nm)

Figure 4.30   *Nitride thickness also affects the shape of the resist profile, causing resist footing, undercuts or vertical profiles. Substrate reflectivity (the square of the magnitude of the reflection coefficient) is shown for comparison*

lithographic requirements for thin film thickness uniformity and control far exceed other device-related restrictions on these thicknesses. As with other swing curve effects, a BARC can be very effective at reducing the sensitivity to underlying film stack variations.

4.7   Imaging in Resist

The aerial image is, quite literally, the image in air. In the world of semiconductor lithography, it is the image of a photomask projected onto the plane of the wafer, but assuming that only air occupies this space rather than the resist-coated wafer. Although aerial images do not really exist in lithography, the aerial image, which is reasonably easy to calculate though very difficult to measure, is a convenient proxy for the final resist image. By picking an exposure-dose dependent intensity threshold, an estimate of the final resist CD is obtained (this essentially assumes that the resist is ideal, with an infinite contrast). Such uses are based on the idea that an aerial image is a good predictor of what the resist image will look like. However, at high numerical apertures the aerial image is in fact a very poor predictor of the final resist image. Due to vector effects, the image in resist can be greatly different from the aerial image, as will be discussed below.

4.7.1   Image in Resist Contrast

At high numerical apertures, the vector nature of light affects the formation of an image as a function of the polarization of the light and the angles of the various diffraction orders that add together to form the image (see Chapter 3). Two plane waves, approaching a wafer at different angles, will interfere to form fringe patterns of light and dark making the simplest of lithographic patterns – an array of lines and spaces. This phenomenon of interference is the key to pattern formation. Without it, light hitting the wafer from different directions would simply add to give a uniform intensity. We can see mathematically the effect of interference by examining how two electric fields combine to form a resultant electric field (magnitude and intensity). Ignoring a few details that will be described in the following section, the intensity of light I is the square of the magnitude of the electric field E. If two electric fields are combined, what is the intensity of the combination? If the two electric fields do not interfere, the total intensity is the sum of the individual intensities (i.e. it is a constant).

$$I = |E_1|^2 + |E_2|^2 \tag{4.106}$$

If, however, the two electric fields interfere completely, the total intensity will be

$$I = |E_1 + E_2|^2 \tag{4.107}$$

Here, the variation in the phase of E_1 relative to E_2 produces a spatial variation in intensity that is our image.

As discussed in detail in Chapters 2 and 3, two electric fields interfere only if two conditions are met: (1) there is a fixed phase relationship between the electric fields, and (2) there is some overlap in electric field direction. The first condition is met when the electric fields arise from diffraction orders created from a single source point. The second condition means that the two electric fields interfere with each other only to the extent that their electric fields oscillate in the same direction. If the electric fields are at right angles to each other, there will be no interference. Thus, to determine the amount of interference between two electric fields, one must first determine the amount of directional overlap between them. Standard vector mathematics gives us some simple tools to calculate directional overlap and thus the amount of interference. When thinking of light as a scalar rather than a vector quantity, we ignore the subtleties discussed above. Essentially, a scalar description of light always assumes that the electric fields are 100% overlapped and all electric fields add together as in Equation (4.107). Since interference is what gives us the patterns we want, a scalar view of light is generally too optimistic. The vector description of light says that there is usually some fraction of the electric fields of our two vectors that don't overlap and thus don't contribute to interference. The noninterfering light is still there, but it adds as a uniform intensity that degrades the quality of the image.

As was done in Chapter 3, consider the interference of two plane waves approaching the wafer at fairly large angles (Figure 4.31). The electric field can point in any direction perpendicular to the direction of propagation. This arbitrary direction can in turn be expressed as the sum of any two orthogonal (basis) directions. The most convenient basis directions are called transverse electric or TE (the electric field pointing out of the page of the drawing) and transverse magnetic or TM (the electric field pointing in the page of the drawing). The TE case means that the electric fields of the two plane waves are always 100% overlapped regardless of the angle between the plane waves. For the TM case, however, the extent of overlap between the two vectors grows smaller as the angle between the plane waves grows larger.

We can calculate how much the two electric fields for the TM case will interfere with each other. Suppose that two rays are traveling at an angle θ with respect to the vertical direction (i.e. the direction normal to the wafer), as shown in Figure 4.31. The electric

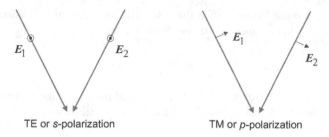

TE or s-polarization   TM or p-polarization

Figure 4.31 *Two plane waves with different polarizations will interfere very differently. For transverse electric (TE) polarization (electric field vectors pointing out of the page), the electric fields of the two vectors overlap completely regardless of the angle between the interfering beams*

fields E_1 and E_2 will have an angle between them of 2θ. The amount of the electric field vector E_2 that points in the same direction as E_1 is just the geometric projection, $E_2\cos(2\theta)$. Thus, the intensity will be given by the coherent (electric field) sum of the parts that overlap plus the incoherent (intensity) sum of the parts that don't overlap.

$$I = |E_1 + E_2\cos(2\theta)|^2 + |E_2\sin(2\theta)|^2 \tag{4.108}$$

Note that for $\theta = 0$, this equation reverts to the perfectly coherent (interfering) sum of Equation (4.107). For $\theta = 45°$, the two electric fields are perpendicular to each other and Equation (4.108) becomes the perfectly incoherent (noninterfering) sum of Equation (4.106). If we consider the simplest case of two unit amplitude plane waves,

$$E_1 = e^{i2\pi(z\cos\theta + x\sin\theta)/\lambda},\ E_2 = e^{i2\pi(z\cos\theta - x\sin\theta)/\lambda}$$
$$I_{TE}(x) = 2 + 2\cos(4\pi x\sin\theta/\lambda) \tag{4.109}$$
$$I_{TM}(x) = 2 + 2\cos(2\theta)\cos(4\pi x\sin\theta/\lambda)$$

Note that for both TE and TM illumination, the resulting images have no z-dependence. In other words, the simple two-beam imaging case has infinite depth of focus.

Two common image metrics, the normalized image log-slope (NILS) and the image contrast will now be calculated for the images in Equation (4.109). For the TE image, the NILS (assuming the desired image is equal lines and spaces) is equal to π, and for the TM image it is $\pi\cos(2\theta)$. The visibility (or contrast) of the resulting fringes is the difference of the maximum and minimum intensities of the interference pattern divided by the sum of these two quantities. For the TE polarization case, the contrast is always exactly 1. For the TM case, the contrast depends on the angle between the two waves and is equal to $\cos(2\theta)$. Thus, as the angle between the plane waves increases, the contrast of the resulting image decreases. One can see why the TM polarization is often called the 'bad' polarization – it provides less interference and reduced image quality.

The resist is not exposed by an aerial image, but by the image in resist. The plane waves that interfere to form the image first propagate into the resist. Once in the resist, they can interfere to form an image in the resist. This image will be different than the aerial image due to refraction. As each plane wave travels from air to resist, refraction lowers the angle of the light according to Snell's law. Letting n be the refractive index of the resist (and assuming the air above the resist has an index of refraction of 1.0), the angle of one of the plane waves inside the resist will be

$$\sin(\theta_{resist}) = \frac{1}{n}\sin(\theta_{air})$$ (4.110)

For the TE light, the interference will be the same and the two plane waves will produce a sinusoidal image of contrast 1. For the TM light, the contrast will be

$$TM\ Contrast = \cos(2\theta_{resist}) = 1 - 2\sin^2(\theta_{resist}) = 1 - \frac{2}{n^2}\sin^2(\theta_{air})$$ (4.111)

Figure 4.32 illustrates the difference between the NILS and the contrast in air versus in resist as a function of angle for TM light when the resist refractive index is 1.7. For an angle of 30° (corresponding to two-beam imaging at the resolution limit of a lens of a modest NA = 0.5), the aerial image for TM illumination has a contrast of 0.5, while the image in resist has a much more acceptable contrast of 0.83. Unpolarized light, which produces an average of the TE and TM images, produces a contrast that is also the average between 1 and the value given by Equation (4.111) (ignoring the difference in transmission of the two polarizations into the resist). Thus,

$$Unpolarized\ Contrast \approx 1 - \frac{1}{n^2}\sin^2(\theta_{air})$$ (4.112)

For the case of unpolarized two-beam imaging at the resolution limit of an NA = 0.9 lens, the aerial image would show a contrast of 0.19 (obviously unacceptable) while the image in resist would have a more reasonable contrast of 0.72. In fact, though, Equation (4.112)

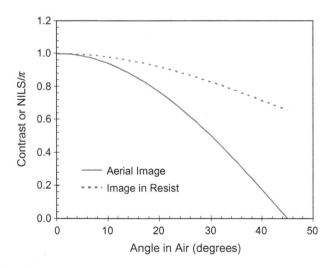

Figure 4.32   *The interference between two TM polarized plane waves produces an image whose contrast and NILS depend on the angle. Since the angle in resist is reduced by refraction, the contrast and NILS of the image in resist are better than those of the aerial image*

is optimistic since TM light will always transmit into the photoresist better than TE light. Using the data in Figure 4.5, for example, at an incident angle of 60° almost 100% of the TM light is transmitted into the resist, whereas only 75% of the high-contrast-producing TE light makes it into the resist.

What does all of this mean? When calculating aerial images, vector effects can dramatically alter the resulting image compared to an approximate scalar calculation. The resist, however, mitigates some of these vector effects and the image in resist can be dramatically different from the aerial image. When trying to approximate a resist feature by a calculated intensity image, only the image in resist (calculated correctly using the vector nature of light) can be expected to give reasonable results.

4.7.2 Calculating the Image in Resist

An aerial image is formed by the interference of plane waves that result from discrete diffraction orders passing through the lens. One of the simplest cases to consider is the imaging of small lines and spaces so that only the zero and the two first orders travel through the lens. For coherent illumination, the zero order will be a plane wave traveling in the z-direction, with magnitude a_0. The two first orders will be plane waves each with magnitudes a_1 and traveling at angles given by

$$n\sin\theta = \pm\lambda/p \tag{4.113}$$

where p is the pitch of the line/space pattern and n is the refractive index of the media. Ignoring for a moment the reflecting substrate, the image will be formed by the interference (sum) of these three plane waves. Assuming that the image is focused at $z = 0$ (the z-position where all three of the plane waves have the same phase), the resulting electric field of the image will be, for s-polarization,

$$E(x,z) = a_0 e^{ikz} + a_1 e^{ikx\sin\theta} e^{ikz\cos\theta} + a_1 e^{-ikx\sin\theta} e^{ikz\cos\theta} \tag{4.114}$$

This sum can be simplified into a more common and convenient form as

$$E(x, z) = e^{ikz}(a_0 + 2a_1 \cos(2\pi x/p)e^{-ikz(1-\cos\theta)})$$

giving

$$I(x,z) = a_0^2 + 2a_1^2 + 4a_0 a_1 \cos(kz(1-\cos\theta))\cos(2\pi x/p) + 2a_1^2 \cos(4\pi x/p) \tag{4.115}$$

Equation (4.115) is the standard s-polarized three-beam image where z can be interpreted as the distance from best focus. A similar expression for p-polarized illumination can also be derived:

$$E_x = e^{ikz}(a_0 + 2a_1 \cos\theta \cos(2\pi x/p)e^{-ikz(1-\cos\theta)})$$
$$E_z = ie^{ikz\cos\theta}(2a_1 \sin\theta \sin(2\pi x/p)) \tag{4.116}$$

giving

$$I(x, z) = a_0^2 + 2a_1^2 + 4a_0 a_1 \cos\theta \cos(kz(1-\cos\theta))\cos(2\pi x/p)$$
$$+ 2a_1^2 \cos(2\theta)\cos(4\pi x/p) \tag{4.117}$$

To determine the aerial image in the presence of a reflecting substrate, the same procedure is followed as above, but including the reflected plane waves as well. The result is, for *s*-polarization,

$$E(x, z) = e^{ikz}(a_0[1 + \rho(0)e^{-ik2z}] + 2a_1 \cos(2\pi x/p)e^{-ikz(1-\cos\theta)}[1 + \rho(\theta)e^{-ik2z\cos\theta}]) \quad (4.118)$$

In this equation, interference between plane waves causes a variation of the electric field in the *x*-direction (the image) and a variation in the *z*-direction (a combination of defocus and standing waves). In fact, one can think of each plane wave as creating its own electric field standing wave, with the final image as the weighted sum of these standing waves:

$$E_{sw}(\theta,z) = e^{ikz\cos\theta} + \rho(\theta)e^{-ikz\cos\theta}$$
$$E(x,z) = a_0 E_{sw}(0,z) + 2a_1 \cos(2\pi x/p)E_{sw}(\theta,z) \quad (4.119)$$

The above image in resist analysis captured the essence of the physics involved, but ignored the air–resist interface and the multiple reflections that would result. However, the final form of Equation (4.119) suggests that the more detailed standing wave expression of Equation (4.32) can be used as E_{sw} in Equation (4.119) to get an accurate image in resist result. Alternately, the diffraction order amplitudes a_0 and a_1 can be replaced by their effective 'in-resist' amplitudes:

$$a_{0r} = a_0 \frac{\tau_{12}(0)}{1 + \rho_{12}(0)\rho_{23}(0)\tau_D^2(0)}$$
$$a_{1r} = a_1 \frac{\tau_{12}(\theta)}{1 + \rho_{12}(\theta)\rho_{23}(\theta)\tau_D^2(\theta)} \quad (4.120)$$

and the simplified standing wave *z*-dependence would become

$$E_{sw}(\theta,z) = e^{ikz\cos\theta} + \rho(\theta)\tau_D^2(\theta)e^{-ikz\cos\theta} \quad (4.121)$$

In these two equations, θ now represents the angle in the resist.

Calculating the intensity in the resist from Equation (4.121) results in cumbersome algebra. However, for the case of a low-reflectivity substrate (through the use of a reasonably good BARC, for example), some of the resulting terms can be ignored, giving

$$I(x,z) \approx |a_{0r}|^2 I_{SW}(0, z) + 4|a_{1r}|^2 I_{SW}(\theta,z)\cos^2(2\pi x/p)$$
$$+ 4|a_{0r}a_{1r}|e^{-\alpha_{eff}z} \cos(2\pi n_2 z(1 - \cos\theta)/\lambda)\cos(2\pi x/p) \quad (4.122)$$

where $I_{sw}(\theta,z) = |E_{sw}(\theta,z)|^2$ and $\alpha_{eff} = \dfrac{\alpha}{2}\left(1 + \dfrac{1}{\cos\theta}\right)$.

Each diffraction order (plane wave) produces a standing wave. The interaction between the zero and first orders produces the interference pattern that forms the image, but the difference in propagation angles also creates a defocus term that accounts for defocusing of the image through the resist thickness. The terms that were ignored in this expression describe the defocusing of the reflected waves.

4.7.3   Resist-Induced Spherical Aberrations

Even with TE-polarized light and no substrate reflection, the resist causes important changes to the nature of the image. Consider Equation (4.122), three-beam imaging for TE illumination, for the case of no substrate reflection:

$$I(x,z) = |a_{0r}|^2 \, e^{-\alpha z} + 4|a_{1r}|^2 \, e^{-\alpha z/\cos\theta} \cos^2(2\pi x/p)$$
$$+ 4|a_{0r}a_{1r}| e^{-\alpha_{\text{eff}} z} \cos(2\pi n_2 \delta(1 - \cos\theta)/\lambda) \cos(2\pi x/p) \tag{4.123}$$

where $\delta = z - z_0$, and z_0 is the position of the plane of best focus. As a comparison, consider the same case for an aerial image:

$$I_{\text{air}}(x,z) = |a_0|^2 + 4|a_1|^2 \cos^2(2\pi x/p)$$
$$+ 4|a_0 a_1| \cos(2\pi n_1 \delta(1 - \cos\theta)/\lambda) \cos(2\pi x/p) \tag{4.124}$$

There are three important and distinct differences between the image in air and the image in resist. First, as given in Equation (4.120), the amplitude of each diffraction order is different in the resist compared to air, and for the case of the first order this difference is angle, and thus pitch, dependent. Second, each diffracted order is absorbed differently also according to its angle, and thus pitch. In fact, an effective z-dependent diffraction order can be defined as

$$a_{0rz} \equiv a_{0r} e^{-\alpha z/2}, \quad a_{1rz} \equiv a_{1r} e^{-\alpha z/(2\cos\theta)} \tag{4.125}$$

so that the image in resist becomes

$$I(x,z) = |a_{0rz}|^2 + 4|a_{1rz}|^2 \cos^2(2\pi x/p)$$
$$+ 4|a_{0rz}a_{1rz}| \cos(2\pi n_2 \delta(1 - \cos\theta)/\lambda) \cos(2\pi x/p) \tag{4.126}$$

As the image propagates further into the resist, the first orders are attenuated due to absorption to a greater degree than the zero order. This difference is more acute for larger angles (smaller pitches).

A more subtle but important difference between the images in air and resist is the defocus term, which includes the normalized optical path difference term $n\delta(1 - \cos\theta)/\lambda$. Both the refractive index and the angle θ are different in resist than in air. The results are depicted graphically in Figure 4.33, where larger angles (corresponding to smaller pitches) are focused further into the resist than smaller angles. First, the plane of best focus is shifted due to refraction. Further, there is a pitch dependence to best focus, a characteristic of spherical aberration.

To compare the defocus term in air to the case in resist, we can express the $\cos\theta$ as a sine and expand the resulting square root in a Taylor series.

$$OPD = n\delta(1 - \cos\theta) = n\delta\left(1 - \sqrt{1 - \sin^2\theta}\right) = \frac{1}{2}n\delta\left(\sin^2\theta + \frac{\sin^4\theta}{4} + \ldots\right) \tag{4.127}$$

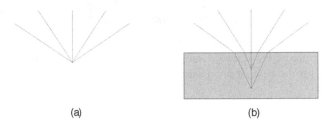

Figure 4.33 *Focusing of plane waves arriving at different angles (a) in air, and (b) in resist, showing that the resist induces spherical aberration*

Using Snell's law to compare the angles and refractive indices in air and resist,

$$OPD_{\text{air}} = \frac{1}{2} n_1 \delta \left(\sin^2 \theta_1 + \frac{\sin^4 \theta_1}{4} + \ldots \right)$$

$$OPD_{\text{resist}} = \frac{1}{2} n_2 \delta \left(\sin^2 \theta_2 + \frac{\sin^4 \theta_2}{4} + \ldots \right) = \frac{1}{2} \frac{n_1}{n_2} n_1 \delta \left(\sin^2 \theta_1 + \frac{n_2^2}{n_1^2} \frac{\sin^4 \theta_2}{4} + \ldots \right)$$

(4.128)

and

$$\frac{OPD_{\text{resist}}}{OPD_{\text{air}}} = \frac{n_1}{n_2} \frac{\left(1 + \frac{\sin^2 \theta_2}{4} + \ldots \right)}{\left(1 + \frac{\sin^2 \theta_1}{4} + \ldots \right)}$$

(4.129)

In the paraxial limit, where only the lowest order term in the series above is kept, the impact of defocus is reduced in resist compared to that in air by the factor n_1/n_2. The distance of the plane of best focus relative to the top of the resist is scaled by this factor. Also, the defocusing of the image through the resist is scaled by this amount. Since $n_2 > n_1$, the plane of best focus is shifted down into the resist, and the thickness of resist causes less defocusing than when the image propagates through the same thickness of air.

To see how the resist induces spherical aberration, we need only keep one more term in the series of Equation (4.129).

$$OPD_{\text{resist}} \approx \frac{n_1}{n_2} OPD_{\text{air}} - \delta \left[\frac{n_1^2 (n_2^2 - n_1^2)}{8 n_2^3} \right] \sin^4 \theta_1$$

(4.130)

The added term varies as the radial position in the pupil to the fourth power, characteristic of 3rd order spherical aberration. Comparing to this Zernike term as given in Chapter 3, an effective amount of 3rd order spherical induced by the resist would be

$$Z_8 \approx -\frac{z}{\lambda} \left[\frac{n_1^2 (n_2^2 - n_1^2)}{24 n_2^3} \right] NA^4$$

(4.131)

Consider a typical case where $n_1 = 1$ and $n_2 = 1.7$:

$$Z_8 \approx -0.016 \frac{z}{\lambda} NA^4 \qquad (4.132)$$

For an NA of 0.9, 10 milliwaves of spherical aberration is induced per wavelength of resist thickness. Since resists tend to be between one and two wavelengths thick for such a high NA, the center of the resist can be expected to show between 5 and 10 milliwaves of effective spherical aberration. Since this resist-induced spherical aberration is systematic, it is possible to correct for it by adding an equivalent amount of oppositely directed spherical aberration into the lens design.

4.7.4   Standing Wave Amplitude Ratio

Our goal is now to understand how reflectivity, as well as the variation of reflectivity with angle, affects standing waves in the image. To begin, we'll define a metric called the standing wave amplitude ratio (SWAR), given by

$$SWAR = \frac{I_{max} - I_{min}}{I_{max} + I_{min}} \approx \frac{I_{max} - I_{min}}{2I(\rho = 0)} \qquad (4.133)$$

Here, $I(\rho = 0)$ denotes the intensity that would be present if there were no reflecting substrate and is approximately the average of the min and max intensities. For a single plane wave, such as Equation (4.31), the SWAR is equal to twice the substrate amplitude reflection coefficient $\rho(\theta)$, and would be zero if the BARC were perfect. In fact, one goal of BARC design is to make the SWAR as close to zero as possible.

For the imaging case, pulling the maximum and minimum values out of Equation (4.118) or Equation (4.122) is complicated by two factors: the defocus dependence of the image, and the fact that each diffraction order produces a standing wave of a different period. If, however, the variation of the image intensity with z caused by defocus is small compared to the standing wave variation (i.e. if the depth of focus is much larger than one standing wave half-period), and best focus is near the position in the resist where the SWAR is being evaluated, this effect can be ignored. Further, by looking at the standing waves at the bottom of the resist, the mismatch in standing wave peak positions will be at their minimum. Under these circumstances, an approximate value for the SWAR can be obtained for this three-beam imaging case. For the simplified case ignoring absorption and multiple reflections in the resist,

$$SWAR \approx 2 \frac{a_o^2 |\rho(0)| + 4a_o a_1 \cos(2\pi x/p)\left(\frac{|\rho(0)| + |\rho(\theta)|}{2}\right) + 4a_1^2 \cos^2(2\pi x/p)|\rho(\theta)|}{a_o^2 + 4a_o a_1 \cos(2\pi x/p) + 4a_1^2 \cos^2(2\pi x/p)} \qquad (4.134)$$

It is very interesting to note that the standing wave amplitude ratio is a function of x. Consider $x = 0$, the center of the space of the line/space pattern. Here, the SWAR is an average of the zero- and first-order reflectivities, weighted in a certain way by the amplitudes of the diffracted orders.

$$SWAR(x = 0) \approx 2 \frac{|\rho(0)|(a_o^2 + 2a_o a_1) + |\rho(\theta)|(4a_1^2 + 2a_o a_1)}{a_o^2 + 4a_o a_1 + 4a_1^2} \qquad (4.135)$$

At $x = p/4$, corresponding to the nominal edge of the feature, the SWAR takes on a very different value.

$$SWAR(x = p/4) \approx 2|\rho(0)| \qquad (4.136)$$

In other words, the standing waves at the edge of the feature are controlled only by the reflectivity of the zero order. Figure 4.34 shows an example of how the standing wave amplitude ratio varies with position along the line/space features. Note that Equations (4.134)–(4.136) and Figure 4.34 all assume a nonabsorbing medium. A real resist, with its reasonably high absorption, will reduce the actual SWAR significantly. Equation (4.134) can be modified to account for absorption and multiple reflections in a real resist if the diffraction order amplitudes a_0 and a_1 are replaced by their in-resist values a_{0r} and a_{1r}, and the substrate reflectivity ρ is replaced by

$$\rho(\theta)\tau_D^2(\theta)e^{-\alpha z/\cos\theta} \qquad (4.137)$$

While the mathematics derived above apply to a fairly simple case, the results are appropriate for considering the approach toward optimizing a bottom antireflection coating for three-beam, high-numerical-aperture imaging (for two-beam imaging there is no ambiguity, since there is only one angle to optimize for). Equation (4.135) shows that standing waves in the middle of the space are controlled by a weighted average of the reflectivities of the zero and first orders (i.e. at zero angle and at the angle of the first orders). However, when considering the more important standing waves at the edge of the feature, only the zero-order reflectivity matters. While Equation (4.136) applies to both s- and p-polarized illumination, Equation (4.135) can be modified to approximate p-polarization by replacing a_1 with $a_1\cos\theta$ (to account for reduced interference creating the image) and by adding a $\cos 2\theta$ term to the weighting of the high-angle reflectivity (to account for reduced interference in the standing wave formation). Thus, for p-polarization,

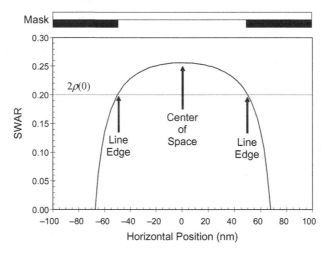

Figure 4.34 *The standing wave amplitude ratio (SWAR) at different positions on the feature for coherent three-beam imaging and s-polarization. For this example of three-beam imaging of 100-nm lines and spaces, $a_o = 0.5$, $a_1 = 0.3183$, $|\rho(0)| = 0.1$ and $|\rho(\theta)| = 0.15$*

even the SWAR in the center of the space becomes more heavily weighted to the normally incident reflectivity due to the reduced interference of the large-angle first diffracted orders. It seems that designing a BARC by reducing the normally incident reflectivity provides a very good starting point for obtaining the best standing wave control. Beyond this initial design, full lithographic simulation will be required.

4.8 Defining Intensity

While seemingly simple in concept, the definition of light intensity is more complicated than expected. In particular, a comparison of intensity values when the light is in different materials and traveling at different angles requires careful consideration. One case where these difficulties become apparent is the simple refraction of a plane wave traveling from one medium to another. Thus, our discussion will begin with a look at electric field and intensity reflection and transmission coefficients. The following derivations are based on the standard treatment given in Born and Wolf.[12]

4.8.1 Intensity at Oblique Incidence

When determining the intensity (also called irradiance) transmitted into a material at an oblique angle, it is very important to understand the exact definition of intensity. By definition (and as was discussed briefly in Chapter 2), the *intensity* of light is the magnitude of the (time averaged) Poynting vector, the energy per second crossing a unit area perpendicular to the direction of propagation of the light. In a medium of refractive index n the intensity I is given by

$$I = n|E|^2 \tag{4.138}$$

where E is the electric field. Note that the definition given in Equation (4.138) may differ by a constant multiplicative factor depending on the units used.

The intensity reflectivity and transmission, for either polarization, are derived by considering a unit area on the interface between two materials labeled 1 and 2. Consider the projected power per unit area, J, of the incident light along the surface of the interface between the materials. If the angle of incidence is θ_1,

$$J_i = I_i \cos(\theta_1) \tag{4.139}$$

Likewise, the projected power per unit area of the reflected and transmitted light along this surface are

$$\begin{aligned} J_r &= I_r \cos(\theta_1) \\ J_t &= I_t \cos(\theta_2) \end{aligned} \tag{4.140}$$

Now the reflectivity and transmission coefficients for each polarization can be defined:

$$\begin{aligned} R_{12} = R_{21} = R &= \frac{J_r}{J_i} = |\rho_{12}|^2 \\ T_{12} = T_{21} = T &= \frac{J_t}{J_i} = \left|\frac{n_2 \cos(\theta_2)}{n_1 \cos(\theta_1)}\right| |\tau_{12}|^2 \end{aligned} \tag{4.141}$$

From these two equations it is easy to show that $R + T = 1$ for each polarization, which is a consequence of conservation of energy. Figure 4.5 shows how the intensity reflectivity varies with incident angle for both *s*- and *p*-polarized illumination for a typical air–resist interface. An alternate form for Equation (4.141), making use of the reverse direction definitions of reflection and transmission coefficients in Equation (4.38), are

$$R = |\rho_{12}\rho_{21}|$$
$$T = |\tau_{12}\tau_{21}|$$

(4.142)

Also note that for randomly polarized light, the overall intensity reflectivity and transmittance can be computed from the values obtained for each polarization:

$$R = R_{\parallel}\cos^2\theta_1 + R_{\perp}\sin^2\theta_1$$
$$T = T_{\parallel}\cos^2\theta_1 + T_{\perp}\sin^2\theta_1$$

(4.143)

Consider a unit intensity plane wave incident on the plane boundary between materials 1 and 2 at an incident angle θ_1 and with intensity I_i. From Equation (4.138), the magnitude of the incident electric field must be

$$|E_i| = \sqrt{\frac{I_i}{n_1}}$$

(4.144)

The transmitted electric field is then

$$|E_t| = |\tau_{12}E_i| = |\tau_{12}|\sqrt{\frac{I_i}{n_1}}$$

(4.145)

The transmitted intensity (i.e. the intensity in material 2) is found by applying the definition of intensity to Equation (4.145).

$$I_t = n_2|E_t|^2 = \frac{n_2}{n_1}|\tau_{12}|^2 I_i$$

(4.146)

By comparing Equation (4.146) with Equation (4.141), the somewhat nonintuitive result below is obtained.

$$\frac{I_t}{I_i} = T\frac{\cos(\theta_1)}{\cos(\theta_2)}$$

(4.147)

As can be seen in Equation (4.147), the transmittance T is not the ratio of the intensities I_t and I_i (see Figure 4.35). The difference comes from the change in the direction of the energy flow caused by refraction. Thus, one might ask the question, which is more important to know inside film 2, the intensity of the plane wave, or its projected power along a surface parallel to the material interface? The answer to this question depends on the task at hand, as discussed in section 4.8.3.

4.8.2   Refraction into an Absorbing Material

The well-known Snell's law of refraction [Equation (4.30)] dictates how a plane wave, traveling across a plane boundary between two materials of different optical properties,

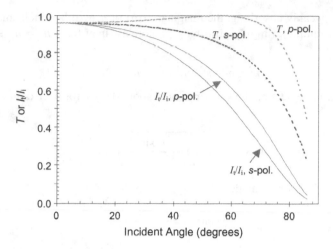

Figure 4.35 *Intensity transmitted into layer 2 relative to the incident intensity (solid lines) and the transmittance* T *(dashed lines) as a function of the angle of incidence for both s- and p-polarization ($n_1 = 1.0$, $n_2 = 1.5$)*

will change its direction. The material boundary conditions required by Maxwell's equations state that any electric field tangential to the material interface must be continuous across that interface. This requirement will lead directly to the derivation of Snell's law. When both refractive indices are real (i.e. neither material is absorbing), the interpretation of Snell's law is quite straightforward. However, if one or both of the materials are absorbing (for example, light traveling from air into photoresist), one or both of the angles will be complex. How is a complex angle of transmittance to be interpreted? To answer this question, a more rigorous treatment of refraction and absorption is required.

Assuming a unit amplitude plane wave, the electric field can be expressed mathematically in two dimensions as

$$E_i(x,z) = \exp(i2\pi n_1(x\sin\theta_1 + z\cos\theta_1)/\lambda) \qquad (4.148)$$

where the interface between the two materials lies in the $x - y$ plane, positive x is up and positive z is to the right (as seen in Figure 4.4). Here we shall assume for simplicity that material 1 is nonabsorbing. The transmitted electric field will be

$$E_t(x,z) = \tau_{12}\exp(i2\pi \boldsymbol{n}_2(x\sin\theta_2 + z\cos\theta_2)/\lambda) \qquad (4.149)$$

where the appropriate expression for τ_{12} is used depending on the polarization of the incident plane wave. If material 2 is absorbing, the angle θ_2 will be complex. The resulting mathematics are quite cumbersome when trying to express all quantities as real numbers. For a weakly absorbing film, however, some simplifications are possible. The optical distance in Equation (4.149) is

$$OD = \boldsymbol{n}_2(x\sin\theta_2 + z\cos\theta_2) \qquad (4.150)$$

Applying Snell's law to eliminate the complex angle θ_2,

$$\boldsymbol{n}_2\cos\theta_2 = \boldsymbol{n}_2\sqrt{1 - \sin^2\theta_2} = \sqrt{\boldsymbol{n}_2^2 - (n_1\sin\theta_1)^2} \qquad (4.151)$$

If material 2 is weakly absorbing, the imaginary part of n_2 will be much smaller than the real part (for example, a highly absorbing photoresist may have $n_2 = n_2 + i\kappa_2 = 1.7 + i0.02$). Thus,

$$n_2^2 = n_2^2 - \kappa_2^2 + i2n_2\kappa_2 \approx n_2^2 + i2n_2\kappa_2 \qquad (4.152)$$

where the error in neglecting the κ_2^2 term is only about 0.01 % for the example photoresist. Thus,

$$n_2 \cos\theta_2 \approx n_2 \sqrt{1 - \left(\frac{n_1 \sin\theta_1}{n_2}\right)^2 + i\frac{2\kappa_2}{n_2}} \qquad (4.153)$$

Now let us define a new angle called θ_2' given by

$$\sin(\theta_2') = \frac{n_1 \sin(\theta_1)}{n_2} \qquad (4.154)$$

This new angle is not the actual transmitted angle, but is the angle of transmission that would be obtained if material 2 were nonabsorbing. For a weakly absorbing material, it is approximately the angle of transmission. Using this angle in Equation (4.153),

$$n_2 \cos\theta_2 \approx n_2 \sqrt{\cos^2\theta_2' + i\frac{2\kappa_2}{n_2}} = n_2 \cos\theta_2' \sqrt{1 + i\frac{2\kappa_2}{n_2 \cos^2\theta_2'}} \qquad (4.155)$$

Expanding the square root into a Taylor series,

$$\sqrt{1 + i\frac{2\kappa_2}{n_2 \cos^2\theta_2'}} = 1 + i\frac{\kappa_2}{n_2 \cos^2\theta_2'} + \frac{1}{2}\left(\frac{\kappa_2}{n_2 \cos^2\theta_2'}\right)^2 - \frac{i}{2}\left(\frac{\kappa_2}{n_2 \cos^2\theta_2'}\right)^3 + \dots \qquad (4.156)$$

If only the first two terms of the Taylor series are kept, both the real and imaginary parts of the result will have a relative error of approximately $\dfrac{1}{2}\left(\dfrac{\kappa_2}{n_2 \cos^2\theta_2'}\right)^2$. Using again the optical properties of photoresist mentioned above and assuming a maximum incident angle of 90° in air, neglecting this term will provide less than 0.02 % error. Even if extreme immersion lithography is considered, the larger angle inside the resist will cause this error to grow to only 0.04 %. Thus, there is very little error in dropping all but the first two terms in Equation (4.156). The final result for Equation (4.155) is

$$n_2 \cos\theta_2 \approx n_2 \cos\theta_2' + i\frac{\kappa_2}{\cos\theta_2'} \qquad (4.157)$$

Thus, the plane wave traveling in material 2 will have an optical path of approximately

$$OD \approx xn_1 \sin\theta_1 + z\left(n_2 \cos\theta_2' + i\frac{\kappa_2}{\cos\theta_2'}\right) = n_2(x\sin\theta_2' + z\cos\theta_2') + i\frac{z\kappa_2}{\cos\theta_2'} \qquad (4.158)$$

and the electric field will be

$$E_t(x,y,z) \approx \tau_{12} \exp(i2\pi n_2(x\sin\theta_2' + z\cos\theta_2')/\lambda)\exp(-2\pi\kappa_2(z/\cos\theta_2')/\lambda) \qquad (4.159)$$

For a weakly absorbing film, the phase term of the electric field is approximately equal to the plane wave that would result for a nonabsorbing film with the same real part of the refractive index. The absorption term in Equation (4.159) provides the somewhat intuitive result that larger angles of propagation with respect to the z-axis give faster attenuation of the light along the z-axis. Because the surfaces of constant phase are not parallel to the surfaces of constant amplitude (which instead are parallel to the interface between the two materials), such a wave is called an *inhomogeneous wave*.

Applying the definition of intensity,

$$I_t(x,y,z) \approx n_2|\tau_{12}|^2 \exp(-\alpha z/\cos\theta_2') \qquad (4.160)$$

where $\alpha = 4\pi\kappa/\lambda$ is the absorption coefficient. Note that the transmitted plane wave has an intensity with only z-dependence.

4.8.3   Intensity and Absorbed Energy

In lithography, the intensity of light inside the resist film is required in order to know the energy absorbed per unit volume in the resist, which in turn determines the amount of chemical change that occurs during exposure. The standard approach to photoresist exposure kinetics makes use of the Dill equation (to be discussed in more detail in Chapter 5),

$$\frac{dm}{dt} = -CIm \qquad (4.161)$$

where m is the relative concentration of photosensitive material, C is the exposure rate constant, and t is the exposure time. Obviously, Equation (4.161) makes use of a specific definition of intensity. What definition is used, and why?

A detailed study of the microscopic mechanism of absorption and exposure can be used to relate the exposure rate constant C to the molar absorptivity of the photosensitive component of the resist (see Chapter 5):

$$C = \frac{\Phi a_m \lambda}{N_A hc} \qquad (4.162)$$

where Φ is the quantum efficiency, a_m is the molar absorptivity of the photosensitive component, λ is the vacuum wavelength, N_A is Avogadro's number, h is Planck's constant, and c is the speed of light. The important aspect of this relationship is the direct proportionality between the exposure rate constant and the molar absorptivity of the photosensitizer (which is the absorption coefficient divided by the concentration of photosensitizer). Thus, the rate of exposure in Equation (4.161) is directly proportional to the absorption coefficient times the intensity of light. However, this same rate of exposure is also directly proportional to the absorbed energy per unit volume of resist per second. After all, it is the absorbed energy that actually causes the chemical reactions to occur. Thus, our use of the intensity of light in the exposure rate Equation (4.161) must be based on the idea that

$$Absorbed\ Energy/volume/time \propto \alpha I \qquad (4.163)$$

Our definition of intensity, for it to be useful in this application, must satisfy Equation (4.163). In fact, this can only be true if intensity is defined as the power per unit area perpendicular to the direction of travel. Consider first Lambert's law of absorption:

$$\frac{dI}{ds} = -\alpha I \tag{4.164}$$

where s is the direction of propagation of light. The change in intensity with s is just the absorbed power per unit volume when I is defined as power per unit area perpendicular to the direction of propagation. Further tying this definition of intensity together with the standard formulations for electric field propagation, absorption is accounted for by the use of a complex refractive index. For a homogeneous film, the solution to Lambert's law is

$$I(x) = I_0 e^{-\alpha s} \tag{4.165}$$

In phasor representation, a plane wave traveling in the s direction is written as

$$E(s) = A e^{i2\pi ns/\lambda} \tag{4.166}$$

Equation (4.166) will yield Equation (4.165) by applying the definition of intensity as given in Equation (4.138) if the absorption coefficient is related to the imaginary part of the refractive index by

$$\alpha = 4\pi\kappa/\lambda \tag{4.167}$$

If the direction s makes an angle θ with respect to the z-axis, then $s = z/\cos\theta$ and

$$I(z) = n|A|^2 \exp(-\alpha z/\cos\theta) = I_0 \exp(-\alpha z/\cos\theta) \tag{4.168}$$

Thus, one can see that by defining intensity as the energy flux perpendicular to the direction of travel of the light, the use of an imaginary refractive index in the standard optics equations provides for a straightforward calculation of energy absorbed per unit volume of resist. It is the intensity within the resist film, not J, the intensity projected along a surface parallel to the resist and substrate, that determines the amount of chemical change during exposure.

Problems

4.1.   Imagine two plane waves moving in opposite directions along the z-axis. Using the time-dependent form of the waves rather than the phasor representation, the $+z$ moving E-field wave is given by

$$U_+(z,t) = A\cos(\omega t - kz)$$

and the $-z$ moving wave is

$$U_-(z,t) = A\cos(\omega t + kz)$$

Show that the sum of these waves is separable in z and t (that is, can be expressed as product of a function of z and a function of t). Show that this results in an intensity 'standing wave', that is, the time average of the square of the electric field varies sinusoidally with position.

4.2. Which of the following situations will result in the formation of standing waves? Explain your reasoning.
  (a) the superposition of identical waves that travel in the same direction;
  (b) the superposition of identical waves that travel in opposite directions;
  (c) the superposition of waves that are nearly identical but of different amplitudes that travel in opposite directions;
  (d) the superposition of nearly identical waves of slightly different frequencies that travel in the same direction.

4.3. Derive Equation (4.19).

4.4. The *optical admittance* of a material, η, at normal incidence is just its refractive index, n. For light traveling through that material at an angle θ, the optical admittance of the material for TE (*s*-polarized) and TM (*p*-polarized) waves are

$$\eta_{TE} = n\cos\theta \quad \eta_{TM} = \frac{n}{\cos\theta}$$

Show that by using optical admittance in the Fresnel formulas [Equation (4.29)], one obtains a single equation for reflection for both polarizations that has the same form as the normally incident definition of Equation (4.8). What are the results for transmittance and how do they compare to Equation (4.7)?

4.5. *Brewster's angle* is defined as the angle at which *p*-polarized light will have zero reflectance at the interface between two materials. Derive an equation for Brewster's angle. For *p*-polarized light incident on a photoresist-coated substrate at Brewster's angle, what is the resulting swing amplitude?

4.6. For *p*-polarized light traveling at 45° with respect to the normal inside of a photoresist, no standing waves will occur since the incident and reflected waves will have no overlap in their electric fields [see Equation (4.37)]. Will the resist swing curve similarly have zero amplitude at this angle? What if the light is traveling at 45° with respect to the normal in air? Explain.

4.7. Derive Equation (4.46) from Equation (4.45).

4.8. Derive an expression for the CD swing curve using Equations (4.51)–(4.53).

4.9. Derive Equation (4.84).

4.10. Calculate the swing ratio for a nominally 400-nm-thick 193-nm resist (assume typical properties) on a substrate of copper. What is the impact on this swing ratio if the maximum is at a greater thickness than the minimum, compared to the minimum at a greater thickness than the maximum?

4.11. Using the normal-incidence BARC solution given in Figure 4.16 for a 40-nm BARC thickness, what BARC thickness control is required to keep the BARC intensity reflectivity below 0.5%?

4.12. At 248 nm, a typical resist on silicon will result in a swing amplitude of about 0.69. Using a BARC with a reflectivity of 1%, the swing amplitude is reduced to about 0.095. Ignoring absorption, what level of resist thickness control is required for each of these two cases (resist on silicon and resist on a 1% reflective BARC) to keep the effective dose error below 1%.

4.13. Derive Equation (4.103), the refractive index of an ideal TARC.

4.14. As an example of the use of Equation (4.100), consider a 193-nm resist with $n_2 = 1.7$, $\alpha = 1.2\,\mu m^{-1}$, $D = 200\,nm$ and $|\rho_{12}| = 0.26$. Suppose that 2% dose error due to resist thickness variations was considered acceptable. Generate a plot of maximum allowed substrate reflectivity ($|\rho_{23}|^2$) as a function of the resist thickness variation Δ.

4.15. What will be the impact of using immersion lithography (versus dry lithography at the same numerical aperture) in terms of the contrast of the image-in-resist for unpolarized illumination? Explain.

4.16. Derive Equation (4.117) from Equation (4.116).

4.17. For light incident on a plane interface between two materials, use Equation (4.141) to show that $R + T = 1$ for each polarization.

References

1   Middlehoek, S., 1970, Projection masking, thin photoresist layers and interference effects, *IBM Journal of Research and Development*, **14**, 117–124.

2   Korka, J.E., 1970, Standing waves in photoresists, *Applied Optics*, **9**, 969–970.

3   Mack, C.A., 1986, Analytical expression for the standing wave intensity in photoresist, *Applied Optics*, **25**, 1958–1961.

4   Macleod, H.A., 1969, *Thin-Film Optical Filters*, American Elsevier Publishing (New York, NY).

5   Widmann, D.W., 1975, Quantitative evaluation of photoresist patterns in the 1 μm range, *Applied Optics*, **14**, 931–934.

6   Brunner, T.A., 1991, Optimization of optical properties of photoresist processes, *Proceedings of SPIE: Advances in Resist Technology and Processing VIII*, **1466**, 297–308.

7   Palik, E.D. (ed.), 1985–1998, *Handbook of Optical Constants of Solids*, Vol. I–III, Academic Press (New York, NY).

8   Brunner, T.A., 1999, Relationship between the slope of the HD curve and the fundamental resist contrast, *Journal of Vacuum Science and Technology B*, **17**, 3362–3366.

9   Tanaka, T., Hasegawa, N., Shiraishi, H. and Okazaki, S., 1990, A new photolithography technique with antireflective coating on resist: ARCOR, *Journal of the Electrochemical Society*, **137**, 3900–3905.

10 Brunner, T.A., 1991, Optimization of optical properties of resist processes, *Proceedings of SPIE: Advances in Resist Technology and Processing VIII*, **1466**, 297–308.

11 Griffing, B.F. and West, P.R., 1983, Contrast enhanced photolithography, *IEEE Electron Device Letters*, **EDL-4**, 14–16.

12 Born, M. and Wolf, E., 1980, *Principles of Optics*, sixth edition, Pergamon Press (Oxford, UK), p. 41.

5

Conventional Resists: Exposure and Bake Chemistry

Photoresists work by converting a spatial distribution of energy (the projected image of a photomask) into a spatial distribution of solubility of the resist in a developer. The first step in this chemical process is exposure, where a light-sensitive component of the resist forms a latent image in response to the aerial image. For so-called 'conventional' resists used for g-line and i-line exposures, the exposure products directly change the solubility of the resist. For a chemically amplified resist (discussed in the next chapter), a post-exposure bake is used to thermally induce a second chemical reaction, the product of which changes the solubility of the resist. In both cases, understanding the nature of the image transfer from a light distribution to a chemical distribution requires an understanding of the chemical kinetics of each of the reactions.

While this chapter deals with conventional resists, which are used only for the older g-line and i-line lithography processes, virtually all of the topics discussed here apply to chemically amplified resists as well. In a sense, chemically amplified resists are a technological superset of conventional resists. Thus, the discussion of chemically amplified resists in the next chapter will assume a familiarity with the material in this chapter.

5.1 Exposure

The kinetics of photoresist exposure is intimately tied to the phenomenon of absorption. The discussion below begins with a description of absorption, followed by the photochemical kinetics of exposure.

5.1.1 Absorption

The first law of photochemistry (also called the Grotthuss–Draper law) states quite simply that only light that is absorbed is effective at causing chemical change. Thus a photochemical reaction, such as the exposure of photoresist, must necessarily begin with the absorption of light. The phenomenon of absorption can be viewed on a macroscopic or a microscopic scale. On the macroscopic level, absorption is described by the familiar Lambert and Beer laws, which give a linear relationship between absorbance and path length multiplied by the concentration of the absorbing species. On the microscopic level, a photon is absorbed by an atom or molecule, promoting an electron to a higher energy state. Both methods of analysis yield useful information needed to describe the effects of light on a photoresist.

The basic law of absorption is an empirical one called Lambert's law. It was expressed by Johann Heinrich Lambert in 1760, though it was in fact first proposed some years earlier by Pierre Bouguer and is sometimes called the Lambert–Bouguer law. It can be expressed in differential form as

$$\frac{dI}{dz} = -\alpha I \tag{5.1}$$

where I is the intensity of light traveling in the z-direction through a medium, and α is the absorption coefficient of the medium and has units of inverse-length. This law is basically a single photon absorption probability equation: the probability that a photon will be absorbed over some small distance traveled is proportional to the photon flux (i.e. the intensity). In a homogeneous medium (i.e. α is not a function of z), Equation (5.1) may be integrated to yield

$$I(z) = I_0 e^{-\alpha z} \tag{5.2}$$

where z is the distance the light has traveled through the medium and I_0 is the intensity at $z = 0$. Note that the integrated form of the Lambert's law requires a homogeneous material. If the medium is inhomogeneous, Equation (5.2) becomes

$$I(z) = I_0 e^{-Abs(z)} \tag{5.3}$$

where $Abs(z) = \int_0^z \alpha(z')dz' = \text{the absorbance}$

When working with electromagnetic radiation, it is often convenient to describe the radiation by its complex electric field vector. The propagation of an electric field through some material can implicitly account for absorption by using a complex index of refraction n for the material such that

$$n = n + i\kappa \tag{5.4}$$

The imaginary part of the index of refraction κ is related to the absorption coefficient by

$$\alpha = 4\pi\kappa/\lambda \tag{5.5}$$

where λ is the vacuum wavelength of the propagating light. Note that the sign of the imaginary part of the index in Equation (5.4) depends on the sign convention chosen for the phasor representation of the electric field (see Chapter 2).

In 1852, August Beer showed that for dilute solutions of an absorbing material in a nonabsorbing solvent, the absorption coefficient is proportional to the concentration of the absorbing species in the solution:

$$\alpha_{\text{solution}} = a_i C_i \tag{5.6}$$

where a_i is the molar absorption coefficient (sometimes called the extinction coefficient) of some absorbing species labeled i (given by $a_i = \alpha_i MW/\rho$, where α_i is the absorption coefficient of the pure material, MW is its molecular weight, ρ is its density) and C_i is the concentration. The stipulation that the solution be dilute expresses a fundamental limitation of Beer's law. At high concentrations, where absorbing molecules are close together, the absorption of a photon by one molecule may be affected by a nearby molecule. Since this interaction is concentration dependent, it causes deviation from the linear relation of Equation (5.6). Also, an apparent deviation from Beer's law occurs if the real part of the index of refraction changes appreciably with concentration. Thus, the validity of Beer's law should always be verified experimentally over the concentration range of interest.

While generally described for solutions, Beer's law can also be applied to solid mixtures. For an N component homogeneous solid, the overall absorption coefficient becomes

$$\alpha_T = \sum_{i=1}^{N} a_i C_i \tag{5.7}$$

The linear addition of absorption terms presumes that Beer's law holds across components, i.e. that the absorption by one material is not influenced by the presence of the other materials. Of the total amount of light absorbed, the fraction of light that is absorbed by component i is given by

$$\frac{I_{Ai}}{I_{AT}} = \frac{a_i C_i}{\alpha_T} \tag{5.8}$$

where I_{AT} is the total light absorbed by the film, and I_{Ai} is the light absorbed by component i.

We will now apply the concepts of macroscopic absorption to a typical positive photoresist. For g-line (436 nm) and i-line (365 nm) lithography, the most common resists are of the diazonaphthoquinone/novolac variety. These positive photoresists are made up of three major components: a base novolac resin that gives the resist its structural properties and etch resistance; a photoactive compound (PAC; this is the light-sensitive component in the resist, also known as the *sensitizer*) called a diazonaphthoquinone (DNQ); and a solvent that renders these components into liquid form for spin coating. (Although photoresist drying during post-apply bake is intended to drive off solvents, a resist may contain up to 10% solvent after a typical post-apply bake – see section 5.2.2.) Denoting the resin R, the sensitizer M and the solvent S, a fourth component appears during exposure: exposure products P generated by the reaction of M with ultraviolet (UV) light.

$$M \xrightarrow{\quad UV \quad} P \tag{5.9}$$

Applying Beer's law, the absorption coefficient α is then

$$\alpha = a_M M + a_P P + a_R R + a_S S \tag{5.10}$$

Since P is being produced and M is disappearing through the exposure process, this equation makes it clear the α will be changing during exposure. If M_0 is the initial sensitizer concentration (i.e. with no exposure), the stoichiometry of the exposure reaction gives

$$P = M_0 - M \tag{5.11}$$

Equation (5.10) may be rewritten as

$$\alpha = Am + B \tag{5.12}$$

where $A = (a_M - a_P)\, M_0$
$\quad\quad B = a_P M_0 + a_R R + a_S S$
$\quad\quad m = M/M_0 =$ the relative sensitizer concentration.

A and B are called the bleachable and nonbleachable absorption coefficients, respectively, and make up the first two Dill photoresist parameters.[1] Other nonbleachable components of the photoresist (such as a dye additive, leveling agents to reduce striations during spin coating, etc.) are added to the B term above. Note that when $A > 0$, the photoresist will become more transparent as it is exposed (that is, the photoresist 'bleaches'). If A is negative, the resist will darken upon exposure.

The quantities A and B are experimentally measurable and can be easily related to typical resist absorbance curves, measured using a UV spectrophotometer. When the resist is fully exposed, $M = 0$ and

$$\alpha_{\text{exposed}} = B \tag{5.13}$$

Similarly, when the resist is unexposed, $m = 1$ $(M = M_0)$ and

$$\alpha_{\text{unexposed}} = A + B \tag{5.14}$$

From this, A may be found by

$$A = \alpha_{\text{unexposed}} - \alpha_{\text{exposed}} \tag{5.15}$$

Thus, $A(\lambda)$ and $B(\lambda)$ may be determined from the UV absorbance curves of unexposed and completely exposed resist (Figure 5.1). A more complete description of the measurement of A and B will be given at the end of this chapter.

As mentioned previously, Beer's law is empirical in nature and consequently should be verified experimentally. In the case of positive photoresists, this means formulating resist mixtures with differing sensitizer to resin ratios and measuring the resulting A parameters. Previous work has shown that Beer's law was valid for a typical conventional photoresist over the full practical range of sensitizer concentrations.[2]

5.1.2 Exposure Kinetics

The kinetics of exposure is best understood by considering absorption on the microscopic scale. On a microscopic level, the absorption process can be thought of as photons being absorbed by an atom or molecule causing an outer electron to be promoted to a higher energy state. The absorption of a photon can have three possible results: the absorbed energy can be dissipated as heat (lattice vibrations); it can be re-emitted as another photon (fluorescence); or it can lead to a chemical reaction. It is the third outcome

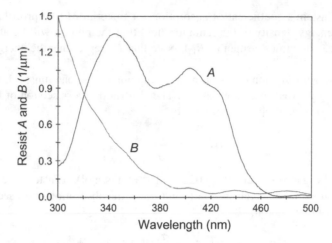

Figure 5.1 *Resist parameters A and B as a function of wavelength measured with a UV spectrophotometer for a typical g-line resist (a 5-arylsulfonate DNQ)*

that is important for the sensitizer since it is the absorption of UV light that leads to the chemical conversion of M to P as seen in Equation (5.9). Einstein first quantified absorption on a microscopic (quantum) scale in 1916–1917 by using probabilities of possible absorption events. Here, we will ignore stimulated absorption by assuming a reasonably low light intensity so that at any given time relatively few molecules will be in the excited state.

Consider a substance M being exposed to quasi-monochromatic radiation. Since the speed of light is constant in a given material, the Lambert law can be converted to a kinetic absorption equation by letting $dz = vdt$ where v is the speed of light (the rate at which new photons arrive). The rate of photon absorption by the molecule M is proportional to N_M (the number density of M), the number of photons, and the molecule's absorption cross section σ_{M-abs}:

$$-\frac{d\varphi_M}{dt} = vN_M\sigma_{M-abs}\varphi_M \tag{5.16}$$

where φ_M is the number density of photons. The intensity I is an energy flux – the amount of energy per unit area per second – and can be related to the photon flux (the photon density times the speed the photons are traveling) multiplied by the energy of one photon:

$$I = v\varphi_M\left(\frac{hc}{\lambda}\right) \tag{5.17}$$

where h is Planck's constant and c is the speed of light in vacuum. Einstein's relation for photon absorption is

$$-\frac{d\varphi_M}{dt} = N_M\sigma_{M-abs}\left(\frac{\lambda}{hc}\right)I = N_M B_E I/v \tag{5.18}$$

where B_E is Einstein's coefficient of absorption for material M (the probability per unit time per unit energy density of the radiation field that the photon will be absorbed by a molecule of M, sometimes written as B_{12}). Note that I/v represents the energy density of the photons.

In order to relate Equation (5.18) to measurable quantities, one must relate the microscopic theory to macroscopic observations. The absorption cross section of the molecule M can be related to its molar absorptivity by

$$\sigma_{M-abs} = \frac{\alpha_M}{N_M} = \frac{a_M}{N_A} \tag{5.19}$$

where N_A is Avogadro's number (6.022×10^{23} atoms/mol). Einstein's coefficient of absorption for material M is the speed of light times the absorption cross section divided by the energy of one photon:

$$B_E = v\sigma_{M-abs}\left(\frac{\lambda}{hc}\right) = \frac{\sigma_{M-abs}}{n}\left(\frac{\lambda}{h}\right) = \frac{a_M\lambda}{nN_A h} \tag{5.20}$$

where n is the refractive index of the resist.

As an example of the use of this microscopic absorption analysis, consider a typical g-line resist of density 1.2 g/ml. For a PAC with molecular weight 400 g/mol formulated at 30 % by weight, the initial PAC concentration will be 0.9 molar (moles/liter). Since a_P is very small for these resists, a_M is about equal to the Dill bleachable absorption coefficient divided by the initial PAC concentration. For $A = 0.63\,\mu m^{-1}$, this gives $a_M = 7 \times 10^6\,cm^2/mol$. The resulting PAC absorption cross section is about 0.12 Å². Since the energy of one photon at this wavelength is 4.56×10^{-19} J, the resulting Einstein coefficient of absorption is about $4.7 \times 10^5\,m^3/J$-s.

The actual chemistry of diazonaphthoquinone exposure is given by[3]

$$\tag{5.21}$$

where R is a moderate to large molecular weight ballast group (in some cases R is connected to the polymer resin). The DNQ converts to an indene intermediate before converting to its final product, a carboxylic acid. It is interesting to note that nitrogen gas is given off (it diffuses quickly out of the film during exposure) and that water is required for this reaction. Water is supplied by the humidity in the atmosphere, which is one reason why clean room humidity levels are tightly controlled (usually at a set point of around 40 % relative humidity). If the relative humidity of the clean room environment drops below about 30 %, the above DNQ exposure reaction will not go to completion. Instead, the intermediate ketene will slowly react with the novolac resin to form an ester (more on this in section 5.2.1).

The chemical reaction in Equation (5.9) can be rewritten in a more general form as

$$M \underset{k_2}{\overset{k_1}{\rightleftharpoons}} M* \overset{k_3}{\longrightarrow} P \tag{5.22}$$

where M is the sensitizer, M^* is the sensitizer molecule in an excited state, P is the indene carboxylic acid (product), and k_1, k_2, k_3 are the rate constants for each reaction. Simple kinetics can now be applied. The proposed mechanism in Equation (5.22) assumes that all reactions are first order. Thus, the rate equation for each species can be written.

$$\frac{dM}{dt} = k_2 M^* - k_1 M$$

$$\frac{dM^*}{dt} = k_1 M - (k_2 + k_3)M^* \tag{5.23}$$

$$\frac{dP}{dt} = k_3 M^*$$

A system of three coupled linear first-order differential equations can be solved exactly using Laplace transforms and the initial conditions

$$M(t=0) = M_0 \tag{5.24}$$
$$M^*(t=0) = P(t=0) = 0$$

However, if one uses the steady-state approximation, the solution becomes much simpler. This approximation assumes that in a very short time the excited molecule M^* comes to a steady state, i.e. M^* is formed as quickly as it disappears. In mathematical form,

$$\frac{dM^*}{dt} = 0 \tag{5.25}$$

The intermediate M^* does indeed come to a steady state quickly, on the order of 10^{-8} seconds or faster.[4] Thus,

$$\frac{dM}{dt} = -KM \tag{5.26}$$

where $K = \dfrac{k_1 k_3}{k_2 + k_3}$

The overall rate constant K is proportional to the intensity of the exposure radiation. The rate constant k_1 will be proportional to the rate of photon absorption (by the Grotthuss–Draper law), which in turn is proportional to the photon flux (by Lambert's law), and thus the intensity. A more useful form of Equation (5.26) is then

$$\frac{dm}{dt} = -CIm \tag{5.27}$$

where the relative sensitizer concentration m $(= M/M_0)$ has been used and C is the standard exposure rate constant and the third Dill photoresist parameter (collectively, the three Dill parameters are also called the *ABC parameters*). A more thorough microscopic analysis of the exposure process allows this exposure rate constant to be broken down into the product of the absorption cross section of the sensitizer and the quantum yield of the reaction (the fraction of absorbed photons that produce the chemical change, Φ):

$$C = \Phi \sigma_{M-abs}\left(\frac{\lambda}{hc}\right) = \frac{\Phi B_E}{\nu} = \frac{\Phi a_M \lambda}{N_A hc} \tag{5.28}$$

Maximum resist sensitivity comes from high quantum efficiency, all absorption coming from the sensitizer, and an absorption coefficient adjusted to be one over the thickness of the resist (see Problem 5.2).

A solution to the exposure rate Equation (5.27) is simple if the intensity within the resist is constant throughout the exposure (i.e. when $A = 0$). Integrating Equation (5.27) gives

$$m = e^{-CIt} \tag{5.29}$$

where $Dose = \int_0^t I dt = It$ for the case of constant intensity during the exposure process.

This result illustrates an important property of first-order kinetics called *reciprocity*. The amount of chemical change is controlled by the product of light intensity and exposure time. Doubling the intensity and cutting the exposure time in half will result in the exact same amount of chemical change. This product of intensity and exposure time is called the *exposure dose* or *exposure energy*. This idea can be made more usefully explicit by relating the exposure dose at some point inside the resist to the exposure dose incident on the resist. For either the simple case of absorption as shown in Equation (5.2) or (5.3), or for the more general case of standing waves in the resist film as described in Chapter 4, the intensity in the resist $I(z)$ is always directly proportional to the incident intensity I_{inc}. Thus,

$$I(x,y,z) = I_{inc} I_r(x,y,z) \tag{5.30}$$

where I_r is the relative intensity, that is, the intensity in the resist assuming a unit incident intensity. Equation (5.29) can now be rewritten as

$$m(x,y,z) = \exp[-C Dose_{inc} I_r(x,y,z)] \tag{5.31}$$

where $Dose_{inc}$ is the incident exposure dose. Reciprocity exists whenever all instances of exposure time and incident intensity can be replaced by incident dose in an expression for the final latent image $m(x,y,z)$.

Reciprocity failures can occur when mechanisms other than the first-order kinetics shown here become significant. For example, an insufficient supply of water during exposure can lead to reciprocity failure if high intensity exposures use up water in the resist film faster than water from the atmosphere can diffuse back into the photoresist. Reciprocity failure can also result from high intensities that heat up the resist during exposure, causing a temperature-dependent change in the molar absorptivity or the quantum yield.

For many resists, especially the conventional DNQ resists popular for g-line and i-line exposure, the intensity within the resist is not constant during exposure. In fact, many resists *bleach* upon exposure, that is, they become more transparent as the sensitizer M is converted to product P. This corresponds to a positive value of A, as seen, for example, in Figure 5.1. Since the intensity varies as a function of exposure time, this variation must be known in order to solve the exposure rate equation. In the simplest possible case, a resist film coated on a substrate of the same index of refraction, only absorption affects the intensity within the resist. Thus, Lambert's law of absorption, coupled with Beer's law, could be applied:

$$\frac{dI}{dz} = -(Am + B)I \qquad (5.32)$$

where Equation (5.12) was used to relate the absorption coefficient to the relative sensitizer concentration. Equations (5.29) and (5.32) are coupled, and thus become first-order nonlinear partial differential equations that must be solved simultaneously. Equations (5.29) and (5.32) were first solved analytically by Herrick:[5]

$$z = \int_{m(z=0)}^{m(z)} \frac{dy}{y[A(1-y) - B\ln(y)]} \qquad (5.33)$$

where y is a dummy variable for the purposes of integration, and

$$I(z) = I(0)\frac{A[1 - m(z)] - B\ln[m(z)]}{A[1 - m(0)] - B\ln[m(0)]} \qquad (5.34)$$

The sensitizer concentration at the top of the resist, $m(0)$, is obtained from the nonbleaching solution [Equation (5.29)] since the intensity at the top of a bleaching resist on a nonreflecting substrate does not change during exposure. For many g-line and i-line resists, $A \gg B$. Thus, at least near the beginning of exposure, B can often be safely neglected (i.e. $Am \gg B$). Under these conditions, the integral in Equation (5.33) can be solved exactly (by setting $B = 0$) giving

$$m(z) = \frac{m(0)}{m(0) + [1 - m(0)]e^{-Az}}$$

$$\frac{I(z)}{I(0)} = \frac{1 - m(z)}{1 - m(0)} = \frac{e^{-Az}}{m(0) + [1 - m(0)]e^{-Az}} = \frac{m(z)}{m(0)}e^{-Az} \qquad (5.35)$$

Although an analytical solution exists for the simple problem of exposure with absorption only, in more realistic problems the variation of intensity with depth in the film is more complicated than Equation (5.32). In fact, the general exposure situation results in the formation of standing waves, as discussed in Chapter 4. Thus, numerical solutions to the time-varying intensity and chemical composition of the film during exposure will almost certainly be required whenever a bleaching photoresist film is used.

The final result of exposure is the conversion of an image intensity in resist $I_r(x,y,z)$ into a latent image $m(x,y,z)$. Figure 5.2 illustrates a one-dimensional case.

5.2 Post-Apply Bake

Baking a resist may have many purposes, from removing solvent to causing chemical amplification. In addition to the intended results, baking may also cause numerous unintended outcomes. For example, the light-sensitive component of the resist may decompose at temperatures typically used to remove solvent. The solvent content of the resist can impact diffusion and amplification rates for a chemically amplified resist. Also, all aspects of baking will probably affect the dissolution properties of the resist. Baking a photoresist remains one of the most complicated and least understood steps in the lithographic process.

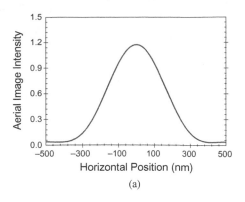

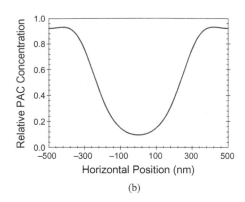

(a)

(b)

Figure 5.2 *The exposure process takes an aerial image (a) and converts it into a latent image (b)*

The post-apply bake (PAB) process, also called a softbake or a prebake, involves drying the photoresist after spin coating by removing excess solvent. There are four major effects of removing solvent from a photoresist film: (1) film thickness is reduced and stabilized (i.e. becomes relatively independent of the time spent waiting for exposure and development); (2) post-exposure bake and development properties are changed; (3) adhesion is improved; and (4) the film becomes less tacky and thus less susceptible to particulate contamination. Unfortunately, there are other consequences of baking many photoresists. For conventional resists, the photosensitive component may begin to decompose at temperatures greater than about $70\,°C$[6] (for chemically amplified resists, decomposition begins at much higher temperatures). Thus, one must search for the optimum post-apply bake conditions that will maximize the benefits of solvent evaporation while maintaining a tolerable level of sensitizer decomposition.

5.2.1   Sensitizer Decomposition

When heated to temperatures above about $70\text{--}80\,°C$, the sensitizer of a DNQ-type positive photoresist begins to decompose to a nonphotosensitive product. The initial reaction mechanism is thought to be identical to that of the sensitizer reaction during ultraviolet exposure.[6,7]

$$\text{(reaction scheme 5.36)} \qquad (5.36)$$

The possible identity of the product or products X will be discussed below.

To determine the concentration of sensitizer M as a function of post-apply bake time and temperature, consider the first-order decomposition reaction,

$$M \xrightarrow{\Delta} X \tag{5.37}$$

If we let M_0' be the concentration of sensitizer before post-apply bake and M_0 the concentration of sensitizer after post-apply bake, simple kinetics tells us that

$$\frac{dM_0}{dt} = -K_T M_0$$

$$M_0 = M_0' \exp(-K_T t_b) \tag{5.38}$$

$$m' = e^{-K_T t_b}$$

where t_b is the post-apply bake time, K_T is the decomposition rate constant at the absolute temperature T and $m' = M_0/M_0'$, the fraction of sensitizer remaining after the bake. The dependence of K_T upon temperature can be described by the Arrhenius equation,

$$K_T = A_r e^{-E_a/RT} \tag{5.39}$$

where A_r is the Arrhenius coefficient, E_a is the activation energy of the reaction, R is the universal gas constant (1.98717 cal/mole-K or 8.31431 J/mole-K), and T is the absolute temperature (K). Thus, the two parameters E_a and A_r allow us to know m' as a function of the post-apply bake conditions, provided Arrhenius behavior is followed. In polymer systems, caution must be exercised since bake temperatures near the glass transition temperature sometimes lead to non-Arrhenius behavior. For normal post-apply bakes of typical photoresists, the Arrhenius model appears well founded.

The effect of this decomposition is a change in the chemical makeup of the photoresist. Thus, any parameters which are dependent upon the quantitative composition of the resist are also dependent upon post-apply bake. The most important of these parameters fall into three categories: (1) optical (exposure) parameters such as the resist absorption coefficient; (2) diffusion parameters during post-exposure bake; and (3) development parameters such as the development rates of unexposed and completely exposed resist. A technique will be described to measure E_a and A_r and thus begin to quantify these effects of post-apply bake.

In the Dill model of exposure described in the previous section, the exposure of a positive photoresist can be characterized by the three parameters A, B and C, as given in Equations (5.12) and (5.29). These expressions do not explicitly take into account the effects of post-apply bake on the resist composition. To do so, we can modify Equation (5.12) to include absorption by the component X:

$$B = a_P M_0 + a_R R + a_X X \tag{5.40}$$

where a_X is the molar absorption coefficient of the decomposition product X (and the absorption term for the solvent has been neglected for simplicity). The stoichiometry of the decomposition reaction gives

$$X = M_0' - M_0 \tag{5.41}$$

Thus,

$$B = a_P M_0 + a_R R + a_X (M_0' - M_0) \tag{5.42}$$

Let us consider two cases of interest, no bake (NB) and full bake (FB). When there is no post-apply bake (meaning no decomposition), $M_0 = M_0'$ and

$$A_{NB} = (a_M - a_P)M_0'$$

$$B_{NB} = a_P M_0' + a_R R \tag{5.43}$$

We shall define 'full bake' as a post-apply bake which decomposes all of the sensitizer. Thus $M_0 = 0$ and

$$A_{FB} = 0$$

$$B_{FB} = a_X M_0' + a_R R \tag{5.44}$$

Using these special cases in our general expressions for A and B, we can show explicitly how these two parameters vary with sensitizer decomposition:

$$A = A_{NB} m'$$

$$B = B_{FB} - (B_{FB} - B_{NB})m' \tag{5.45}$$

The A parameter decreases linearly as decomposition occurs, and B typically increases slightly (since B_{FB} is typically slightly higher than B_{NB}).

The development rate, as we shall see in Chapter 7, is dependent on the concentration of PAC in the photoresist. However, the product X can also have a large effect on the development rate. There are two likely products and the most common outcome of a post-apply bake decomposition is a mixture of the two. The first product is formed via the reaction below and is identical to the product of UV exposure.

$$\tag{5.46}$$

As can be seen, this reaction requires the presence of water. A second reaction, which does not require water, is the esterification of the ketene with the novolac resin.

$$\tag{5.47}$$

Both possible products have a dramatic effect on dissolution rate. The carboxylic acid is very soluble in developer and enhances dissolution. The formation of carboxylic acid can be thought of as a blanket exposure of the resist. The dissolution rate of unexposed resist (r_{min}) will increase due to the presence of the carboxylic acid. The dissolution rate of fully exposed resist (r_{max}), however, will not be affected. Since the chemistry of the dissolution process is unchanged, the basic shape of the development rate function will also remain unchanged. The ester, on the other hand, is very difficult to dissolve in aqueous solutions and thus retards the dissolution process. It will have the effect of decreasing r_{max}, although the effects of ester formation on the full dissolution behavior of a resist are not well known.

If the two mechanisms given in Equations (5.46) and (5.47) are taken into account, the rate Equation (5.38) will become

$$\frac{\mathrm{d}M_0}{\mathrm{d}t} = -k_1 M_0 - k_2 [H_2 O] M_0 \tag{5.48}$$

where k_1 and k_2 are the rate constants of Equations (5.46) and (5.47), respectively. For a given concentration of water in the resist film, $[H_2O]$, this reverts to Equation (5.38) where

$$K_T = k_1 + k_2 [H_2 O] \tag{5.49}$$

Thus, the relative importance of the two reactions will depend not only on the ratio of the rate constants but on the amount of water in the resist film. The concentration of water is a function of atmospheric conditions during the bake and the past history of the resist-coated wafer.

Examining Equation (5.45), one can see that the parameter A can be used as a means of measuring m', the fraction of sensitizer remaining after post-apply bake. Thus, by measuring A as a function of post-apply bake time and temperature, one can determine the activation energy and the corresponding Arrhenius coefficient for the proposed decomposition reaction. Figure 5.3a shows the variation of the resist parameter A with post-apply bake conditions for one particular resist.[8] According to Equations (5.38) and (5.45), this variation should take the form

$$\frac{A}{A_{\mathrm{NB}}} = e^{-K_T t_b}$$

$$\ln\left(\frac{A}{A_{\mathrm{NB}}}\right) = -K_T t_b \tag{5.50}$$

Thus, a plot of $\ln(A)$ versus bake time should give a straight line with a slope equal to $-K_T$. This plot is shown in Figure 5.3b. Knowing K_T as a function of temperature, one can determine the activation energy and Arrhenius coefficient from Equation (5.39). One should note that the parameters A_{NB}, B_{NB} and B_{FB} are wavelength dependent, but E_a and A_r are not.

Figure 5.3 shows an anomaly in which there is a lag time before decomposition occurs. This lag time is the time it took the wafer and wafer carrier to reach the temperature of the convection oven. Equation (5.38) can be modified to accommodate this phenomena,

$$m' = e^{-K_T(t_b - t_{wup})} \tag{5.51}$$

where t_{wup} is the warm-up time. A lag time of about 11 minutes was observed when convection oven baking a 1/4″-thick glass substrate in a wafer carrier. For a silicon wafer on a proximity hot plate, the warm-up time is on the order of 15–20 seconds.

From the data presented in Figure 5.3b, the activation energy is 30.3 Kcal/mol and the natural logarithm of the Arrhenius coefficient (in 1/minute) is 35.3. Thus, a 100 °C,

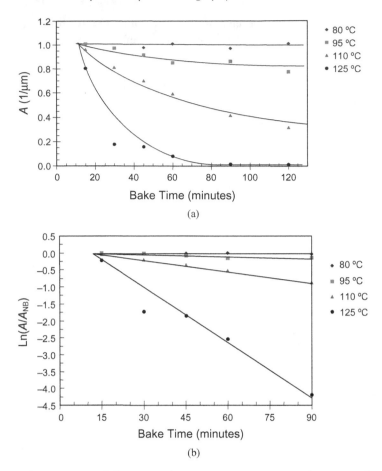

Figure 5.3   *The variation of the resist absorption parameter A with post-apply bake time and temperature for Kodak 820 resist at 365 nm (a convection oven bake was used): (a) linear plot, and (b) logarithmic plot*

30-minute convection oven post-apply bake would decompose 11 % of the photoactive compound. The nature of the decomposition product (the identity of X) will also affect the value of B_{FB}. If thermal decomposition leads to generation of both carboxylic acid and ester, only the ester will affect the change of B from B_{NB} to B_{FB}. In this case, Equation (5.44) must be modified to take into account the fraction of PAC that decomposes into ester, ϕ, by replacing a_X with $a_X\phi + (1 - \phi)a_P$.

The examples given above for photoactive compound decomposition all involve DNQ/ novolac resist systems used at g-line and i-line exposure wavelengths. For these systems, typical bake temperatures always result in some (hopefully small) amount of sensitizer decomposition. However, for chemically amplified resists, the photoacid generators are much more thermally stable . Typical post-apply bake temperatures result in essentially no thermal decomposition of the chemically amplified sensitizer.

5.2.2 Solvent Diffusion and Evaporation

Thermal processing of photoresists (post-apply bake and post-exposure bake) can dramatically influence resist performance in a number of ways. It is well known that residual solvent has a powerful influence on the dissolution rate of thin polymer films and that post-apply bake (PAB) determines the resist's residual solvent content. However, quantitative determination of the influence of bake conditions on residual solvent is difficult. In order to describe solvent evaporation during the baking of a photoresist, a model for how the diffusivity varies during the bake (that is, as a function of the changing photoresist film composition) must be established. In this section, a historical review of the Fujita–Doolittle equation for solvent diffusivity is given. Then, modifications to this equation are proposed to improve its accuracy and dynamic range as needed for its use in describing PAB.

The influence of temperature, molecular weight and chemical composition on the physical and mechanical properties of polymers is a difficult but important topic of study. Since the early 1950s the use of a 'free-volume' description of physical relaxation phenomena in polymers and other materials has proven extremely useful. Although the idea that the free, unoccupied volume in a liquid or solid greatly influences the physical properties of the material is quite old, it was first applied successfully by Arthur Doolittle in 1951.[9]

Doolittle described the influence of temperature on the viscosity of a liquid in two steps: the viscosity depends on the free volume in the liquid, and this free volume depends on temperature. The fraction of the volume of a liquid which is free, that is, unoccupied by the molecules of the liquid, was expressed as

$$v_f = \frac{V - V_0}{V} \tag{5.52}$$

where v_f is the free volume fraction, V is the volume that a specified mass of the liquid occupies at some temperature T, and V_0 is the volume occupied by the liquid extrapolated to zero absolute temperature assuming no phase change. The assumption here is that the 'liquid' extrapolated to absolute zero will contain no free volume and that the thermal expansion of the liquid is due completely to the creation of free volume. Although the first assumption seems quite reasonable, the second is probably less accurate. Thermal expansion of the liquid will undoubtedly lead to free volume formation, but the increased vibrations of the liquid molecules will also lead to an increased 'occupied' volume, the volume that is occupied by the molecule to the exclusion of other molecules. Doolittle's free volume expression can be modified by multiplying by ζ, the fraction of the increased volume which is actually free. Thus,

$$v_f = \zeta \frac{V - V_0}{V} \tag{5.53}$$

where the magnitude of ζ for a liquid is probably close to unity, but for a solid could be considerably smaller.

The difficulty in measuring the free volume is in determining the value of V_0. Doolittle attempted this by measuring the density of a liquid as a function of temperature, fitting this data to an empirical equation, then extrapolating to zero temperature. With this empirically determined temperature dependence for the free volume of the liquid, Doo-little then showed that experimental viscosity data as a function of temperature behaved as

$$\eta = Ae^{B/v_f} \tag{5.54}$$

where η is the viscosity and A and B are empirical constants. Equation (5.54) was found to match the experimental data over a wider range of temperatures much better than any of the other empirical expressions in common use at the time, lending credence to this free volume interpretation. Interestingly, the experimental value of B obtained by Doo-little was very close to unity (0.9995) and the free volume of the liquid ranged from 0.22 to 0.68 over the temperatures studied.

Others began applying the ideas of free volume to polymer systems. Describing both viscosity and diffusion processes in polymers by the same mechanism, Bueche[10] calculated the probability that sufficient free volume would be available to allow movement of the polymer. Relaxation processes (i.e. processes that are limited by the movement of the polymer) were thought to include such phenomena as viscous flow and diffusion. Thus, describing the rate at which a polymer moved was the first step in defining other relaxation rates. Since the polymer segments are vibrating as a function of temperature, the volume occupied by the polymer will have a thermodynamically controlled probability distribution about the average volume per polymer segment. If the volume occupied by the polymer segment exceeds some critical volume, the polymer segment can move or 'jump' to a new configuration. Thus, by integrating the volume probability from this critical volume to infinity, the frequency of jumps can be determined. In turn, the viscosity is almost completely controlled by this jump frequency. By comparing this theory with experimental polymer viscosity data, the temperature dependence of the average volume of a polymer segment was deduced.

For temperatures above the glass transition temperature of the polymer, Bueche described the polymer volume by

$$V = V_{Tg}(1 + (\alpha_1 + \alpha_2)T) \tag{5.55}$$

where V_{Tg} is the volume at the glass transition temperature, α_1 is the coefficient of thermal expansion of the solid-like polymer below the glass transition temperature, and $\alpha_1 + \alpha_2$ is the thermal expansion coefficient of the liquid-like polymer above the glass transition temperature. This abrupt change in the coefficient of thermal expansion is considered one of the fundamental indicators of the glass transition phenomenon. Both α_1 and α_2 typically have magnitudes on the order of $5 \times 10^{-4} K^{-1}$ for most polymers.

Fox and Flory measured the specific volume of polystyrene[11] and polyisobutylene[12] as a function of temperature for many molecular weights and found that Equation (5.55) was quite adequate. Bueche and Fox and Flory went on to speculate that of the total volume expansion given by Equation (5.55), only the α_2 term resulted in the generation of free volume. Essentially, they argued that ζ of Equation (5.53) was very small for the solid-like behavior of the polymer, and that the excess volume that comes from thermal expansion above the glass transition temperature is all free volume. Thus,

$$\begin{aligned} v_f = v_g, \quad T \leq T_g \\ v_f = v_g + \alpha_2(T - T_g), \quad T \geq T_g \end{aligned} \tag{5.56}$$

where T_g is the glass transition temperature and v_g is the fractional free volume at T_g.

The power of the free volume approach to relaxation mechanisms is that the fundamental relationship of the relaxation mechanism to free volume is independent of the mechanism by which free volume changes. Fujita, Kishimoto and Matsumoto[13] used this fact to expand the Doolittle model to include the effect of small amounts of solvent in the polymer. Solvent dissolved in a polymer results in additional free volume which in turn increases diffusivity and decreases viscosity. Fujita *et al.* modified the free volume Equation (5.56) to include solvent content:

$$v_f = v_g + \alpha_2(T - T_g) + \beta\phi_s \tag{5.57}$$

where ϕ_s is the fractional volume of solvent and β describes the fraction of this solvent volume which can be considered free. Here, T_g is usually considered the glass transition temperature of the completely dry (no dissolved solvent) polymer. For the polymethyl acrylate polymer and four different solvents studied by Fujita, β was found to be 0.19.

The Fujita–Doolittle equation has been used extensively to characterize the change in solvent diffusivity with changing solvent concentration. There are, however, constraints to this approach that have led some workers in this field to criticize its use and to look to alternate expressions. In particular, higher solvent concentration causes a dilution of the polymer-induced free volume with the addition of solvent. A correction for this effect gives[14]

$$v_f = (1 - \phi_s)v_g + \alpha_2(T - T_g) + \phi_s\beta \tag{5.58}$$

Equation (5.58) is a more accurate form of the Fujita free volume expression which is valid for higher concentrations of solvent, while still requiring the same number of parameters as the original free volume expression.

The final expression for free volume, Equation (5.58), can now be combined with the Doolittle equation to give a modified Fujita–Doolittle equation for the diffusivity D of a solvent in a polymer matrix as a function of temperature and solvent content.

$$D = Ae^{-B/v_f} \tag{5.59}$$

Letting D_0 be the minimum diffusivity, that is the diffusivity in the limit of no solvent content and $T = T_g$,

$$D_0 = Ae^{-B/v_g} \tag{5.60}$$

Thus, Equation (5.59) can be rearranged as

$$D = D_0 e^{-B(1/v_f - 1/v_g)}$$

where

$$v_f = (1 - \phi_s)v_g + \alpha_2(T - T_g) + \phi_s\beta \tag{5.61}$$

It will often be more convenient to describe the solvent content as a mass fraction rather than a volume fraction. Letting x be the mass fraction and ρ the density,

$$x_s \approx \frac{\rho_s}{\rho_p} \phi_s$$

$$v_f = (1 - x_s)v_g + \alpha_2(T - T_g) + x_s \beta' \tag{5.62}$$

$$\beta' = \frac{\rho_p}{\rho_s} \beta + v_g \left(1 - \frac{\rho_p}{\rho_s} \right) \approx \frac{\rho_p}{\rho_s} \beta$$

where the subscripts 's' and 'p' refer to solvent and polymer, respectively.

Based on previous work, the order of magnitude for each of the terms in Equation (5.62) is known. Doolittle and Williams *et al.*[15] found B to be approximately 1.0, while Fujita *et al.* found $B = 0.73$. Cohen and Turnbull[16] limit B to be between 0.5 and 1 (their so-called geometric factor). The Williams, Landel and Ferry (WLF) equation predicts $v_g = 0.025$ and $\alpha_2 = 4.8 \times 10^{-4}\,\mathrm{K}^{-1}$, but further work by Ferry[17] found values of v_g/B between 0.013 and 0.07, with most of the data between 0.02 and 0.035, and values of α_2 between 1 and $11 \times 10^{-4}\,\mathrm{K}^{-1}$, with most data between 3 and $5 \times 10^{-4}\,\mathrm{K}^{-1}$. Fox and Flory found $\alpha_2 = 3 \times 10^{-4}\,\mathrm{K}^{-1}$. Fujita found β to be 0.19, and in any case it must be ≤ 1.0. The value of β' should be similar. From these typical numbers and the diffusivity model given above, one can see that the diffusivity of solvent changes by four to six orders of magnitude during the course of a typical post-apply bake.

Using this solvent diffusion model, one can predict the impact of PAB time and temperature on the final resist thickness and the resulting solvent distribution as a function of depth into the resist at the end of the bake. Such modeling provides for a very interesting result. Any combination of initial resist thickness and initial solvent content that produces a certain final resist thickness always results in essentially the same solvent distribution in the resist at the end of the bake. Figure 5.4 shows the distribution of solvent left in the resist at the end of a 60-second bake of various temperatures using the parameters from Table 5.1.

Another consequence of using the modified Fujita–Doolittle equation is a prediction of how solvent diffusivity varies with temperature. Using the parameters from Table 5.1,

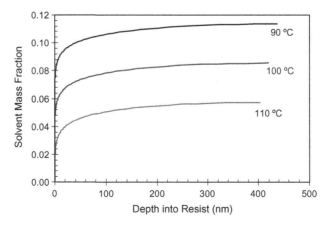

Figure 5.4 *Predicted variation of solvent concentration as a function of depth into the resist at the end of a 60-second post-apply bake*

Table 5.1 Typical solvent diffusion parameters in photoresist

Parameter	Value
B	0.737
β'	0.351
v_g	0.0289
α_2 (1/K)	8.7E-04
T_g (°C)	110.5
D_0 (nm²/s)	0.0967

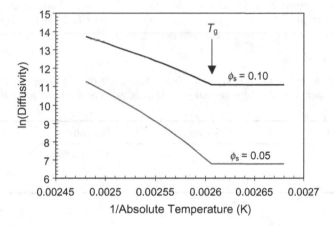

Figure 5.5 *Temperature dependence of solvent diffusivity (using the parameters from Table 5.1 and assuming a solvent mass fractions of 0.05 and 0.1) showing an essentially fixed diffusivity below the glass transition temperature*

and assuming a solvent mass fraction of 0.05, an Arrhenius plot of solvent diffusivity is given in Figure 5.5. There is a large discontinuity at the glass transition temperature, since the model assumes such a discontinuity as a consequence of Equation (5.56). Note that above the glass transition temperature, the behavior is non-Arrhenius, that is, the curve is not perfectly linear in the Arrhenius plot of Figure 5.5.

5.2.3 Solvent Effects in Lithography

Residual solvent in the film after baking can have two significant effects on the lithographic performance of a photoresist. Just as residual solvent increases the free volume and thus the diffusivity of solvent during PAB, any residual solvent at post-exposure bake will increase the diffusivity of exposure products. For a chemically amplified resist, especially when baking in a diffusion-controlled temperature regime, this increase in acid diffusivity can have a quite large impact. Figure 5.6 shows an example where the top portion of the resist exhibits standing waves while the rest of the resist does not. Reduced solvent content at the top of the resist leads to reduced acid diffusion during PEB and (as will be discussed in the following section) less smoothing out of the standing wave pattern.

Figure 5.6   *Lower solvent content at the top of this 248-nm resist leads to reduced acid diffusion during PEB, and thus the presence of standing waves only at the top of the resist (photo courtesy of John Petersen, used with permission)*

Table 5.2   *Development* r_{max} *and* n *values (see Chapter 7) fitted to measured dissolution rate data as a function of PAB temperature (and thus, solvent content) for a thick i-line resist*[18]

Temperature (°C)	wt.% Solvent	r_{max} (nm/s)	Develop n
70	21.1	104.3	1.71
80	19.5	90.3	1.77
90	17.8	78.9	1.96

As Figure 5.5 showed, diffusivity grows almost exponentially as a function of the temperature above T_g. Thus, for the post-exposure bake to cause appreciable diffusion (as will be discussed in the next section), the PEB temperature must be above the T_g of the resist. However, as seen above, the T_g of a resist film is a function of the amount of solvent in the resist, and thus a function of the post-apply bake conditions. For a sufficiently long post-apply bake, the T_g of the resist film approaches the post-apply bake temperature. For shorter bakes, the resist film T_g will be somewhat below the PAB temperature. Thus, as a simple rule of thumb, the PEB temperature must be the same or greater than the PAB temperature in order for appreciable diffusion to occur during PEB.

Secondly, residual solvent can impact resist dissolution rates, affecting both the average dissolution rate and the variation of dissolution rate with exposure dose (Table 5.2). Unfortunately, this behavior is rarely well characterized for any given resist material.

5.3   Post-exposure Bake Diffusion

Many attempts have been made to reduce the standing wave effect and thus increase linewidth control and resolution. One particularly useful method is the post-exposure, predevelopment bake as described by Walker.[19] A 100 °C oven bake for 10 minutes was found to reduce the standing wave ridges on a resist sidewall significantly. This phenomenon can be explained quite simply as due to the diffusion of sensitizer in the resist (which acts as a dissolution inhibitor for a conventional resist, see Chapter 7) during the high-temperature bake. A mathematical model which predicts the results of such a post-exposure bake (PEB) is described below.

In general, molecular diffusion is governed by Fick's Second Law of Diffusion, which states (in one dimension, r)

$$\frac{\partial C_A}{\partial t} = \frac{\partial}{\partial r}\left(D_A \frac{\partial C_A}{\partial r}\right) \tag{5.63}$$

where C_A is the concentration of species A, D_A is the diffusion coefficient of A at some temperature T, and t is the time that the system is at temperature T. A more general form is

$$\frac{\partial C_A}{\partial t} = \nabla(D_A \nabla C_A) \tag{5.64}$$

In many cases, it is quite accurate to assume that the diffusivity is independent of concentration (and thus position). Under these conditions, this differential equation can be solved given a set of boundary conditions (two for each dimension) and an initial distribution of A.

Consider one possible initial condition known as the impulse source (a delta function of concentration). At some point x_0, there are N moles of substance A and at all other points there is no A. Thus, the concentration at x_0 is infinite. For boundary conditions, assume zero concentration as r goes to $\pm\infty$ for all time. Given this initial distribution of A and these boundary conditions, the solution to Equation (5.63) is the Gaussian distribution function,

$$C_A(x) = \frac{N}{\sqrt{2\pi\sigma^2}} e^{-r^2/2\sigma^2} \tag{5.65}$$

where $\sigma = \sqrt{2D_A t}$ is the diffusion length, and $r = x - x_0$.

In practice there are no impulse sources. Instead, we can approximate an impulse source as having some concentration C_0 over some small distance Δx centered at x_0, with zero concentration outside of this range. An approximate form of Equation (5.65) is then

$$C_A(x) \cong \frac{C_0 \Delta x}{\sqrt{2\pi\sigma^2}} e^{-r^2/2\sigma^2} \tag{5.66}$$

This solution is fairly accurate if $\Delta x \ll \sigma$. If there are two 'impulse' sources located at x_1 and x_2, with initial concentrations C_1 and C_2 each over a range Δx, the concentration of A at x after diffusion is

$$C_A(x) = \left[\frac{C_1}{\sqrt{2\pi\sigma^2}} e^{-r_1^2/2\sigma^2} + \frac{C_2}{\sqrt{2\pi\sigma^2}} e^{-r_2^2/2\sigma^2}\right]\Delta x \tag{5.67}$$

where $r_1 = x - x_1$ and $r_2 = x - x_2$.

If there are a number of sources, Equation (5.67) becomes

$$C_A(x) = \frac{\Delta x}{\sqrt{2\pi\sigma^2}} \sum_n C_n e^{-r_n^2/2\sigma^2} \tag{5.68}$$

Extending the analysis to a continuous initial distribution $C_0(x)$, Equation (5.68) becomes

$$C_A(x) = \frac{1}{\sqrt{2\pi\sigma^2}} \int_{-\infty}^{\infty} C_0(x')e^{-(x-x')^2/2\sigma^2} dx' \tag{5.69}$$

where x' is now the distance from the point x. Equation (5.69) is simply the convolution (denoted by $\otimes$) of the original concentration profile with a Gaussian:

$$C_A(x) = C_0(x) \otimes f(x) \tag{5.70}$$

where $f(x) = \dfrac{1}{\sqrt{2\pi\sigma^2}} e^{-x^2/2\sigma^2}$

This equation can now be made to accommodate two-dimensional diffusion:

$$C_A(x,y) = C_0(x,y) \otimes f(x,y) \tag{5.71}$$

where

$$f(x,y) = \frac{1}{2\pi\sigma^2} e^{-r^2/2\sigma^2}$$

$$r = \sqrt{x^2 + y^2}$$

Three-dimensional diffusion can similarly be calculated (as shown below).

From this formulation of diffusion, one can see that the diffusion impulse response function of Equation (5.65) or its higher dimensional equivalents is analogous to the imaging point spread function of Chapter 2. Thus, $f(r)$ is sometimes referred to as the diffusion point spread function (*DPSF*).

We are now ready to apply our solution to the diffusion of sensitizer in a conventional photoresist during a post-exposure bake. For the full three-dimensional case, the sensitizer distribution after exposure can be described by $m(x,y,z)$, where m is the relative sensitizer concentration. According to Equation (5.71) the relative sensitizer concentration after a post-exposure bake, $m*(x,y,z)$, is given by

$$m*(x,y,z) = \frac{1}{(2\pi\sigma^2)^{3/2}} \int_{-\infty}^{\infty} \int \int m(x',y',z')e^{-((x-x')^2+(y-y')^2+(z-z')^2)/2\sigma^2} dx'dy'dz' \tag{5.72}$$

or,

$$m*(x,y,z) = m(x,y,z) \otimes DPSF \tag{5.73}$$

For a given latent image, the only parameter that needs to be specified to solve Equation (5.72) is the diffusion length σ, or equivalently, the diffusion coefficient D_A and the bake time t. In turn, D_A is a function of the bake temperature T (by the Arrhenius equation for temperature ranges which do not traverse the glass transition temperature) and, of course, the resist system used. Unfortunately, it is quite difficult to measure the diffusivity of sensitizer in photoresist. From Walker's work, one can estimate the values of E_a and D_0 for the resist studied to be about 35 Kcal/mol and 3.2×10^{21} nm^2/s, respectively.

The diffusion model can now be used to simulate the effects of a post-exposure bake on standing waves for a conventional resist (Figure 5.7). Obviously, the major effect is

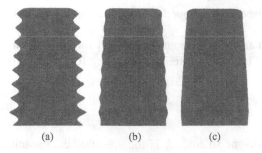

Figure 5.7 *Typical i-line photoresist profile simulations (using PROLITH) for resist on silicon as a function of the PEB diffusion length: (a) 20 nm, (b) 40 nm and (c) 60 nm*

the smoothing out of the standing waves. Less obvious from these pictures is the gradual degradation of the photoresist image itself. The trade-off between these two effects puts a significant constraint on the acceptable values of the diffusion length. In order to remove standing waves, the diffusion length must be on the order of or larger than half of the standing wave period. But to avoid degrading the feature itself, diffusion length must be much smaller than the smallest feature size being printed.

From Chapter 4, recall that the standing wave period is $\lambda/2n_{resist}$ where n_{resist} is the resist refractive index (the period will be slightly less than this for oblique illumination). For i-line exposure of a typical resist, this puts the half period at about 55 nm. Thus, diffusion lengths must be at about this magnitude or larger to completely remove standing waves. Since i-line resists are rarely used to resolve features less than about 300 nm, the requirement that diffusion length is much less than feature size is also realized. However, for 248-nm lithography this is no longer the case. The standing wave half period is about 35–40 nm, but features smaller than 130 nm are regularly imaged. Thus, to remove standing waves effectively, the diffusion length must approach one-third of the feature size, a fraction that is too large to avoid significant image degradation. As a result, for 248- and 193-nm lithography, standing waves are reduced using bottom antireflection coatings rather than relying solely on diffusion during PEB. Additionally, it is important to remember that a post-exposure bake, while capable of reducing or eliminating standing waves, will have no impact on the photoresist swing curve. Since bottom antireflection coatings can be very effective at reducing both standing waves and swing curves, they have become the preferred solution for critical lithography levels.

5.4 Detailed Bake Temperature Behavior

In the discussion of section 5.2.1, a somewhat clumsy approach of adding a 'warm-up time' to the bake analysis was used to account for the noninstantaneous heating of the resist. By modeling the heat transfer from the hot plate to the wafer, a more accurate and detailed picture can be drawn of how the temperature of the wafer varies with time during the bake.

We can approximate the temperature profile of the resist on a typical wafer baking (track) system by the following differential equation (called a lumped capacitance model for heat transfer):[20,21]

$$\rho C_p L \frac{dT}{dt} = -\frac{k_{air}}{\delta}(T - T_{plate}) - h(T - T_{ambient}) \tag{5.74}$$

where ρ is the density of silicon, C_p is the specific heat capacity of silicon, L is the thickness of the wafer, T is the temperature of the resist-coated wafer, k_{air} is the thermal conductivity of air, δ is the thickness of the gap between the hot plate and the wafer, and h is a heat transfer coefficient for the convective heat lost from the top surface of the wafer to the surroundings. In this model, the term on the left side represents the thermal mass of the wafer, and the terms on the right side represent heat transfer on the bottom side of the wafer from the hot plate and heat loss from the top of the wafer to the surroundings. Note that the thermal mass of the photoresist is extremely small compared to that of the wafer, so that it can be ignored. Also, the thermal conductivity of silicon is high enough that the temperature gradient from the bottom to the top of the wafer can also be ignored. Radiative heat transfer mechanisms are small enough to be ignored as well.

The solution to this lumped capacitance heat transfer model is

$$T = T^* - (T^* - T_{initial})e^{-t/\tau} \tag{5.75}$$

where $T_{initial}$ is the initial temperature of the wafer, $T*$ is the equilibrium (infinite time) temperature given by

$$T^* = \frac{\frac{k_{air}}{\delta}T_{plate} + hT_{ambient}}{\frac{k_{air}}{\delta} + h} \tag{5.76}$$

and τ is the time constant for heating (or possibly cooling) the wafer,

$$\tau = \frac{\rho C_p L}{\frac{k_{air}}{\delta} + h} \tag{5.77}$$

Some typical values for the constants in these thermal model equations are given in Table 5.3.

We can now describe the full temperature profile of the resist on a typical wafer track system with a three-stage model for heat transfer to the resist-coated wafer: a proximity hot plate bake as described above; a short time for transfer from the hot plate to a chill plate; and then proximity cooling on the chill plate. During the second stage where the wafer is transferred from the hot plate to the chill plate, the gap δ is assumed to be very large so that the first term on the right side of Equation (5.74) is unimportant, and the second term on the right side is assumed to represent heat loss from the top

Table 5.3 *Typical values for silicon wafer properties*

Parameter	Value	Units
L (wafer thickness), 300-mm-diameter wafer	0.775	mm
L (wafer thickness), 200-mm-diameter wafer	0.725	mm
L (wafer thickness), 150-mm-diameter wafer	0.675	mm
L (wafer thickness), 100-mm-diameter wafer	0.525	mm
C_p (silicon molar heat capacitance) @ 25 °C	19.8	J/K-mole
C_p (silicon molar heat capacitance) @ 125 °C	22.3	J/K-mole
ρ (silicon density)	2.33	g/cm^3
Atomic weight of silicon	28.09	g/mole
k_{air} (air thermal conductivity) at sea level, 20 °C	0.025	W/m-K
k_{air} (air thermal conductivity) at sea level, 100 °C	0.031	W/m-K
h (convection heat transfer coefficient)	5–10	W/m^2-K

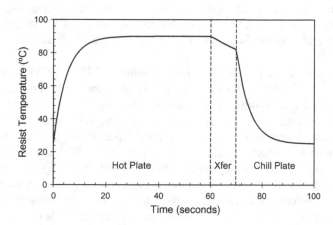

Figure 5.8 *Typical wafer bake profile (60-second bake followed by a 10-second transfer to a chill plate)*

and bottom wafer surfaces. An example of such a full temperature profile is shown in Figure 5.8.

The impact of the time variation of temperature will depend of course on the activation energy of the reactions in question. For a nonzero activation energy, the result will be a time variation of the rate 'constant'. For each rate constant k, an effective rate constant can be defined as

$$k_{\text{eff}} = \frac{1}{t_b} \int_0^{t_b} k(t)\mathrm{d}t \tag{5.78}$$

where t_b is the overall bake cycle time.

It is interesting to note that the equilibrium temperature $T*$ is not equal to the temperature of the hot plate, as shown by Equation (5.76). The fact that the equilibrium temperature of the wafer is usually very close to the hot plate temperature indicates that heating across the gap is the dominant heat transfer mechanism. This is because the gap spacing δ is very small, typically about 0.15–0.2 mm, so that

$$\text{when } \frac{k_{air}}{\delta} \gg h, \quad T^* \approx T_{plate} - \frac{\delta h}{k_{air}}(T_{plate} - T_{ambient}) \tag{5.79}$$

Even if significant variations in δ across the wafer are present, the value of T^* will still be dominated by the temperature of the hot plate, and good temperature uniformity can be achieved. By contrast, variations in the gap spacing can have a large effect on the time constant τ, as seen in Equation (5.77). Again, assuming h is small, the hot plate rise time will be directly proportional to δ.

$$\text{when } \frac{k_{air}}{\delta} \gg h, \quad \tau \approx \frac{\delta \rho C_p L}{k_{air}} \tag{5.80}$$

Variations in the gap spacing can be due to misalignment of the ceramic pins that control the gap spacing in the proximity plate, or due to nonflatness of the silicon wafer – the warp of a 200- and 300-mm wafers can be up to 0.075 and 0.1 mm, respectively (see Figure 5.9).

The lumped capacitance model solution of Equations (5.75)–(5.77) has several limitations. When solving the differential Equation (5.74), all of the terms in the equation except the wafer temperature are assumed to be constant. Obviously, the accuracy of the solution will depend on the accuracy of these assumptions. As Table 5.3 indicates, the heat capacity of silicon will vary by about 10% during a typical bake (a 100 °C temperature range or so), and the thermal conductivity of air will vary by 25%. With a coefficient of linear expansion of about 2.5×10^{-6}/K, the density of silicon will change by less than 0.1% over a 100 °C temperature range. Wafer thickness will typically vary between 1 and 2% across the wafer. The convective heat transfer coefficient can change significantly with air flow, which will change as the air above the wafer heats up and depends on the design of the hot plate chamber. Likewise, the ambient temperature is ill-defined when the hot plate is not in an open environment and can change from wafer to wafer if the hot plate chamber lid is not temperature controlled. But possibly the most significant problem is the assumption that the hot plate temperature is constant. Just as the wafer temperature is affected by the proximity of the hot massive plate beneath it, the hot plate is affected by the cold wafer. When a new wafer is placed on a hot plate, the surface temperature of the plate can drop by 1–2 °C, requiring 5–10 seconds to heat back up to its set point temperature.

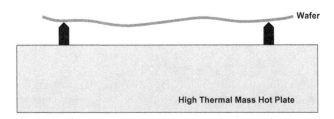

Figure 5.9 *Proximity bake of a wafer on a hot plate showing (in a highly exaggerated way) how wafer warpage leads to a variation in proximity gap (drawing not to scale)*

The thermal model here also assumes a one-dimensional geometry, so that variations in temperature in the plane of the wafer are ignored. Besides edge effects that can cause a temperature difference at the edge of the wafer versus the middle, wafer warpage as depicted in Figure 5.9 will lead to thermal gradients and thus heat transfer in the plane of the wafer, somewhat mitigating the temperature variations that would be predicted from a 1D model.[22]

5.5   Measuring the ABC Parameters

Dill proposed a single, simple experiment for measuring the *ABC* parameters.[1] The photoresist to be measured is coated in a nonreflecting transparent substrate (e.g. glass, quartz or similar material). The resist is then exposed by a normally incident parallel beam of light at the wavelength of measurement. At the same time, the intensity of the light transmitted through the substrate is measured continuously. The output of the experiment, transmitted intensity as a function of exposure time, is then analyzed to determine the resist *ABC* parameters. A typical experimental setup is shown in Figure 5.10. By measuring the incident exposing light intensity, the output of the experiment becomes overall transmittance as a function of incident exposure dose, $T(E)$. Figure 5.11 shows a typical result. Assuming careful measurement of this function, and a knowledge of the thickness of the photoresist, all that remains is the analysis of the data to extract the *ABC* parameters.

Note that the effectiveness of this measurement technique rests with the nonzero value of the resist bleachable absorption parameter *A*. If the photoresist does not change its optical properties with exposure (i.e. if $A = 0$), then measuring transmittance will provide no insight into the exposure reaction, making *C* unobtainable by this method. Other methods for determining *C* for the case when $A = 0$ are discussed briefly in Chapter 6.

Analysis of the bleaching experimental data is greatly simplified if the experimental conditions are adjusted so that the simple exposure and absorption Equations (5.29) and (5.32) apply exactly. This means that light passing through the resist must not reflect at

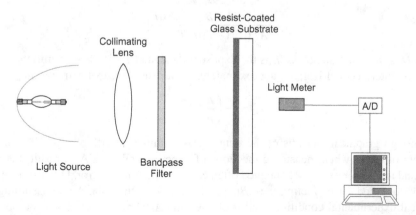

Figure 5.10   *Experimental configuration for the measurement of the ABC parameters*

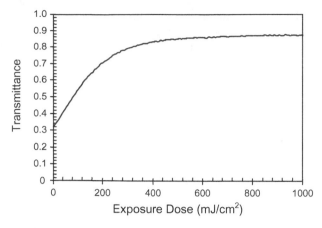

Figure 5.11 *Typical transmittance curve of a positive g- or i-line bleaching photoresist measured using an apparatus similar to that pictured in Figure 5.10*

the resist–substrate interface. Further, light passing through the substrate must not reflect at the substrate–air interface. The first requirement is met by producing a transparent substrate with the same index of refraction as the photoresist. The second requirement is met by coating the backside of the substrate with an interference-type antireflection coating.

Given such ideal measurement conditions, Dill showed that the *ABC* parameters can be obtained from the transmittance curve by measuring the initial transmittance $T(0)$, the final (completely exposed) transmittance $T(\infty)$ and the initial slope of the curve. The relationships are:

$$A = \frac{1}{D}\ln\left[\frac{T(\infty)}{T(0)}\right]$$

$$B = -\frac{1}{D}\ln\left[\frac{T(\infty)}{T_{12}}\right] \qquad (5.81)$$

$$C = \frac{A+B}{AT(0)[1-T(0)]T_{12}}\left.\frac{dT}{dE}\right|_{E=0}$$

where D is the resist thickness, E is the exposure dose, and T_{12} is the transmittance of the air–resist interface and is given approximately, for a resist index of refraction n_{resist}, by

$$T_{12} = 1 - \left(\frac{n_{\text{resist}}-1}{n_{\text{resist}}+1}\right)^2 \qquad (5.82)$$

Although graphical analysis of the data is quite simple, it suffers from the common problem of errors when measuring the slope of experimental data. As a result, the value of *C* (and to a lesser extent, *A*) obtained often contains significant error. Dill also proposed a second method for extracting the *ABC* parameters from the data. Again assuming that the ideal experimental conditions had been met, the *ABC* parameters could be obtained by directly solving the two coupled differential Equations (5.27) and (5.32), and finding

Table 5.4 *Typical values for resist ABC parameters*

Resist Type	A (μm^{-1})	B (μm^{-1})	C (cm²/mJ)	Refractive Index
g-line (436 nm)	0.6	0.05	0.015	1.65
i-line (365 nm)	0.9	0.05	0.018	1.69
248 nm	0.0	0.50	0.05	1.76
193 nm	0.0	1.20	0.05	1.71
Dyed	Unchanged	Increased by 0.3–0.5	Unchanged	Approximately unchanged

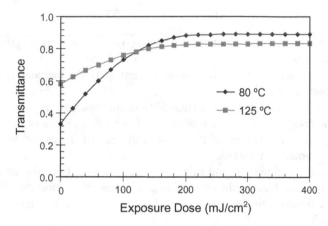

Figure 5.12 *Two transmittance curves for Kodak 820 resist at 365 nm. The curves are for a convection oven post-apply bake of 30 minutes at the temperatures shown*

the values of A, B and C for which the solution best fits the experimental data. Obviously, fitting the entire experimental curve is much less sensitive to noise in the data than taking the slope at one point.

Typical values for *ABC* parameters for several types of resists are given in Table 5.4.

As discussed above, one impact of post-apply baking for conventional resists is the decomposition of sensitizer. Since it is the sensitizer that bleaches, a loss of sensitizer results in a decrease in the value of A. Figure 5.12 shows two example transmittance curves measured for an i-line resist baked at two different temperatures.

Problems

5.1. Assuming a plane wave traveling through a uniform material, derive Equation (5.5).

5.2. For a resist film of a given thickness, there is one value of the absorption coefficient that maximizes the amount of light absorbed by the resist at the bottom. (This can be seen by looking at the extremes. When α is zero, all of the incident light makes it to the bottom, but none is absorbed. When α is infinite, no light makes it to the bottom, so none is absorbed there. Thus, in between these two extremes there must be a maximum.) For a resist of thickness D, and assuming a homogenous material and a nonreflecting substrate, what value of α maximizes the amount of light absorbed by the resist at the bottom?

5.3.   An i-line resist has the following properties:

$$A = 0.85\,\mu m^{-1}$$

$$B = 0.05\,\mu m^{-1}$$

$$C = 0.018\,cm^2/mJ$$

$$\text{Refractive index} = 1.72$$

The resist is coated to a thickness of $1.1\,\mu m$ on a glass substrate optically matched to the photoresist.

(a) At the beginning of exposure, what percentage of the incident light makes it to the bottom of the resist?

(b) The photoresist is exposed so that 55% of the sensitizer is converted at the top of the resist. Give your best estimate of what percentage of the incident light makes it to the bottom of the resist at the end of the exposure. Explain why this estimate is not exact.

5.4.   For most conventional (g- and i-line) resists, $a_M \gg a_P$. For this case, show that a measurement of bleachable absorption coefficient as a function of wavelength, $A(\lambda)$, can be used to determine how the exposure rate constant C varies with wavelength.

5.5.   Derive Equations (5.35).

5.6.   Reciprocity during exposure is seen for the case of a nonbleaching resist by transforming Equation (5.29) to Equation (5.31) – all instances of incident intensity and exposure time can be replaced by incident dose.

(a) Show that the first Equation (5.35) obeys reciprocity.

(b) Show that the general first-order exposure Equation (5.27) results in reciprocity regardless of the effects of bleaching.

5.7.   Suppose that a resist was found to produce the same thickness after PAB for a 95 °C, 300-second bake, a 100 °C, 60-second bake, or a 105 °C, 17-second bake. If the DNQ of that resist decomposes with an activation energy of 31 Kcal/mole and $\ln(Ar) = 37$ (1/minute), which bake is preferable from a thermal decomposition perspective?

5.8.   The modified Fujita–Doolittle Equation (5.61) predicts a temperature dependence for solvent diffusivity which is not the Arrhenius equation. Using the parameters from Table 5.1 and assuming 10% solvent volume fraction, create an Arrhenius plot (log [diffusivity] vs. 1/absolute temperature) for solvent diffusivity from 115 to 130 °C. How close is this to true Arrhenius behavior? Find a best estimate for A_r and E_a based on this plot.

5.9.   Solve the 1D diffusion equation for a 'fixed source' boundary condition (diffusion of a constant concentration of material at the surface of a very thick resist). Assume the diffusivity is constant and

$$C(z,0) = 0$$

$$C(0,t) = C_0$$

$$C(\infty,t) = 0$$

5.10. Solve the 1D diffusion equation for the 'initial source' boundary condition (diffusion of a given amount material at the surface of a very thick resist). Assume the diffusivity is constant and

$$C(z,0) = 0$$

$$dC(0,t)/dt = 0$$

$$C(\infty,t) = 0$$

$$\int_0^\infty C(z, t)dz = Q_0 = \text{constant}$$

5.11. For a typical 300-mm silicon wafer baked at a hot plate temperature of 100 °C (ambient temperature of 20 °C), and assuming a proximity gap of 0.2 mm, calculate:
   (a) The equilibrium wafer temperature ($T*$) and the rise time of the bake (τ).
   (b) The equilibrium wafer temperature and the rise time of the bake if the proximity gap were increased to 0.3 mm.
   (c) At what proximity gap distance would the coefficient of conduction (heat transfer from the hot plate across the gap) equal that of convection (thermal transfer from the top of the wafer)?

5.12. For a typical 300-mm silicon wafer baked at a hot plate temperature of 100 °C and a proximity gap of 0.2 mm, the equilibrium wafer temperature was found to vary by ±0.2 °C across the wafer. How much wafer warpage would be needed to explain all of this temperature variation?

5.13. Derive Equation (5.79).

5.14. From the transmittance curve in Figure 5.11, estimate the values of A, B and C. The resist thickness used was 0.75 μm and the measurement was performed in the standard way. Assume a typical i-line resist.

References

1 Dill, F.H., Hornberger, W.P., Hauge, P.S. and Shaw, J.M., 1975, Characterization of positive photoresist, *IEEE Transactions on Electron Devices*, **ED-22**, 445–452.

2 Mack, C.A., 1988, Absorption and exposure in positive photoresist, *Applied Optics*, **27**, 4913–4919.

3 Dammel, R., 1993, *Diazonaphthoquinone-based Resists*, SPIE Tutorial Texts, Vol. TT 11 (Bellingham, WA).

4 Albers, J. and Novotny, D.B., 1980, Intensity dependence of photochemical reaction rates for photoresists, *Journal of the Electrochemical Society*, **127**, 1400–1403.

5 Herrick, C.E., Jr., 1966, Solution of the partial differential equations describing photo-decomposition in a light-absorbing matrix having light-absorbing photoproducts, *IBM Journal of Research and Development*, **10**, 2–5.

6   Dill, F.H. and Shaw, J.M., 1977, Thermal effects on the photoresist AZ1350J, *IBM Journal of Research and Development*, **21**, 210–218.
7   Shaw, J.M., Frisch, M.A. and Dill, F.H., 1977, Thermal analysis of positive photoresist films by mass spectrometry, *IBM Journal of Research and Development*, **21**, 219–226.
8   Mack, C.A. and Carback, R.T., 1985, Modeling the effects of prebake on positive resist processing, *Proceedings of the Kodak Microelectronics Seminar*, 155–158.
9   Doolittle, A.K., 1951, Studies in Newtonian flow. II. The dependence of the viscosity of liquids on free-space, *Journal of Applied Physics*, **22**, 1471–1475.
10 Bueche, F., 1953, Segmental mobility of polymers near their glass temperature, *Journal of Chemical Physics*, **21**, 1850–1855.
11 Fox, T.G. and Flory, P.J. 1950, Second-order transition temperatures and related properties in polystyrene. I. Influence of molecular weight, *Journal of Applied Physics*, **21**, 581–591.
12 Fox, T.G. and Flory, P.J., 1951, Further studies on the melt viscosity of polyisobutylene, *Journal of Physical and Colloid Chemistry*, **55**, 221–234.
13 Fujita, H., Kishimoto, A. and Matsumoto, K., 1960, Concentration and temperature dependence of diffusion coefficients for systems polymethyl acrylate and n-alkyl acetates, *Transactions of the Faraday Society*, **56**, 424–437.
14 Mack, C.A., 1998, Modeling solvent effects in optical lithography, PhD Thesis, University of Texas at Austin.
15 Williams, M.L., Landel, R.F. and Ferry, J.D., 1955, The temperature dependence of relaxation mechanisms in amorphous polymers and other glass-forming liquids, *Journal of the American Chemical Society*, **77**, 3701–3706.
16 Cohen, M.H. and Turnbull, D., 1959, Molecular transport in liquids and gases, *Journal of Chemical Physics*, **31**, 1164–1169.
17 Ferry, J.D., 1980, *Viscoelastic Properties of Polymers*, 3rd edition, John Wiley & Sons, Ltd (New York, NY).
18 Mack, C.A., Mueller, K.E., Gardiner, A.B., Qiu, A., Dammel, R.R., Koros, W.G. and Willson, C.G., 1997, Diffusivity measurements in polymers, part 1: lithographic modeling results, *Proceedings of SPIE: Advances in Resist Technology and Processing XIV*, **3049**, 355–362.
19 Walker, E.J., 1975, Reduction of photoresist standing-wave effects by post-exposure bake, *IEEE Transactions on Electron Devices*, **ED-22**, 464–466.
20 Mack, C.A., DeWitt, D.P., Tsai, B.K. and Yetter, G., 1994, Modeling of solvent evaporation effects for hot plate baking of photoresist, *Proceedings of SPIE: Advances in Resist Technology and Processing XI*, **2195**, 584–595.
21 Smith, M.D., Mack, C.A. and Petersen, J.S., 2001, Modeling the impact of thermal history during post exposure bake on the lithographic performance of chemically amplified resists, *Proceedings of SPIE: Advances in Resist Technology and Processing XVIII*, **4345**, 1013–1021.
22 Tay, A., Ho, W.K. and Hu, N., 2007, An *in situ* approach to real-time spatial control of steady-state wafer temperature during thermal processing in microlithography, *IEEE Transactions on Semiconductor Manufacturing*, **20**, 5–12.

6

Chemically Amplified Resists: Exposure and Bake Chemistry

Unlike conventional resists, such as the diazonaphthoquinone (DNQ)/novolac systems discussed in the previous chapter, chemically amplified resists require two separate chemical reactions in order to change the solubility of the resists. First, exposure turns an aerial image into a latent image of exposure reaction products. Although very similar to conventional resists, the reaction products of exposure for a chemically amplified resist do not appreciably change the solubility of the resist. Instead, a second reaction during a post-exposure bake is catalyzed by the exposure reaction products. The result of the post-exposure bake reaction is a change in the solubility of the resist. This two-step process has some interesting characteristics and challenges.

In many respects, chemically amplified resists are a conceptual superset of conventional resists. With the exception of thermal decomposition, all of the concepts discussed in the previous chapter apply to chemically amplified resists as well. Familiarity with the concepts of that chapter is thus assumed in this chapter.

6.1 Exposure Reaction

Chemically amplified photoresists are composed of a polymer resin (possibly 'blocked' to inhibit dissolution), a sensitizer called a *photoacid generator* (PAG), and possibly a cross-linking agent, dye or other additive. As the name implies, the PAG forms an acid when exposed to deep-UV light. Ito and Willson at IBM first proposed the use of an aryl onium salt,[1] and triphenylsulfonium salts have been studied extensively as PAGs. The reaction of a very simple triphenylsulfonium salt PAG is:

$$
\begin{array}{c}
\text{Ph} \\
| \\
\text{Ph}-\overset{+}{\text{S}}\;\text{CF}_3\text{COO}^- \xrightarrow{h\nu} \text{CF}_3\text{COOH} + \text{others} \\
| \\
\text{Ph}
\end{array}
\qquad (6.1)
$$

Fundamental Principles of Optical Lithography: The Science of Microfabrication, Chris Mack.
© 2007 John Wiley & Sons, Ltd.

The acid generated in this case (trifluoroacetic acid) is a derivative of acetic acid where the electron-drawing properties of the fluorines greatly increase the acidity of the molecule. In fact, very strong acids will be required for the chemical amplification step described in the next section. The PAG is mixed with the polymer resin at a concentration of typically 5–15% by weight for 248-nm resists, with 10% as a typical formulation (corresponding to a PAG concentration of about 0.2 mol/liter). For 193-nm resists, PAG loading is kept lower at 1–5 wt% in order to keep the optical absorbance of the resist within desired levels.

The kinetics of the exposure reaction are presumed to be standard first order (just as for the exposure of PAC in a conventional resist):

$$\frac{\partial G}{\partial t} = -CIG \tag{6.2}$$

where G is the concentration of PAG at exposure time t (the initial PAG concentration is G_0), I is the exposure intensity, and C is the exposure rate constant. For most chemically amplified resists, the bleachable absorption coefficient A is zero and the intensity in the resist stays constant during exposure. For this case, the rate equation can be solved for G:

$$G = G_0 e^{-CIt} \tag{6.3}$$

The acid concentration H is given by

$$H = G_0 - G = G_0(1 - e^{-CIt}) \tag{6.4}$$

Some 248-nm PAGs react to exposure by a different mechanism. The polymer resin used by the resist (such as polyhydroxystyrene, discussed below) will absorb light, resulting in electrons promoted to higher energy states in the polymer. It is possible for some of this energy to be transferred from the polymer to the PAG, providing an indirect pathway for photon absorption to provide energy to the PAG. If this mechanism exists in a resist, the resulting kinetics will still be well described by the first-order kinetics discussed above.

6.2   Chemical Amplification

Exposure of the resist with an aerial image $I(x)$ results in an acid latent image $H(x)$. A post-exposure bake (PEB) is then used to thermally induce a chemical reaction. This may be the activation of a cross-linking agent for a negative resist or the deblocking of the polymer resin for a positive resist. The defining characteristic of a chemically amplified resist is that this reaction is *catalyzed* by the acid so that the acid is not consumed by the reaction and, to first order, H remains constant. As a result, each exposure event (creating one acid molecule) can cause a large number of dissolution-changing chemical events during PEB. The effects of exposure are said to be 'amplified' by the catalytic nature of this chemical reaction. The *catalytic chain length*, defined to be the average number of amplification chemical events caused by one acid molecule, is in the range of 10 to 100 for most chemically amplified resists, with 20 a typical catalytic chain length. Chemically amplified resists use exposure doses in the range of 20–50 mJ/cm^2, making them 5–10 times more sensitive than the DNQ resists used at i-line.

6.2.1   Amplification Reaction

Willson, Ito and Frechet first proposed the concept of deblocking a polymer to change its solubility.[2] A base polymer, such as polyhydroxystyrene (PHS), is used that is very soluble in an aqueous base developer. It is the acidic hydroxyl (—OH) groups that give the PHS its high solubility, so by 'blocking' these sites (by reacting the hydroxyl group with some longer-chain molecule) the solubility can be reduced. The IBM team employed a *t*-butoxycarbonyl group (*t*-BOC), resulting in a very slowly dissolving polymer. In the presence of a strong acid and heat, the *t*-BOC blocked polymer will undergo acidolysis to generate the soluble hydroxyl group, as shown below.

$$\text{(6.5)}$$

One drawback of this scheme is that the cleaved *t*-BOC is volatile and will evaporate, causing film shrinkage in the exposed areas. Larger molecular weight blocking groups are commonly used to reduce this film shrinkage to acceptable levels (below 10%). Also, the blocking group is such an effective inhibitor of dissolution that nearly every blocked site on the polymer must be deblocked to obtain significant dissolution. Thus, the photoresist can be made more 'sensitive' by only partially blocking the PHS (creating what is called a copolymer, since the polymer chain is now made up of two repeating units – blocked and unblocked PHS). Additionally, fully blocked polymers tend to have poor coating and adhesion properties. Typical photoresists use 10–30% of the hydroxyl groups blocked, with 20% as a typical value. Molecular weights for the PHS run in the range of 3000–6000 giving about 20–40 hydroxyl groups per polymer molecule, about 4 to 10 of which are initially blocked in typical formulations. As an example, the early chemically amplified resist Apex-E has a molecular weight of about 5000, giving about 35 hydroxyl sites per polymer, about 25% of which are initially blocked by *t*-BOC.

This deblocking reaction (also called the deprotection reaction) is often characterized as being either low activation or high activation, referring to the amount of heat that must be supplied to make the reaction go. For a low activation resist, deblocking can occur at room temperature, and so begins during the exposure process. For a high activation resist, a post-exposure bake in the 95–135 °C temperature range is required. While low activation resists have an advantage in not being sensitive to delay time effects (see section 6.2.3), high activation resists tend to be more readily controlled (produce more uniform resist feature sizes) and are thus the most commonly used. Typical high and low activation blocking groups are shown in Figure 6.1. Many resists today in fact include copolymers of both high and low activation blocking groups.

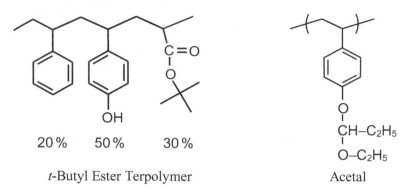

Figure 6.1   *Examples of 248-nm blocking groups: the high activation t-butyl ester; and the low activation acetal*

Using M as the concentration of some reactive site (such as the t-BOC blocking group), these sites are consumed (i.e. are reacted) according to kinetics of first order in H and first order in M:

$$\frac{\partial M}{\partial t_{PEB}} = -k_4 M H \tag{6.6}$$

where k_4 is the rate constant of the amplification reaction (cross-linking, deblocking, etc.) and t_{PEB} is the bake time. Assuming that H is constant (an assumption that will be eliminated later in this chapter), Equation (6.6) can be solved:

$$M = M_0 e^{-k_4 H t_{PEB}} \tag{6.7}$$

(Note: Although H⁺ is not consumed by the reaction, the value of H is not locally constant. Diffusion during the PEB and acid loss mechanisms cause local changes in the acid concentration, thus requiring the use of a reaction–diffusion system of equations as discussed below. The approximation that H is constant is a useful one, however, that gives insight into the reaction as well as accurate results under some conditions.)

It is useful here to normalize the concentrations to some initial values. This results in a normalized acid concentration h and normalized unreacted sites m:

$$h = \frac{H}{G_0} \quad m = \frac{M}{M_0} \tag{6.8}$$

Equations (6.4) and (6.7) become

$$h = 1 - e^{-CIt}$$
$$m = e^{-K_{amp} t_{PEB} h} \tag{6.9}$$

where $K_{amp} = G_0 k_4$. The result of the PEB is an amplified latent image $m(x)$, corresponding to an exposed latent image $h(x)$, resulting from the aerial image $I(x)$. The amount of amplification (the conversion of an exposed latent image of acid into an amplified latent image of blocked and deblocked polymer) is a function of PEB time and temperature. The time dependence in Equation (6.9) is obvious, and the temperature dependence comes through the amplification rate constant K_{amp}.

Table 6.1   *Typical values for amplification reaction activation energies as a function of blocking group*

Blocking Group (248-nm resists)	Activation Energy (kcal/mol)
t-BOC	30
t-Butyl Ester	22
Acetal	15

Like most rate constants, the temperature dependence of K_{amp} can be described by an Arrhenius relation:

$$K_{amp} = A_r e^{-E_a/RT} \tag{6.10}$$

where A_r is the Arrhenius coefficient for this reaction (1/s), E_a is the activation energy (cal/mole or J/mole), R is the universal gas constant (1.98717 cal/mole-K or 8.31431 J/mole-K), and T is the absolute temperature (K). Typical values for activation energies as a function of blocking group are given in Table 6.1. The resulting K_{amp} at the PEB temperature tends to be in the range of 0.02–0.1 s^{-1}.

Equations (6.9) reveal an interesting symmetry between exposure dose, given by It, and a kind of 'thermal dose'. Consider the case of a small exposure dose ($It \ll 1/C$) so that the amount of acid generated is small. Equations (6.9) will become

$$h \approx CIt$$
$$m \approx e^{-K_{amp}t_{PEB}CIt} \tag{6.11}$$

If we define the effective thermal dose as $K_{amp}t_{PEB}/C$, then there exists a reciprocity between exposure dose and thermal dose. Cutting exposure dose in half and doubling the thermal dose will result in the same chemical change at the end of the amplification reaction. As we shall see in the following section, however, the true picture is complicated by the effects of diffusion.

6.2.2   Diffusion

The above analysis of the kinetics of the amplification reaction assumed a locally constant concentration of acid H. Although this could be exactly true in some circumstances, it is typically only an approximation, and is often a poor approximation. In reality, the acid diffuses during the bake. The standard diffusion equation takes the form

$$\frac{\partial H}{\partial t_{PEB}} = \nabla(D_H \nabla H) \tag{6.12}$$

where D_H is the diffusivity of acid in the photoresist. Solving this equation requires a number of things: two boundary conditions for each dimension, one initial condition and knowledge of the diffusivity as a function of position and time.

The solution of Equation (6.12) can now be performed if the diffusivity of the acid in the photoresist is known. Unfortunately, this solution is complicated by two very important factors: the diffusivity is a strong function of temperature and, most probably, the extent of amplification. Since the temperature is changing with time during the bake, the diffusivity will be time dependent. The concentration dependence of diffusivity results from an increase in free volume for typical positive resists: as the amplification reaction proceeds, the polymer blocking group evaporates resulting in a decrease in film thickness but also an increase in free volume (and probably a change in the glass transition temperature as well). Since the acid concentration is time and position dependent, the diffusivity in Equation (6.12) must be determined as a part of the solution of Equation (6.12) by an iterative method. The resulting simultaneous solution of Equations (6.7) and (6.12) is called a *reaction–diffusion* system.

The temperature dependence of the diffusivity can be expressed in a standard Arrhenius form:

$$D_0(T) = A_r e^{-E_a/RT} \tag{6.13}$$

where D_0 is a general diffusivity. A full treatment of the amplification reaction would include a thermal model of the hot plate in order to determine the actual time–temperature history of the wafer (see Chapter 5). To simplify the problem, an ideal temperature distribution can be assumed – the temperature of the resist is zero (low enough for no diffusion or reaction) until the start of the bake, at which time it immediately rises to the final bake temperature, stays constant for the duration of the bake, then instantly falls back to zero.

The concentration dependence of the diffusivity is less obvious. Several authors have proposed and verified the use of different models for the concentration dependence of diffusion within a polymer. Of course, the simplest form (besides a constant diffusivity) would be a linear model. Letting D_0 be the diffusivity of acid in completely unreacted resist and D_f the diffusivity of acid in resist that has been completely reacted,

$$D_H = D_0 + (D_f - D_0)(1 - m) \tag{6.14}$$

Here, diffusivity is expressed as a function of the extent of the amplification reaction $1 - m$. Another common form is the Fujita–Doolittle equation (see Chapter 5), which can be predicted theoretically using free volume arguments. A form of that equation that is convenient for calculations is:

$$D_H = D_0 \exp\left(\frac{D_2(1-m)}{1+D_3(1-m)}\right) \tag{6.15}$$

where $D_2 = (1 + D_3)\ln(D_f/D_0)$ and D_3 is an experimentally determined constant. Other concentration relationships are also possible.

While general solutions to these reaction–diffusion equations require numerical techniques, there are some special cases where analytical solutions are possible. Consider a constant acid diffusivity and an ideal temperature profile (that is, ignore the ramp up to the bake temperature). Further, ignore any sources of acid loss (to be discussed in the next section). Then, the solution to the reaction rate equation given in Equation (6.9) can be used if the relative acid concentration is replaced by an effective acid concentration, given by

$$h_{\text{eff}}(x) = \frac{1}{t_{\text{PEB}}} \int_0^{t_{\text{PEB}}} h(x, t = 0) \otimes DPSF \, dt \tag{6.16}$$

where $h(x, t = 0)$ is the acid concentration at the beginning of the bake and t_{PEB} is the PEB bake time. Diffusion is given by the convolution of the acid profile with the diffusion point spread function (*DPSF*, defined in Chapter 5). The effective acid concentration at a particular x-position is the average concentration at that position over time. This equation can be further arranged to define a reaction–diffusion point spread function, *RDPSF*:

$$h_{\text{eff}}(x) = h(x,0) \otimes RDPSF$$

where

$$RDPSF = \frac{1}{t_{PEB}} \int_0^{t_{PEB}} DPSF \; dt \tag{6.17}$$

The Gaussian diffusion kernel is affected by time integration through the diffusion length, $\sigma_D = \sqrt{2Dt}$, where D is the acid diffusivity in resist. Thus, using the 1D case as an example,

$$RDPSF = \frac{1}{t_{PEB}\sqrt{4\pi D}} \int_0^{t_{PEB}} \frac{e^{-x^2/4Dt}}{\sqrt{t}} \; dt \tag{6.18}$$

The integral is solvable, resulting in an interesting final solution.

$$RDPSF(x) = 2\frac{e^{-x^2/2\sigma_D^2}}{\sqrt{2\pi}\sigma_D} - \frac{|x|}{\sigma_D^2}\text{erfc}\left(\frac{|x|}{\sqrt{2}\sigma_D}\right) \tag{6.19}$$

The first term on the right hand side of Equation (6.19) is nothing more than twice the *DPSF*, and thus accounts for pure diffusion. The second term, the complimentary error function times x, is a reaction term that is subtracted and thus reduces the impact of pure diffusion. Each term, as well as the final *RDPSF*, is plotted in Figure 6.2.

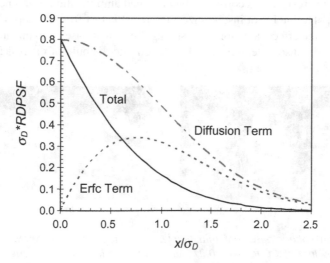

Figure 6.2 *The 1D reaction–diffusion point spread function (RDPSF) and its component terms*

Equation (6.19) gives the *RDPSF* in one dimension. Extension to two or three dimensions is straightforward:

$$2D: \quad RDPSF(r) = -\frac{1}{2\pi\sigma_D^2} Ei\left(-\frac{r^2}{2\sigma_D^2}\right), \quad r = \sqrt{x^2 + y^2}$$

$$3D: \quad RDPSF(r) = \frac{1}{2\pi\sigma_D^3}\left[\frac{\sigma_D}{|r|}\text{erfc}\left(\frac{|r|}{\sqrt{2}\,\sigma_D}\right)\right], \quad r = \sqrt{x^2 + y^2 + z^2}$$

(6.20)

where *Ei* is the exponential integral. Once the effective acid concentration is known, the amount of deblocking can be calculated as before with a simple modification to Equation (6.9):

$$m = e^{-K_{amp}t_{PEB}h_{eff}}$$

(6.21)

6.2.3   Acid Loss

Through a variety of mechanisms, acid formed by exposure of the resist film can be lost and thus not contribute to the catalyzed reaction to change the resist solubility. There are two basic types of acid loss – loss that occurs between exposure and post-exposure bake, and loss that occurs during the post-exposure bake. The first type of loss leads to delay time effects – the resulting lithographic patterns are affected by the delay time between exposure and post-exposure bake. Delay time effects can be very severe and, of course, are very detrimental to the use of such a resist in a manufacturing environment. Acid loss during PEB may be detrimental (if, for example, it is nonuniform or affects only the top or the bottom of the resist profile), or it may serve the beneficial function of stopping the reaction and limiting the sensitivity of the final feature size to bake time or temperature.

The typical mechanism for delay time acid loss is the diffusion of atmospheric base contaminants into the top surface of the resist. The result is a neutralization of the acid near the top of the resist and a corresponding reduced amplification. For a negative resist, the top portion of a line is not insolubilized and resist is lost from the top of the line. For a positive resist, the effects are more devastating. Sufficient base contamination can make the top of the resist insoluble, blocking dissolution into the bulk of the resist (Figure 6.3)

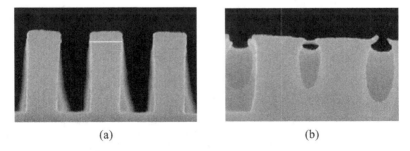

(a)   (b)

Figure 6.3 *Atmospheric base contamination leads to T-top formation. Shown are line/space features printed in APEX-E for: (a) 0.275-µm features with no delay; and (b) 0.325-µm features with 10-minute delay between exposure and post-exposure bake (courtesy of SEMATECH)*

and creating a characteristic 'T-top'-shaped resist profile. In extreme cases, no patterns are observed after development.

The effects of acid loss due to atmospheric base contaminants can be accounted for in a straightforward manner. The base diffuses from the top surface of the resist into the bulk. Assuming that the concentration of base contaminant in contact with the top of the resist remains constant, the diffusion equation can be solved for the concentration of base, B, as a function of depth into the resist film:

$$B = B_0 \, \mathrm{erfc}\left(\frac{z}{\sqrt{2}\sigma_B}\right) \tag{6.22}$$

where B_0 is the base concentration at the top of the resist film, z is the depth into the resist ($z = 0$ at the top of the film) and σ_B is the diffusion length of the base in resist. B_0 will be directly proportional to the concentration of base in the atmosphere by Henry's law (see Problem 6.8). The standard assumption of constant diffusivity has been made here so that the diffusion length goes as the square root of the delay time and as the square root of the diffusivity. Note that this equation assumes that during the post-exposure delay period, the base diffusion length is much greater than the photogenerated acid diffusion length. Since the acid generated by exposure for most resist systems of interest is fairly strong, it is a good approximation to assume that all of the base contaminant will react with acid if there is sufficient acid present.

Atmospheric base contamination can be devastating to chemically amplified resist performance because of the very nature of chemical amplification. Since one photo-generated acid causes dozens of dissolution changing events during PEB, loss of even a small amount of acid before the PEB will have an amplified effect. Several different countermeasures are usually taken to reduce the impact of atmospheric base. As Equation (6.22) shows, base contamination in the resist can be reduced by reducing the atmospheric base (B_0) or by reducing the base diffusion length. In turn, diffusion length can be reduced by reducing the base diffusivity or by reducing the delay time. In practice, all three of these techniques are commonly used. Photolithography equipment is placed as far as possible from sources of base contaminants (resist strippers and ammonia being common examples). Imaging and resist processing tools are then enclosed and the air is filtered with activated charcoal. The delay time between exposure and post-exposure bake is kept to a minimum by directly linking the resist processing track to the step and scan exposure tool. Each wafer is sent to be baked immediately after its exposure is complete. Finally, the diffusivity of base in the resist is reduced by 'annealing' the resist. If the post-apply bake temperature is set to be above the glass transition temperature of the resist polymer, solvent and free volume will be annealed out of the resist (see Chapter 5). It is this free volume that enables the diffusion of base into the resist, so elimination of free volume greatly reduces the contaminant diffusivity.

Another possible acid loss mechanism is base contamination from the substrate, as has been observed on TiN and other nitrogen-containing substrates. These substrates are thought to trap acid that diffuses down to the substrate, resulting in resist footing for a positive resist. Plasma and chemical treatments of nitrogen containing substrates can pacify trapping sites, but the common use of an organic bottom antireflection coating (BARC) generally eliminates the possibility of substrate interactions. Acid can also be lost at the top surface of the resist due to evaporation. The amount of evaporation is a function of the size of the acid and the degree of its interaction with the resist polymer. A small acid (such as the trifluoroacetic acid discussed above) may have very significant evaporation. Thus, most resists today employ fairly large acid molecules to avoid significant acid evaporation.

6.2.4   Base Quencher

Another mechanism for acid loss is intentional rather than accidental. Most modern formulations of chemically amplified resists include the addition of a base quencher (compounds such as tetraoctylammonium hydroxide, for example[3]). Loaded at concentrations of 5–20% of the initial PAG loading, this base quencher (Q) is designed to neutralize any photogenerated acid that comes in contact with it.

$$H + Q \xleftrightarrow{\;K_{eq}\;} HQ \qquad\qquad (6.23)$$

Acid–base neutralization reactions tend to be equilibrium reactions, but in this case the equilibrium constant K_{eq} is large due to the strength of the acid, meaning the reaction favors heavily the formation of the salt HQ. Thus, it is a good approximation to ignore the reversible nature of the reaction and replace the equilibrium constant with a standard forward reaction rate constant, k_Q.

For low exposure doses, the small amount of photoacid generated will be neutralized by the base quencher and amplification will not take place (Figure 6.4). Only when the exposure rises above a certain threshold will the amount of acid be sufficient to completely neutralize all of the base quencher and have leftover acid that can cause deblocking during PEB. The main purpose of the base quencher is to neutralize the low levels of acid that might diffuse into the nominally unexposed regions of the wafer, thus making the final resist linewidth less sensitive to acid diffusion. In addition, base quencher will reduce the sensitivity of the resist to airborne base contaminants.

The simple description of base quenching behavior above is made more complicated by the fact that the quencher will, in general, diffuse during the post-exposure bake. Base quencher diffusion can take on any of the concentration and temperature dependencies described above for acid diffusion. The difference in diffusivity between the acid and the base becomes an important descriptor of lithographic behavior for these types of

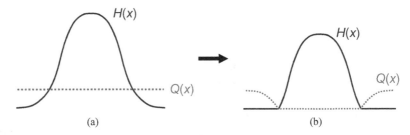

Figure 6.4   *The effect of quencher (Q) on the acid latent image H(x): (a) after exposure but before the quenching reaction; and (b) after the quenching reaction (assuming $K_{quench} \gg K_{amp}$)*

resists. Rate equations for the acid and base, for the case of constant diffusivity, are given by

$$\frac{dh}{dt} = -K_{quench}hq + D_H\nabla^2 h$$

$$\frac{dq}{dt} = -K_{quench}hq + D_Q\nabla^2 q$$

(6.24)

where K_{quench} ($= k_Q G_0$) is the normalized quenching reaction rate constant and q ($= Q/G_0$) is the normalized base quencher concentration. The impact of diffusion can be understood qualitatively by examining Figure 6.4b. If acid diffuses, but base does not, the zero points (the points at which the acid and base completely neutralize each other) will move toward the outside. If the base diffuses but the acid does not, the opposite will occur and the zero points will move in toward the center of the figure. There will be some combination of acid and base diffusion (diffusion lengths about equal) that will leave the zero point stationary. Note that diffusion is driven by the *change* in the gradient of the diffusing species (a second derivative), so that it is a combination of the diffusivities of the acid and the base and the shape of the latent image that determines which direction the zero point will move.

If diffusion is ignored and $K_{quench} \gg K_{amp}$, the impact of a base quencher is to first order a dose offset – one must generate enough acid to overcome the quencher before amplification can occur. Thus, the simple (nondiffusion) kinetic result of Equation (6.9) becomes

$$m = e^{-K_{amp}t_{PEB}(h-q_0)} \quad \text{when} \quad h > q_0$$

$$m = 1 \quad \text{when} \quad h < q_0$$

(6.25)

where the '0' subscript refers to the initial condition (at the beginning of the PEB). Note that in the formulation of the above equation, h represents the relative acid concentration generated upon exposure, but before any quenching takes place.

The impact of base quencher is most readily seen by comparing latent images of resist formulations with and without quencher. When both resist formulations are exposed at the dose-to-size, with all other processing parameters the same, the latent image slope improves dramatically near the resist feature edge when quencher is added (Figure 6.5).

6.2.5 Reaction–Diffusion Systems

The combination of a reacting system and a diffusing system is called a reaction–diffusion system. The solution of such a system is the simultaneous solution of Equations (6.6) and (6.12) using Equation (6.4) as an initial condition and possibly Equation (6.14) or (6.15) to describe the reaction-dependent diffusivity. Of course, any or all of the acid loss mechanisms can also be included.

The coupling of reaction and diffusion can be made more explicit. The Byers–Petersen model for chemically amplified reactions[4] is a superset of the simpler chemically amplified models discussed above. In this model, the amplification reaction is thought to occur by two sequential steps: first, acid diffuses to the reaction site; then, acid reacts at that site (an idea first described by the Polish physicist Marian von Smoluchowski[5] in 1917). If the diffusion step is very fast, the overall rate is controlled by the rate of the reaction

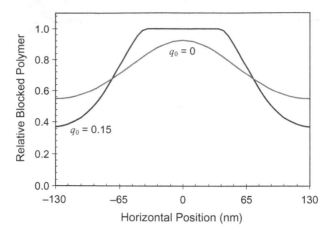

Figure 6.5 *The effect of quencher on the shape of the latent image. Both resists have identical processing, except that the dose for each is adjusted to be the dose-to-size (130-nm lines and spaces)*

itself (the amplification is said to be reaction-controlled). If, however, the diffusion is very slow, the overall rate of amplification will be controlled by the rate of diffusion of acid to the site (a diffusion-controlled amplification rate). If K_{react} is the rate constant for the reaction step, and $K_{diff}D_H$ is the rate constant (equal to the Smoluchowski trap rate normalized in the same way as K_{amp}) for the diffusion step, the overall amplification reaction rate constant will have the form

$$K_{amp} = \frac{K_{react}K_{diff}D_H}{K_{react} + K_{diff}D_H} \tag{6.26}$$

where D_H is the diffusivity of the acid.

Note that if K_{diff} is very large, then even for moderate values of acid diffusivity the kinetics will be reaction-controlled and Equation (6.26) will become

$$K_{amp} = K_{react} \tag{6.27}$$

and the Byers–Petersen model will revert to the original chemically amplified resist model given above. In general, the activation energies for K_{react} and for D_H will be different and it is possible for the overall reaction to switch from diffusion-controlled to reaction-controlled as a function of PEB temperature. For example, if the activation energy for the diffusion portion of the reaction is much smaller than the reaction activation energy, the overall reaction will be reaction-controlled at lower temperatures and diffusion-controlled at higher temperatures (Figure 6.6). K_{diff} represents the likelihood that the acid will be captured by a blocked site so that the reaction can occur. In the diffusion-controlled region, K_{react} is very high and

$$K_{amp} = K_{diff}D_H = 4\pi D_H a G_0 N_A \tag{6.28}$$

where a is the capture distance for the deblocking reaction (once the acid approaches the reaction site closer than the capture distance, a deblocking reaction is possible).

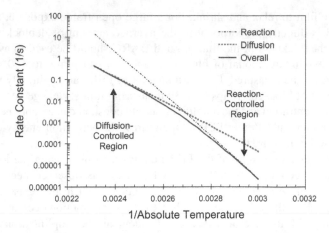

Figure 6.6 *An Arrhenius plot (log of the rate constant versus the inverse of absolute temperature) for K_{amp} when the diffusion and amplification rate constants have different activation energies (diffusion activation energy = 26.5 kcal/mol; reaction activation energy = 45 kcal/mol)*

As was discussed at the end of section 6.2.1, there is a trade-off in chemically amplified resists between exposure dose and thermal dose. Ignoring diffusion and acid loss, there is an approximate reciprocity between the two forms of dose. Generally, thermal dose is increased by an increase in the bake temperature (for production throughput reasons, PEB time is generally varied only over a small window). However, as the above discussion has shown, diffusion also varies exponentially with PEB temperature. Since there is an upper limit to the desired diffusion length of acid (and of base quencher), there is a practical upper limit to the temperature that can be used. And because of the exponential nature of the thermal dose with temperature, there is a practical minimum temperature that still provides for a low-enough dose to give good exposure tool throughput. Typically, the useful range of PEB temperatures for chemically amplified resists is at most 20 °C.

6.3 Measuring Chemically Amplified Resist Parameters

Since most chemically amplified resists do not bleach, the traditional optical transmittance experiment for measuring the *ABC* parameters gives no information about *C*. Other approaches have been devised to measure *C* that take advantage of the strength of the acid being generated in a chemically amplified resist. Direct measurement of the acid concentration is possible using titration, or indirectly using fluorescence. Alternatively, resist formulations with varying amounts of added base can be used to measure acid generation by determining the dose required to overcome the added base.

Deblocking during the chemical amplification reaction leads to the generation of volatile components that evaporate out of the resist during PEB. This slightly unpleasant side effect of amplification can be used as a way to measure the kinetics of the amplification reaction.[6] Evaporation of volatile components at the relatively high PEB temperatures used allows the film to relax and shrink. For a large open-frame exposure, the thickness loss after PEB is directly proportional to the average amount of deblocking that takes place through the thickness of the film (a good BARC should be used to avoid standing waves). If the maximum amount of film shrinkage, corresponding to 100% deblocking of the polymer, can be measured, film shrinkage after PEB can be directly related to the fractional amount of blocked groups left. By measuring film shrinkage versus dose, both the quencher concentration and the amplification rate constant can be obtained. By repeating this measurement at different PEB temperatures, the activation energy of the amplification can also be measured.

Figure 6.7 shows an example of the PEB kinetics measured using the PEB thickness loss measurement technique. Initially, no thickness loss is observed because the doses are not large enough to overcome the base quencher present in the resist. Past the dose required to neutralize all of the base quencher, the blocked fraction decreases as an exponential function, with the rate of decrease dependent on the amplification rate constant (and thus on the temperature). This figure also illustrates one of the major difficulties of this technique. Because the PEB time is fixed, there is no guarantee that even at the highest temperatures and highest exposure doses the resist will approach 100% deblocking. It is quite difficult to ensure that the measurement of the 100% deblocking thickness loss used to calibrate this technique does in fact correspond to 100% deblocking.

A more direct method of measuring chemical kinetics during PEB involves the use of Fourier transform infrared (FTIR) spectroscopy. Absorbance peaks as a function of wavelength indicate specific bonds in the various chemical species in the resist. As bonds are broken and formed during amplification, peaks of the FTIR spectrum appear and disappear, providing a quantifiable measure of the extent of reaction. FTIR equipment mounted directly above the PEB hot plate can measure these spectra in real time during the bake. However, since silicon is mostly transparent to IR light, a backside wafer coating of metal is required to provide a reflected signal to detect,[7] or a hole in the hot plate must be used for measurement in transmission mode.[8]

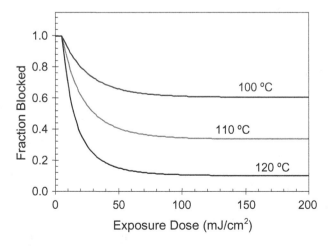

Figure 6.7 *Example of PEB amplification kinetics measured by using thickness loss after PEB to estimate the fraction of polymer that remains blocked*

The most difficult parameters to measure for chemically amplified resists are the acid and base diffusivities (the difficulties increase if these diffusivities vary with extent of amplification). In general, these parameters are empirically determined by matching predictions to experimental linewidth data.

6.4   Stochastic Modeling of Resist Chemistry

The theoretical descriptions of lithography given so far in this book make an extremely fundamental and mostly unstated assumption about the physical world being described: the so-called *continuum approximation*. Even though light energy is quantized into photons and chemical concentrations are quantized into spatially distributed molecules, the descriptions of aerial images and latent images ignore the discrete nature of these fundamental units and use instead continuous mathematical functions. For example, the very idea of chemical concentration assumes that the volume one is interested in is large enough to contain many, many molecules so that an average number of molecules per unit volume can be used. While we can mathematically discuss the idea of the concentration of some chemical species at a point in space, in reality this concentration must be an average extended over a large enough region. While in most cases the volumes of interest are large enough not to worry about this distinction, when trying to understand lithography down to the nanometer level the continuum approximation begins to break down.

When describing lithographic behavior at the nanometer level, an alternate approach, and in a very real sense a more fundamental approach, is to build the quantization of light as photons and matter as atoms and molecules directly into the models used. Such an approach is called *stochastic modeling*, and involves the use of random variables and probability density functions to describe the statistical fluctuations that are expected. Of course, such a probabilistic description will not make deterministic predictions – instead, quantities of interest will be described by their probability distributions, which in turn are characterized by their moments, such as the mean and variance.

6.4.1   Photon Shot Noise

Consider a light source that randomly emits photons at an average rate of L photons per unit time into some area A. Assume further that each emission event is independent. Over some small time interval dt (smaller than $1/L$ and small enough so that it is essentially impossible for two photons to be emitted during that interval), either a photon is emitted or it is not (a binary proposition). The probability that a photon will be emitted during this interval will be Ldt. Consider now some long time T ($\gg dt$). What can we expect for the number of photons emitted during the period T? This basic problem is called a Bernoulli trial and the resulting probability distribution is the well-known *binomial distribution*. If $N = T/dt$, the number of time intervals in the total time, then the probability that exactly n photons will be emitted in this time period, $P(n)$, will be

$$P(n) = \binom{N}{n}(Ldt)^n(1-Ldt)^{N-n} = \frac{N!}{(N-n)!n!}(Ldt)^n(1-Ldt)^{N-n} \qquad (6.29)$$

The binomial distribution is extremely cumbersome to work with as N gets large. If, however, $NLdt = TL$ remains finite as N goes to infinity, the binomial distribution converges to another, more manageable equation called the Poisson distribution:

$$\lim_{N \to \infty} \binom{N}{n} (Ldt)^n (1 - Ldt)^{N-n} = \frac{(TL)^n}{n!} e^{-TL} \tag{6.30}$$

Since there is no limit to how small dt can be made, letting dt go to zero will by default make N go to infinity for any nonzero time interval T and nonzero photon emission rate L.

The Poisson distribution can be used to derive the statistical properties of photon emission. The expectation value of n [that is, the mean number of photons that will be emitted in a time interval T, denoted by the notation $E(n)$ or $\langle n \rangle$] is TL (a very reasonable result since L was defined as the average rate of photon emission):

$$\langle n \rangle \equiv E(n) = \sum_{n=0}^{\infty} n P(n)$$
$$= \sum_{n=0}^{\infty} n \frac{(TL)^n}{n!} e^{-TL} = e^{-TL} \sum_{n=1}^{\infty} \frac{(TL)^n}{(n-1)!} = TL e^{-TL} \sum_{m=0}^{\infty} \frac{(TL)^m}{(n)!} = TL \tag{6.31}$$

The variance (the standard deviation squared) is also TL.

$$\sigma_n^2 \equiv E\left[(n - \langle n \rangle)^2 \right] = E(n^2) - \langle n \rangle^2 = TL \tag{6.32}$$

To use these statistical properties, we must convert from number of photons to a more useful measure, intensity. If n photons cross an area A over a time interval T, the intensity of light will be

$$I = \frac{n}{TA} \left(\frac{hc}{\lambda} \right) \tag{6.33}$$

The mean value of this intensity will be

$$\langle I \rangle = \frac{\langle n \rangle}{TA} \left(\frac{hc}{\lambda} \right) = \frac{L}{A} \left(\frac{hc}{\lambda} \right) \tag{6.34}$$

The standard deviation of the intensity can also be computed from the properties of the Poisson distribution.

$$\sigma_I = \frac{1}{TA} \left(\frac{hc}{\lambda} \right) \sigma_n = \frac{\langle I \rangle}{\sqrt{TL}} = \frac{\langle I \rangle}{\sqrt{\langle n \rangle}} \tag{6.35}$$

As this equation shows, the uncertainty of getting the mean or expected intensity grows as the number of photons is reduced, a phenomenon known as *shot noise*. Perhaps a more useful form is to consider the standard deviation as a fraction of the average value.

$$\frac{\sigma_I}{\langle I \rangle} = \frac{\sigma_n}{\langle n \rangle} = \frac{1}{\sqrt{\langle n \rangle}} = \left[\langle I \rangle TA \left(\frac{\lambda}{hc} \right) \right]^{-1/2} \tag{6.36}$$

The shot noise (the relative uncertainty in the actual intensity that the resist will see) increases with decreasing intensity, exposure time and area of concern.

As an example, consider a 193-nm exposure of a resist with a dose-to-clear of 10 mJ/ cm^2. At the resist edge, the mean exposure energy ($=\langle I \rangle T$) will be on the order of the dose-to-clear. At this wavelength, the energy of one photon, hc/λ, is about 1.03×10^{-18} J. For an area of 1×1 nm, the mean number of photons during the exposure, from Equation (6.34), is about 97. The standard deviation is about 10, or about 10% of the average. For an area of 10 nm $\times$ 10 nm, the number of photons increases by a factor of 100, and the relative standard deviation decreases by a factor of 10, to about 1%. Since these are typical values for a 193-nm lithography process, we can see that shot noise contributes a noticeable amount of uncertainty as to the actual dose seen by the photoresist when looking at length scales less than about 10 nm.

For *Extreme Ultraviolet Lithography* (EUVL), the situation will be considerably worse. At a wavelength of 13.5 nm, the energy of one photon will be 1.47×10^{-17} J, about 15 times greater than at 193 nm. Also, the goal for resist sensitivity will be to have EUV resists that are 2–4 times more sensitive than 193-nm resists (though it is unclear whether this goal will be achieved). Thus, the number of photons will be 30–60 times less for EUV than for a 193-nm lithography. A 1×1 nm area will see only two to three photons, and a 100-nm^2 area will see on the order of 200 photons, with a standard deviation of 7%.

6.4.2 Chemical Concentration

As mentioned above, there really is no such thing as concentration at a point in space since the chemical species is discrete, not continuous. Concentration, the average number of molecules per unit volume, exhibits counting statistics identical to photon emission. Let C be the average number of molecules per unit volume, and dV a volume small enough so that at most one molecule may be found in it (thus requiring that the concentration be fairly dilute, so that the position of one molecule is independent of the position of other molecules). The probability of finding a molecule in that volume is just CdV. For some larger volume V, the probability of finding exactly n molecules in that volume will be given by a binomial distribution exactly equivalent to that for photon counting. And, as before, this binomial distribution will also become a Poisson distribution by letting dV go to zero.

$$P(n) = \frac{(CV)^n}{n!} e^{-CV} \tag{6.37}$$

The average number of molecules in the volume will be CV, and the variance will also be CV. The relative uncertainty in the number of molecules in a certain volume will be

$$\frac{\sigma_n}{\langle n \rangle} = \frac{1}{\sqrt{\langle n \rangle}} = \frac{1}{\sqrt{CV}} \tag{6.38}$$

[The requirement that the concentration be 'dilute' can be expressed as an upper limit to the Poisson distribution – for a given molecule size, saturation occurs at some n_{max} molecules in the volume V. So long as $P(n_{max})$ is small, the mixture can be said to be dilute.[9]]

As an example, consider a 193-nm resist that has an initial PAG concentration of 3% by weight, or a concentration of about 0.07 mole/liter (corresponding to a density of

1.2 g/ml and a PAG molecular weight of 500 g/mole). Converting from moles to molecules with Avogadro's number, this corresponds to 0.042 molecules of PAG per cubic nanometer. In a volume of $(10 \text{ nm})^3$, the mean number of PAG molecules will be 42 (see Figure 6.8). The standard deviation will be 6.5 molecules, or about 15%. For 248-nm resists, the PAG loading is typically 3 times higher or more, so that closer to 150 PAG molecules might be found in a $(10\text{-nm})^3$ volume, for a standard deviation of 8%. Note that when the mean number of molecules in a given volume exceeds about 20, the Poisson distribution can be well approximated with a Gaussian distribution.

As mentioned briefly above, Poisson statistics apply only for reasonably low concentrations. The random distribution of molecules assumes that the position of each molecule is independent of all the others. As concentrations get higher, the molecules begin to 'crowd' each other, reducing their randomness. In the extreme limit, molecules become densely packed and the uncertainty in concentration goes to zero. This saturation condition is a function of not only the concentration, but the size of the molecule as well. To avoid saturation, the volume fraction occupied by the molecules under consideration must be small.

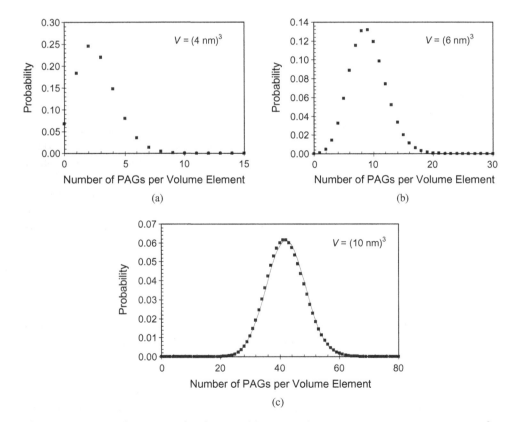

Figure 6.8 *Example Poisson distributions for a typical 193-nm resist ($G_0 N_A = 0.042 / nm^3$): (a) volume of 64 nm³; (b) volume of 216 nm³; and (c) volume of 1000 nm³. For (c), a Gaussian distribution with the same mean and standard deviation as the Poisson distribution is shown as a solid line*

6.4.3 Some Mathematics of Binary Random Variables

The mathematical approach to be taken below in considering the stochastic nature of molecular events (such as absorption, reaction or diffusion) is to define a binary random variable to represent the result of that event for a single molecule. For example, we may let y be a random variable that represents whether a given single PAG molecule was converted to acid or remains unexposed by the end of the exposure process. Thus, $y = 1$ if the event has happened, and $y = 0$ if the event has not happened. Then, by considering the kinetics of the event or some other physical model of how that event might happen, we assign a probability P to the event. Since the event is binary, $P(y = 1) = 1 - P(y = 0)$.

We next consider some properties of the random variable y. In particular, what is the mean and variance of y? The mean value of y uses the definition of the expectation value for a discrete random variable:

$$\langle y \rangle \equiv E(y) = \sum_{i=0,1} y_i P(y_i) = (0)P(0) + (1)P(1) = P(1) \tag{6.39}$$

Likewise, the variance of y is

$$\sigma_y^2 \equiv E((y - \langle y \rangle)^2) = E(y^2) - \langle y \rangle^2 \tag{6.40}$$

But, since y is binary, $y^2 = y$ and

$$\sigma_y^2 = \langle y \rangle - \langle y \rangle^2 = \langle y \rangle (1 - \langle y \rangle) \tag{6.41}$$

Thus, if an appropriate probability for a molecular event can be assigned, the mean and variance of the binary random variable y, representing that event for a single molecule, can easily be calculated. Of course, these results are simply the properties of a binomial distribution, which is the probability distribution of a binary random variable. Interestingly, the random variable y has no uncertainty at the extremes of the event process ($\langle y \rangle = 1$ or $\langle y \rangle = 0$), and maximum uncertainty when $\langle y \rangle = 0.5$.

Now consider a collection of molecules. Suppose a volume V contains some number n molecules. After the event process is complete (e.g. at the end of exposure), the number of molecules Y in that volume that underwent the event will be

$$Y = \sum_{i=1}^{n} y_i \tag{6.42}$$

where y_i is the discrete random variable representing the state of the ith molecule found in this volume (each event in the volume is assumed independent). For a given n, the mean and variance of Y can be readily computed:

$$\langle Y | n \rangle = \sum_{i=1}^{n} \langle y_i \rangle = n \langle y \rangle$$

$$\sigma_{Y|n}^2 = \sum_{i=1}^{n} \sigma_y^2 = n \sigma_y^2 \tag{6.43}$$

where the notation $\langle a | b \rangle$ signifies the mean value of a for a given b.

But n itself is not a fixed quantity and will in general follow a Poisson statistical distribution, as discussed in section 6.4.2. To find the mean value of Y including the statistical variation of n,

$$E(Y) = \langle Y \rangle = \sum_{n=0}^{\infty} \langle Y|n \rangle \frac{\langle n \rangle^n}{n!} e^{-\langle n \rangle} = \sum_{n=0}^{\infty} \langle y \rangle n \frac{\langle n \rangle^n}{n!} e^{-\langle n \rangle} = \langle n \rangle \langle y \rangle \qquad (6.44)$$

Likewise, the variance of Y can be computed from

$$\sigma_Y^2 = E(Y - \langle Y \rangle)^2 = E(Y^2) - \langle Y \rangle^2 \qquad (6.45)$$

Starting the calculation of the expectation value of Y^2,

$$E(Y^2) = E\left[\left(\sum_{i=1}^{n} y_i \right)^2 \right] = E\left(\sum_{i=1}^{n} \sum_{j=1}^{n} y_i y_j \right) = E\left(\sum_{i=1}^{n} \sum_{j=1}^{n} E(y_i y_j) \right) \qquad (6.46)$$

Assuming that the exposure state of each molecule is independent of the states of the others,

$$E(y_i y_j) = E(y^2), \quad i = j$$
$$E(y_i y_j) = E(y_i)E(y_j), \quad i \neq j \qquad (6.47)$$

giving

$$E(Y^2) = E[n(n-1)\langle y \rangle^2 + nE(y^2)] = E[n^2 \langle y \rangle^2 + n\sigma_y^2] = \langle y \rangle^2 E(n^2) + \sigma_y^2 E(n) \qquad (6.48)$$

But, since

$$E(n^2) = \sigma_n^2 + \langle n \rangle^2 \qquad (6.49)$$

we can combine Equations (6.45), (6.48) and (6.49) to give

$$\sigma_Y^2 = \langle y \rangle^2 \sigma_n^2 + \langle n \rangle \sigma_y^2 \qquad (6.50)$$

The variance of n, for a Poisson distribution, is just the mean value of n. Thus,

$$\sigma_Y^2 = \langle n \rangle (\langle y \rangle^2 + \sigma_y^2) \qquad (6.51)$$

Using the variance of y from Equation (6.41),

$$\sigma_Y^2 = \langle n \rangle \left[\langle y \rangle^2 + \langle y \rangle (1 - \langle y \rangle) \right] = \langle n \rangle \langle y \rangle = \langle Y \rangle \qquad (6.52)$$

Like the Poisson distribution of molecules before the event process begins, the variance of the final distribution is equal to its mean. In the sections below, these simple results will be applied to the specific chemical events of exposure, diffusion and reaction.

6.4.4   Photon Absorption and Exposure

What is the probability that a photon will be absorbed by a molecule of light-sensitive material in the resist? Further, what is the probability that a molecule of sensitizer will react to form an acid? As discussed above, there will be a statistical uncertainty in the number of photons in a given region of resist, a statistical uncertainty in the number of PAG molecules, and additionally a new statistical uncertainty in the absorption and exposure event itself.

Consider a single molecule of PAG. First-order kinetics of exposure was used to derive Equation (6.3), the concentration of PAG remaining after exposure (and, as well, the concentration of acid generated) in the continuum approximation (this is also called the *mean-field* solution to the kinetics of exposure). From a stochastic modeling perspective, this kinetic result represents a probability density function for reaction: G/G_0 is the fraction of PAG that is unreacted in some large volume, and by the Law of Large Numbers this must be the probability that any given PAG will remain unexposed. Let y be a random variable that represents whether a given single PAG molecule remains unexposed or was converted to acid by the end of the exposure process. Thus $y = 0$ means an acid has been generated (PAG has reacted), and $y = 1$ means the PAG has not been exposed (no acid generated). A kinetic analysis of exposure gives us the probability for each of these states, given a certain intensity-in-resist I:

$$P(y = 0 | I) = 1 - e^{-CIt}, \quad P(y = 1 | I) = e^{-CIt} \tag{6.53}$$

For a given intensity, the mean value of y can be calculated using the definition of a discrete probability expectation value:

$$\langle y | I \rangle = P(1 | I) = e^{-CIt} \tag{6.54}$$

Likewise, the variance of y for a given intensity is

$$\sigma_{y|I}^2 = \langle y \rangle (1 - \langle y \rangle) = e^{-CIt}(1 - e^{-CIt}) \tag{6.55}$$

Note that the above equations were carefully derived under the assumption of a given intensity of light. However, we know from our discussion of photon counting statistics that I is a probabilistic function [the aerial image intensity $I(x)$ can be thought of as proportional to the probability of finding a photon at position x]. Thus, the mean and variance of y must take into account this probabilistic nature. Letting n be the number of photons exposing a given area A over an exposure time t [related to I by Equation (6.33)], it will be useful to define a new constant:

$$CIt = \psi n \quad \text{which leads to } \psi = \left(\frac{hc}{\lambda} \right) \frac{C}{A} = \frac{\Phi a_M}{N_A A} = \frac{\Phi \sigma_{M-\text{abs}}}{A} \tag{6.56}$$

where the last two equations come from the microscopic absorption analysis presented in Chapter 5. The term ψ is the *exposure shot noise coefficient*, and is equal to the exposure quantum efficiency multiplied by the ratio of the PAG absorption cross section to the area of statistical interest. Since the exposure quantum efficiency is typically in the 0.3–0.7 range and the PAG absorption cross section is on the order of 1 Å^2 for 193-nm resists, for most areas of interest this exposure shot noise coefficient will be much less than 1.

Using this exposure shot noise coefficient to convert intensity to number of photons, the total probability becomes

$$P(y \text{ and } n) = P(y|n)P(n) = \begin{cases} (1 - e^{-\psi n}) \dfrac{\langle n \rangle^n}{n!} e^{-\langle n \rangle} & \text{for } y = 0 \\[3mm] e^{-\psi n} \dfrac{\langle n \rangle^n}{n!} e^{-\langle n \rangle} & \text{for } y = 1 \end{cases} \tag{6.57}$$

$$\langle y \rangle = \sum_{n=0}^{\infty} \sum_{y} y P(y \text{ and } n) = \sum_{n=0}^{\infty} (e^{-\psi n}) \frac{\langle n \rangle^n}{n!} e^{-\langle n \rangle} \tag{6.58}$$

Evaluating the summation (and recognizing the Taylor series expansion of an exponential),

$$\langle y \rangle = \sum_{n=0}^{\infty} \frac{(e^{-\psi} \langle n \rangle)^n}{n!} e^{-\langle n \rangle} = e^{-\langle n \rangle(1-e^{-\psi})} \tag{6.59}$$

When ψ is small, the exact expression (6.59) can be approximated as

$$\langle y \rangle = e^{-\langle n \rangle(1-e^{-\psi})} \approx e^{-\langle n \rangle(\psi - 0.5\psi^2)}$$

$$= \left[e^{-\langle n \rangle \psi} \right]^{1-\psi/2} = \left[e^{-C\langle I \rangle t} \right]^{1-\psi/2} = e^{-C\langle I \rangle t} e^{\psi^2 \langle n \rangle/2} = e^{-C\langle I \rangle t} e^{\frac{1}{2}(Ct\sigma_I)^2} \tag{6.60}$$

This final approximate form can be derived in an alternate way. For a reasonable number of photons (greater than about 20), the Poisson distribution of photons is well approximated by a normal distribution of the same mean and variance. A normal distribution for I will result in a log-normal distribution for y. The final equation on the right-hand side of Equation (6.60) is the well-known result for the mean of a log-normal distribution.

As often happens when taking statistical distributions into account, the mean value of the output of the function is not equal to the function evaluated at the mean value of the input. Another approximate form for the mean value of y is

$$\langle y \rangle \approx y_c + \frac{\psi}{2}(y_c)\ln(y_c) \quad \text{where } y_c = e^{-C\langle I \rangle t} \tag{6.61}$$

Since $(y_c)\ln(y_c)$ is always negative (it goes between zero and about -0.37), the mean value of y is always less than the value of the function y evaluated at the mean value of the intensity. When averaging over a large area, ψ goes to zero, so that y_c can be thought of as the mean value of y when averaged over a large area (that is, in the continuum limit). As the area over which we average our results decreases, the mean value of y (which is the same as the mean value of the relative acid concentration, as we will see below) decreases due to shot noise.

The mean value and uncertainty of the state of *one* acid molecule after exposure can now be translated into a mean and uncertainty of the overall acid concentration after exposure. Consider a volume V that initially contains some number n_{0-PAG} PAG molecules. After exposure, the number of remaining (unexposed) PAG molecules Y will be

$$Y = \sum_{i=1}^{n_{0-PAG}} y_i \tag{6.62}$$

Assuming that each exposure event is independent, the results from section 6.4.3 give the mean of Y:

$$\langle Y \rangle = \langle n_{0-PAG} \rangle \langle y \rangle \tag{6.63}$$

The variance becomes

$$\sigma_Y^2 = \langle Y \rangle + \langle Y \rangle^2 (e^{(Ct\sigma_I)^2} - 1) = \langle Y \rangle + \langle Y \rangle^2 (e^{\psi^2 \langle n \rangle} - 1) \tag{6.64}$$

Since ψ will in general be small,

$$\sigma_Y^2 \approx \langle Y \rangle + \langle Y \rangle^2 (\psi^2 \langle n \rangle) = \langle Y \rangle + \langle Y \rangle^2 (C \langle I \rangle t) \psi = \langle Y \rangle + \frac{\langle Y \rangle^2 (C \langle I \rangle t)^2}{\langle n \rangle} \qquad (6.65)$$

The variance of Y has two components. The Poisson chemical distribution gives the first term, $\langle Y \rangle$. Photon shot noise adds a second term, inversely proportional to the mean number of photons.

At this point it is useful to relate the number of remaining PAG molecules per unit volume Y to the concentration of acid H, and the initial number of PAGs n_{0-PAG} to the initial PAG concentration G_0.

$$H = G_0 - \frac{Y}{N_A V}, \quad G_0 = \frac{n_{0-PAG}}{N_A V} \qquad (6.66)$$

We can also define a relative acid concentration h to be

$$h = \frac{H}{\langle G_0 \rangle} = \frac{G_0}{\langle G_0 \rangle} - \frac{Y}{\langle n_{0-PAG} \rangle} = \frac{n_{0-PAG}}{\langle n_{0-PAG} \rangle} - \frac{Y}{\langle n_{0-PAG} \rangle} \qquad (6.67)$$

The means of these quantities can be related by

$$\langle h \rangle = \frac{\langle H \rangle}{\langle G_0 \rangle} = 1 - \frac{\langle Y \rangle}{\langle n_{0-PAG} \rangle} = 1 - \langle y \rangle \qquad (6.68)$$

Using Equation (6.67),

$$\sigma_h^2 = \frac{\langle h \rangle}{\langle n_{0-PAG} \rangle} + \frac{\sigma_Y^2 - \langle Y \rangle}{\langle n_{0-PAG} \rangle^2} \qquad (6.69)$$

Finally, using Equation (6.65), the variance in acid concentration will be

$$\sigma_h^2 = \frac{\langle h \rangle}{\langle n_{0-PAG} \rangle} + \frac{[(1 - \langle h \rangle) \ln (1 - \langle h \rangle)]^2}{\langle n \rangle} \qquad (6.70)$$

This final result, which accounts for photon fluctuations, uncertainty in the initial concentration of photoacid generator, and the probabilistic variations in the exposure reaction itself, is reasonably intuitive. The first term on the right-hand side of Equation (6.70) is the expected Poisson result based on exposure kinetics – the relative uncertainty in the resulting acid concentration after exposure goes as one over the square root of the mean number of acid molecules generated within the volume of interest. For large volumes and reasonably large exposure doses, the number of acid molecules generated is large and the statistical uncertainty in the acid concentration becomes small. For small volumes or low doses, a small number of photogenerated acid molecules results in a large uncertainty in the actual number within that volume. The second term accounts for photon shot noise. For the case of the $(10\,nm)^3$ of 193-nm resist given above, the standard deviation in initial acid concentration near the resist edge (where the mean acid concentration will be about 0.4) will be >20%. For 193-nm resists, the impact of photon shot noise is minimal compared to variance in acid concentration caused by simple molecular position uncertainty.

6.4.5   Acid Diffusion, Conventional Resist

In Chapter 5, diffusion was treated using a Gaussian diffusion point spread function (DPSF) – the latent image after diffusion is the initial latent image convolved with a Gaussian, whose standard deviation σ is called the diffusion length. From a stochastic modeling perspective, the Gaussian represents a probability distribution: for a particle initially at some position, the Gaussian *DPSF*, given by

$$DPSF = (2\pi\sigma^2)^{-3/2} e^{-r^2/2\sigma^2} \qquad (6.71)$$

represents the probability density of finding that particle some distance r from its original location. This probability distribution is itself derived from a stochastic look at the possible motions of the particle during the bake. Given that the particle can randomly move in any possible direction at a particular speed determined by its diffusivity, and can change directions randomly, the resulting path of the particle is called a *random walk*. Averaging over all possible random walk paths produces the Gaussian probability distribution (when the diffusivity is constant).

How does diffusion affect the statistical fluctuations of concentration? The latent image of acid after exposure, $h(x,y,z)$, as used in the continuum approximation, is actually the mean relative concentration $\langle h \rangle$, with a variance given by Equation (6.70). Ignoring the photon shot noise, the probability distribution for the number of acids in a given volume is Poisson. Now consider a PEB process that causes only diffusion of acid (such as for a conventional resist). The mean acid concentration will 'diffuse' as was described in the continuum approximation of Chapter 5, convolving the initial acid latent image with the *DPSF*. The stochastic uncertainty in acid concentration will still be given by Equation (6.70), but with a new mean 'after diffusion' concentration of acid $\langle h^* \rangle$. Ignoring photon shot noise,

$$\sigma_{h^*} = \sqrt{\frac{\langle h^* \rangle}{\langle n_{o-\mathrm{PAG}} \rangle}}, \quad \langle h^* \rangle = \langle h \rangle \otimes DPSF \qquad (6.72)$$

The proof of this result (which will be sketched briefly here) becomes evident by considering first the diffusion of a single acid molecule, located at some initial position designated by the subscript i. Defining a binary random variable y_i to represent whether that molecule is found in some small volume dV located a distance r_i from its original location,

$$P(y_i = 1) = (2\pi\sigma^2)^{-3/2} e^{-r_i^2/2\sigma^2} dV \qquad (6.73)$$

This binary random variable will follow all of the properties derived in section 6.4.3:

$$\langle y_i \rangle = P(y_i = 1), \quad \sigma_{y_i}^2 = \langle y_i \rangle (1 - \langle y_i \rangle) \qquad (6.74)$$

If, instead of one acid molecule at a certain location which then diffuses, there are n_i acid molecules at this location that then diffuse, the total number of acid molecules in that volume dV will be Y_i, again with the properties derived in section 6.4.3:

$$Y_i = \sum_{k=1}^{n_i} y_{ik}, \quad \langle Y_i \rangle = \langle n_i \rangle \langle y_i \rangle, \quad \sigma_{Y_i}^2 = \langle Y_i \rangle \qquad (6.75)$$

Adding up the contributions from all of the locations that could possibly contribute acid molecules into the volume dV produces the convolution result of Equation (6.72).

$$Y = \sum_i Y_i$$

$$\langle Y \rangle = \sum_i \langle Y_i \rangle = \sum_i \langle n_i \rangle \langle y_i \rangle = \sum_i \langle n_i \rangle (2\pi\sigma^2)^{-3/2} e^{-n^2/2\sigma^2} dV \qquad (6.76)$$

Here, a summation over i is equivalent to an integration over all space, so that the final summation becomes an integral, which in this case becomes a convolution.

$$\langle Y \rangle = \langle n \rangle \otimes DPSF \qquad (6.77)$$

How does diffusion affect the variance of the post-diffusion concentration?

$$\sigma_Y^2 = E(Y^2) - \langle Y \rangle^2$$

$$E(Y^2) = E\left[\left(\sum_i Y_i\right)^2\right] = E\left(\sum_i Y_i \sum_j Y_j\right) = \sum_i \sum_j E(Y_i Y_j) = \sum_i \sum_j E\left(\sum_{k=1}^{n_i} \sum_{l=1}^{n_j} E(y_{ik} y_{jl})\right) \qquad (6.78)$$

Since the diffusion of a given molecule is independent of all of the other molecules,

$$E(y_{ik} y_{jl}) = E(y_{ik}^2), \quad ik = jl$$
$$E(y_{ik} y_{jl}) = E(y_{ik})E(y_{jl}), \quad ik \neq jl \qquad (6.79)$$

As a result, the variance follows in the same way as shown in section 6.4.3:

$$\sigma_Y^2 = \sum_i \sigma_{Y_i}^2 = \langle Y \rangle \qquad (6.80)$$

Thus, diffusion does not intrinsically increase the uncertainty in acid concentration due to the extra stochastic process of the random walk, nor does it intrinsically 'smooth out' any uncertainties through the process of diffusion. Only through a change in the mean concentration does diffusion affect the uncertainty in the concentration. For a typical conventional resist, the concentration of exposure products near the edge of the resist feature does not change appreciably due to diffusion during PEB. Thus, it is unlikely that diffusion will have a significant impact on the concentration statistics near the resist line edge.

6.4.6 Acid-Catalyzed Reaction–Diffusion

Of course, for a chemically amplified resist acid diffusion is accompanied by one or more reactions. In this section and the next, we'll consider only the polymer deblocking reaction (acid-quencher neutralization will be discussed in a following section). In the continuum limit, the amount of blocked polymer left after the PEB is given by

$$M = M_0 e^{-K_{\text{amp}} t_{\text{PEB}} h_{\text{eff}}}$$

$$h_{\text{eff}}(x,y,x) = \frac{1}{t_{\text{PEB}}} \int_0^{t_{\text{PEB}}} (h(x,y,z,t=0) \otimes DPSF)dt = h(x,y,z,t=0) \otimes RDPSF \qquad (6.81)$$

As before, the latent image of acid after exposure, $h(x,y,z, t = 0)$ used in the continuum approximation is actually the mean acid concentration $\langle h \rangle$, with a standard deviation given above. The effective acid concentration, however, has a very specific interpretation: it is the time average of the acid concentration at a given point. The interesting question to be answered, then, is whether this time-averaging effect of diffusion coupled with the acid-catalyzed reaction affects the uncertainty in the effective acid concentration compared to the original acid concentration uncertainty.

To determine the statistical properties of the effective acid concentration, we'll begin as we did for the pure diffusion case in the previous section by looking at the diffusion of a single molecule of acid. Let the binary random variable $y_i(t)$ represent whether that molecule is found in some small volume dV located a distance r_i from its original location, during the interval of time between t and $t + dt$.

$$P(y_i(t) = 1) = (2\pi\sigma^2)^{-3/2} e^{-r_i^2/2\sigma^2} dV, \quad \sigma^2 = 2Dt \tag{6.82}$$

For n_i acid molecules at this location that then diffuse, the total number of acid molecules in that volume dV and over the same time interval will be $Y_i(t)$, again with the properties derived in the previous section:

$$Y_i(t) = \sum_{j=1}^{n_i} y_{ij}(t), \quad \langle Y_i(t) \rangle = \langle n_i \rangle \langle y_i(t) \rangle, \quad \sigma_{Y_i}^2 = \langle Y_i(t) \rangle \tag{6.83}$$

Adding up the contributions from all of the locations that could possibly contribute acid molecules into the volume dV during the interval of time between t and $t + dt$ produces the standard convolution result:

$$Y(t) = \sum_i Y_i(t)$$

$$\langle Y(t) \rangle = \sum_i \langle n_i \rangle \langle y_i(t) \rangle = \langle n \rangle \otimes DPSF(t) \tag{6.84}$$

So far, the results are identical to that for pure diffusion since we have only considered how many acids are in the volume at time t. We now wish to integrate over time, from 0 to t_{PEB}.

$$Y = \frac{1}{t_{PEB}} \int_0^{t_{PEB}} Y(t) dt$$

$$\langle Y \rangle = \frac{1}{t_{PEB}} \int_0^{t_{PEB}} \langle Y(t) \rangle dt = \frac{1}{t_{PEB}} \int_0^{t_{PEB}} \langle n \rangle \otimes DPSF(t) dt = \langle n \rangle \otimes RDPSF \tag{6.85}$$

Thus, as expected, the effective acid concentration used in the continuum approximation is in fact the mean value of a stochastic random variable. The uncertainty of Y, however, involves some extra complications. Proceeding as we have done in the past,

$$\sigma_Y^2 = E(Y^2) - \langle Y \rangle^2$$

$$E(Y^2) = \frac{1}{t_{PEB}^2} E\left[\left[\int_0^{t_{PEB}} Y(t) dt \right]^2 \right] = \frac{1}{t_{PEB}^2} E\left(\int_0^{t_{PEB}} Y(t) dt \int_0^{t_{PEB}} Y(t') dt' \right) \tag{6.86}$$

$$= \frac{1}{t_{PEB}^2} \int_0^{t_{PEB}} \int_0^{t_{PEB}} E(Y(t')Y(t)) dt dt'$$

What is different in this case compared to previous derivations is that $Y(t')$ and $Y(t)$ are not independent. If an acid finds itself in the volume of interest at time t, the probability of finding that same acid in the same volume at time t' will be greater than if the acid had not been found in the volume at time t. Thus, there will be some correlation between $Y(t')$ and $Y(t)$ that will be strongest when t' is close to t, and diminish as the difference between the two times increases.

Applying the definition of $Y(t)$ to Equation (6.86),

$$E(Y^2) = \frac{1}{t_{PEB}^2} \int_0^{t_{PEB}} \int_0^{t_{PEB}} E\left(\sum_i \sum_{j=1}^{n_i} y_{ij}(t) \sum_k \sum_{l=1}^{n_k} y_{kl}(t') \right) dt\ dt' \tag{6.87}$$

For $ij \neq kl$ (that is, for different acid molecules), there can be no correlation through time. Thus, the only time correlation comes from the same acid molecule venturing into the volume at different times. Changing the order of integration to bring the time integration inside and the spatial integration to the outside, we are interested in finding

$$\Lambda = \frac{1}{t_{PEB}^2} \int_0^{t_{PEB}} \int_0^{t_{PEB}} E\left(\sum_{j=1}^{n_i} y_{ij}(t) \sum_{l=1}^{n_i} y_{il}(t') \right) dt\ dt' \tag{6.88}$$

Since the only correlation through time exists when $j = l$,

$$E\left(\sum_{j=1}^n y_j(t) \sum_{l=1}^n y_l(t') \right) = E[n(n-1)\langle y(t)\rangle\langle y(t')\rangle + nE(y(t)y(t'))] \tag{6.89}$$

Following along the same lines as the derivation given in section 6.4.3, we obtain

$$\sigma_Y^2 = \langle n \rangle \otimes CovPSF$$

$$CovPSF = \frac{1}{dV\ t_{PEB}^2} \int_0^{t_{PEB}} \int_0^{t_{PEB}} E(y(t)y(t')) dt\ dt' \tag{6.90}$$

where *CovPSF* is a new function that I call the 'covariance point spread function'.

Consider the unphysical case of $y(t)$ and $y(t')$ being perfectly correlated. In that case, $y(t)y(t') = y(t)$ (since y is a binary variable) and the *CovPSF* is equal to the *RDPSF*. Thus, perfect correlation over the bake time produces the result that

$$\sigma_Y^2 = \langle n \rangle \otimes RDPSF = \langle Y \rangle \tag{6.91}$$

indicating that the final effective acid concentration would retain an approximately Poisson distribution. Without perfect correlation, the *CovPSF* will differ from the *RDPSF* and the variance will be different from the Poisson result.

Since $y(t)$ is a binary random variable, the expectation value is equal to the probability that both conditions are true (that is, the acid found its way into the volume at both time t and time t').

$$E(y(t')y(t)) = P(y(t') = 1 \text{ and } y(t) = 1) = P(y(t') = 1|y(t) = 1)P(y(t) = 1) \tag{6.92}$$

The conditional probability is governed by the difference in time between t and t'.

$$P(y(t') = 1|y(t) = 1) = \lim_{r' \to 0} \frac{e^{-r'^2/2\sigma^2}}{\sqrt{2\pi}\sigma} dV, \quad \sigma^2 = 2D|t - t'| \tag{6.93}$$

where the one-dimensional case is shown for convenience.

Thus, the *CovPSF* can be written as

$$CovPSF = \frac{dV}{4\pi D t_{PEB}^2} \int_0^{t_{PEB}} \frac{e^{-r^2/4Dt}}{\sqrt{t}} dt \int_0^{t_{PEB}} \frac{e^{-r'^2/4D|t-t'|}}{\sqrt{|t-t'|}} dt' \qquad (6.94)$$

where again the 1D version is shown for convenience. The inner integral over t' can be readily evaluated:

$$\lim_{r' \to 0} \frac{1}{\sqrt{4\pi D t_{PEB}}} \int_0^{t_{PEB}} \frac{e^{-r'^2/4D|t-t'|}}{\sqrt{|t-t'|}} dt' = \frac{t_{PEB} - t}{t_{PEB}} RDPSF(0, t_{PEB} - t) + \frac{t}{t_{PEB}} RDPSF(0, t) \quad (6.95)$$

Unfortunately, the next integral does not appear to have an analytical solution. Some properties of the integral, however, can be determined. When $r = 0$, the integral evaluates to

$$CovPSF(r = 0) = \frac{dV}{\sigma^2} \left(\frac{1}{2} + \frac{1}{\pi} \right), \quad \sigma^2 = 2Dt_{PEB} \qquad (6.96)$$

A numerical integration shows that the 1D *CovPSF* is very similar to the 1D *RDPSF*:

$$CovPSF(r) \approx 1.05 \frac{dV}{\sigma} RDPSF(r) \qquad (6.97)$$

A comparison of this approximate form with a numerical integration of Equation (6.94) is shown in Figure 6.9.

The impact of the *CovPSF* can now be determined. The shape of the *CovPSF* is very similar to that of the *RDPSF*. Since the effective acid concentration near the line edge does not differ appreciably from the acid concentration when convolved with

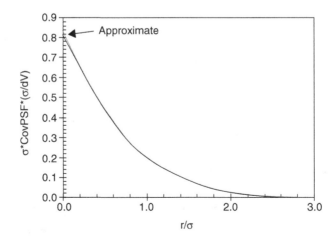

Figure 6.9 *Comparison of the approximate 1D CovPSF of Equation (6.97) with a numerical evaluation of the defining integral in Equation (6.94)*

the *RDPSF*, the same will be true when convolved with this portion of the *CovPSF*. Thus, the effective acid concentration and its standard deviation can be approximated as

$$\langle h_{\text{eff}} \rangle = \langle h \rangle \otimes RDPSF$$

$$\sigma_{h_{\text{eff}}} \approx \sqrt{\frac{dV}{\sigma_D}} \sigma_h \qquad (6.98)$$

The rest of the *CovPSF* simply gives the ratio dV/σ_D, where σ_D is the overall diffusion length. In this context, dV is the (one-dimensional) capture volume for the deblocking reaction. In other words, $dV/2$ represents how close the acid must come to a blocked site before it can potentially participate in a deblocking reaction (called the capture radius, a). As Equation (6.98) indicates, when the acid diffusion length exceeds the capture range of the deblocking reaction, the impact of the diffusing catalyst is to 'smooth out' the statistical fluctuations present in the original acid distribution.

While the above derivation used the one-dimensional form of the diffusion probability, the extension to three dimensions is straightforward. For a problem of dimensionality p ($p = 1, 2$ or 3),

$$\sigma_{h_{\text{eff}}} \approx \left(\frac{2a}{\sigma_D} \right)^{\frac{p}{2}} \sigma_h \qquad (6.99)$$

where a is the deblocking reaction capture distance. In other words, if the acid diffuses a distance less than the reaction capture range, the catalytic nature of the amplification reaction actually increases the stochastic variation in the effective acid concentration compared to the original acid concentration. If, however, the diffusion length is greater than this capture range, the time-averaging effect of the catalytic reaction will smooth out stochastic roughness. It is not diffusion, *per se*, that reduces stochastic uncertainty, but rather the diffusion of a reaction catalyst that does so. Since in real resist systems the diffusion length will invariably be greater than the reaction capture distance, the net affect will always be a reduction in the effective acid concentration standard deviation.

6.4.7   Reaction–Diffusion and Polymer Deblocking

The stochastics of the deblocking of a single blocked site will follow along the same lines as the single PAG exposure analysis of section 6.4.4. Let y be a random variable that represents whether a given single blocked site remains blocked by the end of the PEB. Thus $y = 1$ means the site remains blocked, and $y = 0$ means the site has been deblocked. As before, the continuum kinetic analysis gives us the probability that a single site is deblocked for a given effective acid concentration.

$$P(y = 1 \mid h_{\text{eff}}) = e^{-K_{\text{amp}} t_{\text{PEB}} h_{\text{eff}}}, \quad P(y = 0 \mid h_{\text{eff}}) = 1 - e^{-K_{\text{amp}} t_{\text{PEB}} h_{\text{eff}}} \qquad (6.100)$$

The probability distribution of h_{eff}, however, is not obvious. While the relative acid concentration has a Poisson distribution, the time-averaging effect on the acid diffusion turns the discrete acid random variable into a continuous effective acid random variable.

It will be reasonable to assume that h_{eff} is normally distributed with mean and standard deviations as given in the previous section. Thus, the mean value of y becomes

$$\langle y \rangle = \frac{1}{\sqrt{2\pi}\,\sigma_{h_{eff}}} \int_{-\infty}^{\infty} (e^{-K_{amp}t_{PEB}h_{eff}})e^{-(h_{eff}-\langle h_{eff} \rangle)^2/2\sigma_{h_{eff}}^2}\,dh_{eff}$$

$$\langle y \rangle = e^{-K_{amp}t_{PEB}\langle h_{eff} \rangle}e^{\frac{1}{2}(K_{amp}t_{PEB}\sigma_{h_{eff}})^2}$$

(6.101)

The random variable y has a log-normal probability distribution and Equation (6.101) can be recognized as the standard result for a log-normal distribution.

The total number of blocked groups remaining in a certain small volume will be given by Y.

$$Y = \sum_{i=1}^{n_{0-block}} y_i$$

(6.102)

The mean of Y can be easily computed, as before.

$$\langle Y \rangle = \langle n_{0-block} \rangle \langle y \rangle$$

(6.103)

The variance of Y can be found with a result similar to that for photon shot noise during exposure:

$$\sigma_Y^2 = \langle Y \rangle + \langle Y \rangle^2 (e^{(K_{amp}t_{PEB}\sigma_{h_{eff}})^2} - 1)$$

(6.104)

From the definitions of M and Y,

$$\langle M_0 \rangle = \frac{\langle n_{0-blocked} \rangle}{N_A V}, \quad \langle M \rangle = \frac{\langle Y \rangle}{N_A V}, \quad m \equiv \frac{M}{\langle M_0 \rangle}$$

(6.105)

Thus,

$$\langle m \rangle = \langle y \rangle$$

(6.106)

and

$$\sigma_m^2 = \frac{\sigma_M^2}{\langle M_0 \rangle^2} = \frac{\sigma_Y^2}{\langle n_{0-blocked} \rangle^2}$$

(6.107)

giving

$$\sigma_m^2 = \frac{\langle m \rangle}{\langle n_{0-blocked} \rangle} + \langle m \rangle^2 (e^{(K_{amp}t_{PEB}\sigma_{h_{eff}})^2} - 1)$$

(6.108)

For small levels of effective acid uncertainty,

$$\sigma_m^2 \approx \frac{\langle m \rangle}{\langle n_{0-blocked} \rangle} + \langle m \rangle^2 (K_{amp}t_{PEB}\sigma_{h_{eff}})^2 = \frac{\langle m \rangle}{\langle n_{0-blocked} \rangle} + (\langle m \rangle \ln\langle m \rangle)^2 \left(\frac{\sigma_{h_{eff}}}{\langle h_{eff} \rangle}\right)^2$$

(6.109)

As before, the first term captures the Poisson uncertainty due to the initial distribution of blocked polymer. The second term captures the influence of the effective acid concentration uncertainty. Combining this expression with Equation (6.99) for the variance of the effective acid concentration,

$$\sigma_m^2 = \frac{\langle m \rangle}{\langle n_{0-\text{blocked}} \rangle} + (\langle m \rangle \ln \langle m \rangle)^2 \left(\frac{\sigma_h}{\langle h \rangle} \right)^2 \left(\frac{2a}{\sigma_D} \right)^p \tag{6.110}$$

Finally, using Equation (6.70) for the variance of the acid concentration,

$$\sigma_m^2 = \frac{\langle m \rangle}{\langle n_{0-\text{blocked}} \rangle} + \left(\frac{\langle m \rangle \ln \langle m \rangle}{\langle h \rangle} \right)^2 \left(\frac{2a}{\sigma_D} \right)^p \left(\frac{\langle h \rangle}{\langle n_{0-\text{PAG}} \rangle} + \frac{[(1 - \langle h \rangle) \ln (1 - \langle h \rangle)]^2}{\langle n \rangle} \right) \tag{6.111}$$

Or, in a slightly different form,

$$\left(\frac{\sigma_m}{\langle m \rangle} \right)^2 = \frac{1}{\langle n_{0-\text{blocked}} \rangle \langle m \rangle} + (K_{\text{amp}} t_{\text{PEB}})^2 \left(\frac{2a}{\sigma_D} \right)^p \left(\frac{\langle h \rangle}{\langle n_{0-\text{PAG}} \rangle} + \frac{[(1 - \langle h \rangle) \ln (1 - \langle h \rangle)]^2}{\langle n \rangle} \right) \tag{6.112}$$

While the above equations show how fundamental parameters affect the resulting variance in the final blocked polymer concentration, interpretation is somewhat complicated by the fact that these parameters are not always independent. In particular, the Byers–Petersen model of Equation (6.26) shows a relationship between $K_{\text{amp}} t_{\text{PEB}}$ and $\sigma_D a$. For the case where the PEB temperature puts the deblocking reaction in the diffusion-controlled regime, Equation (6.28) can be expressed as

$$K_{\text{amp}} t_{\text{PEB}} = 2 \pi \sigma_D^2 a \langle G_0 \rangle N_A \tag{6.113}$$

When used in Equation (6.112), for the 3D case ($p = 3$), and ignoring photon shot noise, the result is

$$\left(\frac{\sigma_m}{\langle m \rangle} \right)^2 = \frac{1}{\langle n_{0-\text{blocked}} \rangle \langle m \rangle} + (2 \pi \sigma_D^2 a \langle G_0 \rangle N_A)^2 \left(\frac{2a}{\sigma_D} \right)^3 \left(\frac{\langle h \rangle}{\langle n_{0-\text{PAG}} \rangle} \right)$$

$$= \frac{1}{\langle n_{0-\text{blocked}} \rangle \langle m \rangle} + 32 \pi^2 \left(\frac{\sigma_D a^5}{V^2} \right) \langle h \rangle \langle n_{0-\text{PAG}} \rangle \tag{6.114}$$

$$= \frac{1}{V} \left[\frac{1}{\langle M_0 \rangle N_A \langle m \rangle} + 32 \pi^2 \sigma_D a^5 \langle h \rangle \langle G_0 \rangle N_A \right]$$

While this result is interesting, it is unlikely that resist will be processed at temperatures high enough to put the PEB reaction fully into the diffusion-controlled regime.

Using the example of a typical 193-nm resist, $M_0 N_A = 1.2/\text{nm}^3$, $G_0 N_A = 0.042/\text{nm}^3$ and $K_{\text{amp}} t_{\text{PEB}} = 2$. Consider the case of $\langle h \rangle = \langle h_{\text{eff}} \rangle = 0.3$, $\sigma_D/2a = 5$ and $p = 3$. For a $(10\,\text{nm})^3$ volume, $\sigma_h/\langle h \rangle \approx 0.28$ and $\sigma_{h_{\text{eff}}}/\langle h_{\text{eff}} \rangle \approx 0.025$. The remaining blocked polymer will have $\langle m \rangle = 0.55$ and $\sigma_m = 0.023$, or about 4.3%. For a $(5\,\text{nm})^3$ volume, $\sigma_m = 0.064$, or about 11%. The factors that have the largest impact are the diffusion length and the volume of resist under examination, as shown in Figure 6.10.

6.4.8   Acid–Base Quenching

Interestingly, the acid–base neutralization reaction poses the greatest challenge to stochastic modeling. While acid concentrations in chemically amplified resists are low, base quencher concentrations are even lower, leading to greater statistical uncertainty in

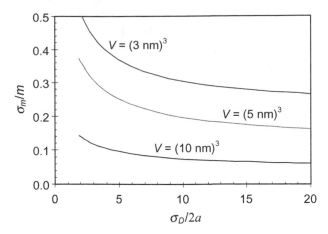

Figure 6.10   *The relative uncertainty of the relative blocked polymer concentration as a function of the volume (= length cubed) under consideration and the diffusion length of acid for a two-dimensional problem. Typical 193-nm resist values were used: $G_0N_A = 0.042/nm^3$, $M_0N_A = 1.2/nm^3$, $K_{amp}t_{PEB} = 2$, and for an exposure such that $<h> = 0.4$*

concentration for small volumes. Further, since the reaction is one of annihilation, statistical variations in acid and base concentrations can lead effectively to acid–base segregation, with clumps of all acid or all base.[10] Such clumping is likely to lead to low-frequency line edge roughness that is easily discernable from the higher-frequency roughness that will be predicted with the stochastic models presented above. Further discussion of this interesting and difficult topic is beyond the scope of this book. Thus, while acid–base quenching is extremely important in its impact on LER, it will not be considered in the model being developed here.

The stochastic descriptions of exposure and reaction–diffusion in chemically amplified resist will be used in Chapter 9 to develop a model for line-edge roughness.

Problems

6.1. Equation (6.9) provides the (no diffusion or acid loss) kinetic solution for the blocking group concentration in a chemically amplified resist. Under the assumption that h is locally constant,
(a) Derive an expression for the relative bake time sensitivity of m (i.e. calculate $dm/d\ln t_{PEB}$).
(b) Derive an expression for the relative temperature sensitivity of m (i.e. calculate $dm/d\ln T$). From this, will a low activation energy resist or a high activation energy resist be more sensitive to temperature variations?
(c) Does the presence of base quencher change the bake time or temperature sensitivity of m?

6.2. Plot the effective thermal dose versus temperature for a resist with an amplification reaction $E_a = 25.0$ kcal/mol and $A_r = \exp(28.0)$ s^{-1}, and an exposure rate constant $C = 0.05$ cm^2/mJ. Use a temperature range from 100 to 140 °C and assume a PEB time of 60 seconds.

6.3. What is the impact of increasing the PAG loading (the initial concentration of PAG) on the effective thermal dose for a given PEB process?

6.4. Derive the simplified Fujita–Doolittle Equation (6.15) from the full Fujita–Doolittle equation given in Chapter 5. What assumptions have to be made?

6.5. Ignoring diffusion and acid loss mechanisms, derive an expression for the catalytic chain length (the number of deblocking events per acid molecule). At what dose will this catalytic chain length be at its maximum? What is the maximum value for the catalytic chain length?

6.6. One possible acid loss mechanism is called acid trapping, where acid is consumed by being trapped within the resist. Ignoring diffusion and all other acid loss mechanisms, this acid loss mechanism is governed by

$$\frac{\partial H}{\partial t_{\mathrm{PEB}}} = -k_5 H$$

Under these conditions, derive an expression for the concentration of blocked sites at the end of the PEB.

6.7. For $x \gg \sigma_D$, derive a simplified 'large argument' approximation to the 1D reaction–diffusion point spread function.

6.8. Henry's law (first formulated in 1803) states that the amount of a gas which will dissolve into a solution is directly proportional to the partial pressure of that gas above the solution, and is generally found to be true only at low concentrations. When applied to a solid, the amount of a substance that is adsorbed onto the surface of the solid (the surface concentration) is proportional to the partial pressure of the substance in the atmosphere. The constant of proportionality, called the Henry's law constant, is dependent on both materials. Suppose ammonia is in the atmosphere at a concentration of 100 ppb (parts per billion), and that the Henry's law constant for ammonia in photoresist is 30 liters-atm/mol at room temperature. What is B_0, the surface concentration of ammonia?

6.9. Derive an expression for the total amount of base contaminant found inside a resist, assuming that Equation (6.22) applies and that $\sigma_B \ll$ resist thickness. If one wanted to reduce this total amount of base contaminant by a factor of 2, how much would the post-exposure delay time have to be reduced (all other factors held constant)?

6.10. Why does the addition of base quencher reduce the sensitivity of the resist to airborne base contaminants?

6.11. For the deblocking kinetics data of Figure 6.7, assume a resist with an exposure rate constant $C = 0.03\,\mathrm{cm^2/mJ}$, and a PEB time of 60 seconds.
  (a) If the resist shows no deblocking until a dose of $8\,\mathrm{mJ/cm^2}$, what is the relative quencher loading for the resist?
  (b) Estimate the amplification rate constant for each temperature shown.
  (c) What is the activation energy for the amplification reaction?

6.12. Assuming that a certain amount of acid is required to achieve a desired lithographic effect (that is, assuming the mean concentration of photogenerated acid is fixed), how low can the mean number of photons go before photon shot noise exceeds the PAG loading shot noise? Assume, for example, that $\langle h \rangle = 0.3$ and $\langle n_{0-\mathrm{PAG}} \rangle / V = 0.05/\mathrm{nm^3}$, and the region of interest is $(10\,\mathrm{nm})^3$.

6.13. Suppose for a given process that $\langle h \rangle = 0.3$ and $\langle m \rangle = 0.5$. Further, assume that photon shot noise can be neglected and that $\langle n_{0-\text{blocked}} \rangle / \langle n_{0-\text{PAG}} \rangle = 30$. Considering the total uncertainty of the final blocked polymer concentration for a 1D problem, how large must σ_D / a be in order to make the contribution of acid uncertainty equal to the Poisson uncertainty of the deblocking reaction?

6.14. As Equation (6.111) shows, increasing acid diffusion during PEB reduces the stochastic uncertainty caused by acid concentration fluctuations. Is there a limit as to how much this uncertainty can be reduced (that is, is there a practical maximum value for the acid diffusion length)? Explain.

6.15. Qualitatively, how will the presence of base quencher affect the variance of final blocked polymer concentration?

References

1   Ito, H. and Willson, C.G., 1984, Applications of photoinitiators to the design of resists for semiconductor manufacturing, in *Polymers in Electronics*, ACS Symposium Series 242, 11–23.

2   Willson, C.G., Ito, H. and Frechet, J.M.J., 1982, L'amplification chimique appliquée au développement de polymères utilisables comme résines de lithographie, *Colloque Internationale sur la Microlithographie: Microcircuit Engineering*, **82**, 261.

3   Miya, Y., Toishi, K. and Hashimoto, K. US Patent 6,893,792.

4   Petersen, J. and Byers, J., 1996, Examination of isolated and grouped feature bias in positive acting, chemically amplified resist systems, *Proceedings of SPIE: Advances in Resist Technology and Processing XIII*, **2724**, 163–171.

5   von Smoluchowski, M., 1917, Versuch Einer Mathematischen Theorie der Koagulationskinetik Kolloier Lösungen, *Z. Phys. Chem.*, **92**, 129–168.

6   Byers, J., Petersen, J. and Sturtevant, J., 1996, Calibration of chemically amplified resist models, *Proceedings of SPIE: Advances in Resist Technology and Processing XIII*, **2724**, 156–162.

7   Gamsky, C.J., Dentinger, P.M., Howes, G.R. and Taylor, J.W., 1995, Quantitative analysis of chemically amplified negative photoresist using mirror-backed infrared reflection absorption spectroscopy, *Proceedings of SPIE: Advances in Resist Technology and Processing XII*, **2438**, 143.

8   Sekiguchi, A., Mack, C.A., Isono, M. and Matsuzawa, T., 1999, Measurement of parameters for simulation of deep UV lithography using a FT-IR baking system, *Proceedings of SPIE: Advances in Resist Technology and Processing XVI*, **3678**, 985–1000.

9   Byers, J., private communication.

10 ben-Avraham, D. and Havlin, S., 2000, *Diffusion and Reactions in Fractals and Disordered Systems*, Cambridge University Press (Cambridge, UK), Chapter 13.

7

Photoresist Development

The chemistry of photoresists is designed to turn a spatial distribution of energy (the aerial image) into a spatial distribution of resist solubility. Ultimately, the dissolution process turns the continuous energy distribution of the projected aerial image into a binary resist image: either the resist is dissolved or it remains on the wafer. Exposure and post-exposure bake create latent images of chemical distributions, but it is the chemistry of dissolution that has the greatest impact on the ability to discern between line and space and to control the dimensions of the final resist features.

A kinetic approach to understanding development will be taken here, with a postulated reaction mechanism that then leads to a development rate equation. Deviations from the expected kinetic development rates at the surface of the resist, called surface induction or surface inhibition, will be related empirically to the expected development rate, i.e. to the bulk development rate. Next, the influence of developer temperature and normality are discussed. Once a basic understanding of what affects the development rate is established, the traditional resist metric 'contrast' will be defined and explored. Finally, to arrive at a resist profile shape, the physical mechanism by which a spatial variation in development rate turns into a final resist profile is examined by way of the development path.

7.1 Kinetics of Development

In order to derive an analytical development rate expression, a simple kinetic model of the development process will be used. This approach involves proposing a reasonable mechanism for the development reaction and then applying standard kinetics to this mechanism in order to derive a rate equation.

Unfortunately, fundamental experimental evidence of the exact mechanism of photoresist development is lacking. The models presented below are reasonable, and the resulting rate equations have been shown to describe actual development rates extremely well.

Fundamental Principles of Optical Lithography: The Science of Microfabrication, Chris Mack.
© 2007 John Wiley & Sons, Ltd.

However, faith in the exact details of the mechanism is limited by this dearth of fundamental studies. In addition, photoresists are not simple, uniform collections of model compounds. Distributions of polymers with different chain lengths (molecular weights) and functional groups mean that macroscopic observations reflect average material properties that are only approximately mapped to microscopic mechanisms.

7.1.1   A Simple Kinetic Development Model

Our discussion here will apply to both conventional and chemically amplified positive photoresists, and will be generalized to negative working resists in a later section. For all resist systems in common use today, the developer is a dilute aqueous solution of a base, most commonly tetramethyl ammonium hydroxide (TMAH).

Photoresist dissolution involves three processes: diffusion of the active component of the developer from the bulk solution to the surface of the resist, reaction of the developer with the resist, and diffusion of the product back into the solution. For this analysis, we shall assume that the last step, diffusion of the dissolved resist into solution, occurs very quickly so that this step may be ignored. Let us now look at the first two steps in the proposed mechanism. The diffusion of developer to the resist surface can be described with the simple diffusion rate equation, given approximately by a simple difference equation:

$$r_D = k_D(D - D_S) \qquad (7.1)$$

where r_D is the rate of diffusion of the developer to the resist surface, D is the bulk developer concentration, D_S is the developer concentration at the resist surface, and k_D is the rate constant (equal to the diffusivity of the developer in the solution divided by the boundary layer thickness).

We shall now propose a mechanism for the reaction of developer with the resist. It is quite likely that this step is in fact a series of more detailed steps, including the diffusion of the developer cation into the resist to form a thin gel layer. However, we will assume a simple surface-limited reaction here. The resist is composed of large macromolecules of resin R along with a dissolution inhibitor M. For the conventional DNQ/novolac resist, the dissolution inhibitor is the photoactive compound itself. For a chemically amplified resist, dissolution is inhibited by the blocking group attached to the PHS (polyhydroxy styrene) resin. The resin is somewhat soluble in the developer solution, but the presence of the inhibitor makes the development rate very slow. Solubility in the resist is changed by removing the inhibitor. For a conventional resist, the sensitizer converts to carboxylic acid upon exposure to UV light. This carboxylic acid is highly soluble in developer, enhancing the dissolution rate of the resin. For a chemically amplified resist, the PEB amplification reaction deblocks the polymer, generating the highly soluble –OH site.

Let us assume that n inhibitor sites must be removed from the influence of one resin in order for the developer to dissolve that resin molecule. This n parameter, the reaction order for this step in the mechanism, will turn out to be the most critical development parameter, representing the number of exposure or deblocking events that work together to get one resin molecule into solution. The rate of the reaction for this step is

$$r_R = k_R D_S (M_0 - M)^n \qquad (7.2)$$

where r_R is the rate of reaction of the developer with the resist, k_R is the rate constant, and M_0 is the initial inhibitor concentration. The two steps outlined above are in series, i.e. one reaction follows the other. Thus, the two steps will come to a steady state such that

$$r_R = r_D = r \qquad (7.3)$$

Equating the rate equations, one can solve for D_S and eliminate it from the overall rate equation, giving the steady-state development rate:

$$r = \frac{k_D k_R D (M_0 - M)^n}{k_D + k_R (M_0 - M)^n} \qquad (7.4)$$

Letting $m = M/M_0$, the relative inhibitor concentration, Equation (7.4) becomes

$$r = \frac{k_D D (1 - m)^n}{k_D / k_R M_0^n + (1 - m)^n} \qquad (7.5)$$

When $m = 1$ (resist unexposed/unreacted), the rate is zero. When $m = 0$ (resist completely exposed/reacted), the rate is equal to r_{max} where

$$r_{max} = \frac{k_D D}{k_D / k_R M_0^n + 1} \qquad (7.6)$$

If we define a constant a such that

$$a = k_D / k_R M_0^n \qquad (7.7)$$

the rate equation becomes

$$r = r_{max} \frac{(a + 1)(1 - m)^n}{a + (1 - m)^n} \qquad (7.8)$$

Note that the simplifying constant a describes the rate constant of diffusion relative to the surface reaction rate constant. A large value of a will mean that diffusion is very fast, and thus less important, compared to the fastest surface reaction (for completely exposed/reacted resist), and the rate is reaction-controlled. When a is small, the rate is diffusion-controlled.

There are three constants that must be determined experimentally: a, n and r_{max}. The constant a can be put in a more physically meaningful form as follows. A characteristic of some experimental rate data is an inflection point in the rate curve at values of m between 0.2 and 0.8. The point of inflection can be calculated by letting

$$\frac{d^2 r}{dm^2} = 0 \quad \text{at} \quad m = m_{th}$$

giving

$$a = \frac{(n+1)}{(n-1)} (1 - m_{th})^n \qquad (7.9)$$

where m_{th} is the value of m at the inflection point, called the *threshold inhibitor concentration*.

This model so far does not take into account the nonzero dissolution rate of unexposed resist (r_{min}). Simply adding this term to Equation (7.8) gives what has been called the *'original' kinetic model* or the 'Original Mack model':[1]

$$r = r_{max}\frac{(a+1)(1-m)^n}{a+(1-m)^n} + r_{min} \qquad (7.10)$$

This approach assumes that the mechanism of development of the unexposed resist is independent of the above-proposed development mechanism. In other words, there is a finite dissolution of resist that occurs by a mechanism that is independent of the presence of exposed photoactive compound (PAC). Note that the addition of the r_{min} term means that the true maximum development rate is actually $r_{max} + r_{min}$. In most cases $r_{max} \gg r_{min}$ and the difference is negligible.

Consider the case when the diffusion rate constant is large compared to the surface reaction rate constant (i.e. the rate is reaction-controlled). If $a \gg 1$, the development rate Equation (7.10) will become

$$r = r_{max}(1-m)^n + r_{min} \qquad (7.11)$$

The interpretation of a as a function of the threshold inhibitor concentration m_{th} given by Equation (7.9) means that a very large a would correspond to a large negative value of m_{th}. In other words, if the surface reaction is very slow compared to the mass transport of developer to the surface there will be no inflection point in the development rate data and Equation (7.11) will apply. It is quite apparent that Equation (7.11) could be derived directly from Equation (7.2) if the diffusion step were ignored.

Figure 7.1 shows some plots of this model for different values of n. The behavior of the dissolution rate with increasing n values is to make the rate function more 'selective'

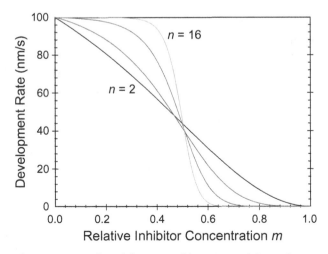

Figure 7.1 *Development rate plot of the original kinetic model as a function of the dissolution selectivity parameter ($r_{max} = 100\,nm/s$, $r_{min} = 0.1\,nm/s$, $m_{th} = 0.5$, and $n = 2, 4, 8$ and 16)*

between resist exposed/reacted above m_{th} and resist exposed/reacted below m_{th}. For this reason, n is called the *dissolution selectivity parameter*. Also from this behavior, the interpretation of m_{th} as a 'threshold' concentration becomes quite evident.

The key parameter in this development model that affects lithographic performance is the dissolution selectivity n. As n goes to infinity, the behavior seen in Figure 7.1 reaches its extreme of a step function. In other words, for $n = \infty$ the resist becomes an ideal threshold resist. Physically, this parameter represents the number of exposure or deblocking reactions that work together to cause one resin molecule to go into solution. Engineering the materials of the resist to increase n is one of the primary tasks of the resist designer. In general, a high n implies a high molecular weight of the resin (for example, a large number of deblocking reactions per resin means a large chain length of the resin). Unfortunately, a resist with all high molecular weight resin will tend to develop more slowly (lower r_{max} and m_{th}), thus degrading resist sensitivity.

The second most important factor in resist performance is the ratio of the maximum to minimum development rates, r_{max}/r_{min}. Increasing this ratio above 100 will definitely improve lithographic performance, though there will be diminishing returns as the ratio exceeds 10 000 or so.

7.1.2 Other Development Models

The previous kinetic model is based on the principle of dissolution enhancement. For a diazonaphthoquinone (DNQ)/novolac resist, the exposure product enhances the dissolution rate of the resin/PAC mixture. For a chemically amplified resist, the deblocked site enhances the solubility of the blocked polymer. In reality this is a simplification – there are really two mechanisms at work. The inhibitor acts to inhibit dissolution of the resin while the product that results from removing that inhibitor enhances dissolution. Thus, a development rate expression could reflect both of these mechanisms. A new model, call the *enhanced kinetic model*, has been proposed to include both effects:[2]

$$r = r_{resin} \frac{1 + k_{enh}(1-m)^n}{1 + k_{inh}(m)^l} \tag{7.12}$$

where k_{enh} is the rate constant for the enhancement mechanism, n is the enhancement reaction order, k_{inh} is the rate constant for the inhibition mechanism, l is the inhibition reaction order, and r_{resin} is the development rate of the resin alone.

For no exposure/reaction, $m = 1$ and the development rate is at its minimum. From Equation (7.12),

$$r_{min} = \frac{r_{resin}}{1 + k_{inh}} \tag{7.13}$$

Similarly, when $m = 0$, corresponding to complete exposure/reaction, the development is at its maximum.

$$r_{max} = r_{resin}(1 + k_{enh}) \tag{7.14}$$

Thus, the development rate expression can be characterized by five parameters: r_{max}, r_{min}, r_{resin}, n and l.

The enhanced kinetic model for resist dissolution is a superset of the reaction-controlled version of the original kinetic model. If the inhibition mechanism is not important, then $k_{inh} = 0$. For this case, Equation (7.12) is identical to Equation (7.11) when

$$r_{min} = r_{resin}, \quad r_{max} = r_{resin} k_{enh} \tag{7.15}$$

The enhanced kinetic model of Equation (7.12) assumes that mass transport of developer to the resist surface is not significant. Of course, a simple diffusion of developer can be added to this mechanism as was done above with the original kinetic model (see Problem 7.3).

Figure 7.2 shows several plots of this model. The five parameters of the enhanced kinetic model give this equation more flexibility in fitting experimental development rate curves compared to the four-parameter original kinetic model. As Figure 7.2 shows, the combination of enhancing and inhibiting mechanisms can produce a sort of double threshold behavior with rates rising and lowering around a plateau at the resin development rate.

Perhaps a more intuitive way to understand the basic mechanism of the enhanced kinetic rate model is with the Meyerhofer plot.[3] This plot, such as the one shown in Figure 7.3, shows how the addition of increasing concentrations of inhibitor in the resist formulation (in this case DNQ added to a novolac resin) changes the original development rate of the resin and increases r_{max} while decreasing r_{min}. Experimentally, the Meyerhofer plot shows an approximately linear dependence of log-development rate on the initial concentration of inhibitor, M_0 (Figure 7.3a). This empirical result can be explained reasonably well if one assumes that k_{enh} has a first-order dependence on M_0 and k_{inh} has a second-order dependence on M_0, as seen in Figure 7.3b.

Some experimentally derived $r(m)$ behavior is not well fit by either kinetic model described above. Figure 7.4 shows the best fit of both the original and the enhanced development rate models to one set of data.[4] In the region around a 0.5 inhibitor concentration, the data exhibits a 'notch' behavior where the actual development rate drops very quickly to a value much less than that predicted by either model. This sudden drop in

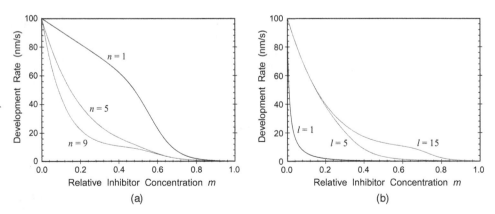

Figure 7.2   *Plots of the enhanced kinetic development model for $r_{max} = 100\,nm/s$, $r_{resin} = 10\,nm/s$, $r_{min} = 0.1\,nm/s$ with: (a) $l = 9$; and (b) $n = 5$*

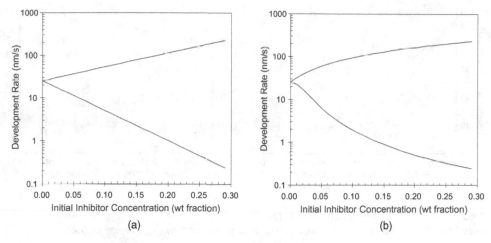

Figure 7.3 *An example of a Meyerhofer plot, showing how the addition of increasing concentrations of inhibitor increases r_{max} and decreases r_{min}: (a) idealized plot showing a log-linear dependence on initial inhibitor concentration, and (b) the enhanced kinetic model assuming $k_{enh} \propto M_0$ and $k_{inh} \propto M_0^2$*

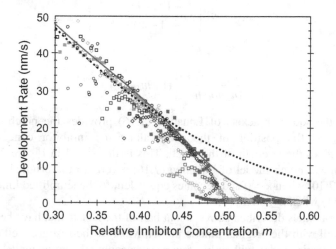

Figure 7.4 *Comparison of experimental dissolution rate data (symbols) exhibiting the so-called 'notch' shape to best fits of the original (dotted line) and enhanced (solid line) kinetic models. The data shows a steeper drop in development rate at about 0.5 relative inhibitor concentration than either model predicts*

rate, resembling a notch in the otherwise slowly varying behavior, is critical to resist performance because it is the region of steepest change in relative development rate versus *m*. Thus, it is critical that a development model properly describes this region of the dissolution rate curve in order to accurately predict the behavior of the resist.

A semiempirical model has been devised in order to better fit the notch behavior described above. This *notch model* begins with the simple version of the original model

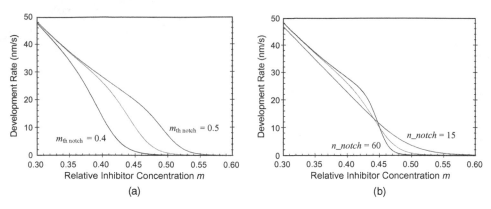

Figure 7.5 *Plots of the notch model: (a) m_{th_notch} equal to 0.4, 0.45 and 0.5 with n_notch equal to 30; and (b) n_notch equal to 15, 30 and 60 with m_{th_notch} equal to 0.45*

given in Equation (7.11) and adds a notch function equivalent to the threshold behavior given by Equation (7.10):

$$r = r_{max}(1-m)^n \left[\frac{(a+1)(1-m)^{n\_notch}}{a+(1-m)^{n\_notch}} \right] + r_{min} \qquad (7.16)$$

where

$$a = \frac{(n\_notch+1)}{(n\_notch-1)}(1 - m_{TH\_notch})^{n\_notch}$$

The term in the square brackets of Equation (7.16) provides the notch-like behavior where m_{th_notch} is the position of the notch along the inhibitor concentration axis and *n_notch* gives the strength of the notch. This is illustrated in Figure 7.5. Note that the five-parameter notch model of Equation (7.16) reverts to the original kinetic model of Equation (7.10) when $n = 0$ and becomes equivalent to the simplified kinetic model of Equation (7.11) when *n_notch* = 1.

Of the three models described above, about half of the resists that have been characterized are fit well with the original kinetic model. Most of the rest are well described by the enhanced kinetic model, with only a few resists needing the notch model to accurately describe their dissolution properties.

7.1.3 Molecular Weight Distributions and the Critical Ionization Model

In each of the models above, the mechanism of surface reaction is based on the idea that *n* dissolution-changing chemical events are required to make one polymer molecule dissolve. Real resists, however, are formulated with polymers that have a range of molecular weights (that is, different polymer chain lengths). It seems unlikely that each polymer molecule, regardless of its chain length, would require the same number of deblocking events (taking chemically amplified resists as an example) to make it soluble. Another approach, called the *critical ionization model*, assumes that the polymer becomes soluble whenever some critical fraction of its blocked groups become deblocked.[5] Thus, lower molecular weight polymers would require fewer deblocking events (lower n) to become soluble compared to higher molecular weight polymers.

High molecular weight polymers (>5000 g/mol, for example) result in high n, but with less than ideal sensitivity. By mixing in a small amount of low molecular weight polymer (<500 g/mol, for example),[6] resist sensitivity can be improved without a significant reduction in average n. Mid-molecular weight resins, however, tend to degrade resolution by lowering n.

7.1.4 Surface Inhibition

The kinetic models given above predict the development rate of the resist as a function of the inhibitor concentration remaining after the resist has been exposed/reacted. There are, however, other parameters that are known to affect the development rate, but that were not included in these models. The most notable deviation from the kinetic theory is the surface inhibition effect (Figure 7.6). The inhibition, or surface induction, effect is a decrease in the expected development rate at the surface of the resist.[7] Thus, this effect is a function of the depth into the resist and requires a new description of development rate.

Several factors have been found to contribute to the surface inhibition effect. Baking of the photoresist can produce surface inhibition and two possible mechanisms are thought to be likely causes. One possibility is an oxidation of the resist at the resist surface, resulting in reduced development rate of the oxidized film. Alternatively, the induction effect may be the result of reduced solvent content near the resist surface, which also results from baking the resist (see Chapter 5). Both mechanisms could be contributing to the surface inhibition. Quite commonly, surface inhibition can be induced with the use of surfactants (surface-acting agents) in the developer. Chemically amplified resists open a wide array of mechanisms for surface inhibition due to reduced amplification near the

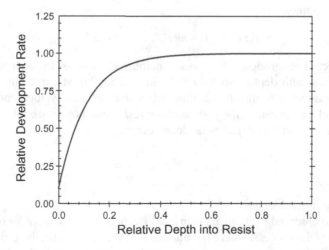

Figure 7.6 *Example surface inhibition with $r_0 = 0.1$ and $\delta = 100$ nm for a 1000-nm-thick resist*

resist surface (caused, for example, by acid evaporation, base contamination, or reduced acid diffusion in the low-solvent surface region).

An empirical model can be used to describe the positional dependence of the development rate. If we assume that the development rate near the surface of the resist exponentially approaches the bulk development rate, the rate as a function of depth, $r(z)$, is

$$r(z) = r_\mathrm{B}[1 - (1 - r_0)e^{-z/\delta}] \tag{7.17}$$

where r_B is the bulk development rate, r_0 is the development rate at the surface of the resist relative to r_B, and δ is the depth of the surface inhibition layer. In several conventional resists, the induction effect has been found to take place over a depth of about 100 nm or less.[8] For chemically amplified resists, 200- to 300-nm inhibition depths can be observed.

A small amount of surface inhibition can be good for photoresist profile control. Slower development of the top of the resist can lead both to less resist loss for small lines, and also profiles with sharper, more square tops. Too much inhibition, however, can lead to T-top-shaped profiles and a loss of linewidth control. In general, moderate values of r_0 and large values of δ lead to improved resist profiles without significant loss of dimensional control.

One of the most useful aspects of surface inhibition is when this effect is used to counter the decrease in development rate caused by absorption in positive resists. For resist sitting on a good bottom antireflection coating, the energy exposing the resist falls off exponentially with depth into the resist. We can approximate the impact of this absorption on the variation of development rate with depth into the resist in the following way. For a conventional resist, ignoring resist bleaching and assuming small exposure doses, the sensitizer will vary with depth into the resist approximately as

$$m(z) = \exp[-CDose_\mathrm{inc}I_\mathrm{r}(z)] \approx 1 - CDose_\mathrm{inc}I_\mathrm{r}(z) = 1 - CDose_\mathrm{inc}e^{-\alpha z} \tag{7.18}$$

Using the simple reaction-limited development rate Equation (7.11) and ignoring the minimum development rate,

$$r(z) \approx r_\mathrm{max}(1 - m(z))^n \approx r(z = 0)e^{-\alpha n z} \tag{7.19}$$

Thus, the effect of absorption is to cause something close to an exponentially decaying development rate with depth into the resist. Surface inhibition can counteract some of this falloff. Reasonable results are obtained when the surface development rate is set to match the development rate at the bottom of the resist and the surface inhibition depth is set to match the absorption exponential decay depth.

$$r_0 \approx e^{-\alpha n D}, \quad \delta \approx \frac{1}{\alpha n} \tag{7.20}$$

where D is the resist thickness.

Note that the observed magnitude of surface effects is often on the order of 50–100 nm. Thus, for thin resist process, where the thickness of the resist is on the order of 100 nm, the very idea that resist response can be separated into surface and bulk behavior is called into question.

7.1.5   Extension to Negative Resists

The development rate models discussed above were derived based on positive resist chemistry. However, with only a slight change in interpretation, these models are perfectly applicable to most negative resist chemistries. For the three models given by Equations (7.10), (7.12) and (7.16), the term m represents the concentration of the species that inhibits dissolution. For a conventional positive resist, this quantity is the photoactive compound. For conventional negative resists, the inhibitor is the exposure product. By interpreting the m in the development rate equations as the inhibitor concentration for the negative resist, whatever it might be, the rate equations can be used without further modification.

7.1.6   Developer Temperature

It is well known that the temperature of the developer solution during development can have a significant impact on resist performance. The speed (i.e. overall development rate) varies in a complicated way with temperature, usually resulting in the counterintuitive result of a 'faster' resist process (i.e. a process requiring lower exposure doses) at lower developer temperatures. The shape of the development rate versus dose (or versus inhibitor concentration) curve will also vary considerably with temperature, leading to possibly significant performance differences. The effect of temperature on dissolution rate shows a complicated behavior where changes in developer temperature give changes in dissolution rate that are dose dependent. At one dose, the effect of temperature on dissolution rate can be very different than at another dose.[9] Use of a dissolution rate model can simplify the description of temperature effects by showing just the change in the model parameters with developer temperature.

Figure 7.7 shows an example of how temperature can affect dissolution rate. In general, one usually expects simple kinetic rate-limited reactions to proceed faster at higher

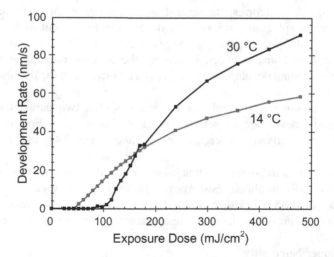

Figure 7.7   *Development rate of THMR-iP3650 (averaged through the middle 20 % of the resist thickness) as a function of exposure dose for different developer temperatures shows a change in the shape of the development dose response*

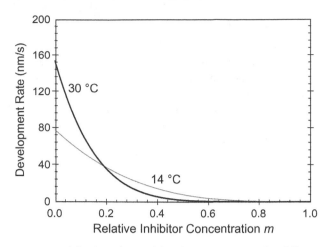

Figure 7.8 *Comparison of the best-fit models of THMR-iP3650 for different developer temperatures shows the effect of increasing r_{max} and increasing dissolution selectivity parameter n on the shape of the development rate curve*

temperatures (indicating a positive activation energy for the reaction). The behavior shown in Figure 7.7 is obviously more complicated than that. At high doses, increasing developer temperature does increase the development rate. But at low doses the opposite is true. Thus, developer temperature has a significant impact on the shape of the dissolution rate curve. Reasons have been proposed for this behavior,[10] but here we will only strive to accurately describe this behavior quantitatively using model parameters.

By fitting the dissolution rate behavior to a development model, the variation of the $r(m)$ curve with temperature can be deduced, as shown in Figure 7.8. Again, the results show that at high doses (corresponding to low concentrations of inhibitor remaining), higher developer temperature increases the development rate. But at low doses (high concentrations of inhibitor remaining), the opposite is true. Using the terminology of the original kinetic development model, increasing the developer temperature caused an increase in the maximum development rate r_{max} and an increase in the dissolution selectivity parameter n.

Figure 7.9 shows the final results of the analysis. The two parameters r_{max} and n are plotted versus developer temperature in an Arrhenius plot. For the i-line resist THMR-iP3650, the activation energies for r_{max} and n are 7.41 and 7.02 Kcal/mole, respectively.[11]

Not all resists exhibit the simple, regular behavior with developer temperature described above. The chemically amplified resist Apex-E, for example, shows a maximum dissolution n value at about 35 °C, as seen in the data of Table 7.1. More work is required to fully understand the impact of developer temperature on a wider array of resists.

7.1.7   Developer Normality

TMAH developer at 0.26 N (2.38 % by weight) is extremely widely used (almost to the point of being a universal standard). However, as one would expect, developer normality

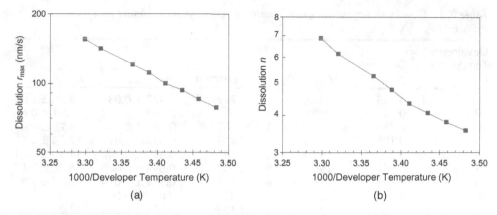

Figure 7.9 *Arrhenius plots of (a) the maximum dissolution rate r_{max} and (b) the dissolution selectivity parameter n for THMR-iP3650*

Table 7.1 *Developer temperature-dependent modeling parameters of Apex-E (0.26 N developer)[12]*

Developer Temperature (°C)	r_{max} (nm/s)	r_{min} (nm/s)	m_{th}	n
5	53.1 ± 9.0	1.18	0.571 ± 0.036	3.93 ± 0.56
10	68.0 ± 10.9	1.283	0.578 ± 0.031	4.17 ± 0.52
15	91.9 ± 13.8	1.35	0.586 ± 0.028	4.25 ± 0.48
20	115.7 ± 14.1	1.48	0.682 ± 0.016	4.79 ± 0.49
25	146.1 ± 15.2	1.55	0.637 ± 0.015	5.56 ± 0.54
30	177.8 ± 16.0	1.62	0.646 ± 0.012	6.36 ± 0.55
35	199.7 ± 14.6	2.56	0.693 ± 0.008	15.58 ± 2.17
40	196.2 ± 13.1	2.14	0.694 ± 0.008	11.25 ± 1.32
45	209.5 ± 22.5	1.83	0.665 ± 0.014	6.64 ± 0.74

(the concentration of TMAH) impacts dissolution rates greatly. In general, higher normality produces higher dissolution rates. Using again the model coefficients as a way of describing the impact of developer normality on dissolution, Equation (7.6) would suggest that the parameters r_{max} and r_{min} increase with an increase concentration of TMAH. The important dissolution selectivity parameter n tends to have an optimum, essentially going to zero at very high and very low normalities. The original kinetic model parameters r_{max} and the dissolution selectivity parameter n vary with developer normality as shown in Table 7.2 for the chemically amplified resist Apex-E. As the developer temperature increases, the impact of different developer normalities on r_{max} is lessened. Although the data here is limited, r_{max} exhibits a well-known phenomenon – there is a critical normality (between about 0.1 and 0.15 N) below which development does not proceed (r_{max} becomes zero). The combined impact of temperature and normality on the dissolution selectivity n is quite complicated.

Table 7.2   *Developer temperature- and normality-dependent modeling parameters for Apex-E*[12]

Developer Temperature (°C)	Developer Normality	r_{max} (nm/s)	r_{min} (nm/s)	m_{th}	n
20	0.195	45.9 ± 7.9	0.272	0.527 ± 0.03	5.26 ± 1.00
20	0.24425	90.9 ± 11.3	0.912	0.607 ± 0.018	5.41 ± 0.61
20	0.26	115.7 ± 14.1	1.482	0.682 ± 0.016	4.79 ± 0.49
35	0.195	78.1 ± 12.9	0.395	0.533 ± 0.01	17.33 ± 2.18
35	0.24425	151.3 ± 13.1	1.191	0.667 ± 0.008	11.10 ± 1.14
35	0.26	199.6 ± 14.6	2.562	0.693 ± 0.008	15.58 ± 2.17
40	0.195	118.5 ± 94.6	0.395	0.511 ± 0.074	8.72 ± 1.93
40	0.24425	182.9 ± 12.4	1.473	0.675 ± 0.005	16.44 ± 1.62
40	0.26	196.2 ± 13.1	2.135	0.694 ± 0.008	11.25 ± 1.32

One important reason for understanding developer normality effects is to better understand the impact of *developer loading*. One of the most popular styles of development is puddle development, where a puddle of developer is formed on top of the wafer which sits stationary through most of the development cycle. One potential issue with this development method, especially for thicker resists, is the accumulation of dissolved resist and the decrease in –OH concentration in the developer as development proceeds in what is called developer loading. Additionally, if an integrated circuit has very different pattern densities across the die, this developer loading effect can vary across the die. Since development will slow down as the developer loading increases, the result can be an additional source of CD nonuniformity. A common solution to this problem is to break the puddle step into a *double puddle*, where each puddle is allowed to sit for half of the development time and fresh developer is used to form the second puddle.

7.2   The Development Contrast

The use of 'contrast' to describe the response of a photosensitive material dates back over 100 years. Hurter and Driffield measured the optical density of photographic negative plates as a function of exposure.[13] The 'perfect negative' was one that exhibited a linear variation of optical density with the logarithm of exposure. A plot of optical density versus log-exposure showed that a good negative exhibited a wide 'period of correct representation', as is shown in the *Hurter–Driffield (H–D) curve* in Figure 7.10. Hurter and Driffield called the slope of this curve in the linear region γ, the 'development constant'. Negatives with high values of γ were said to be 'high contrast' negatives because the photosensitive emulsion quickly changed from low to high optical density when exposed. Of course, high contrast film is not always desirable in photography since it easily saturates.

7.2.1   Defining Photoresist Contrast

Photolithography evolved from the photographic sciences and borrowed many of its concepts and terminology. When exposing a photographic plate, the goal is to change the

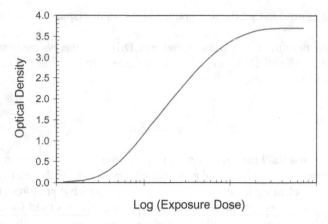

Figure 7.10 *An example Hurter–Driffield (H–D) curve for a photographic negative*

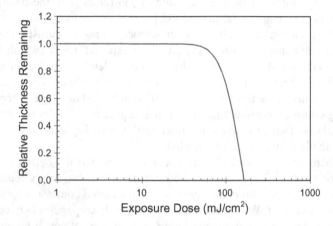

Figure 7.11 *Positive resist characteristic curve used to measure contrast*

optical density of the material. In lithography, the goal is to remove resist. Thus, one analogous H–D curve for lithography plots resist thickness after development versus log-exposure. Experimentally, a wafer is exposed using a blank glass plate as the mask. As each field on the wafer is exposed, the dose is adjusted so that a large array of doses is exposed on one wafer. After going through the normal PEB and development processes, the thickness remaining in this open-frame area is measured for each dose (Figure 7.11). The lithographic H–D curve (often called the *characteristic curve*) is a portion of a complete H–D curve as shown in Figure 7.10. Because the goal is to completely remove unwanted photoresist, there is usually a range of energies for which all of the photoresist is removed and thus the H–D curve would show no response. If, however, a very thick photoresist were used so that it could never be completely removed, the result would be the 'S'-shape of a complete H–D curve. It is common practice to normalize the initial resist thickness to one, so that the lithographic H–D curve displays the relative thickness remaining.

Following the definition of γ from Hurter and Driffield, the *photoresist contrast* has traditionally been defined as the slope of the characteristic curve at the point where the thickness goes to zero. Thus,

$$\gamma_m = \pm \frac{1}{D} \frac{dz}{d\ln E}\Big|_{E=E_0} \tag{7.21}$$

where z is the resist thickness removed during development, D is the resist thickness before development, E is the nominal exposure energy, and E_0 is the energy at which $z = D$. E_0 is called the clearing dose (or *dose-to-clear*) for positive photoresists and the gel dose for negative systems. The positive sign in Equation (7.21) is used for positive resists and the minus sign is used for negative systems in order to keep the value of γ_m positive. For the remainder of this section, positive systems will be used as our example. The results, however, can easily be applied to negative photoresists as well. [Note that sometimes a base-10 logarithm is used in Equation (7.21) rather than the natural logarithm. This book will always employ the natural log.]

Following the tradition of the photographic sciences, a high-contrast photoresist is one that makes a quick transition from being an 'underexposed' resist (which does not dissolve) to an 'overexposed' resist (which dissolves completely). The traditional definition in Equation (7.21) seems to fit this concept. Further, it is reasonably analogous to the photographic contrast as defined by Hurter and Driffield. The slope of optical density (which is a logarithm of transmittance) versus log-exposure is similar in form to the slope of relative thickness (which is like a log-thickness) versus log-exposure. Thus, it would seem that a suitable definition has been used.

However, there are numerous circumstances under which the definition in Equation (7.21) does not meet our expectations of what contrast means. In particular, the use of surfactant-laden developers dramatically increases measured contrast values, but without lithographic improvement. What causes these apparent discrepancies between the behavior of the measured contrast and our concept of how contrast affects lithography? Is there a problem with the definition, or with the measurement technique? These questions can be answered by putting the concept of contrast on a firm theoretical basis and applying rigorous analysis to the observed behavior.

With the advantage of retrospect, let us look at the evolution of the traditional definition of contrast and provide a different definition. In photography, the desired effect of exposure is a change in the optical density of the material. In photolithography, the desired effect is a change in development rate. This change in development rate is manifest as a change in resist thickness after development. Analogous to the photographic H–D curve, let us plot log-development rate versus log-exposure energy. Figure 7.12 shows the results for a typical photoresist. Note that this graph gives a complete H–D curve and does not cut off at some energy E_0. We can give a new definition of contrast, called the *theoretical contrast*, as[14]

$$\gamma_{th} = \frac{d\ln r}{d\ln E} \tag{7.22}$$

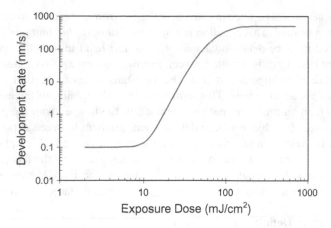

Figure 7.12   *A lithographic H–D curve used to define the theoretical contrast*

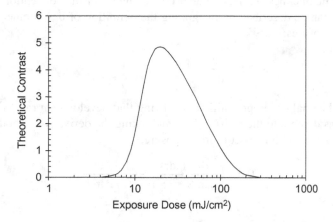

Figure 7.13   *A typical variation of theoretical contrast with exposure dose. For this data, the FWHM dose ratio is about 5.5*

where r is the development rate. Note that the theoretical contrast will be a function of dose, going to zero at very low and very high doses. A typical plot of $\gamma_{th}(E)$ is shown in Figure 7.13. Not only is the maximum of this curve important, but the width of the $\gamma_{th}(E)$ function also affects lithographic performance. The equivalent of the full-width half-maximum (FWHM) on a logarithmic exposure scale would be the ratio of the high and low energies that give contrast equal to half of the maximum value, called the *FWHM dose ratio*.

The goal of lithographic exposure is to turn a gradient in exposure energy (an aerial image) into a gradient in development rate. From Equation 7.22, it is very easy to express this effect as

$$\frac{d\ln r}{dx} = \gamma_{th}\frac{d\ln I}{dx} \tag{7.23}$$

where I is the aerial image intensity and x is the horizontal distance from the center of the feature being printed. This equation is called the *Lithographic Imaging Equation* (and will be explored in more detail in Chapter 9). The left-hand term is the spatial gradient of development rate. To differentiate between exposed and unexposed areas, it is desirable to have this gradient as large as possible. The right-hand side of Equation (7.23) contains the log-slope of the aerial image. This term represents the quality of the aerial image, or alternatively the amount of information contained in the image about the position of the mask edge. In order to achieve a large development gradient between exposed and unexposed parts of the resist, one needs a large exposure gradient and a high photoresist contrast. Also, since contrast is a function of dose, it is important that the theoretical contrast stays high over a wide enough range of doses (that is, the FWHM dose ratio must be large). This is also equivalent to saying that r_{max}/r_{min} must be large (see section 7.2.4).

7.2.2 Comparing Definitions of Contrast

The theoretical definition of contrast can now be compared to the conventional measured contrast, γ_m. The thickness remaining after development for the conventional H–D curve measurement can be described by integrating the definition of development rate for our open-frame exposures:

$$r = \frac{dz}{dt}, \quad \int_0^{t_{dev}} dt = t_{dev} = \int_0^z \frac{dz'}{r(z')} \tag{7.24}$$

where t_{dev} is the final development time. Knowing that development rate is a function of dose as well as depth into the resist, $r(E,z)$, and taking the derivative of both sides of the right-hand equation with respect to log-exposure,

$$0 = -\int_0^z \frac{dr}{d\ln E} \frac{dz'}{r^2(z')} + \frac{1}{r(z)} \frac{dz}{d\ln E} \tag{7.25}$$

Rearranging,

$$\frac{1}{D} \frac{dz}{d\ln E} = \frac{r(z)}{D} \int_0^z \frac{d\ln r}{d\ln E} \frac{dz'}{r(z')} \tag{7.26}$$

Evaluating this expression at $E = E_0$ (i.e. $z = D$) and using the definitions of measured and theoretical contrast,

$$\gamma_m = \frac{r(z=D)}{D} \int_0^D \gamma_{th} \frac{dz}{r(z)} \tag{7.27}$$

If the development rate does not vary through the resist thickness, it is easy to show that Equation (7.27) predicts that measured and theoretical contrast values are equal. But if the development rate varies with depth into the resist, the measured contrast fails to provide a good measurement of the theoretical contrast.

Consider the case of surface inhibition (caused, for example, by surfactants in the developer). The development rate at the top of the resist will be much lower than that at the bottom. Thus, Equation (7.27) shows that the measured contrast will be equal to the theoretical contrast multiplied by some number bigger than 1. As the surface development rate goes to zero, the measured contrast becomes infinite, independent of the value of the theoretical contrast. Likewise, absorption makes the measured contrast value artificially low. Since every resist will exhibit some depth-dependent dissolution behavior, the conventional measured contrast is rarely an accurate measure of the true resist contrast. Both of these cases can be explored in detail using Equation (7.27).

For absorption, assume that the development rate follows the z-dependence given by Equation (7.19). Let's also assume that the theoretical contrast is constant (that is, through the thickness of the resist, all of the doses lie in the range of the approximately linear region of Figure 7.12). Thus,

$$\gamma_m = \frac{r(z=D)}{D} \int_0^D \gamma_{th} \frac{dz}{r(z)} = \frac{r(0)e^{-\alpha nD}}{D} \gamma_{th} \int_0^D \frac{dz}{r(0)e^{-\alpha nz}} \tag{7.28}$$

Solving the integral,

$$\gamma_m = \gamma_{th} \left(\frac{1 - e^{-\alpha nD}}{\alpha nD} \right) \tag{7.29}$$

The behavior of this result can be seen easily by considering the case of relatively small absorption ($\alpha nD \ll 1$) and taking a Taylor expansion of the exponential out to second order:

$$\gamma_m = \gamma_{th} \left(\frac{1 - e^{-\alpha nD}}{\alpha nD} \right) \approx \gamma_{th} \left(1 - \frac{\alpha nD}{2} \right) \tag{7.30}$$

Thus, absorption reduces the measured contrast by an amount approximately proportional to αnD. Also, as will be seen in Chapter 9, the dissolution selectivity parameter n is directly proportional to the theoretical contrast. Thus, in the presence of absorption, any increase in the theoretical contrast will result in only a fractional increase in the measured contrast.

Surface inhibition will increase the measured contrast over the theoretical contrast. Consider again the case where theoretical contrast is constant, but let the depth dependence of the development rate be governed by the surface inhibition function of Equation (7.17). Further, assume that $D \gg \delta$ so that the development rate at the bottom of the resist is about equal to the bulk development rate. The measured contrast will become

$$\gamma_m = \frac{r_B}{D} \gamma_{th} \int_0^D \frac{dz}{r_B[1 - (1 - r_0)e^{-z/\delta}]} \tag{7.31}$$

Solving the integral,

$$\gamma_m = \gamma_{th} \left[1 - \frac{\delta}{D} \ln(r_0) \right] \tag{7.32}$$

As the relative surface rate r_0 goes to zero, the measured contrast goes to infinity. For example, if the surface inhibition depth is 20% of the resist thickness and the relative surface rate is 0.1, the measured contrast will be about 50% higher than the theoretical contrast. If the relative surface rate is 0.01, the measured contrast will be about double the true value.

The problems with the characteristic curve as discussed above are often very severe. While absorption can be easily characterized for a given resist and the measured contrast corrected for absorption effects, surface inhibition is rarely characterized with sufficient accuracy to allow its impact to be removed from the measured contrast. As a result, few lithographers use the characteristic curve for anything but a dose-to-clear measurement. Instead, the theoretical contrast is most often measured directly using a development rate monitor (see section 7.4), if it is measured at all.

7.2.3   The Practical Contrast

One possible improvement in the measurement of photoresist contrast uses a method that Peter Gwozdz calls the '*practical contrast*'. Rather than measuring the resist thickness remaining as a function of dose for open-frame exposures, as for the conventional measurement based on the characteristic curve, the practical contrast measures the development time required to clear each open frame. For reasonably slowly developing resists (true for many g- and i-line resists), the clear time can be measured visually using a stopwatch. Plotting time-to-clear versus dose on a log-log scale gives a curve whose slope is equal to the negative of the theoretical contrast.

To see how this is true, consider a resist where the influence of dose on development rate is separable from the z-dependence.

$$r(E,z) = f(E)g(z) \tag{7.33}$$

Such separable behavior has been assumed for the previously discussed absorption and surface inhibition effects. The theoretical contrast can then be expressed as

$$\gamma = \frac{\partial \ln r}{\partial \ln E} = \frac{\partial \ln f}{\partial \ln E} \tag{7.34}$$

The time to clear will be given by Equation (7.24) when $z = D$:

$$t_{\text{clear}} = \int_0^D \frac{dz}{r(z)} = \frac{1}{f(E)} \int_0^D \frac{dz}{g(z)} \tag{7.35}$$

Since the integral on the right-hand side of Equation (7.35) is a constant (independent of dose), the time-to-clear will be inversely proportional to $f(E)$ and

$$\gamma = \frac{\partial \ln f}{\partial \ln E} = -\frac{\partial \ln t_{\text{clear}}}{\partial \ln E} \tag{7.36}$$

Unfortunately, accurate measurement of time-to-clear is often difficult without automated equipment such as a development rate monitor (see section 7.4). For g- and i-line resists, with maximum development rates of a few hundred nm/s and thicknesses greater than 500 nm, the smallest time-to-clear will be a few seconds. For many 248-nm chemically amplified resists, the maximum development rate is 1000–5000 nm/s and thicknesses are as low as a few hundred nanometers, so that over a fair range of doses the time-to-clear will be less than 1 second. Fortunately, though, the steep 'high contrast' portion of the time-to-clear versus dose curve must necessarily (for any reasonably good resist) have time-to-clear values on the order of the development time (typically 45–90 seconds). Thus, if the goal is to measure the maximum contrast rather than the full contrast versus dose function, the practical contrast approach can work quite well.

7.2.4   Relationship Between γ and r_{max}/r_{min}

As was mentioned briefly at the end of section 7.1.1, a good development rate function not only has a high dissolution selectivity parameter n, but also a large ratio r_{max}/r_{min}. In the context of development contrast, these two requirements are equivalent to saying that the resist must have a high maximum contrast and a large FWHM dose ratio (see Figure 7.13). To more explicitly see the relationships at work, let's assume that our $\gamma(E)$ function is Gaussian in shape (on a log-exposure scale):

$$\gamma(E) = \gamma_{max} e^{-(\ln E - \ln E^*)^2 / 2\sigma^2} \tag{7.37}$$

where γ_{max} is the maximum contrast, and E^* is the dose that gives the maximum contrast. The FWHM dose ratio will be given by

$$\frac{E_{1/2}}{E_{-1/2}} = \exp(2\sigma\sqrt{2\ln 2}) \approx \exp(2.355\sigma) \tag{7.38}$$

Integrating Equation (7.37) with respect to $\ln E$ gives an error function shape to the contrast curve. Using the limits of zero and infinite dose to define r_{max} and r_{min} gives

$$\ln\left(\frac{r}{\sqrt{r_{max}r_{min}}}\right) = \ln\left(\sqrt{\frac{r_{max}}{r_{min}}}\right) erf\left(\frac{\ln(E/E^*)}{\sqrt{2}\sigma}\right) \tag{7.39}$$

where

$$\sigma = \frac{1}{\sqrt{2\pi}\gamma_{max}} \ln\left(\frac{r_{max}}{r_{min}}\right) \tag{7.40}$$

Combining Equations (7.38) and (7.40),

$$\frac{1}{\gamma_{max}} \ln\left(\frac{r_{max}}{r_{min}}\right) = \sqrt{\frac{\pi}{4\ln 2}} \ln\left(\frac{E_{1/2}}{E_{-1/2}}\right) \approx 1.06 \ln\left(\frac{E_{1/2}}{E_{-1/2}}\right) \tag{7.41}$$

The required FWHM dose ratio is determined by the aerial image and is proportional to the ratio of the maximum intensity in the space to the intensity at the line edge (which has typical values of 2–4). Thus, the required maximum to minimum development rate ratio goes as

$$\frac{r_{max}}{r_{min}} \sim e^{a\gamma_{max}}, \quad 1 < a < 2 \quad \text{or} \quad \frac{r_{max}}{r_{min}} \sim \left(\frac{E_{1/2}}{E_{-1/2}}\right)^{\gamma_{max}} \tag{7.42}$$

Equation (7.42) provides the interesting result that resists with higher contrast require a higher r_{max}/r_{min} in order to make effective use of that contrast. A doubling of the contrast requires a squaring of the development rate ratio. For example, if the dose ratio is 2 and the maximum contrast is 10, the maximum to minimum development rate ratio should be at least 1000. For the same dose ratio, however, a maximum contrast of 15 would require that r_{max}/r_{min} be greater than 30000 in order to take full advantage of the higher contrast.

7.3   The Development Path

Most of the discussion in this chapter so far has been concerned with how dose or resist chemical composition is converted to development rate. A spatial distribution of energy turns into a spatial distribution of chemicals which turns into a spatial distribution of dissolution rates. But how does a spatial distribution in development rates turn into an actual resist profile? How does the resist–developer front propagate through time and space during development?

For the one-dimensional case of large open-frame exposures (such as the characteristic curve experiment described in the previous section), propagation of the resist–developer interface is strictly in the z-direction, with the rate of development at a given point equal to the rate at which the interface moves down in z. For the more general case, we can define the path of development by tracing the position of the resist surface through the development time (Figure 7.14). This path will have a direction such that it is always perpendicular to the surface of the resist throughout the development cycle. Letting s be the length along this path at any given time, the development rate will then be

$$r = \frac{ds}{dt} \tag{7.43}$$

The problem then is to follow this path at a rate given by the spatially varying dissolution properties of the resist until the total development time is reached. Integrating this general definition of development rate thus gives a path integral:

$$t_{dev} = \oint_{path} \frac{ds}{r(x,y,z)} \tag{7.44}$$

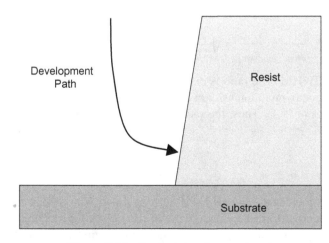

Figure 7.14   *Typical development path*

Before tackling the mathematical problem of solving this path integral, let's consider some of the general properties expected of this path. The path always begins vertically (since the resist surface is horizontal). If the final resist profile is close to our goal of nearly vertical sidewalls, the path must end going nearly horizontally. Thus, we should expect a path that starts out vertically, travels down through the resist thickness, then turns horizontally, stopping at the resist edge (see Figure 7.14).

7.3.1   The Euler–Lagrange Equation

The first step in solving the development path integral will be to parameterize the path in terms of its position. For simplicity, we will consider only the two-dimensional problem $s(x,z)$, though it can be easily extended to three dimensions. Consider a section of that path (Figure 7.15) small enough that it can be approximated as linear. It is clear that

$$\Delta s = \sqrt{\Delta x^2 + \Delta z^2} = \Delta x \sqrt{1 + \left(\frac{\Delta z}{\Delta x}\right)^2} \tag{7.45}$$

Letting the differentials go to zero,

$$ds = dx\sqrt{1 + z'^2} \tag{7.46}$$

where z' is the slope of the development path at any given point. The path integral then becomes

$$t_{dev} = \int_{x_0}^{x} \frac{dx\sqrt{1 + z'^2}}{r(x,z)} \tag{7.47}$$

where the path starts at $(x_0,0)$ and ends at (x,z).

Obviously, to solve Equation (7.47), one must know not only the development rate as a function of position, but the slope of the path at every point along the path as well. But how is one to determine the path of development? Physical insight and mathematical methods must both be exploited. The physical constraint on what path is chosen is called the *principle of least action*: the path taken will be that which goes from the starting

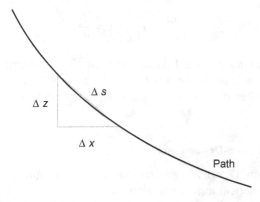

Figure 7.15   *Section of a development path relating path length s to x and z*

point to the finish point in the least amount of time. To see why this must be true, just imagine a different path competing with the least-time path. By the time the competing path made it to the point (x,z), the faster path would have already reached there, and thus would have removed the photoresist from that point! Thus, only the least-time path survives.

The principle of least action means that a path must be found that minimizes the integral in Equation (7.47). A fundamental result from the calculus of variations states that the general integral

$$\int f(x,z,z')dx \tag{7.48}$$

will have a minimum or maximum whenever the Euler–Lagrange equation

$$\frac{\partial f}{\partial z} - \frac{d}{dx}\left(\frac{\partial f}{\partial z'}\right) = 0 \tag{7.49}$$

is satisfied. Applying the Euler–Lagrange approach to the development path, the least-time path will be that which satisfies the second-order nonlinear differential equation:

$$z'' + \left(1 + z'^2\right)\left(\frac{\partial \ln r}{\partial z} - z'\frac{\partial \ln r}{\partial x}\right) = 0 \tag{7.50}$$

(As an interesting aside, there are many problems in physics where the principle of least action is coupled with the Euler–Lagrange technique to solve problems. In optics, techniques in geometrical optics are used to trace rays of light through materials of different refractive indices. Fermat's principle says light will use the path that minimizes the travel time – a principle of least action. The rays follow paths equivalent to the development path discussed here. The optical path length, or equivalently the optical phase, represents the development time in this problem, and a wavefront – a surface of constant phase – is equivalent to the resist profile – a surface of constant development time. The refractive index is equivalent to $1/r$.)

An equivalent formulation of the Euler–Lagrange equation is the Hamilton–Jacobi equation. A surface of constant development time (i.e. the resist surface) is defined by the Hamilton–Jacobi equation:

$$\left(\frac{\partial t}{\partial x}\right)^2 + \left(\frac{\partial t}{\partial z}\right)^2 = \frac{1}{r^2} \tag{7.51}$$

(This equation is equivalent to the Eikonal equation of geometrical optics.) The Hamilton–Jacobi formulation is useful for 'ray-trace'-like calculations, where at each point the dissolution rate is broken down into x- and z-components and the trajectory of the ray is traced out over time.

7.3.2   The Case of No z-Dependence

Solutions to Equation (7.50) for general development problems will almost always require extensive numerical techniques. However, there are a few simplifying cases that yield analytical solutions. Consider the idealized case where the development rate does not vary with z (no absorption, no standing waves, etc.). The Euler–Lagrange equation simplifies to

$$z'' = z'(1+z'^2)\frac{\partial \ln r}{\partial x}$$

(7.52)

This equation can be solved by direct integration with the boundary condition that $z' = \infty$ and $x = x_0$ at $z = 0$ (i.e. the path starts out vertically at $x = x_0$), giving

$$\frac{r(x)}{r(x_0)} = \frac{z'}{\sqrt{1+z'^2}} \quad \text{or} \quad z' = \frac{r(x)}{\sqrt{r^2(x_0)-r^2(x)}}$$

(7.53)

The actual path $z(x)$ can be derived by integrating Equation (7.53),

$$z = \int_{x_0}^{x} \frac{r(x)}{\sqrt{r^2(x_0)-r^2(x)}}\,dx$$

(7.54)

and the development time becomes, by substituting Equation (7.53) into Equation (7.47),

$$t_{dev} = \int_{x_0}^{x} \frac{r(x_0)}{r(x)\sqrt{r^2(x_0)-r^2(x)}}\,dx$$

(7.55)

where the point (x,z) is the end of the path and thus a point on the final resist profile. For a given spatial distribution of development rates $r(x)$, the final resist profile is determined by solving the integrals (7.54) and (7.55) for many different values of x_0 (i.e. for different starting points for the paths), as shown in Figure 7.16, and connecting the paths to form a surface of constant development time.

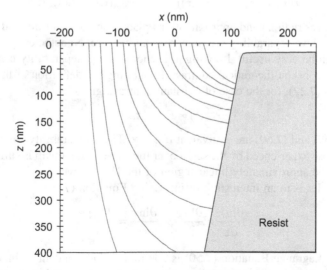

Figure 7.16 *Example of how the calculation of many development paths leads to the determination of the final resist profile*

An interesting form of Equation (7.55) reveals some physical insight into the problem. Combining Equations (7.55) and (7.53) in a different way,

$$t_{\text{dev}} = \frac{1}{r(x_0)} \int_{x_0}^{x} \frac{1+z'^2}{z'} dx = \frac{1}{r(x_0)} \int_{x_0}^{x} \frac{1}{z'} dx + \frac{z(x)}{r(x_0)} \qquad (7.56)$$

The right-hand side of this equation seems to segment the development into horizontal and vertical components. The term $z/r(x_0)$ is the time required to develop vertically, starting at x_0, to the value of z equal to the bottom of the path. The first term on the right-hand side (the integral) is then the effective amount of time spent developing horizontally from x_0 to x.

7.3.3   The Case of a Separable Development Rate Function

Another interesting case to examine is when the development rate function $r(x,z)$ is separable into a function of incident dose E only and a function of the spatial variables independent of dose:

$$r(E,x,z) = f(E)g(x,z) \qquad (7.57)$$

In general, none of the kinetic development models discussed earlier meet this criterion. However, such separability is approximately true for the reaction-controlled original kinetic development model for conventional resists. For moderately low doses and no bleaching, the latent image is approximately

$$m(x,z) = \exp[-CEI_r(x,z)] \approx 1 - CEI_r(x,z) \qquad (7.58)$$

Ignoring the minimum development rate, this approximate latent image gives a development rate proportional to a power of dose:

$$r(x,z) = r_{\text{max}}(1 - m(x,z))^n + r_{\text{min}} \approx r_{\text{max}}(CE)^n(I_r(x,z))^n \qquad (7.59)$$

Obviously, the restrictions and approximations placed on the exposure and development functions to achieve separability are numerous and severe. Nonetheless, the separability assumption can be very useful. Looking at it another way, separability can be achieved by assuming a constant theoretical contrast. Integrating the definition of theoretical contrast, Equation (7.22), for the case of constant contrast, gives

$$r(x,z) \propto E^{\gamma}(I_r(x,z))^{\gamma} \qquad (7.60)$$

Equations (7.59) and (7.60) are equivalent if $n = \gamma$. Thus, separability can be assumed if the range of dose experienced by the resist over the development path is sufficiently small to fall within the approximately linear region of the H–D curve.

Separability leads to an interesting result. Using Equation (7.57),

$$\frac{\partial \ln r}{\partial z} = \frac{\partial \ln g}{\partial z}, \quad \frac{\partial \ln r}{\partial x} = \frac{\partial \ln g}{\partial x} \qquad (7.61)$$

and the Euler–Lagrange Equation (7.50) is completely independent of incident dose. In other words, as the dose is changed, the path of development does not change – the development just follows that path at a different rate. Applying this idea to the defining development path Equation (7.47),

$$f(E)t_{\text{dev}} = \int_{x_0}^{x} \frac{dx\sqrt{1+z'^2}}{g(x,z)} = c, \quad \text{a constant} \tag{7.62}$$

For example, let $f(E) = E^\gamma$. If the dose is increased, the same path is used to get from x_0 to x, but it arrives in a shorter time.

7.3.4 Resist Sidewall Angle

The development path can help to understand even more about the resist profile. Since the path is always perpendicular to the resist profile, knowing the path gives us the sidewall angle of the profile (the angle that the profile makes with the horizontal substrate) – the tangent of the sidewall angle will be one over the development path slope. For the case of no z-variation in development rate, Equation (7.53) leads to

$$\cos\theta = \frac{r(x)}{r(x_0)} \tag{7.63}$$

The goal is to make the sidewall angle close to 90°, and thus make $\cos\theta \approx 0$. This is accomplished by making the development rate at the end of the path very small compared to the rate at the beginning of the path. In other words, good resist profiles are obtained when very little time is spent developing vertically compared to the time spent developing horizontally.

As previously mentioned, numerical methods are required for a general solution for the development path. However, an interesting limit can be observed that sheds further light on the final resist sidewall angle. For the case of no variation in development rate with z, the sidewall angle is given by Equation (7.63) and can be made as close to 90° as possible by making the development rate at the end of the path very small. However, in the presence of a depth dependence to the development rate, there is a different limiting sidewall angle. Consider the Euler–Lagrange Equation (7.50) near the end of the path assuming a reasonably good resist profile. For such a case, the path slope will be small and $z'^2 \ll 1$, so that

$$z'' \approx z' \frac{\partial \ln r}{\partial x} - \frac{\partial \ln r}{\partial z} \tag{7.64}$$

The minimum possible value of z' (and thus the best possible resist sidewall angle) will occur when $z'' = 0$. Thus,

$$z'_{\text{min}} = \frac{\partial \ln r/\partial z}{\partial \ln r/\partial x} \tag{7.65}$$

Consider again the case of absorption causing the only variation in development rate with depth into the resist. Using our simple absorption model in Equation (7.19), the

Lithographic Imaging Equation (7.23), and approximating the dissolution selectivity parameter with the contrast,

$$z'_{min} \approx \frac{\alpha}{\partial \ln I / \partial x} \tag{7.66}$$

The minimum path slope with absorption given in Equation (7.66) is obtained when the development rate at the end of the path goes to zero. For the case of finite development rate, the path slope near the end of path is given approximately by

$$z'_{end} \approx \frac{\alpha}{\partial \ln I / \partial x} + \frac{r(x)}{r(x_0)} \tag{7.67}$$

This result will be derived in the next section for a specific case, but is approximately true in general.

7.3.5   The Case of Constant Development Gradients

Another interesting case that can be explored analytically has constant spatial log-gradients of the development rate. Let

$$\frac{\partial \ln r}{\partial x} = -a, \quad \frac{\partial \ln r}{\partial z} = -b \tag{7.68}$$

where a and b are constants. Interpreting these constants as before, a is something like the resist contrast times image log-slope. The term b is the contrast multiplied by the absorption coefficient. Note that both a and b will be positive. The Euler–Lagrange equation can be solved by direct integration, with the boundary condition that $z' = \infty$ and $x = x_0$ at $z = 0$. After some algebraic manipulation of the result,

$$\exp\left[-a(x - x_0)\left(1 + \frac{b^2}{a^2}\right)\right] = \left[\frac{z' - b/a}{\sqrt{1 + z'^2}}\right] \exp\left[\frac{b}{a} \cot^{-1}(z')\right] \tag{7.69}$$

For a typical process, b/a is in the range of 0.02–0.05 (development gradients in x must be significantly greater than the gradients in z if a good resist profile is to result). Thus, it is reasonable to neglect b^2/a^2 compared to 1. Also, the assumption of constant log-gradient of development rates means that the development rate is separable into x- and z-components and

$$e^{-a(x - x_0)} = \frac{r(x)}{r(x_0)} \tag{7.70}$$

Thus, Equation (7.69) becomes

$$\frac{r(x)}{r(x_0)} \approx \left[\frac{z' - b/a}{\sqrt{1 + z'^2}}\right] \exp\left[\frac{b}{a} \cot^{-1}(z')\right] \tag{7.71}$$

Finally, the inverse cotangent will be zero at the beginning of the path (when $z' = \infty$), going to $\pi/4$ for $z' = 1$, and reaching its maximum of $\pi/2$ as z' goes to zero. Given the small value of b/a, the exponential term can be approximated with a Taylor's series to give

$$\frac{r(x)}{r(x_0)} \approx \left(1 + \frac{b}{a}\cot^{-1}(z')\right)\left(\frac{z'-b/a}{\sqrt{1+z'^2}}\right) \tag{7.72}$$

For a small enough value of b/a,

$$\frac{r(x)}{r(x_0)} \approx \left(\frac{z'-b/a}{\sqrt{1+z'^2}}\right) \tag{7.73}$$

Note that at the end of the path, $z'^2 \ll 1$, and the result given previously in Equation (7.67) is obtained.

Before continuing with the solution, let's consider a simpler case where $b = 0$. Thus, we are solving the no z-dependence case with the development rate function given in Equation (7.70). Substituting this rate expression into Equation (7.54) and integrating gives us the path $z(x)$ for a given starting point x_0:

$$z(x,x_0) = \frac{1}{a}\tan^{-1}\left(\sqrt{e^{2a(x-x_0)}-1}\right) = \frac{1}{a}\cos^{-1}(e^{-a(x-x_0)}) \tag{7.74}$$

Going on to find the development time at each point on the path by carrying out the integration in Equation (7.55),

$$t_{\text{dev}} = \frac{1}{ar(x_0)}\sqrt{e^{2a(x-x_0)}-1} \tag{7.75}$$

The resist profile sidewall angle is simply az.

A convenient way to express this result is to parameterize positions along the path (and thus on the resist profile) as a function of the path starting point and the development time. Rearranging the results in Equations (7.74) and (7.75),

$$z(x_0,t) = \frac{1}{a}\tan^{-1}(ar(x_0)t), \quad x(x_0,t) = x_0 + \frac{1}{a}\ln\left(\sqrt{[ar(x_0)t]^2+1}\right) \tag{7.76}$$

Note that for the very likely case that $ar(x_0)t \gg 1$,

$$x(x_0,t) = x_0 + \frac{1}{a}\ln(ar(x_0)t) \tag{7.77}$$

When absorption is added, the results are in general complicated, but simplify when one doesn't care to know the resist profile at small development times. For normal conditions, the path and develop time can be expressed as the no-absorption solution plus an added term:

$$z(x,x_0) = \frac{1}{a}\tan^{-1}\left(\sqrt{e^{2a(x-x_0)}-1}\right) + \frac{b}{a}(x-x_0)$$

$$t_{\text{dev}} = \frac{1}{ar(x_0)}\sqrt{e^{2a(x-x_0)}-1} + \frac{b}{ar(x_0)}(x-x_0) \tag{7.78}$$

Thus, absorption will slow down the development, making the time required to get to a certain value of x longer. Also, since the path never levels off to be completely horizontal, continued development pushes the profile down (larger z) as well as to the right.

7.3.6   Segmented Development and the Lumped Parameter Model (LPM)

Consider again the case of a constant theoretical contrast. Equation (7.60), for the simple 2D case, can be expressed as

$$r(x,z) = r_0 \left(\frac{E}{E_0} \right)^\gamma (I(x)I_z(z))^\gamma \tag{7.79}$$

where the intensity variation is assumed separable into x- and z-components and E_0 is any arbitrary reference dose that produces the development rate r_0 for an open-frame exposure (so that $I(x)I_z(0) = 1$). It will be convenient to let this dose be the dose-to-clear. Applying Equation (7.24) at this dose,

$$t_{\text{dev}} = \int_0^D \frac{dz}{r(z)} = \frac{1}{r_0} \int_0^D \frac{dz}{(I_r(z))^\gamma} \tag{7.80}$$

An effective resist thickness can be defined as

$$D_{\text{eff}} = r(E_0, z = D)t_{\text{dev}} = \int_0^D \left(\frac{I_z(D)}{I_z(z)} \right)^\gamma dz \tag{7.81}$$

It is clear that for no variation in intensity with depth into the resist this effective resist thickness will equal the actual resist thickness.

As an example, the effective resist thickness can be calculated for the case when only absorption causes a variation in intensity with depth into the resist. For such a case, $I(z)$ will decay exponentially with an absorption coefficient α, and Equation (7.81) can be evaluated to give

$$D_{\text{eff}} = \frac{1}{\alpha\gamma}(1 - e^{-\alpha\gamma D}) \tag{7.82}$$

Equation (7.79) is an extremely simple-minded model relating development rate to exposure energy based on the assumption of a constant resist contrast. In order to use this expression, we still need to determine the path of dissolution. However, rather than calculate the path for some given development variation $r(x,z)$, it will be useful to assume a reasonable development path and then use this path to calculate the final resist CD as a function of exposure dose. The assumed path will be the *segmented development path*:[15] the path goes vertically to a depth z, followed by a lateral development to position x.

A development ray, which traces out the path of development, starts at the point $(x_0, 0)$ and proceeds vertically until a depth z is reached, at which point the development will begin horizontally. Since our interest is to determine the bottom linewidth, consider a path where the depth z is just equal to the resist thickness D. The vertical development time becomes

$$t_z = \frac{1}{r(x_0,D)} \int_0^D \left(\frac{I(z)}{I(D)}\right)^{-\gamma} dz = \frac{D_{\text{eff}}}{r(x_0,D)} \tag{7.83}$$

Similarly, the horizontal development time becomes

$$t_x = \frac{1}{r(x_0,D)} \int_{x_0}^x \left(\frac{I(x')}{I(x_0)}\right)^{-\gamma} dx' \tag{7.84}$$

The total development time is still the sum of these two components.

$$t_{\text{dev}} = \frac{D_{\text{eff}}}{r(x_0,D)}\left(1 + \frac{1}{D_{\text{eff}}} \int_{x_0}^x \left(\frac{I(x')}{I(x_0)}\right)^{-\gamma} dx'\right) \tag{7.85}$$

By applying Equations (7.79) and (7.81),

$$\frac{r(x_0,D)t_{\text{dev}}}{D_{\text{eff}}} = \left(\frac{E(x)}{E(0)}\right)^{\gamma} \tag{7.86}$$

where $E(x)$ is the dose required to create a CD $= 2x$, and $E(0)$ is the dose required to make the CD just equal to zero. Note that $E(0)$ is not the dose-to-clear since in general $I(x_0)$ will not be equal to 1.

Substituting Equation (7.86) into (7.85) gives the Lumped Parameter Model.[16]

$$\left(\frac{E(x)}{E(0)}\right)^{\gamma} = 1 + \frac{1}{D_{\text{eff}}} \int_{x_0}^x \left(\frac{I(x')}{I(x_0)}\right)^{-\gamma} dx' \tag{7.87}$$

Given any aerial image $I(x)$, a CD-versus-dose curve can be calculated by evaluating this integral numerically. For a small space (or small dense lines and spaces), the start of the path x_0 will always be the center of the space. The only parameters that control the relationship between the shape of the aerial image and the shape of the exposure latitude curve are the resist contrast and the effective resist thickness. As the effective resist thickness goes to zero and the resist contrast goes to infinity, the dose required to produce a CD of $2x$ becomes proportional to $1/I(x)$.

7.3.7 LPM Example – Gaussian Image

In the clear region of an image, near the edge of a line, the aerial image (and the image in resist) can quite often be reasonably approximated as a Gaussian function:

$$I(x) = I_0 e^{-x^2/2\sigma^2} \tag{7.88}$$

Since only the space region develops away in a positive resist to form the resist profile, an accurate representation of the image is needed only in this region that develops away. The reason why an aerial image of a space often looks Gaussian in shape can be seen by looking at Taylor expansions of the Gaussian and of a typical cosine Fourier representation of an image. Consider for example an image with two Fourier components (a three-beam imaging case) expanded in a Taylor series about $x = 0$ (the center of the space) for a given pitch p:

$$I(x) = a_o + a_1 \cos(2\pi x/p) + a_2 \cos(4\pi x/p)$$

$$= (a_o + a_1 + a_2) - \left(\frac{a_1 + 4a_2}{2}\right)(2\pi x/p)^2 + \left(\frac{a_1 + 16a_2}{24}\right)(2\pi x/p)^4 - \dots \qquad (7.89)$$

Likewise for the Gaussian image:

$$I(x) = I_0 e^{-x^2/2\sigma^2} = I_0 - \left(\frac{I_0}{2\sigma^2}\right)x^2 + \left(\frac{I_0}{8\sigma 4}\right)x^4 - \dots \qquad (7.90)$$

The zero- and second-order terms of these two series can be made to match exactly when

$$I_0 = a_0 + a_1 + a_2 \quad \text{and} \quad \sigma^2 = \left(\frac{a_0 + a_1 + a_2}{a_1 + 4a_2}\right)\left(\frac{p}{2\pi}\right)^2 \qquad (7.91)$$

In addition, the fourth-order terms for each series are often quite close to each other in magnitude. For example, if $a_0 = a_1$ and $a_2 = 0.2a_0$ (a reasonable set of values), then a Gaussian image with parameters determined by Equation (7.91) will have fourth-order terms in the Taylor series for the aerial image and the Gaussian representation that differ in magnitude by only 5 %. Further, by adjusting σ of the Gaussian slightly from the value given by Equation (7.91), a better overall match to the image in the region of the space can be obtained (see Figure 7.17).

The log-slope of the Gaussian image is

$$ILS = \frac{d \ln I}{dx} = \frac{x}{\sigma^2} \qquad (7.92)$$

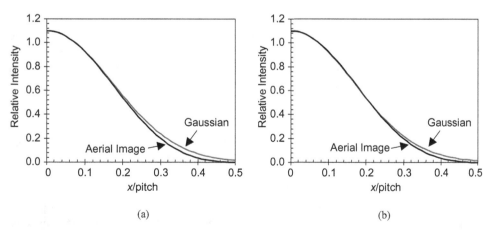

(a)   (b)

Figure 7.17 *Matching a three-beam image ($a_0 = 0.45$, $a_1 = 0.55$, $a_2 = 0.1$) with a Gaussian using (a) a value of σ given by Equation (7.91); and (b) the best-fit σ. Note the goodness of the match in the region of the space (x/pitch < 0.25)*

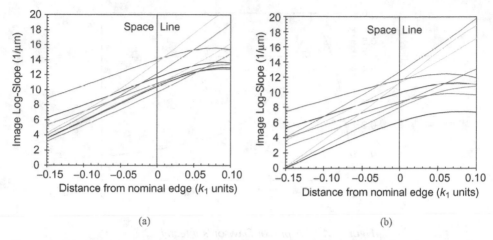

Figure 7.18 *Simulated aerial images over a range of conditions show that the image log-slope varies about linearly with distance from the nominal resist edge in the region of the space: (a)* $k_1 = 0.46$ *and (b)* $k_1 = 0.38$ *for isolated lines, isolated spaces, and equal lines and spaces and for conventional, annular and quadrupole illumination*

Thus, the log-slope varies linearly with x, a result that is approximately true for most images in the region of the space (Figure 7.18). If the nominal CD is CD_1, then the normalized image log-slope (NILS) at this point is

$$NILS_1 = CD_1 \frac{d \ln I}{dx}\bigg|_{x=CD_1/2} = \frac{CD_1^2}{2\sigma^2} \tag{7.93}$$

Putting the Gaussian aerial image into these terms,

$$I(x) = I_0 e^{-NILS_1(x/CD_1)^2} \tag{7.94}$$

Using this Gaussian image in the LPM Equation (7.87) and letting the development path start at $x = 0$,

$$\left(\frac{E(x)}{E(0)}\right)^\gamma = 1 + \frac{1}{D_{\text{eff}}} \int_0^x e^{\gamma NILS_1(x'/CD_1)^2} dx' \tag{7.95}$$

The integral is solvable in terms of the imaginary error function *erfi*. Letting $g = \gamma NILS_1/CD_1^2$,

$$\left(\frac{E(x)}{E(0)}\right)^\gamma = 1 + \frac{1}{2D_{\text{eff}}} \sqrt{\frac{\pi}{g}} erfi\left(\sqrt{g}x\right) \tag{7.96}$$

Since *erfi* goes to infinity as x goes to infinity, a more convenient form can be obtained by relating the imaginary error function to the Dawson's integral $D_w(z)$:

$$erfi(z) = \frac{2}{\sqrt{\pi}} e^{z^2} D_w(z) \quad \text{where} \quad D_w(z) = e^{-z^2} \int_0^z e^{x^2} dx \tag{7.97}$$

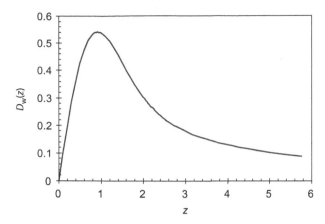

Figure 7.19   *A plot of Dawson's integral, $D_w(z)$*

Dawson's integral is a bit easier to handle since its maximum value is about 0.54 at $z \approx$ 0.92 and goes to zero at large z (see Figure 7.19). The LPM result becomes

$$\left(\frac{E(x)}{E(0)}\right)^{\gamma} = 1 + \frac{1}{D_{eff}\sqrt{g}}e^{gx^2}D_w\left(\sqrt{g}x\right)$$

$$= 1 + \frac{1}{D_{eff}\sqrt{g}}\left(\frac{I(x)}{I(0)}\right)^{-\gamma}D_w\left(\sqrt{g}x\right) \tag{7.98}$$

For large arguments, the Dawson's integral can be approximated reasonably well as

$$D_w(z) = \frac{1}{2z} + \frac{1}{4z^3} + \ldots \approx \frac{1}{2z} \tag{7.99}$$

This final approximation has about 20 % error when $z = 2$, 11 % error at $z = 2.5$, 7 % error at $z = 3$, and 3.5 % error at $z = 4$. Small arguments occur for low g (low contrast resists or low NILS values) and low x (underexposed spaces where the CD is much lower than the nominal). This is also the regime where e^{gx^2} is small so that the impact of this error is much reduced. For typical cases, and where the CD being predicted by Equation (7.98) is within 50 % of the nominal CD, the error in calculating the dose required to achieve that CD caused by using the approximation for the Dawson's integral is usually less than 1–2 %.

In this large argument case, the resulting exposure response expression becomes

$$\left(\frac{E(x)}{E(0)}\right)^{\gamma} \approx 1 + \frac{1}{D_{eff}g(2x)}\left(\frac{I(x)}{I(0)}\right)^{-\gamma} \tag{7.100}$$

Substituting the definition of g and letting $CD = 2x$,

$$\left(\frac{E(x)}{E(0)}\right)^{\gamma} \approx 1 + \frac{1}{\gamma}\left(\frac{1}{D_{eff}}\right)\left(\frac{CD_1^2}{CDNILS_1}\right)\left(\frac{I(x)}{I(0)}\right)^{-\gamma} \tag{7.101}$$

For our Gaussian image [and from Equations (7.92) and (7.93)],

$$ILS = \frac{CD}{2\sigma^2} = \frac{CDNILS_1}{CD_1^2}$$ (7.102)

which gives the final expression:

$$\frac{E(x)}{E(0)} \approx \left[1 + \left(\frac{1}{\gamma D_{\text{eff}} ILS(x)} \right) \left(\frac{I(x)}{I(0)} \right)^{-\gamma} \right]^{\frac{1}{\gamma}}$$ (7.103)

Remembering that log-slope of the aerial image is also a function of x, Equation (7.103) predicts the dose required to achieve a certain CD ($= 2x$) as a function of the aerial image $I(x)$, the resist contrast γ, and the effective thickness of the resist D_{eff}.

Equation (7.103) can be further simplified by assuming that the vertical development time of the LPM can be ignored compared to the horizontal development time (an assumption that is reasonable for a good lithography process, but may not be accurate when trying to understand features that are just able to print, for example at the extremes of focus). In this case, the 1 in the square brackets of Equation (7.103) can be ignored, giving

$$E(x) \approx \frac{E(0)I(0)}{I(x)} \left(\frac{1}{\gamma D_{\text{eff}} ILS} \right)^{\frac{1}{\gamma}} = \frac{E_0}{I(x)} \left(\frac{1}{\gamma D_{\text{eff}} ILS} \right)^{\frac{1}{\gamma}}$$ (7.104)

This interesting result is in fact an example of a threshold resist model: the resist CD is obtained when the local dose $EI(x)$ reaches a certain threshold dose E_{th}.

$$E = \frac{E_{\text{th}}}{I(x)} \quad \text{where} \quad E_{\text{th}} = E_0 \left(\frac{1}{\gamma D_{\text{eff}} ILS} \right)^{\frac{1}{\gamma}}$$ (7.105)

The particular variant of the threshold resist model derived here is called a variable-threshold resist model (sometimes called a VTR model). The threshold varies as a function of the image log-slope of the image. This model also relates the threshold to physically significant resist parameters: E_0, γ and D_{eff}.

Figure 7.20 compares predictions of the CD of a space versus exposure dose for three different models: the LPM Equation (7.98), the LPM using the approximation for the Dawson's integral of Equation (7.103), and the VTR of Equation (7.105). For the conditions here, the approximate form of Dawson's integral is quite accurate and the approximate LPM expression is hard to distinguish from the full LPM results. As expected, the VTR begins to deviate from the LPM at lower doses, where the vertical development time is a significant fraction of the total development time. At the nominal CD, the approximate LPM predicts the dose wrong by 1%, and the VTR is off by 2%. For a CD that is 10 and 20% too small (underexposed), the approximate LPM is still off by only 1% for both, whereas the VTR misses by about 3.5 and 6.5%, respectively. Overexposed (CD of the space getting larger), the VTR and approximate LPM both improve in accuracy.

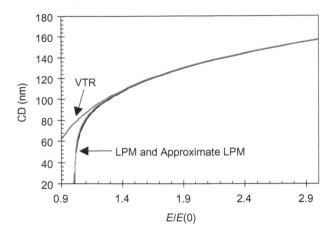

Figure 7.20 *CD-versus-dose curves as predicted by the LPM Equation (7.98), the LPM using the approximation for the Dawson's integral, and the VTR. All models assumed a Gaussian image with g = 0.0025 1/nm² (NILS₁ = 2.5 and CD₁ = 100 nm), γ = 10, and D_{eff} = 200 nm*

An alternate (and usually more convenient) form of this VTR model can be obtained by inserting the Gaussian image intensity [Equation (7.94)] that was used to derive the VTR model directly into Equation (7.104). After some algebraic manipulations, and letting E_1 be the dose that produces the reference spacewidth CD_1,

$$\frac{CD}{CD_1} = \sqrt{1 + \frac{4}{NILS_1}\left[\ln\left(\frac{E}{E_1}\right) + \frac{1}{\gamma}\ln\left(\frac{CD}{CD_1}\right)\right]} \tag{7.106}$$

Unfortunately, the equation is recursive with respect to CD. However, for dimensions close to CD_1, the following approximation can be made:

$$\frac{\ln\left(\frac{CD}{CD_1}\right)}{\ln\left(\frac{E}{E_1}\right)} = \frac{\ln\left(1 + \frac{\Delta CD}{CD_1}\right)}{\ln\left(1 + \frac{\Delta E}{E_1}\right)} \approx \frac{\frac{\Delta CD}{CD_1}}{\frac{\Delta E}{E_1}} \approx \frac{2}{NILS_1} \tag{7.107}$$

Thus,

$$\frac{CD}{CD_1} \approx \sqrt{1 + \frac{4}{NILS_1}\ln\left(\frac{E}{E_1}\right)\left[1 + \frac{2}{\gamma NILS_1}\right]} \tag{7.108}$$

7.4 Measuring Development Rates

In general, development model parameters for a given resist/develop process are obtained by making real-time *in-situ* measurements of resist thickness during development using a development rate monitor (DRM). The resist thickness-versus-develop time curves are measured at many different incident exposure doses. Taking the derivative of this data gives development rate (r) as a function of incident dose (E) and depth into the resist (z).

By modeling the exposure process, a given exposure dose turns into a distribution of inhibitor concentration (m) after exposure as a function of depth. Combining the measured $r(E,z)$ with the calculated $m(E,z)$ produces a resultant $r(m,z)$ data set. These data are then fit with a development model (possibly including a surface inhibition function) and parameters for the model are extracted.

Although the best method for obtaining new parameters is the use of a DRM as described above, there is another approximate approach called the Poor Man's DRM.[17] The Poor Man's DRM involves the measurement of multiple contrast curves – resist thickness remaining as a function of exposure dose for open-frame exposures – at different development times (Figure 7.21).

Analyzing the information in Figure 7.21 to obtain development rate, and eventually development parameters, is not trivial. The thickness as a function of dose and development time is first converted to rate as a function of dose and depth into the resist, $r(E,z)$. From here, analysis is the same as for conventional DRM data. Figure 7.22 shows an example set of results.

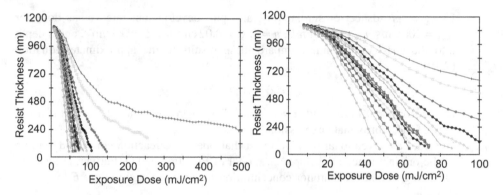

Figure 7.21 *Measured contrast curves[17] for an i-line resist at development times ranging from 9 to 201 seconds (shown here over two different exposure scales)*

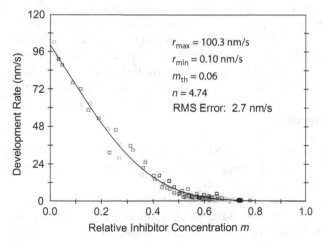

$r_{max} = 100.3$ nm/s
$r_{min} = 0.10$ nm/s
$m_{th} = 0.06$
$n = 4.74$
RMS Error: 2.7 nm/s

Figure 7.22 *Analysis of the contrast curves generates an r(m) data set, which was then fit to the original kinetic development model (best fit is shown as the solid line)*

Problems

7.1. For the original kinetic model of Equation (7.10), what is the value of the development rate at $m = m_{th}$? For $m_{th} > 0$, what is the limit of this value as n becomes large?

7.2. Derive Equation (7.9).

7.3. Using the same approach that was used for the original kinetic development model, add the effect of developer diffusion to the surface of the resist to the enhanced kinetic model. In other words, let Equation (7.12) represent the surface reaction rate and assume that r_{resin} is proportional to D_S.

7.4. Consider a conventional positive resist that obeys the reaction-controlled version of the original kinetic development model of Equation (7.11) Further assume no bleaching, but absorption so that the inhibitor concentration m can be related to incident dose E and depth into resist z by

$$m = e^{-CEI_r(z)} \quad \text{where} \quad I_r(z) = e^{-\alpha z}$$

Using a spreadsheet or similar program, plot development rate versus dose for $r_{max} = 200\,\text{nm/s}$, $r_{min} = 0.05\,\text{nm/s}$, $n = 4$, $C = 0.02\,\text{cm}^2/\text{mJ}$, $\alpha = 0.5\,\mu\text{m}^{-1}$, and a depth into the resist of 500 nm. Compare these results to the approximate Equation (7.19):

$$r(z) \approx r(z = 0)e^{-\alpha n z}$$

Repeat for $\alpha = 0.1\,\mu\text{m}^{-1}$ and $\alpha = 1.5\,\mu\text{m}^{-1}$. What can you conclude about the accuracy of the approximate expression?

7.5. Consider a conventional positive resist that obeys the reaction-controlled version of the original kinetic development model of Equation (7.11). Further, assume no bleaching so that the inhibitor concentration m can be related to dose E by

$$m = e^{-CE}$$

Assuming that the minimum development rate r_{min} can be ignored over the dose range of interest and ignoring any variation in development rate with depth into the resist, derive an expression for the resist characteristic curve (such as that pictured in Figure 7.11). Plot the resulting equation for $r_{max} = 200\,\text{nm/s}$, $n = 7$, $C = 0.02\,\text{cm}^2/\text{mJ}$, an initial resist thickness of 800 nm, and a development time of 60 seconds. What is the value of the dose-to-clear for this case?

7.6. Given a conventionally measured resist characteristic curve, estimate how a small change in resist thickness would change the dose-to-clear.

7.7. Derive Equation (7.32).

7.8. Derive Equation (7.63).

7.9. For the Lumped Parameter Model, consider the case where only absorption causes a variation of development rate with z. In the limit of very large contrast, what is D_{eff} for the two cases of $\alpha = 0$ and $\alpha > 0$?

7.10. The assumption of a segmented development path can lead to a calculation of the resulting CD of a space for a given development rate function $r(x,z)$ and a given development time. How will this CD compare to the CD calculated by rigorously solving the Euler–Lagrange equation for the same conditions?

7.11. Use the Lumped Parameter Model to calculate CD as a function of dose for the case of a space that has a constant log-slope aerial image:

$$I(x) = I_0 e^{-s|x|}$$

where s is positive. Also assume that $x_0 = 0$ and that there is no z-dependence to the development rate. Compare this result to Equation (7.76).

7.12. For the Lumped Parameter Model, consider the case where only absorption causes a variation of development rate with z. Derive an expression that compares the value of resist contrast as measured with a standard photoresist Hurter–Driffield (characteristic) curve to the theoretical (LPM) contrast.

References

1   Mack, C.A., 1987, Development of positive photoresist, *Journal of the Electrochemical Society*, **134**, 148–152.

2   Mack, C.A., 1992, New kinetic model for resist dissolution, *Journal of the Electrochemical Society*, **139**, L35–L37.

3   Meyerhofer, D., 1980, Photosolubility of diazoquinone resists, *IEEE Transactions on Electron Devices*, **ED–27**, 921–926.

4   Arthur, G., Mack, C.A. and Martin, B., 1997, Enhancing the development rate model for optimum simulation capability in the sub-half-micron regime, *Proceedings of SPIE: Advances in Resist Technology and Processing XIV*, **3049**, 189–200.

5   Tsiartas, P.C., Flanagin, L.W., Henderson, C.L., Hinsberg, W.D., Sanchez, I.C., Bonnecaze, R.T. and Willson, C.G., 1997, The mechanism of phenolic polymer dissolution: a new perspective, *Macromolecules*, **30**, 4656–4664.

6   Hanabata, M., Oi, F. and Furuta, A., 1991, Novolak design for high-resolution positive photoresists (IV): tandem-type novolak resin for high-performance positive photoresists, *Proceedings of SPIE: Advances in Resist Technology and Processing VIII*, **1466**, 132–140.

7   Dill, F.H. and Shaw, J.M., 1977, Thermal effects on the Photoresist AZ1350J, *IBM Journal of Research and Development*, **21**, 210–218.

8   Kim, D.J., Oldham, W.G. and Neureuther, A.R., 1984, Development of positive photoresist, *IEEE Transactions on Electron Devices*, **ED-31**, 1730–1735.

9   Garza, C.M., Szmanda, C.R. and Fischer, R.L., Jr., 1988, Resist dissolution kinetics and sub-micron process control, *Proceedings of SPIE: Advances in Resist Technology and Processing V*, **920**, 321–338.

10   Itoh, K., Yamanaka, K., Nozue, H. and Kasama, K., 1991, Dissolution kinetics of high resolution novolac resists, *Proceedings of SPIE: Advances in Resist Technology and Processing VIII*, **1466**, 485–496.

11   Mack, C.A., Maslow, M.J., Carpio, R. and Sekiguchi, A., 1998, New model for the effect of developer temperature on photoresist dissolution, *Proceedings of SPIE: Advances in Resist Technology and Processing XV*, **3333**, 1218–1231.

12   Maslow, M.J., Mack, C.A. and Byers, J., 1999, Effect of developer temperature and normality on chemically amplified photoresist dissolution, *Proceedings of SPIE: Advances in Resist Technology and Processing XVI*, **3678**, 1001–1011.

13   Hurter, F. and Driffield, V.C., 1890, Photochemical investigations and a new method of determination of the sensitiveness of photographic plates, *Journal of the Society of Chemical Industry*, **9**, 455–469.

14   Mack, C.A., Lithographic optimization using photoresist contrast, *Proceedings of the KTI Microlithography Seminar, Interface '90*, 1–12 (1990), and *Microelectronics Manufacturing Technology*, **14**, 36–42 (1991).

15   Watts, M.P.C., 1985, Positive development model for linewidth control, *Semiconductor Inter-national*, **8**, 124–131.
16   Hershel, R. and Mack, C.A., 1987, Chapter 2: lumped parameter model for optical lithography, in R. K. Watts, N.G. Einspruch (eds), *Lithography for VLSI, VLSI Electronics – Microstructure Science Volume 16*, Academic Press (New York, NY) pp. 19–55.
17   Thornton, S.H. and Mack, C.A., 1996, Lithography model tuning: matching simulation to experiment, *Proceedings of SPIE: Optical Microlithography IX*, **2726**, 223–235.

8

Lithographic Control in Semiconductor Manufacturing

Historically, lithography engineering has focused on two key, complimentary aspects of lithographic quality: overlay performance and linewidth control. Linewidth (or critical dimension, CD) control generally means ensuring that the widths of certain critical features, measured at specific points on those features, fall within acceptable bounds. Overlay describes the positional errors in placing one mask layer pattern over an existing pattern on the wafer. Control of both CD and overlay is absolutely essential to producing high-yielding and high-performing semiconductor devices. And achieving adequate control becomes increasingly difficult as the feature sizes of the devices decrease.

8.1 Defining Lithographic Quality

How is lithographic quality defined? This simple question has a surprisingly complex answer. Lithography is such a large component of the total manufacturing cost of a chip and has such a large impact on final device performance that virtually all aspects of the lithography process must be carefully considered. Although somewhat arbitrary, the list below divides lithographic quality into four basic categories: photoresist profile control, overlay, downstream compatibility and manufacturability.

Photoresist Profile Control is a superset of the common CD control metric that is universally thought to be the most important aspect of high-resolution lithography processes. For many lithographic steps, the ability to print features at the correct dimensions has a direct and dramatic influence on device performance. It is typically measured as a mean to target CD difference for one or more specific device features, as well as a distribution metric such as the standard deviation. Spatial variations across the chip, field, wafer or lot are also important and can be characterized together or individually. In addition, the sensitivity of CD to process variations is often characterized and optimized as a method

Fundamental Principles of Optical Lithography: The Science of Microfabrication, Chris Mack.
© 2007 John Wiley & Sons, Ltd.

to improve CD control. Metrics such as resolution, depth of focus and process latitude in general are expressions of CD control. For example, resolution can be defined as the smallest feature of a certain type that provides adequate CD control for a given process.

Profile control, however, recognizes that the printed resist patterns are three-dimensional in nature and a single CD value may not be sufficient to characterize their lithographic quality. Extension of CD control to profile control means taking into account other dimensions. In the 'z-direction' from the top to the bottom of the photoresist, the resist profile shape is usually characterized by a sidewall angle and a final resist height. In the 'x–y direction', patterns more complex than a line or space require a shape characterization that can include metrics such as corner rounding, line-end shortening, area fidelity, line edge roughness, or the critical shape difference (also called edge placement error).

Overlay is the ability to properly position one lithographic level with respect to a previously printed level. In one sense, lithography can be thought of as an effort to position photoresist edges properly on the wafer. But rather than characterize each edge individually, it is more convenient to correlate two neighboring edges and measure their distance from each other (CD) and the position of their midpoint (overlay). One simple reason for this division is that, for the most part, errors that affect CD do not influence overlay, and vice versa. (Unfortunately, this convenient assumption is becoming less and less true as the target feature sizes shrink.) Overlay is typically measured using special targets optimized for the task, but actual device structures can be used in some circumstances. Since errors in overlay are conveniently divided into errors influenced by the reticle and those influenced by the wafer and wafer stage, measurements are made within the exposure field and for different fields on the wafer to separate out these components. While historically CD control has gained the most attention as the limiter to feature size shrinks, overlay control is often just as critical.

Downstream Compatibility describes the appropriateness of the lithographic results for subsequent processing steps, in particular etch and ion implantation. Unlike many other processing steps in the manufacture of an integrated circuit, the handiwork of the lithographer rarely finds its way to the final customer. Instead, the true customers of the lithography process are the etch and implant groups, who then strip off those painstakingly prepared photoresist profiles when finished with them. Downstream compatibility is measured with such metrics as etch resistance, thermal stability, adhesion, chemical compatibility, strippability and pattern collapse.

Manufacturability is the final, and ultimate, metric of a lithographic process. The two major components of manufacturability are cost and defectivity. The importance of cost is obvious. What makes this metric so interesting, and difficult to optimize, is the relationship between cost and other metrics of lithographic quality such as CD control. While buying ultra flat wafers or upgrading to the newest stepper platform may provide an easy improvement in CD and overlay performance, the benefit may be negated by the cost increase. It is interesting to note that throughput (or more correctly overall equipment productivity) is one of the major components of lithographic cost for a fab that is at or near capacity due to the normal factory design that places lithography as the fab bottleneck.

Defectivity in all areas of the fab has been a major contributor to yield loss for most processes throughout the history of this industry. Because lithographic processes are repeatedly applied to make a chip, any improvements in defectivity in the lithography area are multiplied many times over. Finally, concerns such as safety and environmental impact will always play a role in defining the overall manufacturability of a process.

The outline for defining lithographic quality presented above gives a flavor for the complexity of the task. Literally dozens of quality metrics are required to describe the true value of a lithographic result. However, once these metrics have been defined, their relative value to the customer evaluated, and methods for their measurement established, they become powerful tools for the continuous improvement required to remain competitive in semiconductor manufacturing.

8.2 Critical Dimension Control

The shrinking of the dimensions of an integrated circuit is the technological driver of Moore's Law (see Chapter 1). As the dimensions of a transistor shrink, the transistor becomes smaller, lighter, faster, consumes less power, and in many cases is more reliable. In addition, the semiconductor industry has historically been able to manufacture silicon devices at an essentially constant cost per area of processed silicon. Thus, an increased density of transistors per area of silicon results in huge cost advantages in the manufacture of chips. While many of these historical trends are being challenged (shrinking transistor size now often increases power consumption rather than reducing it), the compelling cost advantage of packing more transistors per unit area continues to push feature sizes smaller. However, as features become smaller, the need to control the quality of the lithographic results increases in proportion.

8.2.1 Impact of CD Control

Lithographers work hard to improve the control of the feature dimensions printed on the wafer for two very important reasons. First, and most obviously, the dimensions of device features affect the electrical behavior of those devices. In general, devices have been designed to work optimally over a very limited range of feature sizes. If those sizes become either too big or too small, a variety of undesirable consequences will result. Because of this, the very process of design itself must take into account the variability of the lithography process, resulting in the second important reason why linewidth control is important: its impact on the area of a designed chip.

Design rules are a complex set of geometric restrictions (such as the minimum allowed dimensions and spacings for each circuit element) that are used when creating a circuit layout (the design information that will eventually be translated into a photomask pattern). The overall goal of the design rules is to minimize the silicon area of the circuit while at the same time maintaining sufficient yield and circuit performance in the face of variations in the semiconductor manufacturing processes.

Consider a contact hole making contact to a source or drain of a transistor (Figure 8.1). One of the many design rules would define the minimum layout separation between the contact hole and the active area edge (dimension f in Figure 8.1). In general, this design rule would be determined by the contact to active layer overlay specification ($O_{c\text{-}aa}$)

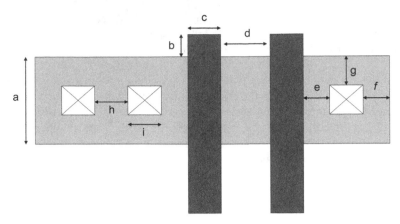

Figure 8.1 *A simple layout showing a two-transistor structure with source/drain contact holes. One design rule would dictate the minimum allowed spacing between the edge of the contact and the edge of the active area (dimension f)*

and the CD control specifications for the contact (ΔCD_c) and the active area width (ΔCD_{aa}).

$$Minimum\ f = O_{c\text{-}aa} + \Delta CD_c/2 + \Delta CD_{aa}/2 \qquad (8.1)$$

Thus, the minimum spacing f can be reduced if the CD control specification is reduced. In general, though, the overlay specification is roughly three times the CD control specification so that overlay control represents about 75 % of the control requirements and thus tends to dominate this design rule (see section 8.4). On the other hand, dimension h is completely determined by CD control considerations.

One of the classic examples of the influence of CD control is at the polysilicon gate level of standard CMOS logic devices. Physically, the polysilicon gate linewidth (paradoxically called the gate *length* by device engineers) controls the electrically important effective gate length (L_{eff}). In turn, L_{eff} is directly related to the transit time τ, the average time required to move an electron or hole from the source to the drain of the MOS device:

$$\tau = \frac{L_{eff}^2}{\mu V_{ds}} \qquad (8.2)$$

where μ is the mobility of the charge carrier for the transistor (either holes or electrons), and V_{ds} is the drain–source voltage. The delay time of an unloaded inverter is about twice the transit time, plus any parasitic delays caused by resistance and capacitance of the device not associated with the gate. These parasitic delays tend to be about equal in magnitude to the delay caused by the gate transit time, but are independent of L_{eff}. Pass transistors have delays about equal to the inverter delay, and a two-input NAND gate has a delay about equal to twice that of an inverter. Given these timing constraints, a circuit's clock period has a minimum of about 10 times the delay given by Equation (8.2), but actual circuits use a slower 15× multiplier to allow for variations in this delay time. Thus, narrower gates tend to make transistors that can switch on and off at higher clock speeds. The squared dependence shown in Equation (8.2) means that small variations in the poly gate CD will cause a 2× larger variation in device delay times.

Obviously, faster chips are more valuable than slower ones, as anyone who has priced a personal computer knows. But smaller is not always better. Transistors are designed (especially the doping levels and profiles) for a specific gate length. As the gate length gets smaller than this designed value, the transistor begins to 'leak' current when it should be off. If this leakage current becomes too high, the transistor is judged a failure. The 'off' current of a transistor, I_{off}, defined as the current when the gate source voltage $V_{gs} = 0$, is given by

$$I_{off} = I_{ds}(@V_{gs} = V_{th})e^{-V_{th}/s} \tag{8.3}$$

where I_{ds} is the drain–source current, V_{th} is the threshold voltage, and s is the temperature-dependent subthreshold swing (typically equal to about 80–90 mV/decade at room temperature, which is about 35–40 mV as used in the above equation). When $V_{gs} = V_{th}$, the device is not turned on and current is caused simply by diffusion. This diffusion current (I_{ds} in the above equation) will in turn be inversely proportional to L_{eff}. However, it is the exponential dependence of I_{off} on the threshold voltage that provides the strongest dependence of leakage current on gate length. Thus, ignoring the less important dependence of I_{ds},

$$\frac{d \ln I_{off}}{dV_{th}} \approx -1/s \tag{8.4}$$

The threshold voltage is dependent on the effective gate length due to what are called *short channel effects*. The change in threshold voltage from its long-channel value V_{th_long} depends on the exact doping profiles used, but a typical simple model is

$$\Delta V_{th} = V_{th\_long} - V_{th} = C_1 e^{-L_{eff}/2l} \tag{8.5}$$

where C_1 depends on the source-to-drain voltage and is on the order of 3–3.5 V, and l is called the *characteristic length* of the short channel effect and depends on the oxide thickness, the junction depth and the depletion layer width. This characteristic length is generally engineered to produce short-channel effects on the order of 0.1 V, so that $l \approx L_{eff}/7$. Thus,

$$\frac{dV_{th}}{dL_{eff}} = \Delta V_{th}/2l \approx 0.35\,V/L_{eff} \tag{8.6}$$

Combining Equations (8.6) and (8.4) using the chain rule,

$$\frac{d \ln I_{off}}{d \ln L_{eff}} \approx -0.35\,V/s \tag{8.7}$$

Assuming a typical subthreshold swing of 35 mV, a 10% reduction in L_{eff} will cause a 100% increase in I_{off}.

When printing a chip with millions of transistor gates, the gate lengths take on a distribution of values across the chip (Figure 8.2a). This across chip linewidth variation (ACLV) produces a range of transistor behaviors that affect the overall performance of

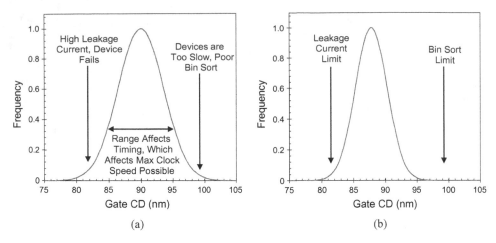

Figure 8.2 *A distribution of polysilicon gate linewidths across a chip (a) can lead to different performance failures. Tightening up the distribution of polysilicon gate linewidths across a chip (b) allows for a smaller average CD and faster device performance*

the chip. Although the specific details can be quite complicated and device specific, there are some very basic principles that apply.[1] As a signal propagates through the transistors of a chip to perform some operation, there will be several paths – connected chains of transistors – that operate in parallel and interconnect with each other. At each clock cycle, transistors are turned on and off, with the results passed to other interconnected transistors. The overall speed with which the operation can be performed (i.e. the fastest clock speed) is limited by the slowest (largest gate CD) transistors in the critical path for that operation (a critical path will typically have between 10 and 20 transistors in the path). Also, when gate lengths become too large, the drive current available to turn on or off the transistor may be too low, resulting in a potential reliability problem. On the other hand, the reliability of the chip is limited by the smallest gate CDs on the chip due to leakage current. If too many transistors on the chip have small gate CDs, the higher leakage current can result in unacceptably high power consumption.

The distribution of linewidths across the chip has another impact besides the limits of the largest and smallest CDs. The range of gate lengths produces a range of switching times for the transistors. If a group of connected transistors in a path all happens to be undersized (and thus faster), while another group of transistors in a path is oversized (and thus slower), a timing error can occur where the results from the two paths don't show up together at the same time. These timing errors, which are very difficult to predict and sometimes difficult to test for, result in device function errors. Today, interconnect delays (the time it takes a signal to travel from one transistor to another) are accounting for an ever larger share of the timing delays of critical paths for chips. Much design focus continues to be on evening out the interconnect delays among competing paths. Still, the gate delay distribution remains a critical part of the timing error problems on chips that are pushing the limits of clock speeds.

So how does improved CD control impact device performance? Obviously from the discussion above, a tighter distribution of polysilicon gate CDs will result in reduced timing errors. This smaller range of linewidths also means that the average linewidth can be reduced without running into the leakage current limit (Figure 8.2b). As a result, the overall speed of the chip can be increased without impacting reliability. The resulting improved 'bin sort', the fraction of chips that can be put into the high clock speed bins upon final test of the device, can provide significant revenue improvements for the fab.

Like transistor gates, the impact of errors in the CD of a contact hole or via is different for oversized versus undersized contacts. If a contact hole is too big, the hole can over-shoot the edge of the pattern below to which it is making contact, possibly causing an electrical short circuit. If a contact hole is too small, the electrical contact resistance will increase. Contact resistance is proportional to the area of the contact, and thus to the square of the contact diameter. A 10% decrease in contact CD causes a roughly 20% increase in contact resistance. If the contact/via resistance gets too high, signals propagating through that contact/via will slow down. For voltage-sensitive parts of the circuit (such as the source/drain contacts), the voltage drop across the contact can change the electrical characteristics of the device.

8.2.2   Improving CD Control

Fundamentally, errors in the final dimension of a feature are the result of errors in the tools, processes and materials that affect the final CD. An error in a process variable (the temperature of a hot plate, for example) propagates through to become an error in the final CD based on the various physical mechanisms by which the variable influences the lithographic result. In such a situation, a propagation of errors analysis can be used to help understand the effects. Suppose the influence of each input variable on the final CD were expressed in a mathematical form, such as

$$CD = f(v_1, v_2, v_3, \ldots) \tag{8.8}$$

where v_i are the input (process) variables. Given an error in each process variable Δv_i, the resulting CD error can be computed from a high-order total derivative of the CD function in Equation (8.8).

$$\Delta CD = \sum_{n=1}^{\infty} \left(\Delta v_1 \frac{\partial}{\partial v_1} + \Delta v_2 \frac{\partial}{\partial v_2} + \ldots \right)^n f(v_1, v_2, \ldots) \tag{8.9}$$

This imposing summation of powers of derivatives can be simplified if the function is reasonably well behaved (and of course we hope that our critical features will be so) and the errors in the process variables are small (we hope this is true as well). In such a case, it may be possible to ignore the higher-order terms ($n > 1$), as well as the cross terms of Equation (8.9), to leave a simple, linear error equation (the first-order total derivative).

$$\Delta CD = \Delta v_1 \frac{\partial CD}{\partial v_1} + \Delta v_2 \frac{\partial CD}{\partial v_2} + \ldots \tag{8.10}$$

Each Δv_i represents the magnitude of a *process error*. Each partial derivative $\partial CD/\partial v_i$ represents the *process response*, the response of CD to an incremental change in the variable. This process response can be expressed in many forms; for example, the inverse of the process response is called *process latitude*. Figure 8.3 shows a simple case where the

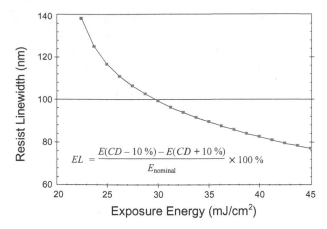

Figure 8.3 *The common CD-versus-exposure dose (E) curve is used to measure exposure latitude (EL)*

process variable is exposure dose and the common-use definition of exposure latitude (EL) is shown.

The linear error Equation (8.10) can be modified to account for the nature of the errors at hand. In general, CD errors are specified as a percentage of the nominal CD. For such a case, the goal is usually to minimize the relative CD error, $\Delta CD/CD$. Equation (8.10) can be put in this form as

$$\frac{\Delta CD}{CD} = \Delta v_1 \frac{\partial \ln CD}{\partial v_1} + \Delta v_2 \frac{\partial \ln CD}{\partial v_2} + \ldots \tag{8.11}$$

Also, many sources of process errors result in errors that are a fraction of the nominal value of that variable (for example, illumination nonuniformity in a stepper produces a dose error that is a fixed percentage of the nominal dose). For such error types, it is best to modify Equation (8.11) to use a relative process error, $\Delta v_i/v_i$.

$$\frac{\Delta CD}{CD} = \frac{\Delta v_1}{v_1} \frac{\partial \ln CD}{\partial \ln v_1} + \frac{\Delta v_2}{v_2} \frac{\partial \ln CD}{\partial \ln v_2} + \ldots \tag{8.12}$$

Although Equations (8.10)–(8.12) may seem obvious, even trivial, in their form, they reveal a very important truth about error propagation and the control of CD. There are two distinct ways to reduce ΔCD: reduce the magnitude of the individual process errors (Δv_i), or reduce the response of CD to that error ($\partial CD/\partial v_i$). The separation of CD errors into these two source components identifies the two important tasks that face the photo-lithography engineer. Reducing the magnitude of process errors is generally considered a *process control* activity and involves picking the right material and equipment for the job and ensuring that all equipment is working in proper order and all materials are meeting their specifications. Reducing the process response is a *process optimization* activity and involves picking the right process settings, as well as the right equipment and materials. Often, these two activities are reasonably independent of each other.

A note of caution: the derivation of Equation (8.10) assumed that the process errors were small enough to be linear and independent in their influence on CD. This will not always be the case in a real lithographic process. One need only consider the two variables of focus and exposure to see that the response of CD is certainly nonlinear and the two variables are highly dependent on each other (see section 8.5). Usually, a linear expansion such as that of Equation (8.10) is most useful as a guide to understanding rather than as a computational tool.

8.2.3 Sources of Focus and Dose Errors

The above sections describe why critical dimension (CD) control is important, and describe a common method for separating the causes of CD errors into two components: the magnitude of a process, material, or tool error, and the response of the lithographic process to that error. This separation of errors allows the process engineer to focus independently on the two methods for improving CD control: reducing the sources of errors and improving process latitude. In general, these approaches can be thought of as independent (though not always). Efforts to improve process latitude usually do not change the magnitude of the process errors, and vice versa. Let's consider the first approach, reducing the magnitude of the errors, using focus errors as an example.

Focus errors arise from many sources, both random and systematic. Tables 8.1 and 8.2 show estimates of the focus errors by source for different lithographic generations: a 1991 0.5-μm i-line process, 1995 0.35-μm i-line and KrF stepper processes, a 0.18-μm KrF scanner process, and a more modern 90-nm ArF process. The values shown are typical, but certainly vary from process to process. Although the tables are reasonably

Table 8.1 *Examples of random focus errors (μm, 6σ) for different lithographic generations*

Error Source	1991[2] i-line 0.50 μm	1995[3] i-line 0.35 μm	1995[3] KrF stepper 0.35 μm	2001 KrF scanner 0.18 μm	2005 ArF scanner 0.09 μm
Lens Heating (Compensated)	0.10	0.10	0.00	0.00	0.00
Environmental (Compensated)	0.20	0.20	0.10	0.10	0.05
Mask Tilt (actual/16)	0.05	0.05	0.10	0.05	0.05
Mask Flatness (actual/16)	0.12	0.12	0.12	0.12	0.07
Wafer Flatness (over one field)	0.30	0.33	0.33	0.15	0.07
Chuck Flatness (over one field)	0.14	0.03	0.03	0.03	0.03
Laser Bandwidth	0.0	0.0	0.20	0.1	0.04
Autofocus Repeatability	0.20	0.08	0.10	0.07	0.04
Best Focus Determination	0.30	0.15	0.10	0.10	0.05
Vibration	0.10	0.10	0.05	0.05	0.03
Total RSS Random Focus Errors	0.60	0.50	0.45	0.28	0.15

Table 8.2   *Examples of systematic (μm, total range) and random focus error estimates combined to determine the Built-in Focus Errors (BIFE) of a process*

Error Source	1991^2 i-line 0.50 μm	1995^3 i-line 0.35 μm	1995^3 KrF stepper 0.35 μm	2001 KrF scanner 0.18 μm	2005 ArF scanner 0.09 μm
Topography	0.5	0.3	0.3	0.10	0.05
Field Curvature and Astigmatism	0.4	0.4	0.3	0.08	0.05
Resist Thickness	0.2	0.2	0.2	0.10	0.05
Total Systematic Errors (range)	1.1	0.9	0.8	0.28	0.15
Total Random Errors (6σ)	0.60	0.50	0.45	0.28	0.15
Range/σ	11	10.8	10.7	6	6
Total BIFE (6σ equivalent)	1.5	1.2	1.1	0.47	0.25

self-explanatory, a few items are worth noting. Errors in flatness or tilt on the mask are reduced by the reduction ratio of the imaging tool squared (assumed to be 4× in this example). Wafer flatness is a total range over the exposure field assuming autoleveling is turned on. One of the primary advantages of the scanner is that the exposure slit sees a smaller region of the wafer and thus has a smaller focus error due to wafer nonflatness compared to a stepper, even for the same wafer. Note also that hot spots (focus errors caused by a particle on the backside of the wafer) and edge die problems (caused by errors in the autofocus system when attempting to focus at the edge of the wafer) are not included in this list.

Systematic focus errors include wafer topography (where chemical mechanical polishing has proven its worth, though is not yet perfect enough to make wafer topography a negligible problem), lens aberrations across the field or slit, and the errors due to focusing through the thickness of the resist. While not considered here, phase errors on a phase shift mask can also act like a focus error. Random and systematic errors can now be combined by assuming the systematic errors have a uniform probability distribution with the given range and convolving this uniform distribution with the Gaussian distribution of the random errors. Since in all cases the uniform width is large compared to the 3σ of the random errors, the result is a probability distribution that looks like the uniform distribution with error function tails. For an error function probability, the equivalent to a 6σ Gaussian range (99.7% probability) depends on the ratio of range/σ. For range/$\sigma =$ 11, the total built-in focus errors (BIFE) of the process, with a probability of 99.7%, is range $+ 3.6\sigma$. For range/$\sigma = 6$, the BIFE is range $+ 4\sigma$.

The exercise to fill in the values in tables such as those presented here serves two important functions. First, an evaluation of the BIFE can be combined with a measurement of the focus–exposure process window for the target process to see if the process capability (the process window) exceeds the process requirements (see section 8.5). Second, a listing of sources of focus errors inevitably leads to ideas where improvements can be made. The largest sources of errors (wafer nonflatness, for example, and wafer topography) offer the greatest potential for improvement.

Interestingly, a historical look at built-in focus errors reveals an interesting trend: the required depth of focus is about three times the minimum half-pitch for critical lithography layers such as gate polysilicon. While certainly not a rigorous analysis, using this rule of thumb can prove useful for projecting BIFE requirements to future generations of optical lithography.

Many errors that are not focus errors, but still affect CD, can be thought of as equivalent to an exposure dose error. Typical examples include:

- across-field (slit) intensity nonuniformity;
- field-to-field and wafer-to-wafer dose control;
- resist sensitivity variations (including development process variations);
- resist/BARC thickness variations (reflectivity variations);
- PEB time and temperature variations;
- flare variations;
- mask CD nonuniformity;
- errors in optical proximity correction (behaves like a mask CD error).

The required dose control has steadily tightened for smaller feature sizes. At the 0.5-μm generation, the built-in dose errors of a well-controlled process were about 15 % (range). At the 90-nm node, dose errors are typically less than about 6 %.

8.2.4 Defining Critical Dimension

A cross section of a photoresist profile has, in general, a very complicated two-dimensional shape (see Figure 8.4, for example). Measurement of such a feature to determine its width has many complications. Suppose, however, that we have been able to measure the shape of this profile exactly so that we have a complete mathematical description of its shape. How wide is it? The answer depends on how one *defines* the width. The original shape of the photoresist profile is simply too complex to be unambiguously characterized by a single width number. The definition of the width of a complex shape requires the definition of a *feature model*.[4]

A feature model is a mathematical function described by a conveniently small number of parameters used to appropriately depict the actual shape of the resist feature. For our application, one of these parameters should be related to the basic concept of the width of the resist profile. The most common feature model used for this application is a trapezoid (Figure 8.4). Thus, three numbers can be used to describe the profile: the width of

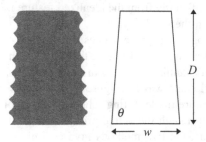

Figure 8.4 *Example photoresist profile and its corresponding 'best-fit' trapezoidal feature model*

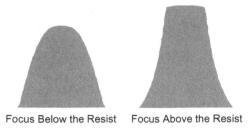

Focus Below the Resist   Focus Above the Resist

Figure 8.5 *Resist profiles at the extremes of focus show how the curvature of a pattern cross section can change*

the base of the trapezoid (linewidth, *w*), its height (profile thickness, *D*) and the angle that the side makes with the base (sidewall angle, θ). To be perfectly general, the position of the feature (defined, for example, by the centroid of the feature model) can be specified and the shape can be made asymmetrical by allowing a different sidewall angle for each side.

Obviously, to describe such a complicated shape as a resist profile with just three numbers is a great simplification. One of the keys to success is to pick a method of fitting this feature model to the profile that preserves the important properties of the profile (and its subsequent use in the device). Thus, one can see that even given an exact knowledge of the actual photoresist profile, there are two potential sources of error in determining the critical dimension: the choice of the feature model and the method of fitting the feature model to the resist profile. Consider Figure 8.5, which shows resist profiles through focus exhibiting different curvatures of their sides (more on this topic in section 8.5). Using a trapezoidal feature model will obviously result in a less than perfect fit, which means that the criterion for best fit will influence the answer.

What is the best feature model and best method of fitting the feature model to measured data for a given application? Since the choice of the feature model is based both on relevance and convenience, and since the trapezoid is so commonly used for CD metrology, the impact of the feature model choice will not be discussed here. When fitting the feature model to the data, there are many possible methods. For example, one could find a best-fit straight line through the sidewall of the profile, possibly excluding data near the top and bottom of the profile. Alternately, one could force the trapezoid to always match the actual profile at some point of interest, for example at the bottom. Whenever the shape of the actual profile deviates significantly from the idealized feature model, the method of fitting can have a large impact on the results.

For example, as a lithographic process goes out of focus, the resist profile and the resulting feature size will change. But because the shape of the resist profile is deviating from a trapezoid quite substantially at the extremes of focus, CD can also be a strong function of how the profile data was fit. Figure 8.6 compares the measured CD through focus for two different feature model fitting schemes: a best-fit line through the sidewall and fitting the trapezoid to match the actual profile at a set threshold (height above the substrate). Near best focus, the two methods give essentially the same value since the resist profile is very close to a trapezoid. However, out of focus there can be a significant difference in the CD values (> 5 %) based only on the fitting method used.

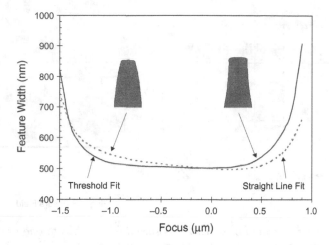

Figure 8.6 *Using resist profiles at the extremes of focus as an example, the resulting measured feature size is a function of how the feature model is fit to the profile*

In real metrology systems, the actual resist profile is never known. Instead, some signal (secondary electrons versus position, scattered light intensity versus wavelength) is measured that can be related to the shape of the resist profile. This signal is then fit to some model, which ultimately relates to the feature size being measured. Although a bit more complicated, the same principles still apply. Both the feature model and how that model is fit to the data will affect the accuracy of the results.

8.3 How to Characterize Critical Dimension Variations

Critical dimension (CD) errors in lithography come from many sources and exhibit many interesting characteristics. The classification of these errors into specific categories (systematic versus random, spatial versus temporal) is often the first step in a root cause analysis that allows the sources of CD variation to be identified and addressed. The nature of step-and-repeat or step-and-scan lithography on many identical wafers in a lot can produce many systematic spatial and temporal processing characteristics that impact CD errors. The discussion below will focus on step-and-scan lithography, with emphasis on the characterization of systematic error sources.

8.3.1 Spatial Variations

The systematic spatial variations of CD within a wafer exhibit four basic signatures: across-wafer, across-field in the scan direction, across-field in the slit direction, and across-field independent of the scan/slit orientation (Figure 8.7). The physical causes of variations in each of these spatial domains can be very different. Thus, dividing up the total spatial variation of CDs among these four components is the first step in identifying causes and remedies for unwanted CD errors.

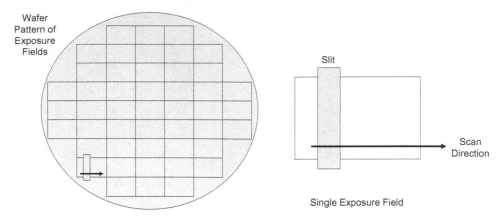

Figure 8.7 *A wafer is made up of many exposure fields, each with one or more die. The field is exposed by scanning a slit across the exposure field*

For those processing steps that result in a full-wafer treatment (resist coating, post-exposure baking, development), errors in that processing can contribute to CD variations with a full-wafer signature. Essentially all wafer processing other than exposure is full-wafer processing and results in an across-the-wafer variation in CD. Examples include:

- underlying filmstack thickness variations due to deposition;
- resist and bottom antireflection coating thickness uniformity;
- prebake temperature uniformity;
- post-exposure bake temperature uniformity;
- development uniformity (due to application, wetting, surface tension, etc.);
- first to last exposure shot variation due to lens heating or post-exposure delay;
- edge shot focus variation due to autofocus/autolevel problems at the edge of the wafer.

Many of these sources of error will exhibit mostly radial signatures, though other signatures (such as a wedge variation – a linear variation from one side of the wafer to the other) are certainly possible. The variation in CD caused by the difference in the first and last exposure shot can have a very different signature and, of course, depends on the specific stepping pattern used. Hot spots, an extended region of the wafer with higher or lower CDs, can be caused by backside defects or wafer chucking errors that produce a localized focus error that cannot be compensated for by autofocus and autoleveling mechanisms.

The process of scanning a slit in one direction across the field results in a distinct difference in the way many across-field error sources are manifest. Spatial variations across the field in the scan direction will average out to produce a fairly uniform signature in that direction. Spatial variations in the slit direction will be systematically reproduced everywhere in the field. There is one major systematic across-field CD variation source that is independent of scan orientation: reticle CD errors. Coupled with the mask error enhancement factor (MEEF), reticle errors will systematically reproduce their signature across the wafer field without any averaging or other scan-induced changes.

Along the long dimension of the slit, there is no averaging effect of scanning to reduce the sources of CD variation. Thus, variations in the slit direction create a unique signature for the exposure tool. Common sources of CD errors with systematic variations in the slit direction include:

- lens aberrations;
- illumination (dose) uniformity;
- source shape (partial coherence) variations;
- flare.

In general, systematic errors in the slit direction dominate the systematic across-field errors.

Most of the systematic sources of CD errors that plague lithographers in the slit direction tend to average themselves out in the scan direction – this is one of the major benefits of step-and-scan technology. And while on-the-fly focusing can greatly reduce the impact of long-range wafer nonflatness, it can also induce some systematic CD errors in the scan direction. Autofocus systems can be fooled by underlying topography on the wafer (for example, when scanning past a logic region onto a dense memory region) and systematically but incorrectly adjust focus as it scans.

Variations from wafer to wafer can take on a spatial component due to systematic differences in how one wafer is processed versus another (it is important to differentiate the spatial variation being considered here from a temporal variation caused by a change in the processing of each wafer over time, to be discussed below). The main source of such variations comes from multiple processing chambers (modules) used per lot. For example, a single photolithography cell may have two developer cups and two post-exposure bake (PEB) hot plates. Any systematic difference between these track modules will give a systematic wafer-to-wafer error grouped by which module was used in the processing. Likewise, incoming wafers may have systematic wafer-to-wafer filmstack variations due to multiple deposition chambers.

Lot-to-lot systematic spatial variations are typically caused by differences in processing tools:

- different photolithography cells;
- different reticles;
- different deposition tools for incoming wafers;
- different metrology tools (tool matching errors).

Again, one should distinguish between those lot-to-lot differences caused by using different tools versus differences caused by temporal drifts in the behavior of a given tool.

8.3.2   Temporal Variations and Random Variations

Temporal variations can be seen both wafer-to-wafer and lot-to-lot, depending only on the timescales involved. Some effects, however, such as lens heating, have different within-lot and lot-to-lot signatures. Temporal variations can be either drifts in lithographic

inputs (a gradual barometric pressure change causing a focus shift) or an abrupt distur-
bance (switching to a new batch of photoresist). Although any lithographic input can
conceivably have a temporal variation, some of the most common are:

- focus drift;
- dose calibration (intensity measurement) drift;
- resist batch change;
- laser center wavelength and bandwidth variations;
- lens heating;
- metrology drift;
- long-term lens degradation.

Of course, almost any lithographic input can have random variations associated with
it. An important distinction between random errors and temporal variations as discussed
above is the time constant of change. If a variable changes randomly but over a period
of time such that many wafers can be processed (that is, between hours and months), such
errors are generally classed as temporal variations. If, however, the variations are rapid
enough to change across a wafer or from wafer to wafer, the errors are described as
random. Some of the most significant random errors are:

- focus;
- dose;
- scan synchronization (x, y and z);
- PEB temperature/heating transients;
- metrology.

Of course, every input to the lithography process has some level of random variation
associated with it.

8.3.3 Characterizing and Separating Sources of CD Variations

Different methods are often employed to characterize and control errors from different
sources. For example, temporal variations are usually characterized using trend chart and
statistical process control (SPC) techniques.[5] Wafer-to-wafer and lot-to-lot variations
caused by differences in processing tools or chambers can be caught using statistical correlation techniques (for example, where the average CD value obtained in one photolithography cell is compared to another and a statistically significant difference is flagged).
The separation of CD errors into random and within-wafer spatially systematic is one of
the most challenging problems in lithography data analysis. Further separation of the
within-wafer systematic errors into their specific components (across-wafer, across-field,
slit direction and scan direction) is both difficult and important. There are two basic
approaches that can be used: characterizing output CD variations and characterizing input
parameter variations.

The basic statistical technique for separating various systematic spatial variations and
random errors is the use of the 'composite' spatial source. For example, the variation of
CD across a wafer can have both a systematic component and a random component. By
measuring a number of wafers at the exact same spatial locations, averaging of the measurements at each spatial location creates a composite wafer with the random errors

mostly averaged out. This composite wafer can then be used to characterize the systematic spatial signature of the error. Subtracting the composite wafer from the actual wafer data produces a residual set of random errors that can be characterized by a mean and standard deviation.

Let $M_i(x,y)$ be the measured CD on wafer i at the location (x,y) on the wafer for a measured group of N wafers. This measured value can be separated into systematic and random errors:

$$M_i(x,y) = CD_0 + S_i(x,y) + R_i(x,y) \tag{8.13}$$

where CD_0 is the nominal (target) CD, S_i is the systematic error for wafer i and R_i is the random error component. If the experiment used to make the measurements is conducted appropriately, the systematic wafer-to-wafer variations can be kept to a minimum (by making sure that each wafer is processed identically and by processing the wafers over a very short period of time). If this wafer-to-wafer variation is much smaller than the across-wafer variation, it is possible to approximate $S_i(x,y) \approx S(x,y) =$ the systematic across-wafer spatial signature. A 'composite' wafer CD error $CCDE(x,y)$ is created by averaging the measurements from the different wafers at each point:

$$CCDE(x,y) = \frac{1}{N}\sum_{i=1}^{N} M_i(x,y) - CD_0 \approx S(x,y) + \frac{1}{N}\sum_{i=1}^{N} R_i(x,y) \tag{8.14}$$

If normally distributed, the random errors will have a mean of zero at each point on the wafer. For a large enough sample N, the average of the sample will approximate the mean of the population with high accuracy. Thus,

$$\frac{1}{N}\sum_{i=1}^{N} R_i(x,y) \approx 0 \quad \text{with an uncertainty of } \sigma_R(x,y)/\sqrt{N} \tag{8.15}$$

where $\sigma_R(x,y)$ is the standard deviation of the random CD error at the point (x,y). By approximating the standard deviation of the random error with the standard deviation of the residuals of the N sample of measurements $s_R(x,y)$, the accuracy of Equation (8.15) can be estimated. By picking N to be sufficiently large, one can use the composite wafer error as an estimate of the systematic across-wafer CD error signature within a known level of error.

$$S(x,y) \approx CCDE(x,y) \quad \text{with an estimated uncertainty of } s_R(x,y)/\sqrt{N} \tag{8.16}$$

In turn, the composite wafer can be analyzed to create a composite field. Averaging the same field measurement points from all fields across the wafer produces a signature that is characteristic of the field error only (with the assumption that the across-wafer errors vary slowly enough across the field as to contribute only negligibly to the composite field). Once the composite field is obtained, it can be subtracted from the composite wafer to create a separate across-wafer-only signature.

Likewise, the composite field can be broken down into a composite slit and composite scan signature. This separation into components is complicated by the fact that the contribution of reticle errors to the composite field is independent of scan direction. For one simplifying case, however, all components can be properly separated. If the reticle CD variations do not exhibit any systematic scan or slit direction components, subtraction of the composite slit and composite scan from the composite field should yield the reticle contribution. It seems likely, however, that this simplifying assumption will often prove false. Any reticle error signature that has a systematic spatial signature (the edges of the reticle might have a lower CD than the middle, for example) will confound the separation of errors. In such a case, the second method of analyzing CD variations – a direct characterization of the sources of variation – can make this spatial decomposition even more meaningful.

While the previous paragraphs described a detailed statistical approach to the characterization of measured CD errors, a second approach would directly look for and measure the variations in lithographic inputs thought to cause these errors. Such a 'sources of variations' study can be coupled with the sensitivities of CD to each error source to predict the resulting CD error distributions. One of the most common examples is the measurement of reticle CD errors to create a spatial map of the reticle. By combining this reticle map with measured or simulated MEEF values for the specific features (mask error enhancement factor, see section 8.7), a predicted wafer CD error map due only to reticle errors can be generated.

The most effective means of characterizing CD errors in a lithographic process is a combination of a sources of variation study and statistical decomposition of measured wafer errors. Using the reticle CD error map described above, for example, the reticle contribution to the across-field CD errors from the scan and slit direction signatures can be subtracted out, making the statistical analysis of spatial variation more accurate. Also, by comparing a 'bottoms up' prediction of wafer CD errors through a sources of variations study with actual measured CD variations, one can check assumptions about which errors are most significant and which fab investments meant to improve CD control are most likely to pay off.

8.4   Overlay Control

Like critical dimension control, overlay control is a vital part of lithography in semiconductor manufacturing. Errors in the overlay of different lithographic levels can directly cause a number of electrical problems. At the gate level, overlay errors can pull a gate line end close to the active area, causing an increase in leakage current. At contact or via, overlay errors can reduce the overlap between the contact hole and the underlying conductor, causing higher contact resistance or, if bad enough, a short to the wrong electrical connection. On the design side, design features are spread out to be insensitive to expected errors in overlay. If overlay control can be improved, designs can be shrunk, allowing smaller die, more die per wafer and lower cost per die. As a result, economics dictates that overlay specifications must shrink in lockstep with device geometries, requiring continuous improvement in overlay measurement and control.

Overlay is defined as the positional accuracy with which a new lithographic pattern has been printed on top of an existing pattern on the wafer, measured at any point on the wafer. This is opposed to the slightly different concept of *registration*, which is the positional accuracy with which a lithographic pattern has been printed as compared to an absolute coordinate grid. While registration can be important (for example, when considering the positional accuracy of features on a photomask), overlay is both much easier to

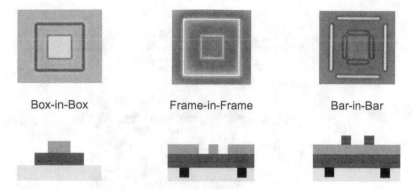

Figure 8.8 *Typical 'box-in-box' style overlay measurement targets, showing top-down optical images along the top and typical cross-section diagrams along the bottom. The outer box is typically 20 μm wide. (Courtesy of KLA-Tencor Corp.)*

measure and more relevant to lithographic quality and yield. In this section we'll examine some basics of how overlay is measured, and how overlay data are used.

8.4.1 Measuring and Expressing Overlay

Overlay measurement involves the design of special patterns used on two different lithographic printing steps such that a metrology tool, when viewing the results of both printed patterns simultaneously, can measure errors in overlay for that point on the wafer. The most common pattern used today is the so-called box-in-box target (Figure 8.8). An outer box is printed during the first lithographic step and an inner box is printed during the second lithography pass.

Overlay measurement looks for differences in the space between the printed boxes (the distances between inner and outer box edges). For example, the *x*-overlay error would be one-half of the difference between the right and left widths between the inner and outer boxes (Figure 8.9). By using both left and right widths, processing errors that affect the dimensions of the boxes symmetrically will cancel out. Repeating this measurement for the top and bottom of the boxes gives the *y*-overlay error. The *x*- and *y*-overlay measurement pair produces a vector of the overlay error at the wafer and field location of the measurement.

While symmetrical errors in CD measurement will cancel when calculating overlay, asymmetrical errors will not. Coma in the overlay measurement optics, for example, can cause a left–right asymmetry in the measured box-in-box CDs, thus giving an error in the overlay measurement. Such an asymmetry can easily be detected by rotating the wafer by 180° and remeasuring the target. In the absence of any measurement asymmetries, the two measurements would be the same, but of opposite signs. Thus, the *tool-induced shift* (TIS) is defined as

$$TIS = Overlay(0°) + Overlay(180°) \tag{8.17}$$

with TIS = 0 for an ideal measurement. If TIS for an overlay measurement tool is fixed, it can be easily measured and subsequent overlay measurement corrected for this

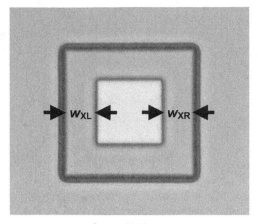

$$x\text{-overlay} = 0.5(w_{XL} - w_{XR})$$

Figure 8.9   *Measuring overlay as a difference in width measurements*

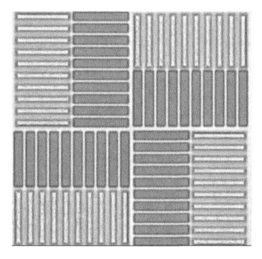

Figure 8.10   *Typical AIM target where the inner bars (darker patterns in this photograph) are printed in one lithographic level and the outer (brighter) bars in another level. (Courtesy of KLA-Tencor Corp.)*

systematic error. Thus, it is the variation of TIS that causes increased measurement uncertainty.

Asymmetries in the measurement target on the wafer can also cause similar shifts, termed *wafer-induced shift* (WIS). Unlike TIS, however, WIS is rarely systematic enough to be simply subtracted out of the data. Additionally, nonideal targets often interact with the measurement tool, especially focus, to produce nonsystematic WIS errors.

Recently, a new type of target called an Advanced Imaging Metrology (AIM) target has shown superior measurement results (Figure 8.10). The AIM bars are less sensitive to chemical mechanical polishing (CMP) errors such as dishing and other processing errors that can damage a box-in-box target, thus making the AIM target more process-robust. Also, the multiple bars essentially allow an averaging of multiple measurements at one time, significantly increasing the precision of the measurement. Special image processing software is used to relate the optical image of the target (determining the center of symmetry of each of the bars) to x- and y-overlay values.

Overlay measurement targets are placed at various spatial positions in the reticle field, minimally at the four corners but often in the streets between the die (and thus within the field) as well. The spatial variation of overlay errors provides extremely important information as to the behavior of the reticle, the wafer and the stepper, as we shall see in the next section. Overlay errors are characterized in terms of x-errors and y-errors separately, rather than being combined into an overlay error length. Since integrated circuit (IC) patterns are mostly laid out on a rectangular grid, the impact of an x-overlay error on device yield or performance is almost independent of any y-overlay errors that might also be present, or vice versa.

8.4.2   Analysis and Modeling of Overlay Data

The goal of overlay data analysis is twofold: assess the magnitude of overlay errors and determine, if possible, their root causes. The overall magnitude of overlay errors is commonly summarized by the mean plus 3σ of the overlay data in both x- and y-directions. Root cause analysis involves explaining the data with a model that assigns a cause to the observed effect.

To see how an overlay model might work, let's examine a simple example of a reticle that is slightly rotated. Figure 8.11a shows how a rectangular reticle field would print on a wafer if the reticle were rotated slightly with respect to the previous layer. If the ideal position of the upper-right corner of the field were (x,y), rotation of the reticle about an angle ϕ would result in that corner printing at a location (x^*,y^*) given by

$$x^* = x\cos\phi - y\sin\phi$$
$$y^* = y\cos\phi + x\sin\phi$$

(8.18)

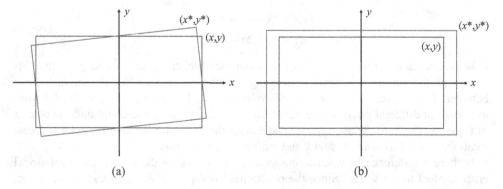

(a)   (b)

Figure 8.11   *Examples of two simple overlay errors: (a) rotation and (b) magnification errors*

where the origin was set to be in the middle of the field. In any reasonable lithography process, these rotational errors will be quite small (on the order of 10 μradians or less) so that the trigonometric functions can accurately be replaced by their small angle approximations: $\cos\phi \approx 1$ and $\sin\phi \approx \phi$. The resulting overlay errors will then be

$$\begin{aligned} dx &\equiv x^* - x = -\phi y \\ dy &\equiv y^* - y = \phi x \end{aligned} \qquad (8.19)$$

Another possible alignment error is translation (also called offset), where the entire reticle field is shifted in x and y by Δx and Δy. In this case, the resulting overlay errors are simply equal to this translation error:

$$\begin{aligned} dx &= \Delta x \\ dy &= \Delta y \end{aligned} \qquad (8.20)$$

Another interesting error that can occur in optical lithography is a magnification error. While lithography tools are designed to reduce the reticle by a specific amount (most commonly 4×), in fact changes in the index of refraction of air and the exact position of the lens or lens components relative to the reticle and wafer can cause small errors in the magnification of the imaging tool. Sometimes these magnification errors can be different in the x- and y-directions (this is especially true for step-and-scan tools, where the ratio of reticle-to-wafer scan speeds controls the magnification in the scan direction). From Figure 8.11b, for x and y relative magnification errors of Δm_x and Δm_y, respectively,

$$\begin{aligned} x^* &= (1 + \Delta m_x) x \\ y^* &= (1 + \Delta m_y) y \\ dx &= \Delta m_x x \\ dy &= \Delta m_y y \end{aligned} \qquad (8.21)$$

The uniform or overall magnification error is the average of Δm_x and Δm_y. The difference between Δm_x and Δm_y is sometimes called *anamorphism*.

Combining the three sources of overlay errors we've discussed so far, the resulting total error in overlay would be

$$\begin{aligned} dx &= \Delta x - \phi y + \Delta m_x x \\ dy &= \Delta y + \phi x + \Delta m_y y \end{aligned} \qquad (8.22)$$

Like magnification, rotational errors are sometimes broken up into x and y rotations as well, with the uniform rotational error being the average of ϕ_x and ϕ_y and the difference between these angles called *skew* or *nonorthogonality*. If the overlay error (dx, dy) were measured at different points in the field [different (x,y) points], one could find the best fit of Equation (8.22) to the overlay data and extract the five unknown overlay error sources: rotation, x and y translation, and x and y magnification errors.

Is there a difference between a rotational error applied to the reticle and a rotational error applied to the wafer? Since the reticle field is repeated many times on one wafer, rotating the reticle is very different from rotating the wafer, as can be seen more clearly in Figure 8.12. (As an aside, the difference between wafer rotation in the x- and y-

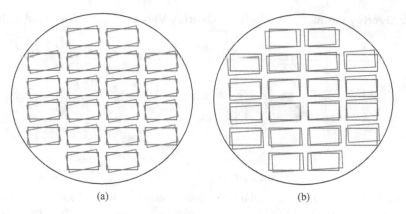

(a)   (b)

Figure 8.12 *Different types of rotation errors as exhibited on the wafer: (a) reticle rotation and (b) wafer rotation*

directions is called skew or orthogonality errors, and results from errors in the stepping motion of the wafer stage. That is why wafer errors are sometimes called stage errors.) Similarly, a reticle magnification error (caused by the imaging tool) would yield a different signature than a wafer magnification (also called scale) error (caused, for example, by a thermal expansion of the wafer). Translation errors, however, are exactly the same regardless of whether the reticle has been offset or the wafer has been offset.

Wafer rotation and magnification errors can be separated from reticle rotation and magnification errors by properly defining the coordinate axes used in Equation (8.22). If x and y represent the coordinates relative to the center of the field being measured, then the model fitting will yield reticle errors. If the coordinates are for the entire wafer (with an origin at the center of the wafer, for example), the resulting model fit will give wafer errors. By combining reticle and wafer terms in the full model (using both wafer and field coordinates in the full equation), both reticle and wafer error terms can be determined. Remember that translation errors cannot be separated into reticle and wafer, so only one set of translation terms should be used. The final model equations are

$$dx = \Delta x - \theta_x y_{\text{wafer}} + \Delta M_x x_{\text{wafer}} - \phi_x y_{\text{field}} + \Delta m_x x_{\text{field}}$$
$$dy = \Delta y + \theta_y x_{\text{wafer}} + \Delta M_y y_{\text{wafer}} + \phi_y x_{\text{field}} + \Delta m_y y_{\text{field}}$$

(8.23)

where θ and ΔM represent the wafer (stage) rotation and magnification errors, and ϕ and Δm represent the reticle (field) rotation and magnification terms.

In the linear model developed here, there are a total of 10 model coefficients: two magnification and two rotation errors each for the reticle and the wafer, and two translation terms. A typical sample plan would measure overlay at the four corners of an exposure field for five different fields (in the middle and four 'corners' of the wafer), and for three wafers in a lot. This number of measurements is generally enough to get good modeling statistics, averaging random errors so that the model coefficients are reasonably precise. The residuals, the difference between the modeled overlay errors (the systematic errors) and the actual measured data, are a measure of the random component of the errors (Figure 8.13).

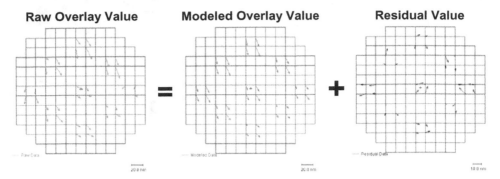

Figure 8.13   *Separation of raw overlay data into modeled + residual values. The sampling shown here, four points per field and nine fields per wafer, is common for production monitoring*

8.4.3   Improving Overlay Data Analysis

As discussed above, the primary overlay errors to consider are rotation of the wafer about an angle θ, rotation of the reticle about an angle ϕ, translation where the entire reticle field or wafer is shifted in x and y by Δx and Δy, relative wafer magnification errors of ΔM_x and ΔM_y in x and y, and relative field magnification errors of Δm_x and Δm_y in x and y, respectively. Combining these five sources of overlay errors gave the model in Equation (8.23). While this basic overlay model captures the majority of systematic errors typically found in stepper or step-and-scan imaging, additional model terms and other approaches can be used to improve the analysis. The goal is to properly account for all systematic overlay errors such that the model residuals, the difference between the best-fit model values and the actual data, are purely random. These additional terms are especially important when the first and second lithography levels are printed on different lithographic tools.

The linear terms in Equation (8.23) can be expanded to include higher-order terms. Second-order terms can model a trapezoidal error in the shape of the printed field, caused by a tilt in the reticle plane for nontelecentric imaging tools. Thus, these terms provide additional 'correctables' that can be used to adjust reticle tilt and reduce subsequent overlay errors. Third- and fifth-order terms are added to the field analysis to account for systematic lens distortion. While these terms are often not correctable, their inclusion in the model can improve the accuracy of the resulting lower-order model terms. A common higher-order model (including only the reticle field terms) is shown here:

$$dx = -\phi_x y + \Delta m_x x + t_{1x}x^2 + t_{2x}xy + d_{3x}x(x^2 + y^2) + d_{5x}x(x^2 + y^2)^2$$
$$dy = \phi_y x + \Delta m_y y + t_{1y}y^2 + t_{2y}xy + d_{3y}y(x^2 + y^2) + d_{5y}y(x^2 + y^2)^2$$
(8.24)

where t_1 and t_2 are the trapezoidal error terms, d_3 are the third-order distortion terms, and d_5 represent fifth-order distortion. Figure 8.14 shows an overview of reticle model terms.

The purpose of using third- and fifth-order distortion terms in modeling overlay data is to account for the very real impact of systematic distortions caused by lens aberrations.

Error Term	Picture	Coefficients

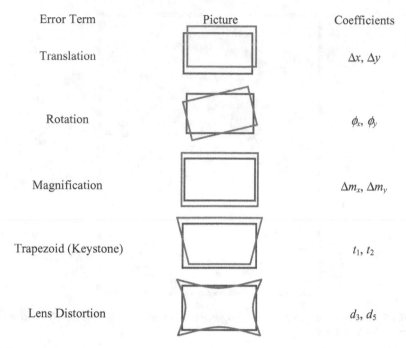

Error Term		Coefficients
Translation		$\Delta x, \Delta y$
Rotation		ϕ_x, ϕ_y
Magnification		$\Delta m_x, \Delta m_y$
Trapezoid (Keystone)		t_1, t_2
Lens Distortion		d_3, d_5

Figure 8.14 *Illustration of field (reticle) model terms including higher-order trapezoid and distortion*[6]

More correctly, such overlay errors are the result of *differences* in the distortion charac-teristics of the lenses used to print the first and second lithographic levels. If the same exposure tool was used for both levels, these high-order distortion terms should disappear (since lens distortion changes only slowly with time), regardless of the distortion of the lens relative to an absolute grid.

While modeling with high-order terms is a reasonable approach, it would be better to characterize the lens distortion more completely. Since distortion caused by lens aberra-tions varies slowly compared to the correctable error terms, it can be characterized infre-quently with a thorough analysis. First, one exposure tool is picked as the reference or 'golden' tool. A 'golden' artifact wafer is printed on this tool and the resulting first-layer overlay marks are etched into the silicon. A test reticle is used that puts many, many overlay measurement targets throughout the exposure field (a 10×10 array, for example). Then, this artifact wafer is printed with the second layer on each exposure tool to be characterized. Measurement results are analyzed to subtract out the low-order correctables and multiple fields are measured to average out random errors. The remaining residuals form a *lens fingerprint* for each tool relative to the reference tool (Figure 8.15).

Once a database of lens fingerprints for each tool has been collected, these fingerprints can be used in different ways. In the simplest application, the lens fingerprint can be subtracted from the raw overlay data before modeling those data. Suppose stepper A was used as the reference and the current lot used stepper D for the first layer and stepper G for the second layer. The measured fingerprints in the database are D-A and G-A. The

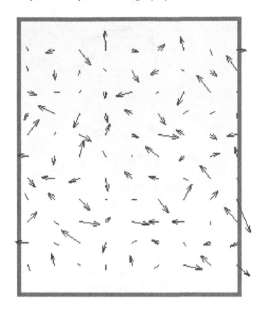

Figure 8.15 *Example of a stepper lens fingerprint showing in this case nearly random distortion across the lens field*

lens fingerprint for this lot would be G-A–D-A. In a second application of lens fingerprint data, every possible combination of steppers would be used to calculate the resulting fingerprints, which then are rank ordered by the maximum error in the fingerprint. This lens matching exercise results in a list of the best and worst combinations of steppers/scanners to use.

Like the imaging lens, every reticle contains fixed errors in registration (overlay relative to an absolute grid). Differences between the *reticle fingerprint* of the first and second lithography layer result in systematic overlay errors across the field that are independent of both lens distortion and stepper/scanner correctables. And just like the lens fingerprint, if this reticle fingerprint were to be measured, it could be used in the data analysis to improve the resulting modeling. Unfortunately, reticle fingerprints are much harder to measure and generally require special registration measurement equipment found only in the mask-making shop. Thus, measurement of a reticle fingerprint must generally be specified at the time of mask purchase.

Awareness of the reticle fingerprint makes measuring the lens fingerprint more difficult. If two reticles are used for the lens fingerprint measurement, the result will be a combined lens and reticle fingerprint measurement. Thus, typically one reticle is used for a lens fingerprint measurement, with both lithography pass patterns (inner and outer boxes, for example) located next to each other to minimize reticle effects.

The switch from pure scanning or pure stepping motion of the wafer stage to a combined step-and-scan stage motion adds several new possible overlay error terms. One such scanner term that affects the wafer overlay model is called *backlash*. With backlash, the x and y translation errors depend on the direction of the scan. Step-and-scan tools work by scanning the wafer and reticle past the exposure slit until a whole field is exposed. Then the wafer is stepped to a new location and the scanning repeats in the opposite direction. Besides an overall translation error between reticle and wafer, backlash adds a translation error that switches sign depending on the direction of the scan. Backlash terms are often added to the standard overlay models [the 10-parameter model of Equation (8.23) becomes a 12-parameter model], and while most common for step-and-scan tools, it is sometimes used for steppers as well.

Scanning also adds the possibility of higher-order stage 'wobble'. If a stage were scanning in the y-direction, for example, that stage could slowly wobble back and forth in the x-direction as it scans. Thus, a y-direction scan adds error terms to the x-overlay error dx that go as powers of the field coordinate y. A y^2 term is called *scan bow*. A y^3 term adds a snake-like signature to the field overlay. Note that a linear variation in y is already included as a reticle rotation.

Like the reticle and the lens, a specific scanner stage can have its own signature. However, unlike the reticle or the lens, the scanner stage signature varies more quickly over time than the other signatures. Thus, ideally each of these signatures should be measured independently. The lens fingerprint can be measured independent of the scan signature by using the lens fingerprint measurement method above for a static exposure (that is, with no scanning motion of stage and reticle). This lens signature within the slit can then be collapsed to a one-dimensional slit signature by averaging all the errors across the slit (that is, in the scan direction). Finally, a dynamic lens/scan signature can be measured by applying the lens fingerprint method for a full scanning exposure (see Figure 8.16). Subtracting out the slit signature leaves the scan signature remaining.

8.4.4 Using Overlay Data

Once overlay on a production lot of wafers has been measured and modeled, how is this data and analysis to be used? In general, production lot measurements of any kind can

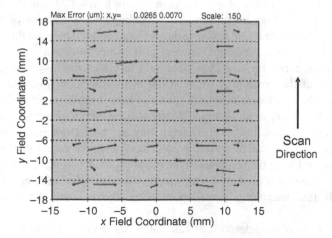

Figure 8.16 *Example of a scanner lens/scan fingerprint[7] (figure used with permission). Note that errors in the scan direction are mostly averaged out by the scan*

be thought of as 'health monitors'. The goal is to answer three questions: How healthy is this lot? How healthy are the process and process tools? How healthy is the metrology? For overlay, this translates into three basic uses of the metrology data: (1) lot disposition-ing – should this lot be reworked or sent on to etch; (2) stepper correctables – have we detected errors in the stepper or scanner that can be adjusted for improved overlay; and (3) metrology diagnostics – do we have a problem with the targets, sample plan, or metrol-ogy tool that should be taken into account.

If overlay errors on a wafer are too large, final electric yield for that wafer can suffer. Since lithography offers the unique option to rework wafers that have not been properly processed, it is very cost-effective to try to catch yield-limiting overlay errors while the option for rework still exists. Since wafers are processed in batches called lots (typically made up of 25 wafers, though lot sizes can vary), and since each lot is processed in an essentially identical manner (common tool settings and materials, very small elapsed time between processing of wafers within a lot), these lots are almost always dispositioned together as a group.

After a lot has been processed through resist development, two or three wafers from that lot are randomly selected for overlay measurement. (For dual wafer stage scanners, care must be taken to sample both stages and analyze the stage/wafer errors separately.) Data analysis leads to a number of statistical and model-based metrics from those mea-surements. This collection of metrics must then be distilled into a binary go-no go deci-sion: Do we predict that the cost of yield losses due to overlay errors, should the lot be passed, will exceed the cost of reworking the wafers? Some basic cost analysis can trans-late this question into a yield-loss threshold: if the predicted yield loss due to overlay errors exceeds a certain value, the lot should be reworked.

But standard overlay data analysis does not directly provide a prediction of overlay-limited yield. Instead, certain statistical or modeled proxies for yield are used. In the simplest approach, if the maximum measured error, the mean value of the error magni-tude, or the standard deviation of the measured data distribution (in either x or y) exceed some predefined thresholds, the lot is to be reworked. These thresholds are generally determined by a combination of experience and simple rules of thumb that have been scaled with each technology node. Similarly, modeling results can also be used to assess the value of reworking the lot. If applying the modeled correctables would reduce overlay errors for this lot by some predefined amount, it becomes worthwhile to rework.

The above approaches for assessing yield have a serious flaw – they only predict the impact on yield of the errors that occur at the measurement points. But measurement sampling on production lots is generally sparse to keep measurement costs low. Thus, it is most likely that errors at nonsampled points will be the yield limiters. Modeling can be used to overcome this limitation by predicting the maximum error expected on the wafer. The best-fit model can be used to predict the systematic variation of overlay at any point on the wafer. Using the residuals of the model fit as an estimate of the random error contribution, a maximum predicted error can be calculated for an average field and for the wafers. If this maximum predicted error is above some threshold, the lot is reworked.

Even the more rigorous approach of calculating the predicted maximum error on the wafer has a flaw: wafers don't fail due to overlay errors, die fail. In general, each exposure field contains several die (2–4 microprocessors or 8–16 memory chips in a field are common). Knowing the number of die per field and the pattern of fields printed on the wafer, the max predicted error approach can be used to find the maximum error that occurs in each die on the wafer. By predicting that a catastrophic yield-loss error will occur whenever the overlay exceeds some threshold at any point in the die, the max predicted error calculation can be used to predict the overlay-limited yield – the number of die that are expected to fail due to overlay errors from the current lithography step. This overlay-limited yield metric translates measured overlay data into a lot pass/fail decision that is directly related to final die yield.

As discussed before, one important use of overlay data modeling is to calculate stepper correctables. Correctables are defined as corrections to the exposure tool settings (affecting reticle and wafer position and motion) that, if they had been made before the measured lot was processed, would have resulted in improved overlay for that lot. Thus, technically these correctables are backward-looking. But their true value comes from applying them in a forward-looking manner. There are several ways in which overlay correctables are used. The most direct and obvious is during lot rework: if the measurements indicate that the lot should be reworked, what stepper/scanner settings should be changed to best ensure reworking will result in significantly improved overlay? Since reworking is typically done immediately after that lot was first processed, without unloading the reticle or changing the basic settings of the tool, the correctables tend to work extremely well at improving the reworked overlay.

Correctables can also be applied even if the current lot is passed. Often an exposure tool is used to process numerous identical lots in a row. Like the rework case, the reticle is not being reloaded and the basic stepper settings are not being changed between lots. Thus, correctables can be used to keep the overlay errors of each subsequent lot at their minimum. A system of feeding back correctables based on current measurements to keep the next batch of wafers in control is termed APC, *advanced process control*. Finally, correctables can be tracked over time for a given exposure tool to look for trends. SPC (statistical process control) techniques can determine if the variation in correctables from lot to lot is statistically unexpected and thus cause for alarm.

The third 'health monitor' use of production lot overlay measurements is to diagnose the health of the metrology itself. Overlay measurement begins with a signal acquisition, which is then analyzed to produce the raw overlay data. The signal can also be analyzed for noise and nonideal characteristics (such as asymmetries) in what is called *target diagnostics*. Wafer processing, film deposition and chemical mechanical polishing (CMP) in particular, can reduce the visibility of target to an optical measurement tool both decreasing the signal and increasing the noise. CMP and its variation across the wafer can also cause an asymmetric 'smearing' of the topography that makes up the overlay measurement target.

Target diagnostics can be used to assess the uncertainty of a given measurement. This uncertainty can be used to weight the data for subsequent modeling, or as a data removal mechanism (if the target is too bad, measurement results from that target can be ignored). But the modeling itself offers another diagnostic tool for the metrology. Fitting the data to a model results in model coefficients plus residuals. If the residuals are too high, this could be an indication that something has gone wrong with the measurement. Additionally, the data fitting process results not only in model coefficients but in estimates of the uncertainties of those coefficients (technically termed the covariance matrix of the fit). If the uncertainty of a correctable is consistently too high, there is a good chance that the sampling plan needs to be improved.

8.4.5 Overlay Versus Pattern Placement Error

Overlay is measured using special measurement targets which, in general, are large compared to the resolution limits of the lithographic imaging tool. As was discussed in Chapter 3, however, aberrations can cause errors in the positions of printed features that are a function of the feature size and pitch of the patterns. The positional errors specific to the feature being printed are called *pattern placement errors* (PPE) and are above and beyond (and thus can be added directly to) the overlay errors that are the same for any pattern. Using the Zernike polynomial terminology, the variation of *x*- and *y*-tilt with field position, called distortion, is an overlay error since the same error is experienced by every feature. However, the higher-order odd aberrations (such as coma) cause additional positional errors that depend on the feature. These additional errors are called pattern placement errors.

Pattern placement errors can be measured by using special overlay targets that have the same (or nearly the same) structure as the device patterns of interest. Since practically this can very difficult to accomplish, pattern placement errors are more commonly predicted by calculating the various images of interesting patterns given measured lens aberrations as a function of field position (see Chapter 3).

8.5 The Process Window

As discussed in section 8.2.2, there are two ways to improve CD control: reduce the variation in CD with a given process error (that is, increase process latitude); and reduce the magnitude of process errors. In order to increase process latitude, one must first measure and characterize that process latitude. Unfortunately, there are a very large number of potential process errors in the fab, from variations in the wafer filmstack to batch-to-batch variations in resist properties, from scanner stage vibrations to PEB hot plate temperature nonuniformities. To fully characterize the response of CD to each individual process error seen in manufacturing would be a daunting task, let alone to consider the interactions of these errors.

Fortunately, the task is simplified by recognizing that essentially all errors in the fab fall into two basic categories: errors that behave like dose errors, and errors that behave like focus errors. Once this bifurcation of error sources is understood, the job of measuring process latitudes is simplified to measuring the focus and dose responses of the process. The sections below will address the importance of dose and focus by explaining the *focus–exposure matrix*, and using it to provide definitions of the *process window* and *depth of focus* (DOF).

8.5.1 The Focus–Exposure Matrix

In general, depth of focus can be thought of as the range of focus errors that a process can tolerate and still give acceptable lithographic results. Of course, the key to a good definition of DOF is in defining what is meant by tolerable. A change in focus results in two major changes to the final lithographic result: the photoresist profile changes, and the sensitivity of the process to other processing errors is increased. Typically, photoresist profiles are described using three parameters: the linewidth (or critical dimension, CD), the sidewall angle and the final resist thickness (that is, the profile is modeled as a trapezoid, see section 8.2.4). The variation of these parameters with focus can be readily determined for any given set of conditions (although CD is the easiest of the three to measure). The second effect of defocus is significantly harder to quantify: as an image goes out of focus, the process becomes more sensitive to other processing errors such as exposure dose, PEB temperature and develop time. Exposure dose is commonly chosen to represent these other process responses.

The effect of focus on a resist feature is dependent on exposure. Thus, the only way to judge the response of the process to focus is to simultaneously vary both focus and exposure in what is known as a *focus–exposure matrix*. Figure 8.17 shows a typical example of the output of a focus–exposure matrix using linewidth as the response (sidewall angle and resist loss can also be plotted in the same way) in what is called a Bossung plot.[8]

Unfortunately, experimental data collected from a focus–exposure matrix will rarely be as smooth and clean as that shown in Figure 8.17. Experimental error is typically several percent for CDs at best focus and exposure, but the error grows appreciably when out of focus. To better analyze focus–exposure CD data, it is common to fit the data to a reasonable empirical equation in order to reduce data noise and eliminate flyers. A reasonable expression can be derived from a basic understanding of how an image is printed in photoresist.

Consider an ideal, infinite contrast positive resist and simple 1D patterns. For such a resist, any exposure above some threshold dose E_{th} will cause all of the resist to be removed (see Chapter 3). For an aerial image $I(x)$ and an exposure dose E, the CD will equal $2x$ when

$$E_{th} = EI(x) \tag{8.25}$$

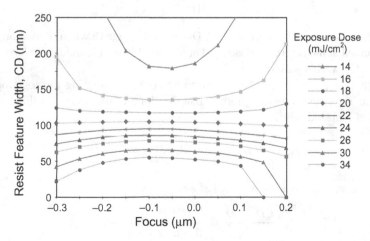

Figure 8.17 *Example of the effect of focus and exposure on the resulting resist linewidth. Here, focal position is defined as zero at the top of the resist with a negative focal position indicating that the plane of focus is inside the resist*

Consider some reference dose E_1 producing $x_1 = CD_1/2$. For some small difference from this x value, let us assume the aerial image is linear (that is, let us use only the first two terms in a Taylor series expansion of I about x_1).

$$I(x) = I(x_1) + \frac{dI}{dx}(x - x_1) \tag{8.26}$$

Combining Equations (8.25) and (8.26),

$$E_{th} = E_1 I(x_1) = EI(x) = E\left[I(x_1) + \frac{dI}{dx}(x - x_1)\right] \tag{8.27}$$

Dividing both sides of the equation by $EI(x_1)$,

$$\frac{E_1}{E} = \frac{I(x)}{I(x_1)} = 1 + \frac{d\ln I}{dx}(x - x_1) \tag{8.28}$$

where the log-slope of the aerial image is evaluated at x_1. Since $x = CD/2$, this equation can be rearranged to give

$$\frac{E_1}{E} - 1 = \frac{d\ln I}{d\ln CD}\left(\frac{CD - CD_1}{CD_1}\right) \tag{8.29}$$

Since we have assumed an infinite contrast resist, it is easy to show from Equation (8.25) that

$$\frac{d\ln I}{d\ln CD} = -\frac{d\ln E}{d\ln CD} \tag{8.30}$$

Combining Equations (8.29) and (8.30),

$$\frac{CD - CD_1}{CD_1} = \left.\frac{d\ln CD}{d\ln E}\right|_{E=E_1}\left(1 - \frac{E_1}{E}\right) \tag{8.31}$$

The term $d\ln CD/d\ln E$ is the slope of the CD-versus-dose curve when plotted on a log-log scale, evaluated at the reference dose E_1. Thus, from this simple model we have a prediction of how CD varies with dose.

An alternate exposure dependence can be derived by assuming the log-slope of the aerial image is constant around x_1 rather than the slope of the image. In this case, Equation (8.26) is replaced with

$$I(x) = I(x_1)e^{s(x - x_1)} \quad \text{where} \quad s = \left.\frac{d\ln I}{dx}\right|_{x=x_1} \tag{8.32}$$

Following the same derivation path as before,

$$\frac{CD - CD_1}{CD_1} = \left.\frac{d\ln CD}{d\ln E}\right|_{E=E_1}\ln\left(\frac{E}{E_1}\right) \tag{8.33}$$

Note that Equation (8.33) reverts to Equation (8.31) with a Taylor expansion of the natural logarithm when E is near E_1.

Finally, the most accurate aerial image dose variation model assumes the log-slope of the image varies linearly with x. As was discussed in Chapter 7, this leads to an aerial image of the space with a Gaussian shape:

$$I(x) = I_0 e^{-NILS_1(x/CD_1)^2} \quad \text{where} \quad NILS_1 = CD_1 \left. \frac{d\ln I}{dx} \right|_{x=x_1} \tag{8.34}$$

Proceeding as before,

$$\frac{CD}{CD_1} = \sqrt{1 + 2 \left. \frac{d\ln CD}{d\ln E} \right|_{E=E_1} \ln\left(\frac{E}{E_1}\right)} \tag{8.35}$$

where in this case the CD is for the space. For a line, the equivalent expression for CD would be

$$\frac{p - CD}{p - CD_1} = \sqrt{1 - 2 \left(\frac{CD_1}{p - CD_1}\right) \left. \frac{d\ln CD}{d\ln E} \right|_{E=E_1} \ln\left(\frac{E}{E_1}\right)} \tag{8.36}$$

where p is the pitch. Note that a Taylor series expansion of the square root when E is near E_1 leads to Equation (8.33). Equation (8.34) for a Gaussian-shaped image has its center at $x = 0$, the center of the space. This describes typical aerial images very well for small spaces, but for larger spaces there will be an offset between the peak of the Gaussian and the center of the space. In this case, Equation (8.36) can still be used if the pitch is replaced by an empirically determined effective pitch.

Equation (8.31), (8.33) or (8.35) describes how CD varies with dose for any given aerial image [though in practice Equation (8.33) doesn't work nearly as well as Equation (8.31) or (8.35), and Equation (8.35) matches data best of all]. Thus, these expressions apply to an out-of-focus image just as well as to an in-focus image. The only thing that changes through focus is the CD-versus-dose slope. To first order, this slope varies quadratically with focus.

$$\frac{d\ln CD}{d\ln E} = \left. \frac{d\ln CD}{d\ln E} \right|_{F=F_0} \left[1 + \left(\frac{F - F_0}{\Delta}\right)^2 \right] \tag{8.37}$$

where F is the focal position, F_0 is best focus, and Δ is the focus error that causes the log-CD-versus-log-dose slope to double.

As an example, consider the case of pure three-beam coherent imaging of equal lines and spaces through focus as discussed in Chapter 3. The aerial image was shown to be

$$I(x) = \frac{1}{4} + \frac{2}{\pi^2} + \frac{2}{\pi} \cos(\Delta\Phi)\cos(2\pi x/p) + \frac{2}{\pi^2}\cos(4\pi x/p) \tag{8.38}$$

where $\Delta\Phi = 2\pi n(F - F_0)\left(1 - \sqrt{1 - (\lambda/p)^2}\right)/\lambda$.

The normalized log-slope of the image evaluated at the nominal line edge is

$$NILS = 8\cos(\Delta\Phi) \tag{8.39}$$

Recalling Equation (8.30) and assuming that the defocus amount is small enough that the cosine can be replaced by the first two terms of its Taylor series expansion,

$$\frac{d\ln CD}{d\ln E} = \frac{d\ln CD}{d\ln E}\bigg|_{F=F_0}\left[1+\frac{\Delta\Phi^2}{2}\right] \tag{8.40}$$

This gives the same result as Equation (8.37) with

$$\Delta = \frac{\lambda}{\sqrt{2}\pi n}\left(1-\sqrt{1-(\lambda/p)^2}\right)^{-1} \approx \frac{\sqrt{2}p^2}{\pi n\lambda} \tag{8.41}$$

This method of adding the focus dependence to the dose dependence gives extra significance to the dose E_1. When $E = E_1$, Equations (8.31), (8.33) and (8.35) require that $CD = CD_1$ regardless of the value of focus. As will be discussed in section 8.5.3, this is the definition of the *isofocal dose* and *isofocal CD*. Thus, the exposure latitude term $d\ln CD/d\ln E$ used in the above expressions will be the value of this slope at the isofocal dose (see Figure 8.18). At other exposure energies, this log-CD-versus-log-dose slope will of course be different. For the simplest model, Equation (8.31), the result is

$$\frac{d\ln CD}{d\ln E} = \frac{d\ln CD}{d\ln E}\bigg|_{E=E_1}\left(\frac{E_1}{E}\right)\left(\frac{CD_1}{CD}\right) \tag{8.42}$$

For the more exact model of Equations (8.35) or (8.36),

$$\text{Space:}\ \frac{d\ln CD}{d\ln E} = \frac{d\ln CD}{d\ln E}\bigg|_{E=E_1}\left(\frac{CD_1}{CD}\right)^2$$

$$\text{Line:}\ \frac{d\ln CD}{d\ln E} = \frac{d\ln CD}{d\ln E}\bigg|_{E=E_1}\left(\frac{CD_1}{CD}\right)\left(\frac{p-CD_1}{p-CD}\right) \tag{8.43}$$

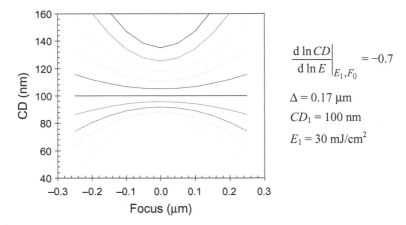

$$\frac{d\ln CD}{d\ln E}\bigg|_{E_1,F_0} = -0.7$$

$$\Delta = 0.17\ \mu m$$

$$CD_1 = 100\ nm$$

$$E_1 = 30\ mJ/cm^2$$

Figure 8.18 *Plot of the simple Bossung model of Equations (8.31) and (8.37) shows that it describes well the basic behavior observed experimentally*

Over reasonable ranges of focus (the magnitudes of focus errors normally encountered in semiconductor lithography), the impact of focus on the blurring of an aerial image is symmetric. That is, going out of focus in each direction (the wafer too close or too far from the projection lens) produces the same result on the aerial image for the same magnitude of focus error. And yet, when we measure the lithographic response to focus, in the form of a focus–exposure matrix or Bossung plot, we can observe an obvious asymmetry in the response of critical dimension (CD) to focus. Figure 8.17 shows a small but noticeable tilt in the shape of the Bossung curves and a decided asymmetry between plus and minus focus.

Since the aerial image responds to focus symmetrically, the cause of the observed asymmetry must involve the way that the aerial image interacts with the photoresist to produce the final CD. In fact, it is the thickness of the photoresist that gives rise to this asymmetry. As the aerial image travels through space, it becomes more 'in focus' as it approaches the focal plane, then defocuses as it travels away from this plane of best focus. Because the photoresist has a nonzero thickness, different parts of the resist will experience different amounts of defocus. Suppose the focal plane were placed just above the top of the resist (Figure 8.19a). At the top, the resist sees an aerial image that is near best focus and thus very sharp. As this image propagates toward the bottom of the resist, however, it goes further out of focus. As a consequence, the top of the resist profile looks sharp but the bottom is curved or 'blurred' by the defocused image. If, on the other hand, the plane of best focus were placed below the bottom of the resist (Figure 8.19b), a very different resist shape will result. The bottom of the resist will see a sharp image, while the top of the resist will experience a more defocused image. The result will be a resist profile with straight sidewalls at the bottom, but a very rounded top.

As Figure 8.19 shows, the shape of the resist profile will be very different depending on the direction of the defocus (note that 'best' focus will typically place the focal plane somewhere near the middle of the resist thickness). As one might expect, the critical dimensions of the two resist profiles shown in Figure 8.19 will also be different. The result is an asymmetric response to focus as shown in the Bossung curves. The asymmetry becomes apparent whenever the thickness of the photoresist becomes an appreciable fraction of the depth of focus of that feature.

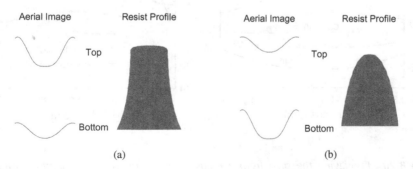

Figure 8.19 *Positioning the focal plane (a) above the top of the resist, or (b) below the bottom of the resist results in very different shapes for the final resist profile*

Asymmetric response to focus can be added to the semiempirical Bossung model by allowing other powers of dose and focus (in particular, allowing odd powers of focus). Thus, a generalized version of the Bossung model becomes

$$\frac{CD - CD_1}{CD_1} = \sum_{n=1}^{N} \sum_{m=0}^{M} a_{nm} \left(1 - \frac{E_1}{E}\right)^n (F - F_0)^m \tag{8.44}$$

In general, very good fits to experimental data are obtained with very few terms of the summation.[9]

8.5.2   Defining the Process Window and DOF

Of course, one output as a function of two inputs can be plotted in several different ways. For example, the Bossung curves could also be plotted as exposure latitude curves (linewidth versus exposure) for different focus settings. Another very useful way to plot this two-dimensional data set is a contour plot – contours of constant linewidth versus focus and exposure (Figure 8.20a).

The contour plot form of data visualization is especially useful for establishing the limits of exposure and focus that allow the final image to meet certain specifications. Rather than plotting all of the contours of constant CD, one could plot only the two CDs corresponding to the outer limits of acceptability – the CD specifications. Because of the nature of a contour plot, other variables can also be plotted on the same graph. Figure 8.20b shows an example of plotting contours of CD (nominal ±10%), 80° sidewall angle and 10% resist loss all on the same graph. The result is a *process window* – the region of focus and exposure that keeps the final resist profile within all three specifications.

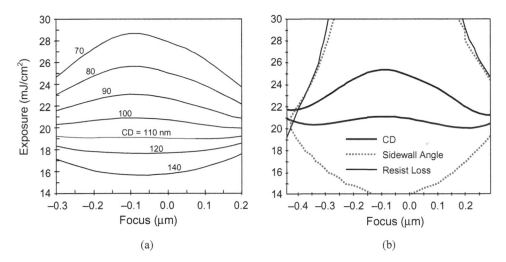

Figure 8.20 *Displaying the data from a focus–exposure matrix in an alternate form: (a) contours of constant CD versus focus and exposure, and (b) as a focus–exposure process window constructed from contours of the specifications for linewidth, sidewall angle and resist loss*

The focus–exposure process window is one of the most important plots in lithography since it shows how exposure and focus work together to affect linewidth (and possibly sidewall angle and resist loss as well). The process window can be thought of as a *process capability* – how the process responds to changes in focus and exposure. An analysis of the error sources for focus and exposure in a given process will give a *process require-ment* (see Table 8.2). If the process capability exceeds the process requirements, yield will be high. If, however, the process requirement is too large to fit inside the process capability, yield will suffer.

What is the maximum range of focus and exposure (that is, the maximum process requirement) that can fit inside the process window? A simple way to investigate this question is to graphically represent errors in focus and exposure as a rectangle on the same plot as the process window. The width of the rectangle represents the built-in focus errors of the processes, and the height represents the built-in dose errors. The problem then becomes one of finding the maximum rectangle that fits inside the process window. However, there is no one answer to this question. There are many possible rectangles of different widths and heights that are 'maximal', i.e. they cannot be made larger in either direction without extending beyond the process window (Figure 8.21a). (Note that the concept of a 'maximum area' is meaningless here.) Each maximum rectangle represents one possible trade-off between tolerance to focus errors and tolerance to exposure errors. Larger DOF can be obtained if exposure errors are minimized. Likewise, exposure latitude can be improved if focus errors are small. The result is a very important trade-off between exposure latitude and DOF.

If all focus and exposure errors were systematic, then the proper graphical representa-tion of those errors would be a rectangle. The width and height would represent the total ranges of the respective errors. If, however, the errors were randomly distributed, then a probability distribution function would be needed to describe them. It is common to assume that random errors in exposure and focus are caused by the summation of many small sources of error, so that by the central limit theorem the overall probability

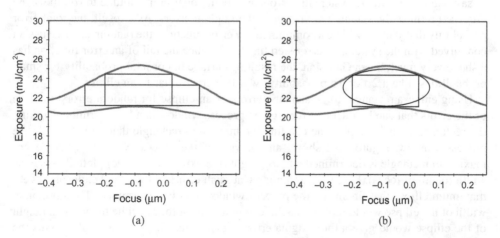

(a)

(b)

Figure 8.21 *Measuring the size of the process window: (a) finding maximum rectangles; and (b) comparing a rectangle to an ellipse*

distributions for focus and dose will be Gaussian (a normal distribution). Given standard deviations in exposure and focus, σ_E and σ_F respectively, the probability density function will be

$$p(\Delta E, \Delta F) = \frac{1}{2\pi\sigma_E\sigma_F}\exp(-\Delta E^2/2\sigma_E^2)\exp(-\Delta F^2/2\sigma_F^2) \tag{8.45}$$

where focus errors and exposure errors are assumed to be independent. In order to graphically represent the errors of focus and exposure, one should describe a surface of constant probability of occurrence. All errors in focus and exposure inside the surface would have a probability of occurring which is greater than the established cutoff. What is the shape of such a surface? For fixed systematic errors, the shape is a rectangle. For a Gaussian distribution, the surface can be derived by setting the probability of Equation (8.45) to a constant, p^*.

$$p^* = \frac{1}{2\pi\sigma_E\sigma_F}\exp(-\Delta E^2/2\sigma_E^2)\exp(-\Delta F^2/2\sigma_F^2)$$

$$-\ln(2\pi\sigma_E\sigma_F p^*) = \frac{\Delta E^2}{2\sigma_E^2} + \frac{\Delta F^2}{2\sigma_F^2} \tag{8.46}$$

Equation (8.46) is that of an ellipse. Suppose, for example, that one wishes to describe a 'three-sigma' surface, where p^* corresponds to the probability of having an error equal to 3σ in one variable. The resulting surface would be an ellipse with major and minor axes equal to $3\sigma_E$ and $3\sigma_F$ (Figure 8.21b).

$$1 = \frac{\Delta E^2}{(3\sigma_E)^2} + \frac{\Delta F^2}{(3\sigma_F)^2} \tag{8.47}$$

As seen in Table 8.2 for focus errors, the errors that actually occur in manufacturing are a combination of random and systematic errors. In fact, it is a reasonable rule of thumb to say that the full-range systematic errors are about equal in magnitude to (on the same order as) 6 times the standard deviation of the random errors. As a result, the total error probability distribution will be a complimentary error function (the random error Gaussian convolved with the systematic error step function). Since the tail of an error function has a shape very much like a Gaussian, the resulting surface of constant probability for small probabilities (like a 2σ or 3σ probability) will be very similar to an ellipse.

Using either a rectangle for systematic errors or an ellipse for random errors, the size of the errors that can be tolerated for a given process window can be determined. Taking the rectangle as an example, one can find the maximum rectangle that will fit inside the process window. Figure 8.22 shows an analysis of the process window where every maximum rectangle is determined and its height (the exposure latitude) plotted versus its width (depth of focus). Likewise, assuming Gaussian errors in focus and exposure, every maximum ellipse that fits inside the process window can be determined. The horizontal width of the ellipse would represent a three-sigma error in focus, while the vertical height of the ellipse would give a three-sigma error in exposure. Plotting the height versus the width of all the maximum ellipses gives the second curve of exposure latitude versus DOF in Figure 8.22.

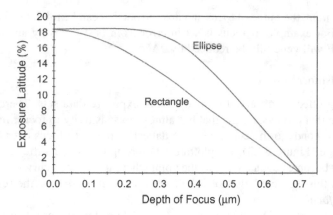

Figure 8.22 *The process window of Figure 8.20b is analyzed by fitting all the maximum rectangles and all the maximum ellipses, then plotting their height (exposure latitude) versus their width (depth of focus)*

The exposure latitude-versus-DOF curves of Figure 8.22 provide the most concise representation of the coupled effects of focus and exposure on the lithography process. Each point on the exposure latitude-DOF curve is one possible operating point for the process. The user must decide how to balance the trade-off between DOF and exposure latitude. One approach is to define a minimum acceptable exposure latitude, and then operate at this point; this has the effect of maximizing the DOF of the process. In fact, this approach allows for the definition of a single value for the DOF of a given feature for a given process. The depth of focus of a feature can be defined as *the range of focus that keeps the resist profile of a given feature within all applicable profile specifications (linewidth, sidewall angle and resist loss, for example) over a specified exposure range.* For the example given in Figure 8.22, a minimum acceptable exposure latitude of 10%, in addition to the other profile specifications, would lead to the following depth of focus results:

- DOF (rectangle) = 0.40 μm
- DOF (ellipse)   = 0.52 μm

As one might expect, systematic errors in focus and exposure are more problematic than random errors, leading to a smaller DOF. Most actual processes would have a combination of systematic and random errors. Thus, one might expect the rectangle analysis to give a pessimistic value for the DOF, and the ellipse method to give an optimistic view of DOF. As mentioned above, however, the ellipse method gives a DOF very close to the true value for a common mix of systematic and random errors.

The definition of depth of focus also leads naturally to the determination of best focus and best exposure. The DOF value read off from the exposure latitude-versus-DOF curve corresponds to one maximum rectangle or ellipse that fits inside the process window. The center of this rectangle or ellipse would then correspond to best focus and exposure for this desired operating point. Knowing the optimum focus and dose values is essential to being able to use the full process window. If the process focus and dose settings deviate from this optimum, the range of focus and dose errors that can be tolerated will be reduced accordingly. For example, if focus is set in error from best focus by an amount Δf, the resulting DOF will generally be reduced by $2\Delta f$.

8.5.3   The Isofocal Point

An interesting effect is observed in many focus–exposure data sets. There is often one exposure dose that produces a CD that has almost no sensitivity to focus. In the Bossung plot, this corresponds to the curve with the flattest response to focus over the middle of the focus range (Figure 8.23). By plotting CD through dose for different focuses, this isofocal point becomes even clearer as the point where the various curves cross. The dose that produces this isofocal behavior is called the *isofocal dose*, and the resulting CD is called the *isofocal CD*.

Ideally, the isofocal CD would equal the target CD of the process. Unfortunately, this is rarely the case without specifically designing the process for that result. The difference between the isofocal CD and the target CD is called the *isofocal bias*. A process with a large isofocal bias will produce smaller usable process windows because the right and left sides of the process window will bend either up or down, making the maximum possible ellipse or rectangle much smaller. In general, more isolated features exhibit greater isofocal bias than dense features, as seen in Figure 8.24.

As was mentioned in section 8.5.1, the isofocal CD and the isofocal dose play an important role in the semiempirical Bossung model of Equation (8.44). The terms E_1 and CD_1 are in fact identical to the isofocal dose and CD. For the simplest version of this model, Equations (8.31) and (8.37), the role of the isofocal point in influencing the shape of the process window can be made explicit. Since the model takes the form

$$\frac{CD - CD_1}{CD_1} = s\left(1 - \frac{E_1}{E}\right)\left[1 + \left(\frac{F - F_0}{\Delta}\right)^2\right], \quad \text{where} \quad s = \frac{d\ln CD}{d\ln E}\bigg|_{E_1, F_0} \qquad (8.48)$$

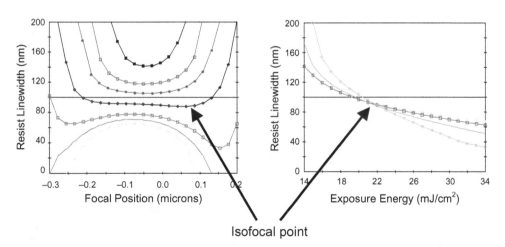

Isofocal point

Figure 8.23   *Two ways of plotting the focus–exposure data set showing the isofocal point – the dose and CD that have minimum sensitivity to focus changes (for the left graph, each curve represents a different exposure dose; for the right graph each curve is for a different focus)*

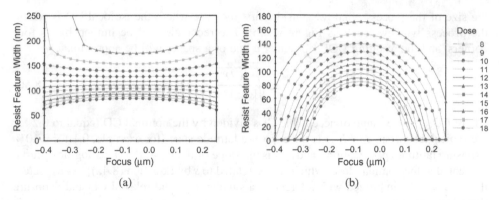

(a)   (b)

Figure 8.24 *Bossung plots for (a) dense and (b) isolated 130-nm lines showing the difference in isofocal bias*

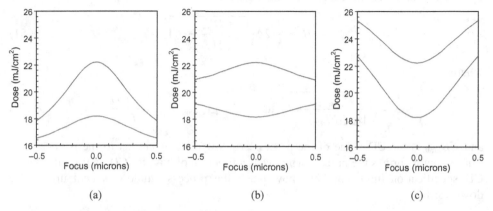

(a)   (b)   (c)

Figure 8.25 *Process windows calculated using Equation (8.49) for a nominal line CD of 100 ± 10 nm, nominal dose of 20 mJ/cm², s = −1, and Δ = 0.45 microns: (a) isofocal CD₁ = 130 nm, (b) isofocal CD₁ = 100 nm and (c) isofocal CD₁ = 70 nm*

contours of constant CD can be found by solving for $E(F,CD)$:

$$E(F,CD) = E_1 \frac{1 + \left(\dfrac{F - F_0}{\Delta}\right)^2}{1 + \left(\dfrac{F - F_0}{\Delta}\right)^2 - \left(\dfrac{CD - CD_1}{sCD_1}\right)} \tag{8.49}$$

(It is useful to remember that s will be negative for a line pattern, and positive for a space.) Figure 8.25 shows three example process windows, showing how the shape of the process window changes depending on whether the nominal CD is smaller, the same, or larger than the isofocal CD.

The process window defined by two contours as given by Equation (8.49) can be analytically measured to give the depth of focus. Using the rectangle approach to measuring the size of the process window, consider first the case where the isofocal CD is within the process window (such as Figure 8.25b). The rectangle will be limited by its four corners, and the resulting DOF (for either a line or a space) will be approximately

$$DOF = 2\Delta\sqrt{\frac{CD_{spec}}{|s_{nom}|EL_{spec}} - 1} \tag{8.50}$$

where CD_{spec} is the range of acceptable CDs divided by the nominal CD (equal to 0.2 for the typical $\pm10\%$ spec), EL_{spec} is the exposure latitude spec (for example, 0.1 for a 10% exposure latitude specification), and s_{nom} is the slope of the log-CD-versus-log-dose curve evaluated at the nominal dose [which can be related to s by Equation (8.42)]. As expected, the DOF can be improved with a larger Δ, a smaller s_{nom}, and relaxed CD and exposure latitude specifications.

When the isofocal CD is not within the process window (Figure 8.25a and c), the bending of the process window further limits the maximum rectangle that can fit inside the window and thus the DOF. For these cases, the DOF becomes

$$DOF \approx \frac{1}{K}2\Delta\sqrt{\frac{CD_{spec}}{|s_{nom}|EL_{spec}} - 1} \tag{8.51}$$

where

$$K = \sqrt{\frac{CD_{rel-iso-limit}}{|s_{nom}|EL_{spec}} + 1} \quad \text{and} \quad CD_{rel-iso-limit} = \frac{|CD_1 - CD_{limit}|}{CD_{nom}} \tag{8.52}$$

and CD_{limit} is the CD value of the process window contour closest to the isofocal CD. This K factor shrinks the process window as a function of how far CD_1 is from its closest CD specification limit, and thus how much the process window is bent upward or downward.

8.5.4   Overlapping Process Windows

Although all of the above results describe the focus and exposure response of one critical feature, in reality a number of mask features must be printed simultaneously. Using the most common example, isolated lines typically occur on the same design as dense lines. Thus, a more important measure of performance than the DOF of each individual feature is the *overlapping* DOF of multiple critical features. Just as multiple profile metrics were overlapped to form one overlapping process window in Figure 8.20b, process windows from different features can be overlapped to determine the DOF for simultaneously printing those multiple features. Figure 8.26 shows such an example for isolated and dense features, where subresolution assist features (SRAFs) are used to reduce the isofocal bias of the isolated line and achieve a significantly better overlapping DOF. Ideally, process windows for all critical feature sizes and pitches would be overlapped to find out which of the features represent the process window limiters.

Systematic variations in patterning as a function of field position can also be accounted for with overlapping process windows. If the same feature were printed at different points in the field (typically the center and the four corners of the field are sufficient), process

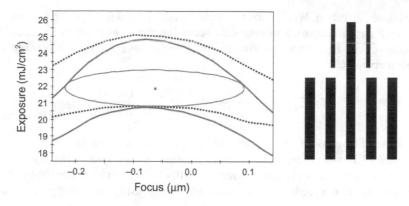

Figure 8.26   *The overlapping process window for the dense (dashed lines) and isolated (solid lines) features shown to the right of the graph*

windows can be overlapped to create the *usable* process window for that feature. The resulting depth of focus is called the *useable depth of focus* (UDOF).

8.5.5   Dose and Focus Control

Controlling focus and dose is an important part of keeping the critical dimensions of the printed patterns in control. The first step in understanding how to control focus and dose is to characterize the response of critical features to variations using the focus–exposure process window, as discussed above. Proper analysis of CD versus focus and dose data allows for the calculation of the process window, measurement of the process window size to produce the exposure latitude–DOF plot, and determination of a single value for the depth of focus. Best focus and dose are also determined with this analysis as the process settings that maximize the tolerance to focus and dose errors. Once best focus and dose are determined, the next goal is to keep the process centered at this best focus and dose condition as thousands of production wafers pass through the lithography cell. And since almost all errors that occur in lithography act like either a dose error or a focus error, properly adjusting dose and focus on a lot-by-lot basis can provide for much tighter CD control. Dose and focus monitoring and feedback control has become a quite common technique for living within shrinking process windows.

The most common way to monitor and correct for drifts in exposure dose grew out of standard product-monitoring metrology applications. Critical dimension test structures (generically called *metrology targets*) are placed in the kerf (streets separating each die) and are designed to mimic the size and surroundings of some critical feature of interest within the die. For example, on the polysilicon level of a CMOS device, the most common critical gate may be a 90-nm line on a 250-nm pitch. The CD targets might then be an array of five 90-nm lines on a 250-nm pitch. These types of targets are called *device-representing*.

By characterizing the change in CD versus dose of this specific device-representing target, any change in measured CD can be directly related to an equivalent change in dose (see Figure 8.3). Often the data are fit to a function, and the function used to predict the required dose correction. While a linear function is often used (everything becomes linear over a small enough range), a significantly better function was given in Equation (8.31). This equation can be inverted to give the dose correctable ΔE for a measured critical dimension error ΔCD.

$$\Delta E = E_1 \left(\frac{\Delta CD}{sCD_1 - \Delta CD} \right) \quad \text{where} \quad s = \left. \frac{\mathrm{d}\ln CD}{\mathrm{d}\ln E} \right|_{E=E_1} \tag{8.53}$$

Multiple measurements and a proper sampling plan help to eliminate random errors in the measured CD and ensure that the changes observed are systematic. By monitoring this target on product lots over time, temporal drifts in CD can be corrected by changing the exposure dose in a feedback loop (often automated using *advanced process control*, APC).

The advantages of using device-representing targets are their simplicity and multipurpose use. Device-representing CD targets are often measured for the purpose of lot monitoring: do the measured CD values indicate that this lot will yield high or low? If these structures are being measured anyway, it only makes sense to use them for dose monitoring and corrections as well. The disadvantage of using device-representing targets, as we shall see below, is that they may have reduced sensitivity to the variable being controlled (exposure dose, in this case), compared to other possible test structures, and may be too sensitive to other variables not being measured or controlled (focus being the most important of these).

Monitoring exposure dose is relatively easy since all features vary monotonically with dose. However, as we saw above, the sensitivity of any given feature to dose depends on focus. In other words, the exposure latitude slope term in Equation (8.53) is focus dependent. In general, s varies quadratically with focus, reaching a minimum at best focus. Further, focus errors most often cause CD errors. If focus drifts, the use of an exposure monitoring strategy like that outlined above will make exposure dose adjustments to compensate for focus errors. If sufficiently large, the combined focus drift and exposure adjustment can render the lithography process unstable (meaning the useable process window becomes too small to ensure adequate CD control) even though the target CD appears to be in control. How can this situation be avoided?

As we saw above, the two variables focus and dose must be characterized together in a focus–exposure matrix (Figure 8.17). Note that at some doses, changes in focus make this CD larger, while at other doses the same change in focus will make the CD smaller. At the isofocal dose, changes in focus have the least impact on CD. This observation suggests a way to improve exposure monitoring: If a CD target is found which exhibits isofocal behavior at the nominal dose, this test structure will have minimum sensitivity to focus errors while maintaining reasonable dose sensitivity. Such a test structure will be called an *isofocal target*. An isofocal target will have minimum sensitivity to focus errors, so that only dose errors will be detected and corrected. However, focus errors will still change the exposure latitude term s in the exposure correction equation. Thus, if both focus and exposure errors are present, the predicted dose correction from Equation (8.53) will be too large. To address this deficiency, focus monitoring must be added to the overall control strategy.

Focus presents an extremely difficult control problem since CD tends to vary quadratically with focus. As a result, it becomes extremely difficult to determine the *direction* of a focus error – was the focal plane too high or too low compared to best focus? Thus, the focus monitoring problem is often broken up into two parts: determining the magnitude of a focus error, and its direction.

Since the responses of CD to dose and to focus are coupled, the goal is to find two targets with different responses to these variables. If, for example, two targets had the same response to dose, but different responses to focus, measuring them both would allow focus errors to be separated from dose errors. A simple approach would be to find a dense line whose isofocal dose matched the nominal dose for the process and an isolated line far removed from its isofocal dose. Further, if the isolated line were sized so that it had approximately the same exposure latitude as the dense line [same value of *s* in Equation (8.53)], then the isofocal dense feature could be used as the dose monitor while the difference in CD between dense and isolated features could be used to monitor focus. Figure 8.27 shows an example of a reasonably well-designed target. The dense feature (which is the dose monitor) varies by only about 2 nm over the full range of focus errors of interest, but varies by over 20 nm over the exposure range of interest. The CD difference (the focus monitor) varies by only 2 nm over the dose range of interest but changes by over 20 nm over the focus range of interest. Thus each monitor has 10× *discrimination*, that is, 10 times more sensitivity to the variable being monitored compared to the other variable. Further refinement of the target design can often lead to even better discrimination between focus and dose errors, while maintaining high sensitivity.

It is possible to improve the analysis of this dual-target dose and focus monitoring strategy by creating models for the combined focus–exposure response of each target. Inverting both models using the two CD target measurements can provide for good discrimination between dose and focus errors even when the two features do not have identical exposure latitude and the dense feature is not perfectly isofocal. It is important to remember, however, that the nearly quadratic response of CD with focus means that focus direction cannot be determined from such data.

To monitor the direction of a focus error, one must measure a response on the wafer that is sufficiently asymmetric with respect to focus direction. CD does not provide such a response in general. Of course, at any time the lithography engineer can stop running product wafers and print a focus–exposure matrix wafer in order to determine best focus and adjust the focus on the stepper or scanner. However, the goal of product monitoring is to make measurements on product wafers that can detect and correct for dose and focus errors quickly without reducing stepper productivity or throughput. Some novel approaches have been developed, each with their own advantages and disadvantages. Two important examples are described below.

The phase-shift focus monitor developed by Tim Brunner[10] uses a 90° phase shift pattern on the mask to turn focus errors into pattern placement errors on the wafer. Thus, a measurement of overlay for this pattern is directly proportional to the focus error, direction included. Unfortunately, the need for a 90° phase shift region on the mask means that this focus monitor cannot in general be placed on product reticles – it is a test mask approach that doesn't meet product monitoring requirements.

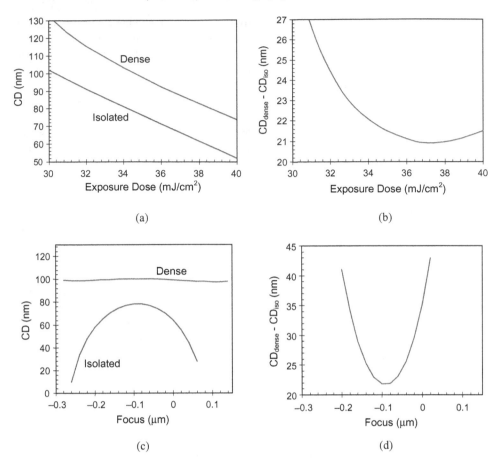

Figure 8.27 *A dual-target approach to monitoring dose and focus using a 90-nm dense line on a 220-nm pitch and an 80-nm isolated line: (a) dense and isolated lines through dose; (b) the iso-dense difference through dose; (c) dense and isolated lines through focus; (d) the iso-dense difference through focus*

Scatterometry-based measurements offer a product-measurement solution to the focus monitoring problem. Unlike scanning electron microscopes, scatterometry allows for the simultaneous measurement of linewidth, sidewall angle and resist height for a pattern of long lines and spaces. While CD varies approximately quadratically with focus, resist height and sidewall angle have a strong linear component to their responses, making detection of focus direction possible. Like Figure 8.19 above, Figure 8.28 shows an example of how a typical resist cross section varies with focus. The strong asymmetry of the sidewall angle can provide sensitive detection of focus direction.

The sensitivity of a monitor target is defined as the smallest error that can be reliably detected. The goal of focus and exposure monitoring is to have sensitivities of detecting focus and exposure errors equal to about 10–20 % of the process window size or less. For example, if the depth of focus is 400 nm, the focus sensitivity must be at least 80 nm, but

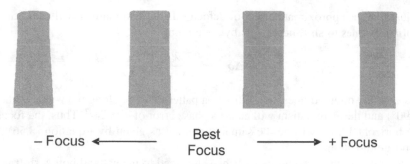

$$- \text{Focus} \longleftarrow \quad \begin{array}{c} \textbf{Best} \\ \textbf{Focus} \end{array} \quad \longrightarrow + \text{Focus}$$

Figure 8.28 *The asymmetric response of resist sidewall angle to focus provides a means for monitoring focus direction as well as magnitude. Here, + focus is defined as placing the focal plane above the wafer (moving the wafer further away from the lens)*

40 nm is preferable. For an 8% exposure latitude specification, the dose sensitivity must be at least 1.6%, but 0.8% is the preferred sensitivity.

8.6   H–V Bias

One interesting effect of aberrations is called horizontal–vertical (H–V) bias. Quite simply, H–V bias is the systematic difference in linewidth between closely located horizontally and vertically oriented resist features that, other than orientation, should be identical. There are two main causes of H–V bias: astigmatism and related aberrations; and source shape aberrations.

8.6.1   Astigmatism and H–V Bias

The aberration of astigmatism results in a difference in best focus as a function of the orientation of the feature. Using the Zernike polynomial description of aberrations (see Chapter 3), 3rd order 90° astigmatism (which affects horizontally and vertically oriented lines) takes the form

$$phase\ error = 2\pi Z_4 R^2 \cos 2\phi \tag{8.54}$$

Consider a vertically, y-oriented pattern of lines and spaces. The diffraction pattern will spread across the x-axis of the pupil, corresponding to $\phi = 0°$ and 180°. Thus, the phase error will be $2\pi Z_4 R^2$ for this feature. Recalling the description of defocus as an aberration (see Chapter 3), the phase error due to defocus is

$$phase\ error \approx \frac{\pi \delta NA^2}{\lambda} R^2 \tag{8.55}$$

where δ is the defocus distance, λ is the wavelength, and NA is the numerical aperture. [Equation (8.55) is approximate because it retains only the first term in a Taylor series. While this approximation is progressively less accurate for higher numerical apertures, it will be good enough for our purposes.] Immediately, one sees that 3rd order astigmatism looks just like the approximate effect of defocus. Thus, astigmatism will cause the vertically oriented lines to shift best focus by an amount

$$\Delta\delta_{\text{vert}} \approx \frac{2Z_4\lambda}{NA^2} \tag{8.56}$$

For horizontally oriented lines, the diffraction pattern will be along the y-axis of the pupil ($\theta = \pm 90°$) and the astigmatism will cause a phase error of $-2\pi Z_4 R^2$. Thus, the focus shift for the horizontal lines will be the same magnitude as given by Equation (8.56), but in the opposite direction.

To see how astigmatism causes H–V bias, we need to understand how a shift in focus might affect the resist feature CD. To first order, CD has a quadratic dependence on focus.

$$CD \approx CD_{\text{best focus}} + a\delta^2 \tag{8.57}$$

where a is the dose-dependent curvature of the CD through focus curve (see section 8.5.1) and is given by

$$a = CD_1\left(\left.\frac{d\ln CD}{d\ln E}\right|_{E_1, F_0}\right)\left(1 - \frac{E_1}{E}\right)\left(\frac{1}{\Delta^2}\right) \tag{8.58}$$

Recalling the typical shapes of Bossung curves, a can vary from positive to negative values as a function of dose (that is, depending on whether E is greater than or less than the isofocal dose, E_1). If best focus is shifted due to astigmatism, we can calculate the CD of the vertical and horizontal features by adding the focus shift of Equation (8.56) to Equation (8.57).

$$CD_{\text{vert}} \approx CD_{\text{best focus}} + a\left(\delta - \frac{2Z_4\lambda}{NA^2}\right)^2$$

$$CD_{\text{horiz}} \approx CD_{\text{best focus}} + a\left(\delta + \frac{2Z_4\lambda}{NA^2}\right)^2 \tag{8.59}$$

From, here a straightforward subtraction gives us the H–V bias:

$$H-V\ bias \approx \frac{8a\delta Z_4\lambda}{NA^2} \tag{8.60}$$

The H–V bias is directly proportional to the amount of astigmatism in the lens (Z_4) and to the curvature of the CD-through-focus curve (a). But it is also directly proportional to the amount of defocus. In fact, a plot of H–V bias through focus is a sure way to identify astigmatism (that is, when not using the isofocal dose, where $a \approx 0$, for the experiment). Figure 8.29 shows some typical results. Note that the isolated lines show a steeper slope than the dense lines due to a larger value of the CD through focus curvature. In fact, if a is determined by fitting Equation (8.57) to CD through focus data, a reasonable estimate of Z_4 can be made using an experimentally measured H–V bias through focus curve. Note also that the true shape of the curves in Figure 8.29 is only approximately linear, since both Equations (8.56) and (8.57) ignore higher-order terms.

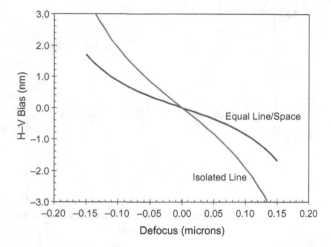

Figure 8.29 *PROLITH simulations of H–V bias through focus showing approximately linear behavior (λ = 193 nm, NA = 0.75, σ = 0.6, 150-nm binary features, 20 milliwaves of astigmatism). Simulations of CD through focus and fits to Equation (8.57) gave the CD curvature parameter a = −184 μm⁻² for the dense features and −403 μm⁻² for the isolated lines*

Assume that the maximum possible value of the defocus is one-half of the depth of focus (DOF). At this defocus, Equation (8.57) would tell us that at the worst case dose, the term $a\delta^2$ will be about 10% of the nominal CD (by definition of the process window with a ±10% CD specification). Thus, we can say that within the process window, the worst case H–V bias due to astigmatism will be

$$\left. \frac{H-V\ bias}{CD_{nominal}} \right|_{max} \approx \frac{1.6Z_4\lambda}{DOF\ NA^2} \qquad (8.61)$$

Using typical numbers for a 65-nm process, for a wavelength of 193 nm, an NA of 0.9, and assuming a depth of focus of 200 nm, the fractional H–V bias will be about $2Z_4$. If we are willing to give 1% CD error to H–V bias, our astigmatism must be kept below 5 milliwaves. In general, Equation (8.61) shows that as new, higher resolution scanners are designed and built, the astigmatism in the lens must shrink as fast or faster than the depth of focus of the smallest features to be put into production.

8.6.2 Source Shape Asymmetry

The second major cause of H–V bias is illumination aberrations (that is, source shape asymmetries). In Chapter 2, a discussion of partial coherence showed that source points must be symmetrically placed about the axis of symmetry for a mask pattern in order to avoid telecentricity errors (which cause pattern placement errors as a function of defocus). For example, for y-oriented lines and spaces, a source point at (f_x, f_y) must be accompanied by a corresponding source point of the same intensity at $(-f_x, f_y)$ in order to avoid telecentricity errors. The avoidance of telecentricity errors for x-oriented lines will likewise

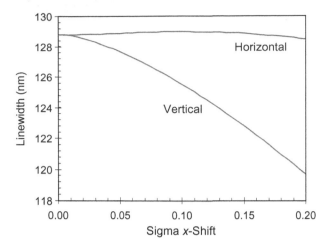

Figure 8.30   *Example of how an x-shift in the center of a conventional source (σ = 0.6) affects mainly the vertical (y-oriented) lines and spaces (CD = 130nm, pitch = 650nm, PROLITH simulations)*

require additional points at $(f_x, -f_y)$ and $(-f_x, -f_y)$, providing the general guidance that all source shapes must have fourfold symmetry. If both horizontal and vertical lines must be printed simultaneously, identical printing will occur only if the source shape is invariant to a 90° rotation. This provides an additional constraint that for every source point (f_x, f_y), there must be a corresponding source point of the same intensity at $(-f_y, f_x)$.

Consider conventional illumination where the center of the disk-shaped source is not perfectly aligned with the center of the optical path (called an illumination telecentricity error). The impact of such a source telecentricity error is dependent on the partial coherence, the pitch, and of course on the amount of telecentricity error. Figure 8.30 shows that, in general, an *x*-shift of the center of the illumination source shape affects vertically oriented (*y*-oriented) lines significantly, but horizontal (*x*-oriented) lines very little.

For the simple case of dense line/space patterns where only the zero and the two first diffraction orders are used in the imaging, the image will be made up of combinations of one-, two- and/or three-beam interference. The change in CD for the vertical features for an *x*-shift in the source center is caused by a change in the ratio of two-beam to three-beam imaging. Figure 8.31 shows an example case where the total three-beam imaging area (light grey) and the total two-beam imaging area (dark grey) do not appreciably change as the center of the source is shifted by about 0.1 sigma. Thus, for this pitch and sigma, one would not expect to see much change in the vertical feature CD. Consider a different pitch, as shown below in Figure 8.32a. At this particular pitch, all of the first order is inside the lens, so that all of the imaging is three-beam (note that the second diffracted order is not shown in the diagram for clarity's sake). However, when the source is shifted in *x* by 0.1 sigma (Figure 8.32b), the amount of three-beam imaging for the vertical lines is reduced and two-beam imaging is introduced. By contrast, the horizontal lines (which spread the diffraction pattern vertically in the pupil) have only an

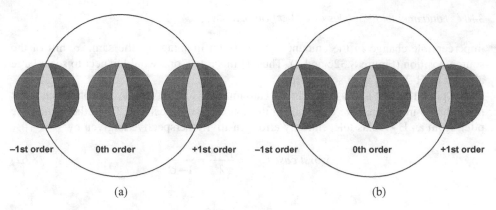

-1st order   0th order   +1st order   -1st order   0th order   +1st order

(a)   (b)

Figure 8.31 *Example of dense line/space imaging where only the zero and first diffraction orders are used (k_{pitch} = 1.05). The middle segment of each source circle represents three-beam imaging, the outer areas are two-beam imaging. (a) source shape is properly centered, (b) source is offset in x (to the right) by 0.1. Note that the diffraction pattern represents vertical (y-oriented) features*

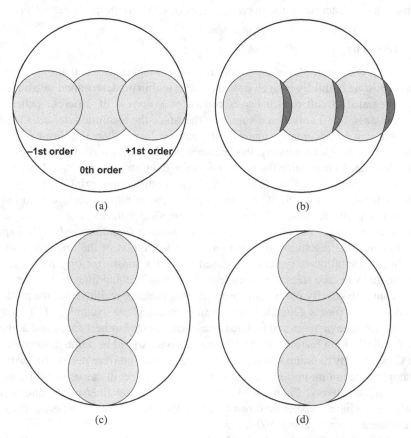

Figure 8.32 *Examples of how a telecentricity error affects the ratio of two-beam to three-beam imaging at the worst case pitch (σ = 0.4, x-shift = 0.1, k_{pitch} = 1.65). (a) vertical lines, no telecentricity error, (b) vertical lines, with telecentricity error, (c) horizontal lines, no telecentricity error, and (d) horizontal lines, with telecentricity error*

imperceptible change in the amount of three-beam imaging for the same x-shift of the source position (Figure 8.32c and d). Thus, at this pitch, one would expect to see a large amount of H–V bias with source shift.

The pitch that just allows only three-beam imaging for a given partial coherence is the pitch that is most sensitive to a shift in the source center position. Thus, the worst case pitch from an H–V bias telecentricity error sensitivity perspective is given by

$$worst\ case\ k_{\text{pitch}} = \frac{pNA}{\lambda} \approx \frac{1}{1-\sigma} \qquad (8.62)$$

8.7   Mask Error Enhancement Factor (MEEF)

Errors in the critical dimensions of the features on a mask have always been a significant cause of critical dimension errors on the wafer. As optical lithography has pushed features to lower and lower k_1 values (= feature size $*NA/\lambda$), however, mask CD errors have taken up a much larger fraction of the overall sources of CD variations on the wafer.

8.7.1   Linearity

An important constraint placed on any lithographic imaging task is that *every* feature on the mask must be faithfully imaged onto the wafer within predetermined tolerances. Most typical integrated circuit device layers consist of a myriad of different pattern types, shapes and sizes. It is a common assumption that since the resolution defines the smallest pattern that can be acceptably imaged, all features larger than this limit will also be acceptably imaged. Unfortunately, this assumption may not always be true. For imaging systems designed to maximize the printability of a given small feature (e.g. using off-axis illumination), often some larger features will be more difficult to print. One way to insure that larger features print well at the same time as the minimum feature is to build this requirement into the definition of resolution: *the smallest feature of a given type such that it and all larger features of the same type can be printed simultaneously with a specified depth of focus*. This resolution is called the *linear resolution* of the imaging system.

The linear resolution is typically assessed using the mask linearity plot. Consider a mask with many feature sizes of a given type, for example, equal lines and spaces, isolated lines, or contact holes. By plotting the resulting resist feature width versus the mask width, the mask linearity plot is generated. Shown in Figure 8.33 are examples of linearity plots for equal line/space patterns and isolated lines, both imaged at best focus and at the dose-to-size for the 350-nm features. Perfect, linear behavior would be a line through the origin with a slope of 1. By defining specifications for any deviation from perfect linearity ($\pm 5\%$, for example), the minimum feature that stays within the specification would be the linear resolution. Qualitatively, Figure 8.33 shows that the equal line/space patterns of this example have a linear resolution down to about 350 nm ($k_1 = 0.54$), whereas the isolated lines are linear down to about 300 nm ($k_1 = 0.46$).

Of course, to be truly practical, one should include variations in exposure and focus as well. The most rigorous approach would begin with the focus–exposure process window

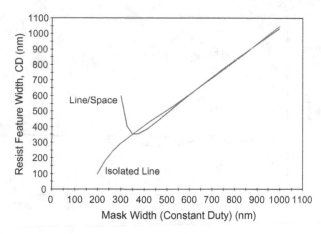

Figure 8.33 *Typical mask linearity plot for isolated lines and equal lines and spaces (i-line, NA = 0.56, σ = 0.5)*

for the largest feature size. The process window from each smaller feature is then overlapped with that from the larger features and the depth of focus calculated. Smaller and smaller features are added until the overlapped depth of focus drops below the specified limit, indicating the linear resolution limit. Thus, although mask linearity plots do not provide a rigorous, general method for determining the linear resolution, they are qualitatively useful.

As lithography for manufacturing continues to push toward its ultimate resolution limits, linearity is playing a decidedly different role in defining the capabilities of low k_1 imaging. Consider, using Figure 8.33 as an example, manufacturing at the linear resolution limits: 350-nm lines and spaces and 300-nm isolated lines. Although these features may be 'resolvable' by the definitions provided above, critical dimension (CD) control may be limited by a new factor: how do errors in the dimensions of the feature on the mask translate into errors in resist CD on the wafer?

8.7.2 Defining MEEF

For 'linear' imaging, mask CD errors would translate directly into wafer CD errors (taking into account the reduction factor of the imaging tool, of course). Thus, a 3-nm CD error on the mask (all CDs on the mask will be expressed here in wafer dimensions) would result in a 3-nm CD error on the final resist feature. If, however, the features of interest are at the very edge of the linear resolution limit, or even beyond it, the assumption of linear imaging falls apart. How then do mask CD errors translate into resist CD errors?

Consider the examples shown in Figure 8.33. If an isolated line is being imaged near its resolution limit, about 300 nm, a 10-nm mask CD error would give a 14-nm resist CD error. Thus, at this feature width, isolated line mask errors are amplified by a factor of 1.4. This amplification of mask errors is called the *mask error enhancement factor*

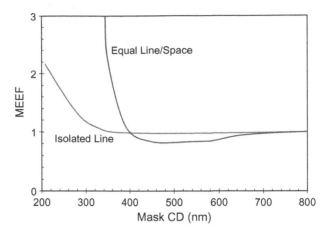

Figure 8.34   *The mask error enhancement factor (MEEF) for the data of Figure 8.33*

(MEEF). First discussed by Wilhelm Maurer,[11] the MEEF (also called MEF by some authors) is defined as the change in resist CD per unit change in mask CD:

$$MEEF = \frac{\partial CD_{\text{resist}}}{\partial CD_{\text{mask}}} \qquad (8.63)$$

where again the mask CD is in wafer dimensions. If we assume a 'zero bias' process (where the target CD on the wafer is the same as the mask CD), then the MEEF can also be expressed as

$$MEEF = \frac{\partial \ln CD_{\text{resist}}}{\partial \ln CD_{\text{mask}}} \qquad (8.64)$$

This form of the MEEF equation will be useful later. Figure 8.34 shows how the MEEF varies with feature size for the mask linearity data of Figure 8.33. Regions where the MEEF is significantly greater than 1 are regions where mask errors may come to dominate CD control on the wafer. Optical proximity correction techniques (discussed in Chapter 10) allow us to lower the linear resolution, but generally without improving the MEEF. As a result, the mask has begun to take on a much larger portion of the total CD error budget.

8.7.3   Aerial Image MEEF

A MEEF of 1.0 is the definition of a linear imaging result. Although a MEEF less than one can have some desirable consequences for specific features, in general a MEEF of 1.0 is best. Fundamentally, anything that causes the overall imaging process to be non-linear will lead to a nonunit valued MEEF. In lithography, every aspect of the imaging process is nonlinear to some degree, with the degree of nonlinearity increasing as the dimensions of the features approach the resolution limits. Consider the first step in the imaging process, the formation of an aerial image. One might judge the linearity of this first step by approximating the resist CD with an image CD, defined to be the width of the aerial image at some image threshold intensity value. It is important to note that the image CD is only an approximate indicator of the resist CD. For real, finite-contrast resists the differences between these two quantities can be substantial. Nonetheless, the image CD will be used here to elucidate some general principles about imaging and the MEEF.

For two simple cases of projection imaging, coherent and incoherent illumination, analytical expressions for the aerial image can be defined. Assuming a pattern of many long lines and spaces with a spacewidth w and pitch p such that only the 0 and ±1 diffraction orders pass through the lens, the TE coherent and incoherent in-focus aerial images would be

$$\text{Coherent Illumination: } I(x) = \left[\frac{w}{p} + \frac{2\sin(\pi w/p)}{\pi}\cos(2\pi x/p)\right]^2 \tag{8.65}$$

$$\text{Incoherent Illumination: } I(x) = \frac{w}{p} + \frac{2\sin(\pi w/p)}{\pi}(MTF_1)\cos(2\pi x/p) \tag{8.66}$$

where MTF_1 is the value of the incoherent Modulation Transfer Function at the spatial frequency corresponding to the first diffraction order. The requirement that no orders higher than the first diffraction order be used to form the image means that the coherent image equation is valid for a limited range of pitches such that $1 < pNA/\lambda < 2$, and the incoherent expression is valid for $0.5 < pNA/\lambda < 1$.

Using these expressions to define the image CD, exact expressions for the *image MEEF* can be derived for these repeating line/space patterns under the conditions given above:

$$image\ MEEF = \frac{\partial CD_{\text{image}}}{\partial CD_{\text{mask}}} = \frac{\partial CD_{\text{image}}}{\partial w} \tag{8.67}$$

$$\text{Coherent Illumination: } image\ MEEF = \frac{2 + \cos(2\pi w/p)}{1 - \cos(2\pi w/p)} \tag{8.68}$$

$$\text{Incoherent Illumination: } image\ MEEF = \frac{\dfrac{1}{MTF_1} + 1 + \cos(2\pi w/p)}{1 - \cos(2\pi w/p)} \tag{8.69}$$

An interesting observation can be made immediately. Over the range of valid pitches, the coherent image MEEF is only dependent upon the duty cycle (w/p), not on the pitch itself. The incoherent image MEEF, on the other hand, has a direct pitch dependence through the value of the MTF (which is approximately equal to $1 - \lambda/[2NAp]$). Figure 8.35 shows how both image MEEFs vary with spacewidth-to-linewidth ratio.

The extreme nonlinearity of the imaging process is evident from the results shown in Figure 8.35. For coherent illumination, a pattern of equal lines and spaces will have an image MEEF of 0.5. A spacewidth twice the linewidth produces a MEEF of 1.0, and a spacewidth three times the linewidth results in a coherent image MEEF of 2.0. Obviously, different duty cycles can have wildly different sensitivities to mask errors. While the

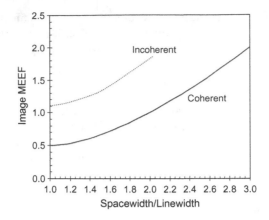

Figure 8.35   *The impact of duty cycle (represented here as the ratio of spacewidth to line-width for an array of line/space patterns) on the image CD-based MEEF for both coherent and incoherent illumination. For the incoherent case, an MTF_1 of 0.45 was used*

approximations used do not apply to truly isolated lines, it is clear that such features will also deviate from unit MEEF. A spacewidth/linewidth ratio less than unity also enhances the effect.

Although neither purely coherent nor purely incoherent illumination is ever used in real lithographic imaging, these two extremes tend to bound the behavior of typical partially coherent imaging tools. Thus, we would expect the image MEEF of a partially coherent imaging system to vary with both duty cycle and pitch, and to vary by about a factor of 2 as a function of the partial coherence.

8.7.4   Contact Hole MEEF

Possibly the most challenging mask layer to print in high-end lithography processes is the contact or via layer. These small features are subwavelength in two dimensions, making them exceptionally sensitive to everything that makes low k_1 lithography difficult. For example, contact holes suffer from the largest MEEFs of any feature type (as we shall see below). Assuming that the source of the mask error is isotropic (that is, it affects both the height and width of the contact in the same way), then such an error is affecting two dimensions of the mask simultaneously. In other words, errors in the area of the mask feature go as errors in the CD squared.

When printing a small contact hole, the aerial image projected onto the wafer is essentially the same as the point spread function (PSF) of the optical system. The point spread function is defined as the normalized aerial image of an infinitely small contact hole on the mask. For small contact holes (about $0.6\lambda/NA$ or smaller), the aerial image takes on the shape of the PSF. If the contact hole size on the mask is smaller than this value, the printed image is controlled by the PSF, not by the dimensions of the mask. Making the contact size on the mask smaller only reduces the intensity of the image peak. This results in a very interesting relationship: a change in the mask size of a small contact hole is essentially equivalent to a change in exposure dose.

It is the area of the contact hole that controls the printing of the contact. The electric field amplitude transmittance of a small contact hole is proportional to the area of the contact, which is the CD squared. The intensity is then proportional to the area squared. Thus, the effect of a small change in the mask CD of a contact hole is to change the effective dose reaching the wafer (E) as the CD to the fourth power.

$$E \propto CD_{mask}^4$$
$$d \ln E \approx 4d \ln CD_{mask} \tag{8.70}$$

where the approximate sign is used because the fourth power dependence is only approximately true for typically sized contact holes.

The above equation relates mask errors to dose errors. Combining this with the MEEF Equation (8.64) allows us to relate MEEF to exposure latitude.

$$MEEF \approx 4 \frac{\partial \ln CD_{wafer}}{\partial \ln E} \tag{8.71}$$

The term $\partial \ln CD / \partial \ln E$ can be thought of as the percent change in CD for a 1 % dose change and is the inverse of the common exposure latitude metric. Thus, anything that improves the exposure latitude of a contact hole will also reduce its MEEF. Anything that reduces exposure latitude (like going out of focus) will result in a proportional increase in the MEEF. The importance of exposure latitude as a metric of printability is even greater for contact holes due to the factor of four multiplier for the MEEF. Additionally, the above expression could be used to lump mask errors into effective exposure dose errors for the purpose of process window specifications. Mask errors could be thought of as consuming a portion of the exposure dose budget for a process.

8.7.5 Mask Errors as Effective Dose Errors

The concept of mask errors as effectively exposure dose errors, first discussed above in the context of contact hole MEEF, can be used as a general concept for understanding the nature of the mask error enhancement factor. Consider again the image MEEF based on an aerial image CD. For a given mask feature size w, the image CD will occur at $x = CD/2$ when the intensity equals the image threshold value.

$$I(x = CD/2, w) = I_{th} \tag{8.72}$$

Consider the total differential of the intensity with x and w as variables:[12]

$$dI = \frac{\partial I}{\partial x} dx + \frac{\partial I}{\partial w} dw = 0 \tag{8.73}$$

where the total differential must be zero since the intensity is constrained to be at the threshold intensity. Thus,

$$\frac{dx}{dw} = -\frac{\partial I/\partial w}{\partial I/\partial x} \quad or \quad image\ MEEF = -2 \frac{\partial I/\partial w}{\partial I/\partial x} \tag{8.74}$$

In this expression, the numerator can be thought of as the change in dose reaching the wafer caused by a change in the mask CD (Figure 8.36). The denominator is the image

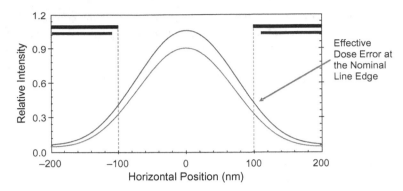

Figure 8.36 *Mask errors can be thought of as creating effective dose errors near the edge of the feature*

slope, which is proportional to exposure latitude. An alternate form of this expression is

$$image\ MEEF = \frac{2}{NILS} \frac{\partial \ln I}{\partial \ln w}\bigg|_{x=CD/2} \tag{8.75}$$

As an example, consider the out-of-focus coherent three-beam image for TE illumination:

$$I(x) = \left[\frac{w}{p} + \frac{2\sin(\pi w/p)}{\pi} \cos(\pi\lambda\delta/p^2)\cos(2\pi x/p) \right]^2 \tag{8.76}$$

(The low NA expression for defocus is used here, but the results will be identical if the more exact expression were used.) Taking the derivative with respect to the mask spacewidth *w*,

$$\frac{\partial \ln I}{\partial \ln w}\bigg|_{x=w/2} = \frac{2w}{p} \frac{[1+2\cos^2(\pi w/p)\cos(\pi\lambda\delta/p^2)]}{\sqrt{I(w/2)}} \tag{8.77}$$

For the case of equal lines and spaces, this derivative becomes a constant (since the edge of the feature is at the image isofocal point):

$$\frac{\partial \ln I}{\partial \ln w}\bigg|_{x=CD/2} = 2 \tag{8.78}$$

Thus, the coherent three-beam equal line/space image MEEF, including defocus, becomes

$$image\ MEEF = \frac{4}{NILS} \tag{8.79}$$

As will be explained in greater detail in Chapter 9, 2/*NILS* is equal to the slope of a log-CD-versus-log-dose curve for the case of an infinite contrast resist. Thus, one can generalize this image MEEF by replacing 2/*NILS* with the actual CD-versus-dose slope.

$$MEEF \approx \left(\frac{\partial \ln CD_{\text{wafer}}}{\partial \ln E} \right) \left(\frac{\partial \ln I}{\partial \ln w} \bigg|_{x=CD/2} \right) \tag{8.80}$$

For the case of a small contact hole, we saw in the previous section that the fractional dose change per fractional mask CD change was 4. For a small isolated space, the value is about 2. For other features, the change in intensity at the nominal line edge can be computed and an estimate of MEEF made using Equation (8.80).

Another way to interpret the above result is to consider that mask errors result in a reduction of the effective exposure latitude. If *EL* is the exposure latitude when no mask errors are present, then the effective exposure latitude in the presence of mask errors of range ΔCD_{mask} will be

$$EL_{\text{eff}} \approx EL \left(1 - MEEF \frac{\Delta CD_{\text{mask}}}{\Delta CD_{\text{spec}}} \right) \tag{8.81}$$

where ΔCD_{spec} is the CD tolerance used to define the exposure latitude (for example, 0.2 times the nominal CD). This expression can be used to scale an entire exposure latitude versus depth of focus plot, such as Figure 8.22, and thus can also be used to determine the effect of mask errors on the available depth of focus.

8.7.6   Resist Impact on MEEF

For a given aerial image response, the response of a real, finite-contrast resist will change the value of the MEEF, sometimes dramatically. Although it is difficult to systematically and controllably vary resist contrast experimentally, simulation provides an effective tool for exploring its impact on MEEF theoretically. Consider a simple baseline process of 248-nm exposure with a 0.6 NA, 0.5 partial coherence imaging tool using 500-nm-thick resist on BARC on silicon. The contrast of the resist is controlled by essentially one simulation parameter, the dissolution selectivity parameter n of the original kinetic dissolution rate model (see Chapter 7). High values of n correspond to high values of resist contrast. For this example, $n = 5$ will be a low-contrast resist, $n = 10$ is a mid-contrast resist and $n = 25$ will be called a high-contrast resist.

The variation of MEEF with nominal mask feature size for these three virtual resists is shown in Figure 8.37 for a mask pattern of equal lines and spaces. For larger feature sizes, the MEEF is near 1.0 for all resists, controlled by the aerial image MEEF. However, as the feature sizes approach the resolution limit, the characteristic skyrocketing MEEF is observed. (In fact, this dramatic increase in MEEF for smaller features can be used as one definition of resolution: the smallest feature size that keeps the MEEF below some critical value, say 3.) As can be seen from the figure, the major impact of resist contrast is to determine at what feature size the MEEF begins its dramatic rise. Above k_1 values of about 0.6, MEEF values are near 1.0 or below, and resist contrast has little impact. Below this value, the MEEF rises rapidly. Resist contrast affects the steepness of this rise.

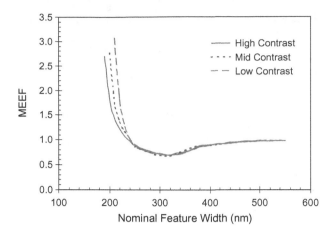

Figure 8.37   *Resist contrast affects the mask error enhancement factor (MEEF) dramatically near the resolution limit*

Obviously, resist contrast has a huge effect on the MEEF for features pushing the resolution limits. At a k_1 value of 0.5 for the example shown here, the MEEF for the high-contrast resist is 1.5, for the mid-contrast resist it is 2.0, and for the low-contrast resist the MEEF has grown to greater than 3.0. As is seen in so many ways, improvements in resist contrast can dramatically improve the ability to control high-resolution linewidths on the wafer, in this case in the presence of mask errors.

8.8   Line-End Shortening

As lithography pushes to smaller and smaller features, single number metrics such as the critical dimension (CD) of a feature are not adequate. The three-dimensional shapes of the final printed photoresist features can, in fact, affect the performance of the final electrical devices in ways that cannot be described by variations in a single width parameter of those features. In such cases, more information about the shape of a photoresist pattern must be measured in order to characterize its quality. One very simple example is known as *line-end shortening* (LES).

Consider a single, isolated line with width near the resolution limit of a lithographic process. Considerable effort is usually required to develop a process that provides adequate CD control over a range of processing errors (focus, exposure, mask errors, etc.). Although such a feature is generally considered to be one-dimensional (with CD, measured perpendicular to the long line, as the only important dimension), it must, by necessity, have a two-dimensional character at the line end. An important question then arises: for a process where control of the linewidth is adequate, will the shape of the line end also behave acceptably? Often, the answer to this question is *no* due to line-end shortening.

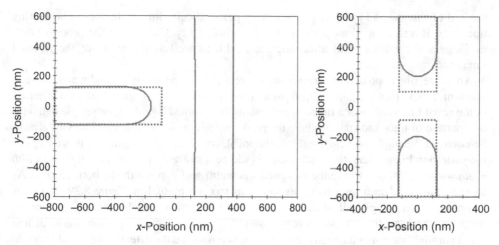

Figure 8.38 *Outline of the printed photoresist pattern (solid) superimposed on an outline of the mask (dashed) shows two examples of line-end shortening ($k_1 = 0.6$)*

Figure 8.38 illustrates the problem. When the process is adjusted to give the correct CD along the length of the line, the result at the line end will be a pull-back of the resist to produce a foreshortened end. The degree of line-end shortening is a strong function of the line width, with effects becoming noticeable for k_1 less than about 0.8. At first glance, it may seem that a solution to this problem is straightforward. If the printed image of a line end is shorter than the drawn pattern on the mask, simply extending the mask by the amount of line-end shortening would solve the problem. Of course, since the degree of LES is feature size dependent, proper characterization would be required. Most commercial optical proximity correction or design rule checking software today can automatically perform such corrections on the design before the photomask is made (see Chapter 10). This solution ignores two very important problems, however. First, what happens when the end of the line is in proximity to another feature, as in Figure 8.38, and second, how does the degree of LES vary with processing errors? For the first problem, a simple extension of the line may not work. Thus, more complicated corrections (such as increasing the width of the line near the end) are required. For the second problem, a more complete characterization of line-end shortening is needed.

8.8.1 Measuring LES

In order to characterize line-end shortening, the first step is to find a way to measure it. Since LES is fundamentally an error of the resist pattern relative to the design, it cannot be independently measured. Instead, it must be measured as the difference between two measurements, such as a measure of the line-end position relative to another feature. A very simple approach is to use a test structure such as those shown in Figure 8.38 where the line-end shortening is considered to be proportional to the width of the gap between the end of the line and the edge of the nearby perpendicular line, or between two butting line ends. But this gap alone does not tell the whole story. Changes in the process (such as focus and exposure) will affect the gap width as well as the width of the isolated line.

An interesting approach to normalizing the relationship between linewidth and line-end shortening is to plot the gap width from a structure like that in Figure 8.38 as a function of the resist linewidth over a range of processing conditions. Based on the simple behavior of a pattern of lines and spaces where the resist linewidth plus the spacewidth will always be equal to the pitch, one can establish the ideal, linear imaging result for this case. For the pattern in Figure 8.38, the ideal result should be a straight line with equation *linewidth + gapwidth* = 500 nm, where the designed linewidth and gap width are both 250 nm. As an example, the data from a focus–exposure matrix are plotted in Figure 8.39 using this technique. Interestingly, all of the data essentially follow a straight line which is offset from the ideal, no line-end shortening result. The vertical offset between the ideal line and a parallel line going through the data can be considered the effective line-end shortening over the range of processing conditions considered. Although not perfect, this result shows that the variables of focus and exposure do not influence the effective LES to first order. The fact that the data form a line which is not exactly parallel to the ideal line simply indicates that, to second order, the LES does not exhibit the same process response to these variables as does the linewidth.

The gap width-versus-linewidth approach to characterizing the effective line-end shortening still ignores the three-dimensional nature of line-end effects. Processing changes, especially focus, can alter the shape as well as the size of a photoresist feature. For the

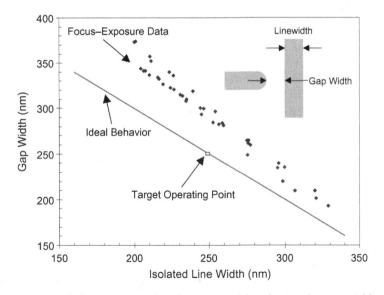

Figure 8.39 *Line-end shortening can be characterized by plotting the gap width of a structure like that in the insert as a function of the isolated linewidth under a variety of conditions. As shown here, changes in focus and exposure produce a linear gap width versus linewidth behavior*

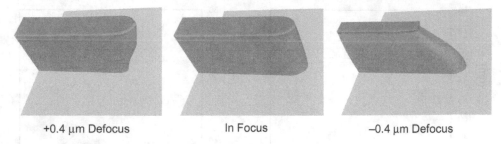

+0.4 µm Defocus   In Focus   −0.4 µm Defocus

Figure 8.40 *Simulated impact of focus on the shape of the end of an isolated line (250-nm line, NA = 0.6, σ = 0.5, λ = 248, positive focus defined as shifting the focal plane up)*

case of a resist line cross section, the sidewall angle of the resist pattern is reduced when out of focus. What is less obvious is that the end of a line is even more sensitive to focus errors than the line itself. Figure 8.40 shows how errors in focus can change the three-dimensional shape of a line end. Obviously, any metrology designed to measure line-end shortening will almost certainly be affected by the shape changes depicted in Figure 8.40.

8.8.2 Characterizing LES Process Effects

Any line whose width is near the resolution limit of a lithographic process will usually exhibit significant line-end shortening. The end of the line is made up of two 90° corners. But as these corners pass through an imaging process of limited resolution, the corners will, by necessity, round. If the feature width is less than the sum of the two corner-rounding radii, the end of the line will pull back due to this corner rounding. There are actually three major causes of this corner rounding, all of which contribute to the final LES magnitude.

The primary corner rounding, and thus line-end shortening, mechanism is the diffraction limitation of aerial image formation. To first order, the aerial image corner-rounding radius is about equal to the radius of the point spread function (about $0.35\lambda/NA$). Thus, significant line-end shortening is expected for feature sizes smaller than about $k_1 = 0.7$. As an example, a 180-nm isolated line imaged at 248 nm with a numerical aperture (NA) of 0.688 (giving a scaled feature size of $k_1 = 0.5$) and with a partial coherence of 0.5 will produce an aerial image with nearly 50 nm of LES. The amount of corner rounding is a strong function of k_1. Figure 8.41a shows the effect of NA on the line-end shortening of the aerial image of a 180-nm-sized structure like that of Figure 8.38. Obviously, a higher NA (just like a lower wavelength or larger feature size) produces less LES. The influence of partial coherence (Figure 8.41b) is less dramatic but still significant.

LES is also influenced by corner rounding on the reticle. Mask making involves lithography processes that, due to their inherent resolution limits, inevitably produce rounded corners. At the end of a line, rounded corners on the mask reduce light intensity transmitted through the mask in a manner nearly identical to a foreshortening of the mask line. Equating the area lost due to mask corner rounding of radius R (wafer dimensions) with

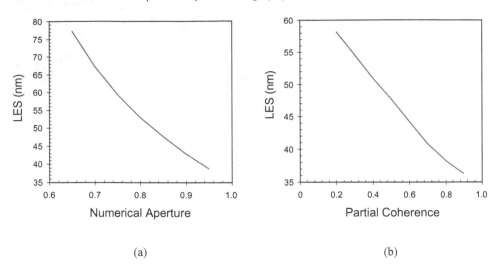

Figure 8.41   *Response of line-end shortening (LES) to imaging parameters (130-nm isolated line, λ = 248 nm): (a) numerical aperture (σ = 0.5), and (b) partial coherence (NA = 0.85)*

an equivalent area due to a foreshortened mask, an approximate LES due to mask corner rounding can be estimated, for a space of width w, as

$$LES_{\text{mask}} \approx \frac{2R^2\left(1 - \dfrac{\pi}{4}\right)}{w} \approx \frac{0.43R^2}{w} \tag{8.82}$$

A similar result is expected for lines. Thus, mask corner rounding leads to an effectively shorter mask line, which of course will lead directly to a shorter line printed in resist. As an example, consider a mask-making process with a corner-rounding radius of 20 nm (wafer dimensions). For an 80-nm feature, the amount of LES due to the mask corner rounding can be estimated at 2 nm. If the mask corner rounding is increased to 40 nm, the impact of the mask on LES grows to 9 nm. If the maximum allowed LES due to mask corner rounding is set to be a fixed fraction of the minimum feature size, Equation (8.82) shows that the maximum allowed mask corner rounding must also be a fixed fraction of the minimum feature size.

The resist also affects the final degree of line-end shortening. Interestingly, development contrast seems to have only a small influence on the amount of LES of the final resist patterns. Diffusion during PEB, on the other hand, has a very significant effect. Figure 8.42 shows the simulated result of increasing diffusion length (for an idealized conventional resist and for a chemically amplified resist, both with constant diffusivity) on the line-end shortening of a completely isolated line. The diffusion length must be kept fairly small in order for diffusion to only marginally impact the LES. In both cases, the increase in line-end shortening goes as about the diffusion length squared. The reason for the large effect of diffusion on line-end shortening is the three-dimensional nature of

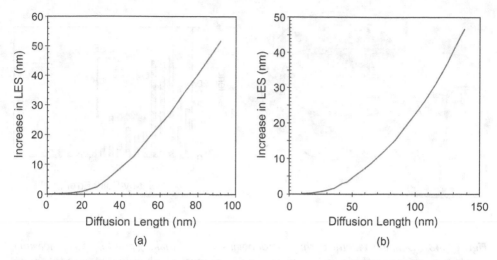

Figure 8.42 *Diffusion can have a dramatic effect on line-end shortening (LES) of an isolated line: (a) 180-nm line, λ = 248 nm, NA = 0.688, σ = 0.5, conventional resist, and (b) 130-nm line, λ = 248 nm, NA = 0.85, σ = 0.5, chemically amplified resist*

the diffusion process: at the end of an isolated line, chemical species from the exposed areas can diffuse from all three sides of the line end (as opposed to one direction for a line edge).

8.9   Critical Shape and Edge Placement Errors

A discussion of CD control assumes that a convenient and accurate dimensional measure of the feature in question is available. For long lines and spaces, and symmetric contact holes, a simple, single metric can be used to describe the difference between the actual result on the wafer and the desired result: the critical dimension error (CDE). The optimum process is that which drives the CDE to zero. But real integrated circuit patterns include numerous complex shapes that defy easy one-dimensional characterization. How does one judge the error in the shape of an arbitrary 2D pattern? For this we must define a *critical shape error* (CSE), the 2D analog to the critical dimension error.

The critical shape error is determined by finding the point-by-point difference between the actual printed resist shape and the desired shape.[13] For example, Figure 8.43a takes a desired shape (equal to the layout design with an acceptable amount of corner rounding added) and compares it to a top-down image of the printed resist shape. The desired shape is broken down into a collection of measurement points (in this case, equally spaced points are placed around the entire desired shape). Then vectors are drawn from these measurement points to the actual wafer shape, collecting their lengths in a frequency distribution of errors as shown in Figure 8.43b. Once such a distribution is determined, some characterization of the distribution can be used to describe the overall shape error. For example, the average error could be used (CSE_{avg}) or the error which is greater than 90% of the

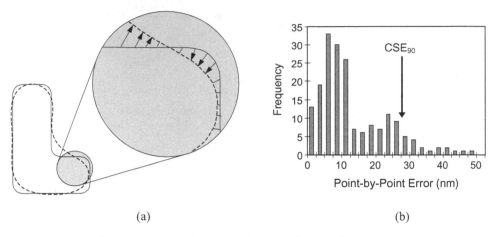

(a)   (b)

Figure 8.43   *Defining a metric of shape error begins with (a) making point-by-point measurements comparing actual (dashed) to desired (solid) shapes, which produces (b) a frequency distribution of errors (unsigned lengths of the vectors are used here). One possible critical shape error (the CSE_{90}) is shown as an example of the analysis of this distribution*

point-by-point measurements (CSE_{90}), or some other percentage could also be used. For the distribution in Figure 8.43, some results are given below:

- $CSE_{avg} = 13.5\,\text{nm}$
- $CSE_{80} = 23\,\text{nm}$
- $CSE_{90} = 28\,\text{nm}$
- $CSE_{95} = 32\,\text{nm}$
- $CSE_{99.7} = 46\,\text{nm}$

Several different definitions of the point-by-point error are possible. In the example above, the unsigned length of the vector is used to generate the error histogram. The signed vector could also be used (a mean of zero for the signed vectors would be an indication of a properly sized feature). An alternate approach, useful for features on a 'Manhattan' layout, constrains the error vectors to be wholly vertical or horizontal. These Manhattan error vectors are typically called *edge placement errors* (EPE) and are commonly used by optical proximity correction software when characterizing pattern fidelity (see Chapter 10).

8.10 Pattern Collapse

Photoresist profile control has traditionally been defined by making a trapezoidal model of the shape of the resist profile and defining the width, height and sidewall angle of the trapezoid as parameters to be controlled. In general, this approach works well for 1D patterns (long lines and spaces). However, under some circumstances, a different problem can plague the control of profile shape – *pattern collapse*, where seemingly well-formed resist profiles fail mechanically and fall over (Figure 8.44).

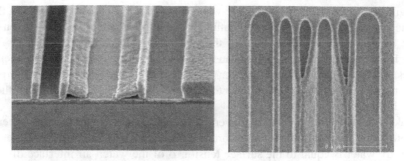

Figure 8.44 *Examples of photoresist pattern collapse, both in cross section (left) and top down* (Courtesy of Joe Ebihara of Canon)

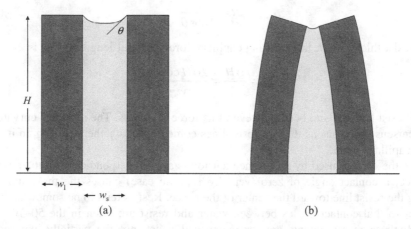

Figure 8.45 *Pattern collapse of a pair of isolated lines: (a) drying after rinse leaves water between the lines, and (b) capillary forces caused by the surface tension of water pull the tops of the lines toward each other, leading to collapse*

Pattern collapse occurs for tall, narrow resist lines when some force pushing against the top of the line causes the profile to bend and eventually break or peal off the substrate. As one might expect, the taller and narrower the line, the easier it would be to push it over. But where does this pushing force come from? After development, the wafer is rinsed with deionized water and then air-dried. As the water dries, surface tension from the receding water will pull on the top of the resist feature. Of course, if the water is symmetrically receding from both sides of the line, the two forces on each side will cancel. If, however, there is an asymmetry and the water from one side of the line dries off faster than the other side, the resultant force will not be zero.

Figure 8.45a shows what is probably the worst case for generating asymmetric surface tension forces during after-rinse drying: two isolated parallel lines with a small gap between them. The large open areas to each side of the pair quickly dry while the small space between them requires a much longer time for removal of the water. As a result, there will be a point in the drying cycle where the top meniscus of the water in the gap will line up with the top of the resist features, producing a capillary force, driven by the surface tension of water, that will pull the two resist lines toward each other (Figure 8.45b). If the force is great enough to cause the two resist tops to touch each other, these features are said to have collapsed. Other patterns of lines and spaces (a five-bar pattern, for example) will behave very similarly to this two-bar test pattern for at least the most outside lines.

Tanaka[14] developed a simple cantilever-beam mechanical model for this pattern collapse situation. The capillary pressure (force per unit area) along the side of the resist line covered by water is equal to the surface tension σ of the water–air interface divided by the radius of curvature R of the meniscus. The radius of curvature in turn is determined by the contact angle θ of the water–resist interface and the spacewidth w_s. Before any bending occurs, this curvature will be

$$R = \frac{w_s}{2\cos\theta} \tag{8.83}$$

For a resist thickness H, the resulting capillary force per unit length of line is

$$F = \frac{\sigma H}{R} = \frac{2\sigma H \cos\theta}{w_s} \tag{8.84}$$

As the resist line starts to bend, however, this force increases. The radius of curvature of the meniscus decreases as the two resist lines come closer together, causing an increase in the capillary force.

Since the force caused by the surface tension is always perpendicular to the air–water interface, a contact angle of zero (very hydrophilic case) causes the maximum force pulling the resist line toward the center of the space. Resists tend to be somewhat hydrophobic, so that contact angles between water and resist are often in the 50–70° range (though these measurements are for unexposed resist, not the partially exposed and deblocked resist along the profile edge where the contact angle should be less). Note that the capillary force increases with the aspect ratio of the space ($A_s = H/w_s$) and with the surface tension of water, which at room temperature is about 0.072 N/m (72 dyne/cm).

The bending of the resist line can be described as an elastic cantilever beam. The important resist material property of mechanical strength is the Young's modulus, E, which is a measure of the stiffness of the resist and is higher for resists with high glass transition temperatures. Resists have been measured to have Young's modulus values in the range of 2–6 GPa with the high end of the range corresponding to novolac resists and the low end of the range for ArF resists.[15] Applying a force F to a line of width w_l causes that line to move (sway) into the space by an amount δ given by

$$\delta = \frac{3}{2}\left(\frac{F}{E}\right)\left(\frac{H}{w_l}\right)^3 \tag{8.85}$$

Note that the amount of bending is proportional to the cube of the aspect ratio of the resist line, and thus will be very sensitive to this aspect ratio.

As the resist line bends, the pulling of the line by the capillary force is countered by a restoring force caused by the stiffness of the resist. Eventually a 'tipping point' is reached where the increasing capillary force exceeds the restoring force of the resist and the line collapses. Tanaka calculated this critical point to occur when

$$\frac{E}{\sigma} \le \frac{4}{w_s} A_i^3 \Big[3A_s \cos\theta + \sin\theta + \sqrt{9A_s^2 \cos^2\theta + 6A_s \cos\theta\sin\theta} \, \Big] \tag{8.86}$$

where A_s is the aspect ratio of the space and A_l is the aspect ratio of the line (H/w_l). If the contact angle is less than about 80° and the aspect ratio of the space is high (that is, we are in an interesting regime where pattern collapse is likely to be a problem), the square root can be approximated with the first terms of a Taylor series to give a somewhat more approachable result:

$$\frac{E}{\sigma} \le \frac{8}{w_s} A_i^3 [3A_s \cos\theta + \sin\theta] \tag{8.87}$$

Equation (8.87) shows what affects pattern collapse and therefore what can be done to try to reduce pattern collapse. Resist chemists could make the resist stiffer by increasing the Young's modulus E (similar to increasing T_g, the glass transition temperature), something that is hard to do without altering other very important resist properties. The water–air surface tension can be reduced by adding surfactant into the rinse liquid. Surfactants can easily reduce the surface tension by a factor of 2 or 3, but may pose contamination problems and can also soften the resist, reducing its Young's modulus. Contact angle of water to the resist also influences collapse. The worst case (maximum value of $3A_s\cos\theta + \sin\theta$) occurs at an angle of $\tan^{-1}(1/3A_s)$ which is generally less than 10°. Thus, hydrophilic resists produce greater pattern collapse. Making the resist more hydrophobic will help (this, incidentally, is also a goal for immersion lithography). The aspect ratio of the space is important, but is less so for hydrophobic resists (small $\cos\theta$). The space width has a direct impact on collapse, with smaller spaces increasing the capillary force and thus the likelihood of collapse. By far the most critical factor is the aspect ratio of the line. Since the tendency to collapse increases as the line aspect ratio cubed, small changes in this factor can have big consequences.

Putting some numbers into the equation, ArF resists will have E/σ in the range of about 35 nm^{-1} assuming no surfactants in the water. Assuming a 60° water–resist contact angle, consider a 100-nm space with a space aspect ratio of 3. The maximum aspect ratio of the line will be 4.3 (a 70-nm linewidth before the line collapses). If the space were shrunk to 50 nm keeping the space aspect ratio at 3, the maximum line aspect ratio drops to 3.4 (a 44-nm linewidth). As another example, when printing a pair of 45-nm equal lines and spaces in a resist of thickness 141 nm, overexposure will cause the patterns to collapse before they reach a 10% CD error. Pattern collapse can become a process window limiter before CD, sidewall angle, or resist loss reach their specifications.

Equation (8.87) can also predict the tendency toward pattern collapse as a function of line/space duty cycle. To simplify, assume that the contact angle is small enough that ignoring the $\sin\theta$ term in Equation (8.87) produces very little error (this approximation is not so bad, even for a 60° contact angle). For this case, a given pitch equal to $w_s + w_l$ will have the minimum tendency to collapse when the linewidth is 50% bigger than the spacewidth (a 3:2 linewidth to spacewidth ratio). All other duty cycles will have an increased chance of pattern collapse. For a fixed duty cycle, the likelihood of collapse goes as the resist thickness to the fourth power, and as one over the pitch to the third power.

The two-dimensional model used here is really a worst case since it assumes very long lines and spaces. For shorter lines that are connected to another resist feature at one or both ends, the other features can give support to the line, inhibiting the twisting that would be required to make the line collapse. However, it is likely that CMOS device levels like poly and metal 1 will always have some patterns that will behave close to the worst case presented here. And the trend toward smaller feature sizes and constant or even greater aspect ratios will only make the problem of pattern collapse worse in the future.

Problems

8.1. A given 90-nm lithography process level exhibits a 0.1-ppm (parts per million) wafer magnification error due to wafer expansion. For a 300-mm wafer, what is the maximum overlay error due to this term if it goes uncorrected? Is the magnitude of this error significant (assume the overlay specifications is 1/3 of the technology node feature size)?

8.2. Silicon has a coefficient of linear expansion of about 3×10^{-6}/K. Assume that 3 nm of overlay error due to lack of wafer temperature control is acceptable. How well must the temperature of the wafer be controlled for a 200-mm wafer? For a 300-mm wafer?

8.3. What are the maximum x- and y-overlay errors due to a 0.3 arc-second reticle rotation error for a 24×28 mm field?

8.4. It is possible that the overlay targets, when printed on the wafer, may be asymmetric, as shown below for a frame-in-frame target. What will be the impact of such asymmetry on overlay measurement?

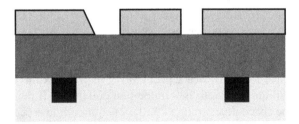

8.5. Derive Equations (8.42) and (8.43).

8.6. Equations (8.31) and (8.35) both predict CD versus dose, with the later equation a bit more accurate. Plot CD versus dose for these two models on one graph for the case of

$$CD_1 = 80\,\text{nm}, \quad E_1 = 32\,\text{mJ/cm}^2, \quad \text{and} \quad \left.\frac{d\ln CD}{d\ln E}\right|_{E=E_1} = 0.6$$

At what dose values do the two models disagree with each other in predicted CD by 2%?

8.7. For the Bossung plot given below,
   (a) Approximately where is best focus? Why?
   (b) What is the approximate isofocal bias (numerical value) for a target CD of 180 nm?

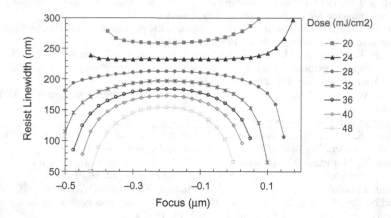

8.8. Using Equations (8.50)–(8.52), plot the depth of focus of a 65-nm line as a function of isofocal bias: let $CD_{spec} = 0.2$, $El_{spec} = 0.08$, $s_{nom} = 0.75$, and $\Delta = 0.2\,\mu m$.

8.9. Use the general expression (8.75) to derive the image MEEF for a coherent equal line/space image that uses only the 0th and ±1st diffraction orders [that is, derive Equation (8.68)].

8.10. For the two-line pattern collapse problem, what linewidth-to-spacewidth ratio (for a fixed pitch) minimizes the tendency for pattern collapse under the following conditions:
   (a) Assume that the water–resist contact angle is small enough that the $\sin\theta$ term in Equation (8.87) can be ignored.
   (b) Assume that the water–resist contact angle is large enough that the $\cos\theta$ term in Equation (8.87) can be ignored.

References

1 Sturtevant, J., Allgair, J., Fu, C., Green, K., Hershey, R., Kling, M., Litt, L., Lucas, K., Roman, B., Seligman, G. and Schippers, M., 1999, Characterization of CD control for sub-0.18 μm lithographic patterning, *Proceedings of SPIE: Optical Microlithography XII*, **3679**, 220–227.

2 Mack, C.A., 1992, Understanding focus effects in submicron optical lithography, part 3: methods for depth-of-focus improvement, *Proceedings of SPIE: Optical/Laser Microlithography V*, **1674**, 272–284.

3 Sethi, S., Barrick, M., Massey, J., Froelich, C., Weilemann, M. and Garza, F., 1995, Lithography strategy for printing 0.35 um devices, *Proceedings of SPIE: Optical/Laser Microlithography VIII*, **2440**, 619–632.

4 SEMI Standard SEMI P35-1106, 2006, *Terminology for Microlithography Metrology*, Semiconductor Equipment and Materials International, San Jose, CA.

5 Levinson, H.J., 1999, *Lithography Process Control*, SPIE Press Vol. TT28, Bellingham, CA.

6 Ausschnitt, C., 1988, Chapter 8: electrical measurements for characterizing lithography, in *VLSI Electronics: Microstructure Science, Vol. 16*, Academic Press, p. 330.

7 DeMoor, S., Brown, J., Robinson, J.C., Chang, S. and Tan, C., 2004, Scanner overlay mix and match matrix generation; capturing all sources of variation, *Proceedings of SPIE: Metrology, Inspection, and Process Control for Microlithography XVIII*, **5375**, 66–77.

8 Bossung, J.W., 1977, Projection printing characterization, *Proceedings of SPIE: Developments in Semiconductor Microlithography II*, **100**, 80–84.

9 Ausschnitt, C.P., Rapid optimization of the lithographic process window, *Proceedings of SPIE: Optical/Laser Microlithography II*, **1088**, 115–133.

10 Brunner, T., Martin, A., Martino, R., Ausschnitt, C., Newman, T. and Hibbs, M., 1994, Quantitative stepper metrology using the focus monitor test mask, *Proceedings of SPIE: Optical/Laser Microlithography VII*, **2197**, 541–549.

11 Maurer, W., 1996, Mask specifications for 193-nm lithography, *Proceedings of SPIE*, **2884**, 562–571.

12 The derivation was first suggested to me by Mark D. Smith.

13 Tsudaka, K., *et al.*, 1995, Practical optical mask proximity effect correction adopting process latitude consideration, *MicroProcess '95 Digest of Papers*, 140–141.

14 Tanaka, T., Morigami, M. and Atoda, N., 1993, Mechanism of resist pattern collapse during development process, *Japanese Journal of Applied Physics*, **32**, 6059–6064.

15 Simons, J., Goldfarb, D., Angelopoulos, M., Messick, S., Moreau, W., Robinson, C., de Pablo, J. and Nealey, P., 2001, Image collapse issues in photoresist, *Proceedings of SPIE: Advances in Resist Technology and Processing XVIII*, **4345**, 19–29.

9

Gradient-Based Lithographic Optimization: Using the Normalized Image Log-Slope

There are many approaches for optimizing a lithographic process. Very commonly, the focus–exposure matrix is used to measure depth of focus (which includes a sensitivity to both dose and focus errors) and the process is optimized to maximize the depth of focus for select critical features. Such experimental optimization efforts are most effective and efficient when guided by appropriate theory. Here, the gradient-based approach to lithographic optimization, using the well-known image log-slope, is presented. By calculating and then optimizing the gradients of the image that is projected and recorded into a photoresist through focus, a clear direction for maximizing depth of focus is seen. This approach can also be thought of as a first-order propagation of errors analysis – a Taylor series approach to how input variations affect the output of CD. Small errors in exposure dose affect CD depending on the focus-dependent slope of the image at the nominal line edge.

9.1 Lithography as Information Transfer

This chapter will focus on CD control using exposure latitude and the variation of exposure latitude through focus as the basic metrics of CD control. The approach taken here will view lithography as a sequence of information transfer steps (Figure 9.1). A designer lays out a desired pattern in the form of simple polygon shapes. This layout data drive a mask writer so that the information of the layout becomes a spatial variation of transmittance (chrome and glass, for example) of the photomask. The information of the layout has been transferred, though not perfectly, into the transmittance distribution of the mask. Next, the mask is used in a projection imaging tool to create an aerial image of the mask.

Fundamental Principles of Optical Lithography: The Science of Microfabrication, Chris Mack.
© 2007 John Wiley & Sons, Ltd.

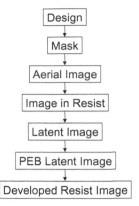

Figure 9.1   *The lithography process expressed as a sequence of information transfer steps*

However, due to the diffraction limitations of the wavelength and lens numerical aperture, the information transmitted to the wafer is reduced (see Chapter 2). The aerial image is an imperfect representation of the information on the mask. Aberrations, flare and other nonidealities degrade the image quality further (see Chapter 3). This aerial image is projected into the resist where the variation of transmittance with angle, resist refraction and substrate reflection makes the image in resist different than the aerial image (see Chapter 4). Exposure then produces a latent image of exposed and unexposed sensitizer concentration, which now contains some portion of the information originally stored as an aerial image (see Chapter 5). Post-exposure bake changes this latent image into a new latent image through diffusion (for a conventional resist) or reaction and diffusion (for a chemically amplified resist), with the result being a loss of information in the latent image (see Chapter 6). Finally, development convents the latent image into an image of development rates, which then is transformed into the final resist profile via the development path (see Chapter 7).

At each stage of this sequence, there is potential for information loss. The final developed resist image is not a perfect representation of the original design data. By defining metrics of information content at each stage of this sequence of information transfer steps, a systematic approach toward maximizing the information contained in the final image can be found. The information contained in each image propagates through the lithography process, from the formation of an aerial image $I(x)$ and image in resist, to the exposure of a photoresist by that image to form a latent image of chemical species $m(x)$ or $h(x)$, to post-exposure bake where diffusion and possibly reactions create a new latent image $m^*(x)$, and finally to development where the latent image produces a development rate image $r(x)$ that results in the definition of the feature edge (a one-dimensional example is used throughout this chapter for simplicity).

9.2   Aerial Image

Projection imaging tools, such as scanners, steppers, or step-and-scan tools, project an image of a mask pattern into air, and then ultimately into the photoresist. The projected

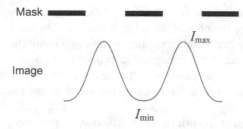

Figure 9.2 *Image contrast is the conventional metric of image quality used in photography and other imaging applications, but is not directly related to lithographic quality*

image in air is called the *aerial image*, a distribution of light intensity as a function of spatial position within (or near) the image plane. The aerial image is the source of the information that is exposed into the resist, forming a gradient in dissolution rates that enables the three-dimensional resist image to appear during development. The quality of the aerial image dictates the amount of information provided to the resist, and subsequently the quality and controllability of the final resist profile.

How do we judge the quality of an aerial image? If, for example, aerial images are known for two different values of the partial coherence, how do we objectively judge which is better? Historically, the problem of image evaluation has long been addressed for applications such as photography. The classical metric of image quality is the *image contrast* (Figure 9.2). Given a mask pattern of equal lines and spaces, the image contrast is defined by first determining the maximum light intensity (in the center of the image of the space) and the minimum light intensity (in the center of the line) and calculating the contrast as

$$Image\ Contrast = \frac{I_{max} - I_{min}}{I_{max} + I_{min}} \tag{9.1}$$

Since the goal is to create a clearly discernible bright/dark pattern, ideally I_{min} should be much smaller than I_{max}, giving a contrast approaching 1.0 for a high-quality ('high-contrast') image.

Although this metric of image quality is clear and intuitive, it suffers from some problems when applied to lithographic images. First of all, the metric is only strictly defined for equal lines and spaces. Although it is possible to modify the definition of image contrast to apply, for example, to an isolated line or to a contact hole, it is not clear that these modified definitions are useful or comparable to each other. It is clearly impossible to apply this metric to an isolated edge. Secondly, the image contrast is only useful for patterns near the resolution limit. For large features, the image contrast is essentially 1.0, regardless of the image quality. Finally, and most importantly, the image contrast is not directly related to practical metrics of lithographic quality, such as resist linewidth control.

Fundamentally, the image contrast metric samples the aerial image at the wrong place. The center of the space and the center of the line are not the important regions of the image to worry about. What is important is the shape of the image near the nominal line edge. The edge between bright and dark determines the position of the resulting photoresist edge. This transition from bright to dark within the image is the source of the information as to where the photoresist edge should be. The steeper the intensity transition, the better the edge definition of the image, and as a result the better the edge definition of the resist pattern. If the lithographic property of concern is the control of the photoresist linewidth (i.e. the position of the resist edges), then the image metric that affects this lithographic result is the slope of the aerial image intensity near the desired photoresist edge.

The slope of the image intensity I as of function of position x (dI/dx) measures the steepness of the image in the transition from bright to dark. However, to be useful it must be properly normalized. For example, if one simply doubles the intensity of the light, the slope will also double, but the image quality will not be improved. (This is because intensity is a reciprocal part of exposure dose, which is adjusted to correctly set the final feature size in photoresist.) Dividing the slope by the intensity will normalize out this effect. The resulting metric is called the image log-slope (ILS):

$$ILS = \frac{1}{I}\frac{dI}{dx} = \frac{d\ln(I)}{dx} \tag{9.2}$$

where this log-slope is measured at the nominal (desired) line edge (Figure 9.3). Note that a relative change in image intensity is the same as a relative change in exposure dose, so that the ILS is also a relative exposure energy gradient. Since variations in the photoresist edge positions (linewidths) are typically expressed as a percentage of the nominal linewidth, the position coordinate x can also be normalized by multiplying the log-slope by the nominal linewidth w, to give the normalized image log-slope (NILS).

$$NILS = w\frac{d\ln(I)}{dx} \tag{9.3}$$

The NILS is the best single metric to judge the lithographic usefulness of an aerial image.[1-3]

Since the NILS is a measure of image quality, it can be used to investigate how optical parameters affect image quality. One of the most obvious examples is defocus, where the NILS decreases as the image goes out of focus. Figure 9.4a shows the aerial image of a space at best focus, and at two levels of defocus. The 'blurred' images obviously have a lower ILS at the nominal line edge compared to the in-focus image. By plotting the log-slope or the NILS as a function of defocus, one can quantify the degradation in aerial

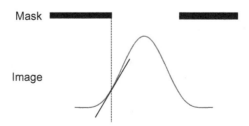

Figure 9.3 *Image Log-Slope (or the Normalized Image Log-Slope, NILS) is the best single metric of image quality for lithographic applications*

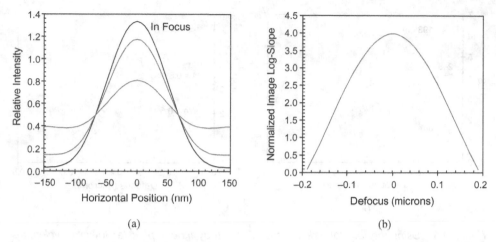

Figure 9.4 *The effect of defocus is to (a) 'blur' an aerial image, resulting in (b) reduced log-slope as the image goes out of focus (150-nm space on a 300-nm pitch, NA = 0.93, λ = 193 nm)*

image quality as a function of defocus (Figure 9.4b). This log-slope defocus curve provides a very important tool for understanding how focus affects a lithographic process. For example, suppose one assumes that there is a minimum acceptable NILS value, below which the aerial image is not good enough to provide adequate resist images or linewidth control. In Figure 9.4b, for example, a minimum acceptable NILS value of 1.5 would mean that this imaging process can tolerate about $\pm 0.14\,\mu$m of defocus and still get aerial images of acceptable quality. Thus, an estimate of the minimum acceptable NILS can lead to an estimate of the depth of focus. Note that, in general, the focus response of an aerial image is symmetric about best focus. When this is the case, the log-slope defocus curve is usually plotted only for positive values of defocus.

To see how the log-slope defocus curve can be used to understand imaging, consider the effects of wavelength and numerical aperture on the focus behavior of an aerial image. Figure 9.5a shows how the NILS of a 0.25-μm line/space pattern degrades with defocus for three different wavelengths (365, 248 and 193 nm), all other lithographic parameters held constant. It is clear from the plot that the lower wavelength provides better image quality for the useful range of defocus. For a given minimum acceptable value of NILS, the lower wavelength will allow acceptable performance over a wider range of focus. One could conclude that, for a given feature being imaged, a shorter wavelength provides better in-focus performance and better depth of focus.

The impact of numerical aperture (NA) is a bit more complicated, as evidenced in Figure 9.5b. Here, the log-slope defocus curves for three different numerical apertures (again, for a 0.25-μm line/space pattern) cross each other. If one picks some minimum acceptable NILS value, there will be an optimum NA which gives the maximum depth of focus (for example, a minimum NILS value of 2.5 has the best depth of focus when $NA = 0.6$). Using a numerical aperture above or below this optimum reduces the depth of focus.

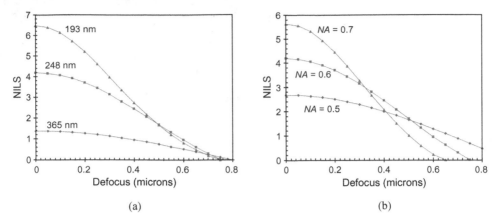

Figure 9.5   *Using the log-slope defocus curve to study lithography: (a) lower wavelengths give better depth of focus (NA = 0.6, σ = 0.5, 250-nm lines and spaces), and (b) there is an optimum NA for maximizing depth of focus (λ = 248 nm, σ = 0.5, 250-nm lines and spaces)*

NILS values are easy and fast to calculate and provide a simple yet valuable metric of image quality. As an example of using this metric, the log-slope defocus curve is one of the easiest ways to quantify the impact of defocus on image quality. By using this tool, we have quickly arrived at two fundamental imaging relationships: when imaging a given mask pattern, (1) lower wavelengths give better depth of focus, and (2) there is an optimum NA that maximizes the depth of focus. But to make the best use of the NILS as an image metric, one must relate the NILS numerical value to lithographically measurable quantities. How does one determine the minimum acceptable NILS? If the NILS is increased from 2.0 to 2.5, what is the lithographic impact? More fundamentally, why is NILS a good image metric?

The answers to these questions lie with the fact that NILS is directly related to the printed feature's exposure latitude (as will be shown in subsequent sections of this chapter). Exposure latitude describes the change in exposure dose that results in a given change in linewidth. Mathematically, it can be expressed as one over the slope of a critical dimension (CD) versus exposure dose (E) curve, $\partial CD/\partial E$. Since both CD and dose variations are most usefully expressed as a percentage of their nominal values, the more useful expression of exposure latitude is $\partial \ln CD/\partial \ln E$. For the simplifying case of a perfect, infinite contrast photoresist the exposure latitude can be related to NILS by

$$\frac{\partial \ln E}{\partial \ln CD} = \frac{1}{2} NILS \qquad (9.4)$$

To put this in more familiar terms, if we define exposure latitude to be the range of exposure, as a percentage of the nominal exposure dose, that keeps the resulting feature width within ±10 % of the nominal size, then for an infinite contrast resist the exposure latitude can be approximately related to NILS by

$$\% Exposure\ Latitude \approx 10 * NILS \qquad (9.5)$$

(the approximation comes from the assumption that NILS is constant over the ±10 % CD range). Thus, in a perfect world (i.e. a perfect photoresist), the impact of NILS can be easily related to a lithographically useful metric: each unit increase in NILS gives us 10 % more exposure latitude. Unfortunately, the real world is not so perfect and infinite contrast photoresists have yet to enter the commercial market. The real impact of NILS on exposure latitude is somewhat reduced from the above ideal. In general, Equation (9.5) can be modified to account for the nonideal nature of photoresists to become

$$\%Exposure\ Latitude \approx \alpha(NILS - \beta) \qquad (9.6)$$

where α and β are empirically determined constants. β can be interpreted as the minimum NILS required to get an acceptable image in photoresist to appear. α then is the added exposure latitude for each unit increase in NILS above the lower limit β. By the end of this chapter, a more accurate and physically meaningful relationship between NILS and exposure latitude will be derived.

The values of α and β can be determined by comparing a calculated NILS-versus-defocus curve to experimentally measured exposure latitude-versus-defocus data. Figure 9.6 shows a simulation of such an experiment for a typical case. Once calibrated, a minimum acceptable exposure latitude specification (say, 10 %) can be translated directly into a minimum acceptable NILS value (in this case, 1.7). Since α and β are resist and process dependent, the minimum acceptable NILS must be also. And of course, the requirements for the minimum acceptable exposure latitude will impact the required NILS directly. Thus, either using Equation (9.5) for the ideal case, or Equation (9.6) for a calibrated resist case, a quantitative valuation of the importance of NILS can readily be made.

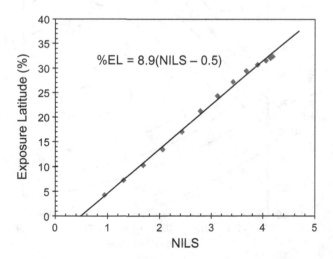

Figure 9.6 *Typical correlation between NILS and exposure latitude (simulated data, λ = 248 nm, NA = 0.6, σ = 0.5, 500 nm of UV6 on ARC on silicon, printing 250-nm lines and spaces through focus)*

The NILS is a generic metric of aerial image quality. Thus, it can be used to optimize any parameter that affects image quality. Some typical use cases include:

- Optimizing numerical aperture (NA) and partial coherence (σ)
- Investigating off-axis illumination parameters
- Comparing phase-shifting mask designs
- Understanding the impact of optical proximity correction (OPC) on image quality and optimizing OPC
- Examining the impact of aberrations, polarization and flare.

As an example, consider the first use case, optimizing NA and σ (the so-called optimum stepper problem). Most steppers and scanners made today have software-controlled variable numerical aperture and variable partial coherence. For a given feature or features, the goal is to pick the settings that maximize lithographic quality. Using the NILS as a proxy for lithographic quality, there are two basic optimization approaches: (1) for a given level of defocus, find the settings that produce the maximum NILS, or (2) for a given minimum acceptable NILS (that is, a given minimum allowed exposure latitude), find the settings that maximize depth of focus.

Figure 9.7 shows an example where the defocus is set to equal the built-in focus errors of the process (0.2 μm in this case), and contours of constant NILS are plotted as a function of NA and σ ($\lambda = 248$ nm printing 130-nm lines on a 360-nm pitch). The maximum NILS of 1.24 is achieved when $NA = 0.72$ and $\sigma = 0.72$. This optimum is sensitive to the amount of defocus used. If the defocus requirement can be lowered to 0.16 μm, the optimum stepper settings for this problem become $NA = 0.76$ and $\sigma = 0.64$ producing a maximum NILS of 1.43.

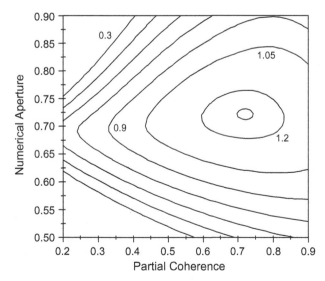

Figure 9.7 *One approach to the optimum stepper problem is to pick a fixed amount of defocus (0.2 μm) and find the settings that maximize the NILS ($\lambda = 248$ nm, 130-nm lines on a 360-nm pitch, contours of constant NILS)*

9.3   Image in Resist

As described in Chapter 4, the image in resist will differ from the aerial image due to the different efficiencies of coupling plane wave energy into the resist as a function of angle and polarization. For a reflective substrate, each incident angle will produce a different reflectivity swing curve, so that for a given resist thickness each incident angle will have a different transmittance into the photoresist. For a nonreflective substrate, the transmittance of the air–resist interface creates an angular and polarization dependence. Figure 9.8 illustrates these two cases.

The electric field aerial image can be thought of as the sum of planes waves, each with magnitude a_n, combining to form the image. For the case of TE-polarization,

$$E(x,z) = \sum_{n=-N}^{N} a_n e^{ikx\sin\theta_n} e^{ikz\cos\theta_n} \tag{9.7}$$

where z is the distance from the plane of best focus, θ_n represents the angle of the nth diffraction order with respect to the optical axis, and $\pm N$ diffraction orders are used to form the image. The image in resist can be calculated in the same way, but with the magnitude of each plane wave modified by the influence of the photoresist (see Chapter 4):

$$a_{nr} \equiv a_n \frac{\tau_{12}(\theta_n)}{1 + \rho_{12}(\theta_n)\rho_{23}(\theta_n)\tau_D^2(\theta_n)} e^{-\alpha_{eff} z/2} \tag{9.8}$$

where θ_n now represents the angle of the nth diffraction order in resist, α_{eff} is the effective absorption coefficient, as defined in Chapter 4, and z is the depth into the resist. For the

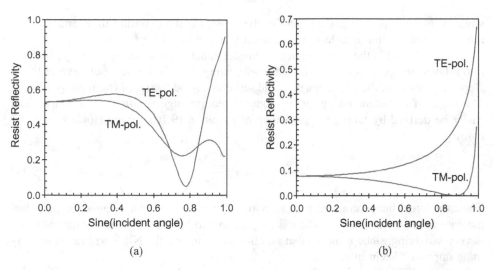

(a)

(b)

Figure 9.8 *Intensity reflectivity between air and resist of plane waves as a function of incident angle and polarization for (a) resist on silicon, and (b) resist on an optically matched substrate ($\lambda = 248\,nm$, resist n = 1.768 + i0.009868)*

case when a reasonably good BARC is used ($|\rho_{23}| \ll 1$), the impact of the film stack is dominated by the transmittance of the air–resist interface and absorption in the resist:

$$a_{nr} \approx a_n \tau_{12}(\theta_n)e^{-\alpha z/(2\cos\theta_n)} \tag{9.9}$$

Once the impact of the resist film is properly accounted for and the image in resist has been calculated, the image log-slope concept can readily be applied to the image in resist. As before, the image-log slope and the NILS, evaluated at the nominal resist edge for the image in resist, will serve as the metric of information content of the image in resist. Additionally, defocus of the image as it propagates through the thickness of the resist means that the location of the image must also be specified. In general, the bottom of the resist will be used for calculating NILS, though sometimes the middle of the resist will prove more convenient.

9.4   Exposure

The NILS is, fundamentally, an aerial image or image in resist metric. It can be thought of as a measure of the amount of information contained in the image that defines the proper position of the desired photoresist edge. The information in this image propagates through exposure into a latent image of exposed and unexposed resist, where a similar metric must be defined to judge the quality of the latent image. Consider a simple yet common case: a resist with first-order kinetics (almost always the case) whose optical properties do not change with exposure dose (commonly the case for chemically amplified resists). For such a case, the latent image $m(x,y,z)$ is related to the intensity in the resist $I_r(x,y,z)$ by

$$m(x,y,z) = \exp(-CI_r(x,y,z)t) \tag{9.10}$$

where C is the exposure rate constant of the resist, t is the exposure time, and m is the relative concentration of light-sensitive resist material (see Chapter 5).

Equation (9.10) is the exposure image transfer function, translating an image in resist into a latent image. From our experience with using the NILS, one would expect that a slope or gradient of the latent image would serve as a good metric of latent image quality. The slope of the latent image (at the nominal feature edge position, for example) can easily be derived by taking the derivative of Equation (9.10), giving (for a simple 1D case)[4]

$$\frac{\partial m}{\partial x} = m\ln(m)\frac{\partial \ln I}{\partial x} \tag{9.11}$$

Thus, the latent image gradient is directly proportional to the image log-slope (and thus the normalized latent image gradient is proportional to NILS). This entirely logical result is very satisfying, since it means that all efforts to improve the NILS will result directly in an improved latent image gradient.

Equation (9.11) reveals another important factor in latent image quality. The term $m\ln(m)$, which relates the image log-slope to the latent image gradient, is exposure dependent (m being the relative amount of resist sensitizer that has not been exposed at the point where the latent image gradient is being described). A simple thought experiment reveals the necessity for a dose dependence to latent image quality. At the two extremes of zero dose and infinite dose, the final latent image becomes a uniform chemical distribution [$m(x,y,z) = 1$ for zero dose, and $m(x,y,z) = 0$ for infinite dose]. Thus, the gradient of the latent image is zero at these extremes and no information is transferred to the wafer. But obviously the gradient is nonzero at doses in between these extremes. By the *Mean Value Theorem* of calculus, there must be a dose which maximizes the latent image gradient.

A plot of $-m\ln(m)$ versus m shows that there is one exposure dose (one value of m) that will maximize the latent image quality (Figure 9.9). When $m = e^{-1} \approx 0.37$, the value of $-m\ln(m)$ reaches its maximum and the full information of the aerial image is transferred into the resist during exposure. It is interesting to note that when $m = 1$ (no exposure) and $m = 0$ (complete exposure of the resist), the latent image gradient is zero and no information is transferred from the aerial image into the resist, as expected.

There are many interesting implications that come from the simple observation of the existence of an optimum exposure dose. Often, dose is used as a 'dimension dial', adjusting dose to obtain the desired feature size without regard to any process latitude implications. If the dose is near the optimum, this approach is valid. If, however, the dose used is significantly off from optimum (say, very underexposed compared to the peak of Figure 9.9), changing dose will affect both dimension and overall latent image quality.

The above discussion assumes a resist that does not bleach. Since g-line and i-line resist materials generally exhibit a large amount of bleaching, some attempts to describe how bleaching affects the latent image gradient will be useful. In general, the combined effects of bleaching with the propagation of the image from top to bottom of the resist yield a very complex picture. However, a few approximations can greatly simplify this picture without greatly sacrificing the accuracy of the derived latent image gradient.

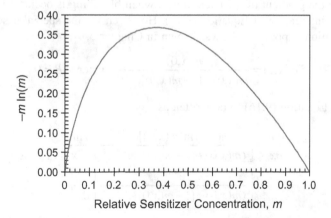

Figure 9.9 *Plot revealing the existence of an optimum exposure, the value of m at which the latent image gradient is maximized. Note that m = 1 corresponds with unexposed resist, while m = 0 is completely exposed resist*

When only absorption affects how the intensity varies with depth into a bleaching resist (that is, when there are no standing waves and when the defocusing of the image through the resist is ignored), Lambert's law of absorption coupled with Beer's law gives

$$\frac{dI}{dz} = -(Am + B)I \tag{9.12}$$

The coupled Equations (9.10) and (9.12) were solved in Chapter 5 to give

$$z = \int_{m(z=0)}^{m(z)} \frac{dy}{y[A(1-y) - B\ln(y)]} \tag{9.13}$$

where y is a dummy variable for the purposes of integration, and

$$I(x,z) = I(x,0) \frac{A[1 - m(x,z)] - B\ln[m(x,z)]}{A[1 - m(x,0)] - B\ln[m(x,0)]} \tag{9.14}$$

Differentiating Equation (9.13) with respect to x gives the interesting result that

$$\frac{\partial m}{\partial x} = \left[\frac{m(x,z)I(x,z)}{m(x,0)I(x,0)} \right] \frac{\partial m}{\partial x}\bigg|_{z=0} \tag{9.15}$$

But at the top of the resist, bleaching does not affect the solution to the kinetic rate equation so that Equations (9.10) and (9.11) still apply at the top of the resist. Thus,

$$\frac{\partial m}{\partial x} = m(x,z)\ln[m(x,z)] \frac{\partial \ln I}{\partial x}\bigg|_{z=0} \tag{9.16}$$

This derivation shows that Equation (9.16) coupled with Equation (9.13) is a more general result for the latent image gradient after exposure and that Equation (9.11) is a special case of (9.16) when $z = 0$ or when $A = 0$ (since the image log-slope is not a function of z when there is no bleaching).

Calculating the gradient of the latent image when bleaching is occurring still involves the numerical integration of Equation (9.13). However, for the special case of $B = 0$ an analytical solution is possible. As was shown in Chapter 5,

$$m(x,z) = \frac{m(x,0)}{m(x,0) + [1 - m(x,0)]e^{-Az}} \quad \text{when} \quad B = 0 \tag{9.17}$$

which allows Equation (9.16) to be written as

$$\frac{\partial m}{\partial x} = \left[\frac{m(x,0)\ln[m(x,0)]e^{-Az}}{(m(x,0) + [1 - m(x,0)]e^{-Az})^2} \right] \frac{\partial \ln I}{\partial x}\bigg|_{z=0}$$

$$= \left[\frac{e^{-Az}}{(m(x,0) + [1 - m(x,0)]e^{-Az})^2} \right] \frac{\partial m}{\partial x}\bigg|_{z=0} \tag{9.18}$$

By plotting Equation (9.18), Figure 9.10 shows that bleaching results in an increased latent image gradient. Like the contrast enhancement layers discussed in Chapter 4, bleaching allows greater transmittance of light in the bright spaces relative to the edges

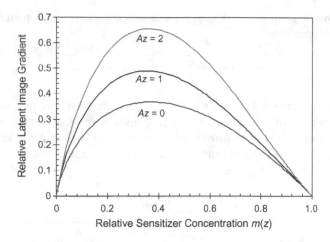

Figure 9.10 *Bleaching (increasing values of Az) results in increased latent image gradient at the bottom of the resist (shown here is the special case where B = 0)*

between lines and spaces, which necessarily have a lower image intensity. As a result, the light near the edge of the feature is more highly absorbed than light in the center of the space. The transmitted image $I(x,z)$ will have a higher slope than the image at the top of the resist. Interestingly, the optimum level of exposure remains very close to the non-bleaching case: $m(x,z) \approx e^{-1}$.

9.5 Post-exposure Bake

The post-exposure bake takes the latent image of exposure products and creates a new latent image at the end of the bake, $m^*(x)$. For a conventional resist, diffusion will spread out the latent image. For a chemically amplified resist, diffusion is accompanied by reaction to form a new latent image.

9.5.1 Diffusion in Conventional Resists

Fickean diffusion, where the diffusivity of the diffusing material remains constant, can be treated as a simple convolution problem (as described in Chapter 5). If $m(x)$ is the original latent image (the spatial distribution of chemical species in the resist), the new latent image after diffusion $m^*(x)$ can be found by convolving the original chemical distribution with a Gaussian sometimes referred to as the diffusion point spread function (*DPSF*):

$$m^*(x) = m(x) \otimes DPSF = \frac{1}{\sqrt{2\pi}\sigma_D} \int_{-\infty}^{\infty} m(\tau)e^{-(x-\tau)^2/2\sigma_D^2}d\tau \qquad (9.19)$$

where σ_D is the diffusion length. While a one-dimensional case is shown here, it is easy to extend the convolution to two or three dimensions (though things do get more complex when boundary conditions are applied).

Consider now a generic latent image for a repeating line/space pattern of pitch p described as a Fourier series:

$$m(x) = \sum_{n=0}^{N} a_n \cos(2\pi nx/p) \tag{9.20}$$

where a pattern symmetrical about $x = 0$ is assumed so that there are no sine terms in the series. Larger values of n represent higher-frequency terms (harmonics) in the image, though a typical minimum pitch pattern will have an upper limit of $N = 2$ or 3. The effect of diffusion on this latent image can be calculated by plugging Equation (9.20) into the convolution integral (9.19) giving

$$m^*(x) = \sum_{n=0}^{N} a_n^* \cos(2\pi nx/p) \tag{9.21}$$

where $a_n^* = a_n e^{-2(\pi n\sigma_D/p)^2}$.

Diffusion can be thought of as simply a reduction of each of the $n > 0$ Fourier coefficients of the latent image. Obviously, a greater diffusion length leads to a greater degradation of the latent image (a_n is reduced). But it is the diffusion length relative to the pitch that matters. Thus, for the same diffusion length, smaller pitch patterns are degraded more than larger pitch patterns. Also, the higher-frequency (larger n) terms degrade faster than the lower-frequency terms. In fact, each frequency term can be said to have an effective diffusion length equal to $n\sigma_D$ and it is the ratio of this effective diffusion length to the pitch that determines the amount of damping for that frequency component (Figure 9.11).

The effect of diffusion on the latent image gradient (LIG) is complicated by the fact that the rate of diffusion is driven by the change in the concentration gradient. Thus, in general the latent image gradient after diffusion will not be directly proportional to the gradient before diffusion – it will be influenced by the shape of the latent image to a

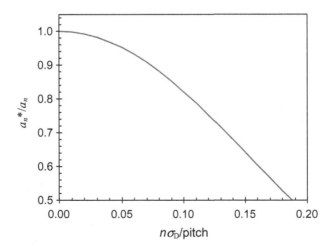

Figure 9.11 *Effect of diffusion on the latent image frequency components for a dense line*

distance of a few diffusion lengths away. However, an approximate response can be obtained by assuming that the latent image for a small pitch pattern, and thus the effect on the gradient, is dominated by the $n = 1$ Fourier component of the latent image. The change in the latent image gradient due to diffusion can then be approximated as the change in a_1:

$$\frac{\partial m^*/\partial x}{\partial m/\partial x} \approx \frac{a_1^*}{a_1} \tag{9.22}$$

This simple result can be generalized for any pattern (not just patterns with a very small pitch) to become

$$\frac{\partial m^*}{\partial x} \approx \frac{\partial m}{\partial x} e^{-\sigma^2\pi^2/2L^2} = \frac{\partial m}{\partial x} e^{-t_{PEB}D\pi^2/L^2} = \frac{\partial m}{\partial x} e^{-t_{PEB}/\tau} \tag{9.23}$$

where L is a characteristic length related to the width of the edge region (the range over which the original latent image gradient is nonzero). For a pattern of small lines and spaces, L is equal to the half-pitch of the pattern. Obviously, increased diffusion (indicated by a larger diffusion length) results in a greater degradation of the latent image gradient (Figure 9.12). Also, sharper edges (smaller values of L) are more sensitive to diffusion, showing a greater fractional decline in the latent image gradient for a given diffusion length. The second (middle) form of Equation (9.23) replaces the diffusion length with $\sqrt{2Dt_{PEB}}$ where D is the diffusivity and t_{PEB} is the bake time. Finally, the left-hand form of the equation defines a characteristic pattern diffusion time $\tau = L^2/(D\pi^2)$. The bake time must be kept much less than this characteristic pattern diffusion time in order to prevent significant latent image degradation due to diffusion.

9.5.2 Chemically Amplified Resists – Reaction Only

For chemically amplified resists, diffusion during PEB is accompanied by a reaction that changes the photogenerated acid latent image into a latent image of blocked and deblocked

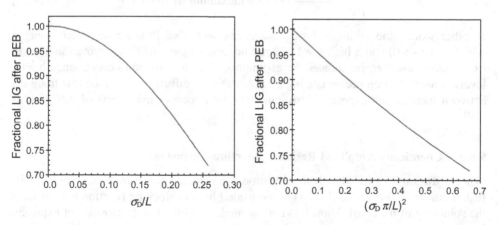

Figure 9.12 *Increased diffusion (shown by the dimensionless quantity σ_D/L, the diffusion length over the width of the edge region) causes a decrease in the latent image gradient (LIG) after PEB*

polymer. Since reaction and diffusion occur simultaneously, rigorous evaluation of the impact on the latent image gradient is required. However, as a start we will look at the impact of the reaction without diffusion. Ignoring the possibilities of acid loss before or during the PEB, a simple mechanism for a first-order chemical amplification would give (from Chapter 6)

$$m^*(x) = e^{-\alpha_f h(x)} \tag{9.24}$$

where $\alpha_f = K_{amp}t_{PEB}$ is the amplification factor, proportional to the PEB time and exponentially dependent on PEB temperature. This simple expression points out the trade-off between exposure dose and thermal dose. A higher exposure dose generates more acid (larger value of h), requiring less PEB (lower value of α_f) to get the same result (same value of m^*). For a given level of required amplification, thermal and exposure doses can be exchanged so long as $\alpha_f h$ is kept constant.

The gradient of this new latent image after amplification is then

$$\frac{\partial m^*}{\partial x} = m^* \ln(m^*)\left(\frac{1}{h}\frac{\partial h}{\partial x}\right) = -m^* \ln(m^*)\left(\frac{(1-h)\ln(1-h)}{h}\right)\frac{\partial \ln I}{\partial x} \tag{9.25}$$

The gradient of the deprotection latent image relative to the image log-slope is often called the chemical contrast:

$$\frac{\partial m^*}{\partial x} \bigg/ \frac{\partial \ln I}{\partial x} = \frac{\partial m^*}{\partial \ln E} = \text{chemical contrast} \tag{9.26}$$

For a given latent image after exposure [that is, given $h(x)$], the optimum latent image after amplification occurs when $m^* = e^{-1}$, giving $\alpha_f h = 1$. For a given required level of amplification (given value of m^*), the trade-off between thermal and exposure dose can be optimized to give the maximum latent image gradient after PEB. This occurs when

$$\frac{1}{h}\frac{\partial h}{\partial x} \propto \frac{(1-h)\ln(1-h)}{h} = \text{maximum when } h \to 0 \tag{9.27}$$

In other words, the optimum latent image gradient after PEB occurs when using a low dose ($h \to 0$) and a high level of amplification (Figure 9.13). Carrying this idea to its extreme, however, invalidates the assumption that no diffusion occurs, since higher levels of amplification necessitate higher levels of acid diffusion. Thus, the true trade-off between thermal and exposure dose must take into account the effects of diffusion as well.

9.5.3   Chemically Amplified Resists – Reaction–Diffusion

When diffusion is included with amplification, the situation becomes more complex. Diffusion during post-exposure bake is accompanied by a deblocking reaction that changes the solubility of the resist. Thus, it is not the final, postdiffusion distribution of exposure reaction products that controls development but rather the integral over time of these exposure products. Things become even more complicated when acid loss is accounted

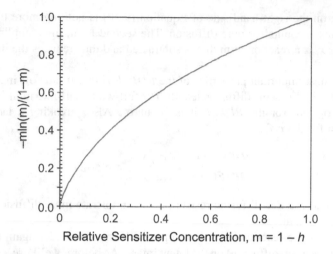

Figure 9.13 *For a chemically amplified resist with a given required amount of amplification, the exposure dose (and thus relative sensitizer concentration m) is optimum as the dose approaches zero (m → 1), assuming negligible diffusion*

for. In particular, the presence of base quencher (that also may diffuse) leads to both complexity and advantage in tailoring the final latent image shape. However, for the simplified case of no acid loss, an analytical solution is possible.

Letting $h(x, t = 0)$ be the concentration of acid (the exposure product) at the beginning of the post-exposure bake (PEB), an effective acid latent image can be defined as

$$h_{\text{eff}}(x) = \frac{1}{t_{\text{PEB}}} \int_0^{t_{\text{PEB}}} h(x,t=0) \otimes DPSF dt = h(x,0) \otimes \frac{1}{t_{\text{PEB}}} \int_0^{t_{\text{PEB}}} DPSF dt \qquad (9.28)$$

$$= h(x,0) \otimes RDPSF$$

where *RDPSF* is the reaction–diffusion point spread function (see Chapter 6). This effective acid concentration distribution $h_{\text{eff}}(x)$ can be used to calculate the reaction kinetics of the PEB as if no diffusion had taken place. In other words, the effects of diffusion are separable from the reaction for the case of no acid loss. The *RDPSF* then becomes analogous to the *DPSF* of a conventional resist.

For the 1D case, the Gaussian diffusion kernel is affected by time integration through the diffusion length, $\sigma_{\text{D}} = \sqrt{2Dt}$, where D is the acid diffusivity in resist. Thus,

$$RDPSF = \frac{1}{t_{\text{PEB}} \sqrt{4\pi D}} \int_0^{t_{\text{PEB}}} \frac{e^{-x^2/4Dt}}{\sqrt{t}} dt \qquad (9.29)$$

The integral is solvable, resulting in an interesting final solution:

$$RDPSF(x) = 2 \frac{e^{-x^2/2\sigma_{\text{D}}^2}}{\sqrt{2\pi}\sigma_{\text{D}}} - \frac{|x|}{\sigma_{\text{D}}^2} erfc\left(\frac{|x|}{\sqrt{2}\sigma_{\text{D}}}\right) \qquad (9.30)$$

The first term on the right-hand side of Equation (9.30) is nothing more than twice the *DPSF*, and thus accounts for pure diffusion. The second term, the complimentary error function times x, is a reaction term that is subtracted and thus reduces the impact of pure diffusion.

One of the most important properties of the *RDPSF* is that it falls off in x much faster than the *DPSF* with the same diffusion length. The full-width half-maximum for the *DPSF* is about $2.35\sigma_D$, but for the *RDPSF* it is about σ_D. Also, invoking a large argument approximation for the *erfc*,

$$\frac{RDPSF(x)}{DPSF(x)} \approx 2\left(\frac{\sigma_D}{x}\right)^2, \quad x \gg \sigma_D \tag{9.31}$$

In other words, a reaction–diffusion resist system can tolerate more diffusion than a conventional resist system.

To further compare reaction–diffusion to pure diffusion, consider again the impact of diffusion (and reaction–diffusion) on the latent image. As before, we'll use a generic latent image for a repeating line/space pattern of pitch p described as a Fourier series:

$$h(x,0) = \sum_{n=0}^{N} a_n \cos(2\pi nx/p) \tag{9.32}$$

where a pattern symmetrical about $x = 0$ is assumed so that there are no sine terms in the series. The effect of pure diffusion, calculated as a convolution with the *DPSF*, is simply a reduction in the amplitude of each harmonic, as shown in Equation (9.21). In the reaction–diffusion case, we convolve Equation (9.32) with the *RDPSF*, which results in an analytical solution:[5]

$$h^*(x) = \sum_{n=0}^{N} a_n^* \cos(2\pi nx/p)$$

where

$$a_n^* = a_n\left(\frac{1-e^{-2(\pi n\sigma_D/p)^2}}{2(\pi n\sigma_D/p)^2}\right) \tag{9.33}$$

Consider the special case of $x = p/4$ and $N = 3$ (that is, at the nominal line edge of a very small line/space pattern), a case that will prove useful later:

$$h(p/4, 0) = a_0 - a_2$$
$$h^*(p/4) = a_0 - a_2^* \tag{9.34}$$

Figure 9.14 compares the effects of pure diffusion with reaction–diffusion on the amplitudes of the Fourier coefficients. As can be seen, pure diffusion causes a much faster degradation of the Fourier components than reaction–diffusion for the same diffusion length. For example, if one is willing to allow a particular Fourier component to fall in amplitude by 20%, a reaction–diffusion system with no acid loss can tolerate about 50% more diffusion than a pure diffusion resist. In fact, for reasonably small $n\sigma_D/p$, one can

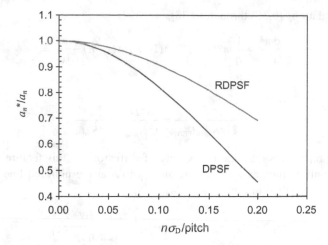

Figure 9.14 *Effect of diffusion on the latent image frequency components for a dense line, comparing pure diffusion (DPSF) to reaction–diffusion (RDPSF)*

compare the diffusion-only case to a reaction–diffusion system using a Taylor series expansion of the exponential:

$$\text{Diffusion:} \quad \frac{a_n^*}{a_n} = e^{-2(\pi n \sigma_D/p)^2} \approx 1 - 2(\pi n \sigma_D/p)^2$$

$$\text{Reaction-Diffusion:} \quad \frac{a_n^*}{a_n} = \frac{1 - e^{-2(\pi n \sigma_D/p)^2}}{2(\pi n \sigma_D/p)^2} \approx 1 - (\pi n \sigma_D/p)^2 \tag{9.35}$$

The two are roughly equivalent if the reaction–diffusion system has a diffusion length $\sqrt{2}$ larger than the diffusion-only case (or a diffusivity that is twice as large).

As with pure diffusion for conventional resist, we can approximate the impact of reaction–diffusion on the final latent image gradient by assuming the $n = 1$ order of the latent image Fourier series dominates. Thus,

$$\frac{\partial h_{\text{eff}}}{\partial x} \approx \frac{\partial h}{\partial x}\left(\frac{1 - e^{-\sigma_D^2 \pi^2/2L^2}}{\sigma_D^2 \pi^2/2L^2}\right) \tag{9.36}$$

Combining now all of the terms of exposure, reaction–diffusion impact on $h(x)$, and reaction to give m^*, the final gradient of the deblocked concentration can be calculated. The exposure term gives

$$h = 1 - e^{-Clt} \quad \text{which leads to} \quad \frac{\partial h}{\partial x} = (1-h)\ln(1-h)\frac{\partial \ln I}{\partial x} \tag{9.37}$$

Reaction converts the effective acid concentration to a blocked polymer concentration:

$$m^* = e^{-\alpha_f h_{\text{eff}}} \quad \text{which leads to} \quad \frac{\partial m^*}{\partial x} = -\alpha_f m^* \frac{\partial h_{\text{eff}}}{\partial x} \tag{9.38}$$

Combining Equations (9.36) through (9.38),

$$\frac{\partial m^*}{\partial x} \approx \frac{1}{\eta} m^*(1-e^{-\eta\alpha_f})(1-h)\ln(1-h)\frac{\partial \ln I}{\partial x} \tag{9.39}$$

where

$$\eta = \frac{\pi^2 \sigma_D^2}{2L^2 K_{amp} t_{PEB}} = \frac{\pi^2 D}{L^2 K_{amp}} = \frac{1}{\tau K_{amp}}$$

The term η represents the ratio of the rate of diffusion for this feature to the rate of reaction. The only remaining step is to relate h_{eff} to h at the nominal line edge. Taking the simple case described in Equation (9.34),

$$h_{eff}(p/4) = a_0 - a_2^* = a_0 - a_2\left(\frac{1-e^{-2(2\pi\sigma_D/p)^2}}{2(2\pi\sigma_D/p)^2}\right) \tag{9.40}$$

Expanding the exponential as a Taylor's series,

$$h_{eff} = a_0 - a_2\left(1-(2\pi\sigma_D/p)^2 + \frac{4}{3}(2\pi\sigma_D/p)^4 - \ldots\right) \tag{9.41}$$

To avoid excessive degradation of the latent image due to diffusion, a typical value for $2\pi\sigma_D/p$ would be 0.5 or less. Typical latent images will have a_2 about one order of magnitude smaller than a_0. Thus, one can approximate h_{eff} for the typical case as

$$h_{eff} \approx a_0 - a_2 + a_2(2\pi\sigma_D/p)^2 \approx h \text{ to an error of } \sim 5\% \text{ or less} \tag{9.42}$$

In other words, at the line edge the acid concentration doesn't change much due to diffusion. As a result, a final expression for the latent image gradient becomes

$$\frac{\partial m^*}{\partial x} \approx \frac{1}{\eta} e^{-\alpha_f h_{eff}}(1-e^{-\eta\alpha_f})(1-h_{eff})\ln(1-h_{eff})\frac{\partial \ln I}{\partial x} \tag{9.43}$$

Figure 9.15 uses Equation (9.43) to show how the latent image gradient (LIG) varies with α_f (which is proportional to PEB time). At low levels of amplification (small PEB time), the gradient is amplification controlled and increased bake produces an improved gradient. At long PEB times, the gradient becomes diffusion controlled and increased bake reduces the gradient due to diffusion. Thus, there is an optimum bake time (optimum α_f) that produces the maximum LIG of blocked polymer concentration.

The goal now becomes one of finding the optimum final deblocked concentration m^* (equivalently, the optimum h_{eff}) and the optimum level of amplification α_f for a given value of η. The optimum α_f can be determined for a given value of m^* or for a given value of h_{eff}:

$$optimum\ \alpha_f = -\left[\frac{1+\ln(1-h_{eff})}{(1-h_{eff})\ln(1-h_{eff})}\right] = \frac{1}{\eta}\ln\left(\frac{h_{eff}+\eta}{h_{eff}}\right) \tag{9.44}$$

Figure 9.16 provides numerical results of the values of α_f and m^* that produce the maximum latent image gradient as a function of η. Figure 9.17 shows the resulting

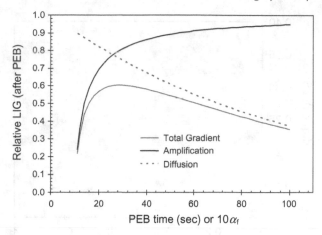

Figure 9.15 *Including diffusion with amplification, there is an optimum PEB to maximize the latent image gradient (LIG), shown here relative to the maximum possible LIG. For this example, $K_{amp} = 0.1\,s^{-1}$, $\tau = 200$ seconds and the exposure dose is chosen to give the maximum gradient for each PEB time*

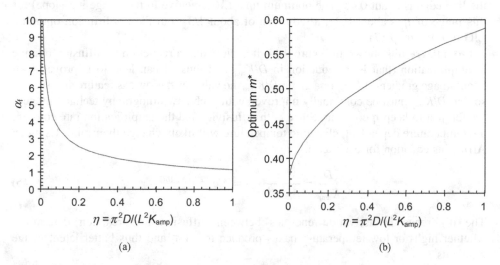

Figure 9.16 *The optimum value of (a) the amplification factor, and (b) the deblocked concentration in order to maximize the final latent image gradient, as a function of η*

maximum latent image gradient. For the case of $\eta = 0$ (no diffusion), the solution matches the results given in the previous section. As the relative amount of diffusion increases, the optimum amplification factor quickly falls. A knee in this response at about $\eta = 0.1$ (and $\alpha_f \approx 3$) marks a transition from high to low sensitivity of the optimum amplification factor to η. Typical values of η for commercial resists are on the order of 0.1–0.2, so that optimum amplification factors are near 3 and the optimum deblocking concentration at

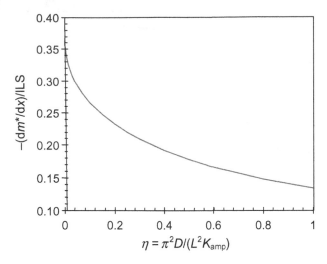

Figure 9.17 *The optimum value of the final latent image gradient (relative to the image log-slope), as a function of η*

the line edge is about 0.45. The optimum final LIG (relative to the image log-slope) for this range of η is about 0.25, a reduction of about 1/3 from the no-diffusion optimum LIG (see Figure 9.17).

From the results shown in Figure 9.17, it is clear that a reduction in diffusion relative to amplification (that is, a reduction in D/K_{amp} and thus η) can lead to improved final latent image gradient. Also, for a given D/K_{amp} the value of η grows as feature size shrinks so that D/K_{amp} must be continually improved with each new lithography technology generation just to keep η constant. Since both diffusivity and the amplification rate constant are temperature dependent, changing temperature will likely change their ratio. Using an Arrhenius equation for each term,

$$\frac{D}{K_{amp}} = \frac{A_{rD}e^{-E_{aD}/RT}}{A_{rK}e^{-E_{aK}/RT}} = A'_r e^{-(E_{aD}-E_{aK})/RT} \tag{9.45}$$

The difference in activation energies between diffusion and reaction determines whether high- or low-temperature bakes produce lower η and thus better latent image gradients.

The dependence of η with temperature is more complicated when the Byers–Petersen model is used. Recalling from Chapter 6, the overall amplification rate constant depends on the acid diffusivity as

$$K_{amp} = \frac{K_{react}K_{diff}D}{K_{react} + K_{diff}D} \tag{9.46}$$

Thus, the ratio of diffusivity to amplification becomes

$$\frac{D}{K_{amp}} = \frac{1}{K_{diff}} + \frac{D}{K_{react}} \tag{9.47}$$

In general, K_{diff} is not expected to vary much with temperature. Thus, if the resist operates in the diffusion-controlled regime ($K_{react} \gg K_{diff}D$), η will be relatively insensitive to temperature. However, in the reaction-controlled regime ($K_{react} \ll K_{diff}D$), η will have the temperature dependence shown in Equation (9.45).

9.5.4 Chemically Amplified Resists – Reaction–Diffusion with Quencher

The analysis of the previous section is very helpful for understanding the role of diffusion in a simple reaction–diffusion system with no acid loss. However, all serious chemically amplified resists include base quenchers – a beneficial source of acid loss. In general, quencher greatly complicates any efforts to analytically solve for the final latent image or the final latent image gradient. However, for one special case, the above analysis applies even when quencher is present. If the acid–quencher reaction is very fast compared to diffusion (typically a very good assumption), and the base quencher has exactly the same diffusivity as the acid, then the initial acid concentration at the start of the PEB $h(x,0)$ is simply reduced by the background quencher concentration. [Note that for a Fourier series representation of the initial acid latent image such as Equation (9.32), the impact of quencher is then just a reduction of the value of a_0.] Negative acid concentrations represent the concentration of base quencher, so that both acid and base diffusion are properly calculated using the convolution of Equation (9.28). Of course, negative values of $h_{eff}(x)$, the effective acid concentration, should be set to zero before performing any amplification calculations.

For quencher of relative initial concentration q_0, the acid gradient after exposure and initial quenching becomes, in regions of excess acid,

$$h = 1 - e^{-CIt} - q_0 \quad \text{which leads to} \quad \frac{\partial h}{\partial x} = (1 - h - q_0)\ln(1 - h - q_0)\frac{\partial \ln I}{\partial x} \quad (9.48)$$

The resulting final latent image gradient becomes

$$\frac{\partial m^*}{\partial x} \approx \frac{1}{\eta}e^{-\alpha_f h_{eff}}(1 - e^{-\eta \alpha_f})(1 - h_{eff} - q_0)\ln(1 - h_{eff} - q_0)\frac{\partial \ln I}{\partial x} \quad (9.49)$$

The value of α_f to maximize the final latent image gradient is

$$optimum\ \alpha_f = -\left[\frac{1 + \ln(1 - h_{eff} - q_0)}{(1 - h_{eff} - q_0)\ln(1 - h_{eff} - q_0)}\right] = \frac{1}{\eta}\ln\left(\frac{h_{eff} + \eta}{h_{eff}}\right) \quad (9.50)$$

The impact of quencher on the optimum gradient is seen in Figure 9.18. Quencher allows the process to be run at higher η values while still providing a good latent image gradient. For the $q = 0.1$ case in this figure, an η value that is about 0.23 higher will give the same optimum gradient compared to the no-quencher case. Alternately, for a given η, higher q_0 allows for a higher optimum gradient. Thus, the presence of quencher significantly relaxes the constraints placed on diffusion, allowing higher-diffusing resists to work well for smaller features. Figure 9.19 shows the values of α_f and m^* that produce the optimum gradient.

Note that in Figure 9.18, the two curves for nonzero quencher are only plotted down to some critical value of η (called η_c). When $\eta < \eta_c$, the optimum m^* approaches 1 (see

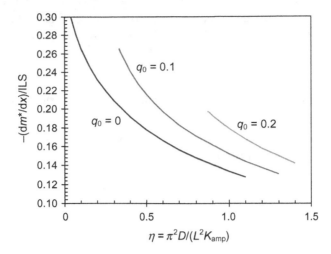

Figure 9.18 *The optimum value of the final latent image gradient (relative to the image log-slope), as a function of η for cases with and without quencher*

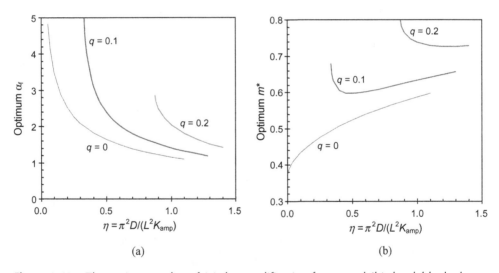

Figure 9.19 *The optimum value of (a) the amplification factor, and (b) the deblocked concentration in order to maximize the final latent image gradient, as a function of η for different quencher loadings*

Figure 9.19). For this case, the best latent image gradient occurs when the point at which quencher just eliminates all acid (that is, $h = 0$) is placed at the nominal line edge. The resulting LIG is quite high (see Problem 9.8):

$$\left.\frac{\partial m^*/\partial x}{ILS}\right|_{max} \approx \frac{1}{\eta}(1-q_0)\ln(1-q_0) \tag{9.51}$$

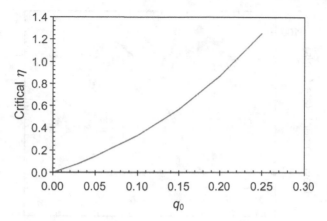

Figure 9.20 *Numerically determined values of the critical η value (η_c) for different quencher loadings*

Of course, real development processes require some level of deblocking at the line edge so that this maximum gradient cannot be practically attained. Nonetheless, operating below η_c and putting the $h = 0$ point very near the nominal line edge result in very steep latent image gradients. A plot of η_c as a function of quencher loading is shown in Figure 9.20.

9.6 Develop

In one sense, the goal of selective exposure of resist with an aerial image is to create a solubility differential: exposed and unexposed regions of the photoresist give rise to regions of higher and lower solubility, measured as a development rate (resist removal rate in nanometers per second). The information contained in the aerial image (or image in resist) $I(x)$ is used to expose the photoresist to form a latent image $m(x)$ or $h(x)$, which is modified by the post-exposure bake to create a new latent image $m^*(x)$, and finally developed based on a development rate 'image' $r(x)$ that results in the definition of the feature edge.

The fundamental chemical response of interest is the change in dissolution rate as a function of the exposure dose seen by the resist. A plot of development rate r versus exposure dose E on a log-log scale is called a Hurter–Driffield (H–D) curve (Figure 9.21) and allows for the definition of the *photoresist contrast*, γ. Quite simply, the photoresist contrast is the slope of the development rate H–D curve (see Chapter 7).

$$\gamma \equiv \frac{\partial \ln r}{\partial \ln E} \qquad (9.52)$$

Note that photoresist contrast is actually a function of exposure, $\gamma(E)$.

The photoresist contrast is a measure of the discrimination of the resist with respect to exposure. Higher contrast means that a given change in dose will result in a greater change

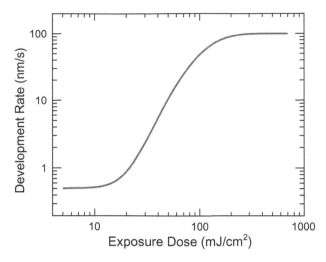

Figure 9.21   *Typical development rate function of a positive photoresist (one type of Hurter–Driffield curve)*

in development rate. This point can be seen clearly using the *lithographic imaging equation*, derived from the definition of photoresist contrast:

$$\frac{\partial \ln r}{\partial x} = \gamma(E)\frac{\partial \ln I}{\partial x} \quad \text{or} \quad w\frac{\partial \ln r}{\partial x} = \gamma NILS \tag{9.53}$$

The quality of the development rate image, described by a gradient in the dissolution rate, is determined by the product of the image log-slope and the photoresist contrast function. To create a large solubility differential, one would like a good aerial image (large image log-slope), a good photoresist (large photoresist contrast) and an optimized process (an exposure dose chosen to use the maximum of the photoresist contrast function).

Breaking up the response of development to exposure into separate latent and development rate images allows us to relate the development rate gradient to the latent image gradient:

$$\frac{\partial \ln r}{\partial x} = \frac{\partial \ln r}{\partial m^*}\frac{\partial m^*}{\partial x} \tag{9.54}$$

where m^* is the concentration of the chemical species (after post-exposure bake) that affects the dissolution rate. For example, for a chemically amplified resist m^* would represent the concentration of blocked polymer. From Equations (9.53) and (9.54), one can see that the definition of photoresist contrast encompasses the exposure, post-exposure bake and development steps in order to relate the final development rate gradient to the original source of the information being imprinted in the resist, the image log-slope.

The variation of development rate with m^* can take on many forms, but a simple one can be used here to illustrate the expected response. The model of development shown here (called the reaction-controlled version of the original kinetic development model) provides for a typical nonlinear development rate function:

$$r = r_{max}(1 - m^*)^n + r_{min} \qquad (9.55)$$

where r_{max} and r_{min} represent the maximum and minimum development rates, respectively ($r_{max} \gg r_{min}$ is assumed), and n is called the dissolution selectivity parameter and controls how nonlinear the development response will be. Using this model of dissolution rate,

$$\frac{\partial \ln r}{\partial m^*} = -\frac{n}{1 - m^*}\left(1 - \frac{r_{min}}{r}\right) \qquad (9.56)$$

Figure 9.22 shows a plot of Equation (9.56) as a function of m^* for $n = 5$ (a low to medium contrast resist) and $n = 10$ (a medium to high contrast resist). Adjusting the process to set the peak of this curve at the nominal feature edge is a key part of process optimization.

From Equation (9.56), it is easy to show that the maximum gradient occurs when

$$-\frac{\partial \ln r}{\partial m^*} \text{ is a maximum} \quad \text{when} \quad r = n r_{min} \qquad (9.57)$$

The maximum gradient will occur at a value of m^* given by

$$-\left.\frac{\partial \ln r}{\partial m^*}\right|_{max} \quad \text{when} \quad m^* = 1 - \left[\frac{(n-1)r_{min}}{r_{max}}\right]^{\frac{1}{n}} \qquad (9.58)$$

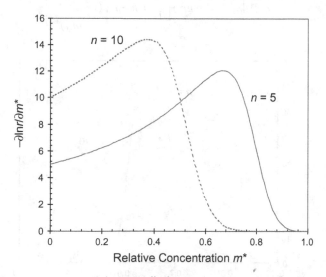

Figure 9.22 *One component of the overall photoresist contrast is the variation in development rate r with chemical species m*, shown here for the reaction-controled version of the original kinetic development rate model (r_{max} = 100 nm/s, $r_{min\bullet}$ = 0.1 nm/s)*

Thus, the maximum gradient is

$$-\frac{\partial \ln r}{\partial m^*}\bigg|_{max} = \frac{(n-1)}{1-m^*} = (n-1)\left[\frac{r_{max}}{(n-1)r_{min}}\right]^{-\frac{1}{n}}$$

(9.59)

For the full kinetic rate model,

$$r = r_{max}\frac{(a+1)(1-m^*)^n}{a+(1-m^*)^n} + r_{min}$$

(9.60)

and the equivalent gradient is

$$\frac{\partial \ln r}{\partial m^*} = -\frac{n}{1-m^*}\left(1-\frac{r_{min}}{r}\right)\left(\frac{a}{a+(1-m^*)^n}\right)$$

(9.61)

Figure 9.23 shows that again there will be an optimum inhibitor concentration to maximize the development rate gradient, and that this optimum will depend on the value of the dissolution selectivity parameter. The maximum gradient occurs when the development rate is given by

$$r = \frac{nr_{min}}{1+(n-1)\dfrac{(1-m^*)^n}{(1-m_{th})^n}}$$

(9.62)

The maximum gradient is then

$$-\frac{\partial \ln r}{\partial m^*}\bigg|_{max} = \frac{(n-1)}{(1-m^*)}\left(\frac{1-\dfrac{(1-m^*)^n}{(1-m_{th})^n}}{1+\left(\dfrac{n-1}{n+1}\right)\dfrac{(1-m^*)^n}{(1-m_{th})^n}}\right)$$

(9.63)

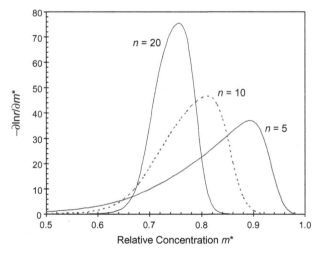

Figure 9.23 *Development rate gradient with inhibitor concentration for the original kinetic development rate model* (r_{max} = 100 nm/s, r_{min} = 0.1 nm/s, m_{th} = 0.7)

As n increases, the optimum inhibitor concentration approaches the threshold inhibitor concentration, m_{th}. For large n, the maximum gradient occurs when

$$1 - m^* \approx (1 - m_{th})\left(1 + \frac{1}{2n}\ln\left(\frac{r_{min}}{r_{max}}\right)\right) \quad \text{and} \quad \frac{(1-m^*)^n}{(1-m_{th})^n} \approx \sqrt{\frac{r_{min}}{r_{max}}} \tag{9.64}$$

giving

$$-\frac{\partial \ln r}{\partial m^*}\bigg|_{max} \approx \frac{(n-1)}{(1-m_{th})\left(1 + \frac{1}{2n}\ln\left(\frac{r_{min}}{r_{max}}\right)\right)}\left(\frac{1 - \sqrt{\frac{r_{min}}{r_{max}}}}{1 + \left(\frac{n-1}{n+1}\right)\sqrt{\frac{r_{min}}{r_{max}}}}\right) \tag{9.65}$$

If the resist is a reasonably good one, so that $r_{max}/r_{min} > 1000$, then this large n approximation can be further simplified:

$$-\frac{\partial \ln r}{\partial m^*}\bigg|_{max} \approx \frac{n}{(1-m_{th})} + \frac{1}{2(1-m_{th})}\ln\left(\frac{r_{max}}{r_{min}}\right) \tag{9.66}$$

This final result describes the basic behavior that we wish to know: the maximum log-development rate gradient is roughly proportional to n, inversely proportional to $1 - m_{th}$, and increases with the log of the ratio of the maximum to minimum development rates.

9.6.1 Conventional Resist

While Equation (9.54) related the development rate gradient to the after-PEB latent image gradient, in fact all of the lithographic process steps can be chained together to relate the final development response to the initial source of the imaging information, the aerial image. For a conventional resist,

$$\gamma = \left(\frac{\partial \ln r}{\partial m^*}\right)\left(\frac{\partial m^*}{\partial m}\right)\left(\frac{\partial m}{\partial \ln E}\right) \tag{9.67}$$

As described above, we have seen that for a resist that does not bleach,

$$\frac{\partial m}{\partial \ln E} = m\ln(m) \tag{9.68}$$

The impact of diffusion can approximated by

$$\frac{\partial m^*}{\partial m} \approx e^{-\sigma^2 \pi^2/2L^2} \tag{9.69}$$

where σ is the diffusion length and L is a characteristic width of the edge region (and is roughly equal to the resolution of the lithography process). Also, diffusion does not have a significant impact on the concentration of the sensitizer at the line edge so that $m \approx m^*$

at the feature edge. Combining each of these expressions with Equation (9.61) gives the overall contrast:

$$\gamma = -\frac{n}{1-m}\left(1-\frac{r_{min}}{r}\right)\left(\frac{a}{a+(1-m)^n}\right)(e^{-\sigma^2\pi^2/2L^2})m\ln m \qquad (9.70)$$

Our goal will now be to find the maximum value of this photoresist contrast. Since diffusion always degrades the contrast, the optimum will always be zero diffusivity. Thus, without loss of generality, we shall omit the diffusion term from the analysis that follows. The exact value of the maximum is most easily determined numerically (Figure 9.24), but it will be useful to derive some general trends. Since the $m\ln m$ term varies slowly, the contrast will be dominated by the development piece $(\partial\ln r/\partial m^*)$. Thus, for reasonably large n the optimum value of m^* will be approximately that given by Equation (9.64).

$$m^* \approx m_{th} + \frac{(1-m_{th})}{n}\ln\left(\sqrt{\frac{r_{max}}{r_{min}}}\right) \qquad (9.71)$$

Thus, at this optimum

$$m^*\ln m^* \approx m_{th}\ln m_{th} + \frac{(1+\ln m_{th})(1-m_{th})}{n}\ln\left(\sqrt{\frac{r_{max}}{r_{min}}}\right) \qquad (9.72)$$

The maximum contrast will be approximately

$$\gamma_{max} \approx -\frac{m_{th}\ln m_{th}}{(1-m_{th})}\left[n+\left(\frac{(1-m_{th}+\ln m_{th})}{m_{th}\ln m_{th}}\right)\ln\left(\sqrt{\frac{r_{max}}{r_{min}}}\right)\right](e^{-\sigma^2\pi^2/2L^2}) \qquad (9.73)$$

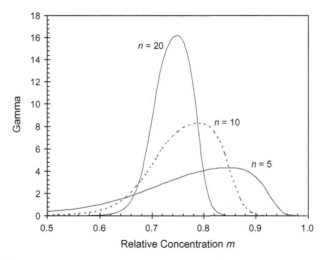

Figure 9.24 *Photoresist contrast as a function of inhibitor concentration for the original kinetic development rate model, assuming no diffusion ($r_{max} = 100\,nm/s$, $r_{min} = 0.1\,nm/s$, $m_{th} = 0.7$) and different values of the dissolution selectivity parameter n*

or, since n is reasonably high,

$$\gamma_{max} \approx -n\frac{m_{th}\ln m_{th}}{(1-m_{th})}(e^{-\sigma^2\pi^2/2L^2}) \tag{9.74}$$

This final approximate form matches well the full numerical calculations for the maximum contrast shown in Figure 9.24 even for n as low as 5. The overall contrast is maximized by minimizing diffusion, maximizing n and letting m_{th} approach 1.

9.6.2 Chemically Amplified Resist

For a chemically amplified resist, the PEB term $\partial m^*/\partial m$ is a function of the amplification factor α_f and the diffusion factor η, as described previously. Figure 9.25 shows a plot of $\gamma(E,\alpha_f)$ for the $n = 5$ case assuming no diffusion and the simple reaction-controlled development rate equation. Without diffusion, the best performance comes from a low dose and a high amplification factor, as described previously.

When diffusion is included in the analysis, the best resist will be the one that matches the optimum blocked concentration to produce a good latent image gradient (for example, see the curves in Figure 9.19b) with the optimum blocked concentration to maximize $\partial\ln r/\partial m^*$ (for example, see the curves in Figure 9.23). Thus, the m_{th} of the development function must be tuned to the value of η and quencher concentration for the resist. And,

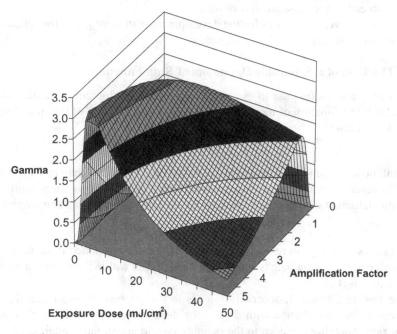

Figure 9.25 *The overall photoresist contrast (gamma) as a function of exposure dose (E) and amplification factor (α_f) for a chemically amplified resist with no diffusion ($\eta = 0$, $r_{max} = 100\,nm/s$, $r_{min} = 0.1\,nm/s$, n = 5, C = 0.05\,cm^2/mJ$)*

if the quencher level is set high enough so that η is near its critical level, the maximum resist contrast can approach its maximum possible level:

$$\gamma_{max} \approx n \frac{m_{th}}{(1-m_{th})} \left[\frac{(1-q_0)\ln(1-q_0)}{\eta} \right], \quad m_{th} \approx e^{-\alpha_f h_{eff}}. \quad (9.75)$$

9.7   Resist Profile Formation

We have now arrived at the final 'image' gradient, the gradient in development rate. As we shall see next, the last step will be to relate this gradient to the control of the resist feature edge by knowing the development path. The development path traces the surface of the resist through the development cycle from the top of the undeveloped resist to a point on the final resist profile (see Chapter 7). The basic equation which defines the physical process of development is an integral equation of motion:

$$t_{dev} = \oint \frac{ds}{r(x,y,z)} \quad (9.76)$$

where t_{dev} is the development time, $r(x,y,z)$ is the development rate at every point in the resist, ds is a differential length along the development path, and the path integral is taken along the development path. The endpoint of the development path defines the position of the final resist profile and, consequently, the final critical dimension. There is only one path possible, determined by the principle of least action: the path will be that which goes from start to end in the least amount of time.

In Chapter 7, several solutions for the development path were given for certain special cases. Two of these cases are described below.

9.7.1   The Case of a Separable Development Rate Function

An interesting and useful case to examine is when the development rate function $r(x,z)$ is separable into a function of incident dose E only and a function of the spatial variables independent of dose:

$$r(E,x,z) = f(E)g(x,z) \quad (9.77)$$

In general, none of the kinetic development models discussed in Chapter 7 meet this criterion. However, separability results from assuming a constant theoretical contrast. Integrating the definition of theoretical contrast for the case of constant contrast gives

$$r(x,z) \propto E^\gamma (I_r(x,z))^\gamma \quad (9.78)$$

where $I_r(x,z)$ is the image in resist. Thus, separability can be assumed if the range of dose seen over the development path is sufficiently small to fall within the approximately linear region of the H–D curve.

As we saw in Chapter 7, separability leads to the interesting result that the path of development does not change with dose – the development just follows this path at a different rate. Applying this idea to the defining development path Equation (9.76),

$$f(E)t_{dev} = \int_{x_0}^{x} \frac{dx\sqrt{1+z'^2}}{g(x,z)} \quad (9.79)$$

For example, let $f(E) = E^\gamma$. If the dose is increased, the same path is used to get from x_0 to x, but it arrives in a shorter time.

Equation (9.79) allows us to calculate exposure latitude: how a change in dose effects the final position on the path x for a given development time. Taking the derivative of this equation with respect to log-exposure dose,

$$\frac{\partial f}{\partial \ln E} t_{dev} = \frac{\sqrt{1+z'^2}}{g(x,z)} \frac{\partial x}{\partial \ln E} \tag{9.80}$$

Solving for the exposure latitude term, $\partial x/\partial \ln E$,

$$\frac{\partial x}{\partial \ln E} = \frac{\partial \ln f}{\partial \ln E} \frac{r(x,z)t_{dev}}{\sqrt{1+z'^2}} = \frac{\partial \ln f}{\partial \ln E} r(x,z)t_{dev} \sin\theta \tag{9.81}$$

where θ is the sidewall angle of the resist profile (which generally is close to 90°). For the case of $f(E) = E^\gamma$,

$$\frac{\partial x}{\partial \ln E} = \gamma r(x,z)t_{dev} \sin\theta \tag{9.82}$$

Similarly, the sensitivity of the final edge position x to changes in development time can also be calculated. Taking the derivative of Equation (9.79) with respect to log-development time,

$$\frac{\partial x}{\partial \ln t_{dev}} = r(x,z)t_{dev} \sin\theta \tag{9.83}$$

While Equations (9.82) and (9.83) were derived to give us insight into exposure latitude and development latitude, they also provide for an interesting approach to understanding the development time/exposure dose trade-off. If a longer development time is used, a lower dose can be used. But what combination of dose and development time will maximize CD control? For our separable development rate assumption, Equation (9.79) shows us that any process that gives the correct edge position (correct CD) must have $r(x,z)t_{dev}$ = constant. But Equations (9.82) and (9.83) show us that $r(x,z)t_{dev}$ = constant will produce a constant sensitivity to relative dose errors and relative development time errors. In other words, under the assumption of constant γ, any dose/develop time combination that produces the correct CD will also allow for the control of that CD with the same sensitivity to relative dose errors and relative development time errors.

9.7.2 Lumped Parameter Model

In Chapter 7, the lumped parameter model (LPM) was used to derive a general way to relate the dose $E(x)$ required to achieve a certain CD ($=2x$) to the shape of the aerial image $I(x)$, using resist contrast γ and the effective resist thickness D_{eff} to describe the influence of the photoresist on this image:

$$\left(\frac{E(x)}{E(0)}\right)^\gamma = 1 + \frac{1}{D_{eff}} \int_{x_0}^{x} \left(\frac{I(x')}{I(x_0)}\right)^{-\gamma} dx' \tag{9.84}$$

where the starting point in the development path, x_0, is generally assumed to be the middle of the space for a pattern with narrow spaces. From this, it is easy to differentiate and calculate the exposure latitude of the resist edge position x:

$$\frac{\partial x}{\partial \ln E} = r(x)\gamma t_{\text{dev}} = \gamma D_{\text{eff}}\left[\frac{E(x)I(x)}{E(0)I(0)}\right]^{\gamma} \tag{9.85}$$

Note that the result matches the general expression (9.82) since the LPM makes the explicit assumption of segmented development so that the development path will always be horizontal and $\sin\theta = 1$.

Equation (9.85) can be used for a specific case also derived in Chapter 7. Assuming that the shape of the aerial image in the region being developed can be approximated as a Gaussian (that is, assuming the log-slope of the image varies linearly with position in this region), the LPM predicts

$$\left(\frac{E(x)}{E(0)}\right)^{\gamma} \approx 1 + \left(\frac{1}{\gamma D_{\text{eff}}ILS}\right)\left(\frac{I(x)}{I(0)}\right)^{-\gamma} \tag{9.86}$$

where the approximate equal sign is used since the actual equation yields a Dawson's integral that is approximated in Equation (9.86). Note that the image log-slope, ILS, is used in this equation as a function of x rather than being defined at the nominal line edge. Substituting this specific result into the general LPM expression for exposure latitude gives

$$\frac{\partial x}{\partial \ln E} \approx \frac{1}{ILS} + \gamma D_{\text{eff}}\left[\frac{I(x)}{I(0)}\right]^{\gamma} \tag{9.87}$$

Often it is convenient to substitute $CD = 2x$ and compute the log-log slope of the CD-versus-exposure dose curve:

$$\frac{\partial \ln CD}{\partial \ln E} \approx \frac{2}{NILS} + \frac{2\gamma D_{\text{eff}}}{CD}\left[\frac{I(CD/2)}{I(0)}\right]^{\gamma} \tag{9.88}$$

There are two distinct terms on the right-hand side of this expression. The first, $2/NILS$, is a pure aerial image term and is the limiting value of the exposure latitude for the case of an infinite contrast resist (Figure 9.26). The second term is a 'development path factor' that includes the aspect ratio of the resist (D_{eff}/CD) and the ratio of the aerial image intensity at the edge of the pattern relative to that in the center of the space (that is, the exposure dose at the end of the path relative to the exposure dose at the beginning of the path). This development path factor is reduced (giving better CD control) by lowering the aspect ratio of the resist, increasing the resist contrast, and reducing the aerial image intensity at the dark line edge (i.e. $x = CD/2$) relative to the bright space center ($x = 0$).

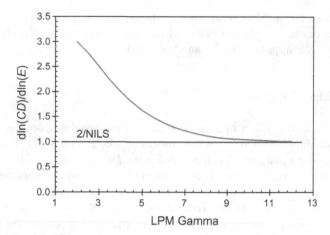

Figure 9.26 *A plot of Equation (9.88) showing how the exposure latitude term approaches its limiting value of 2/NILS as the lumped photoresist contrast increases. In this case, the resist aspect ratio is 2, the ratio I(CD/2)/I(0) is 0.5 and the NILS is 2*

A slightly more accurate form of Equation (9.88) can be obtained by using two terms in the polynomial expansion of the Dawson's integral. The result is

$$\frac{\partial \ln CD}{\partial \ln E} \approx \frac{2}{NILS}\left(1 + \frac{2}{\gamma NILS} + \gamma NILS \frac{D_{eff}}{CD}\left[\frac{I(CD/2)}{I(0)}\right]^{\gamma}\right) \qquad (9.89)$$

The relative difference between this equation and Equation (9.88) is generally very small except for the case where the resist is very thin.

Equation (9.88) finally brings all the pieces of the NILS puzzle together, describing the information transfer from the aerial image through development to the final resist image. It relates the fractional change in CD to the fractional change in exposure dose and thus its inverse defines the exposure latitude. The aerial image affects CD control in two ways, through the NILS directly and through the development path factor. The exposure, PEB and development effects can be lumped together into a photoresist contrast term, or can be separated out into individual components as described earlier.

Since the above expressions were derived specifically for a Gaussian aerial image, that specific image can be used in Equation (9.88) or (9.89).

$$\frac{I(CD/2)}{I(0)} = e^{-NILS/4} \qquad (9.90)$$

giving

$$\frac{\partial \ln CD}{\partial \ln E} \approx \frac{2}{NILS}\left(1 + \frac{2}{\gamma NILS} + \gamma NILS \frac{D_{eff}}{CD} e^{-\gamma NILS/4}\right) \qquad (9.91)$$

In this form, the exposure latitude is determined by just three factors: NILS, resist contrast and the feature aspect ratio. For large $\gamma NILS$, Equation (9.91) approaches the result for an ideal threshold resist. Also, recalling the lithographic imaging equation, $\gamma NILS$ is the normalized log-development rate gradient. Thus, optimizing this gradient helps make the photoresist response approach that of an ideal resist.

9.8   Line Edge Roughness

When variations in the width of a resist feature occur quickly over the length of the feature, this variation is called *line width roughness* (see Figure 9.27). When examining these variations along just one edge, it is called *line edge roughness* (LER). LER becomes important for feature sizes on the order of 100 nm or less, and can become a significant source of linewidth control problems for features below 50 nm. LER is caused by a number of statistically fluctuating effects at these small dimensions such as shot noise (photon flux variations), statistical distributions of chemical species in the resist such as photoacid generators (PAGs), the random walk nature of acid diffusion during chemical amplification, and the nonzero size of resist polymers being dissolved during development.

LER is usually characterized as the 3σ deviation of a line edge from a straight line, though a more complete frequency analysis of the roughness can be valuable as well. For 193-nm lithography, LER values of 4 nm and larger are common. The impact of LER on device performance depends on the specific device layer and specific aspects of the device technology. For lithography generations below 100 nm, typical specifications for the 3σ LER are about 5% of the nominal CD. It is possible that LER will become the main limiter of CD control below 65-nm production.

In Chapter 6, a stochastic model for exposure and reaction–diffusion of chemically amplified resists was developed. This stochastic model will now prove useful for the prediction of line edge roughness. As a review, for a given volume of resist under consideration, the statistical variance in the final blocked polymer concentration is given by

$$\sigma_{m*}^2 = \frac{\langle m^* \rangle}{\langle n_{0-\text{blocked}} \rangle} + (\langle m^* \rangle \ln \langle m^* \rangle)^2 \left(\frac{2a}{\sigma_D} \right)^p \left(\frac{1}{\langle n_{0-\text{PAG}} \rangle \langle h \rangle} + \frac{[(1 - \langle h \rangle) \ln(1 - \langle h \rangle)]^2}{\langle h \rangle^2 \langle n \rangle} \right) \qquad (9.92)$$

where $\langle h \rangle$ is the mean value of the acid concentration, $\langle m^* \rangle$ is the mean value of the blocked polymer concentration after PEB, σ_D/a is the ratio of acid diffusion length to the

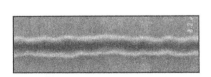

Figure 9.27   *SEM pictures of photoresist features exhibiting line edge roughness*

capture range of the deblocking reaction, $\langle n \rangle$ is the mean number of photons (to account for photon shot noise), $\langle n_{0-PAG} \rangle$ is the mean initial number of PAGs in the control volume at the start of exposure, and $\langle n_{0-blocked} \rangle$ is the mean initial number of blocked polymer groups in the volume before PEB.

While development can also be included, for the sake of simplicity, we will assume an infinite contrast development process so that the line edge will be determined by the blocked polymer latent image. Thus, a simple threshold model for the latent image will determine the resist critical dimension. And, as we saw in Chapter 2 for an aerial image threshold model, a Taylor series expansion of the blocked polymer concentration, cut off after the linear term, allows us to predict how a small change in blocked polymer concentration (Δm^*) will result in a change in edge position (Δx):

$$\Delta x = \frac{\Delta m^*}{dm^*/dx} \tag{9.93}$$

From this, we can devise a simple qualitative model for line edge roughness. The standard measure of line edge roughness, from a top-down SEM, will be proportional to the standard deviation of blocked polymer concentration divided by its gradient perpendicular to the line edge:

$$LER \propto \frac{\sigma_{m^*}}{dm^*/dx} \tag{9.94}$$

Thus, to achieve a low LER it will be necessary to make the standard deviation of the deprotection small and make the gradient of deprotection large. Of course, a main topic of this chapter was how process parameters can be used to maximize the latent image gradient. From Equation (9.92) we can see how to minimize the statistical uncertainty in deprotection. There is one interesting variable in common to both: acid diffusion. Increasing acid diffusion will reduce σ_{m^*}, but will reduce the latent image gradient. One would expect, then, an optimum level of diffusion to minimize the LER.

To investigate the impact of diffusion on LER, we can combine Equations (9.92) and (9.43) into (9.94). Thus, for the no-quencher case, and ignoring photon shot noise,

$$\sigma_{m^*} = \sqrt{\frac{\langle m^* \rangle}{\langle n_{0-blocked} \rangle} + (\langle m^* \rangle \ln \langle m^* \rangle)^2 \left(\frac{2a}{\sigma_D} \right)^p \left(\frac{1}{\langle n_{0-PAG} \rangle \langle h \rangle} \right)}$$

$$\frac{\partial m^*}{\partial x} \propto \frac{1}{\sigma_D^2} (1 - e^{-\pi^2 \sigma_D^2 / 2L^2}) \approx \frac{\pi^2}{2L^2} \left(1 - \frac{\pi^2 \sigma_D^2}{4L^2} \right) \text{(since } \sigma_D \ll L\text{)} \tag{9.95}$$

so that

$$LER \propto \frac{\sigma_D^2}{1 - e^{-\pi^2 \sigma_D^2 / 2L^2}} \sqrt{\frac{\langle m^* \rangle}{\langle n_{0-blocked} \rangle} + (\langle m^* \rangle \ln \langle m^* \rangle)^2 \left(\frac{2a}{\sigma_D} \right)^p \left(\frac{1}{\langle n_{0-PAG} \rangle \langle h \rangle} \right)} \tag{9.96}$$

and

$$LER \propto \frac{\sigma_D^2}{1 - e^{-\pi^2 \sigma_D^2 / 2L^2}} \sqrt{1 - (K_{amp} t_{PEB}) \langle m^* \rangle \ln \langle m^* \rangle \left(\frac{2a}{\sigma_D} \right)^p \frac{\langle n_{0-block} \rangle}{\langle n_{0-PAG} \rangle}} \tag{9.97}$$

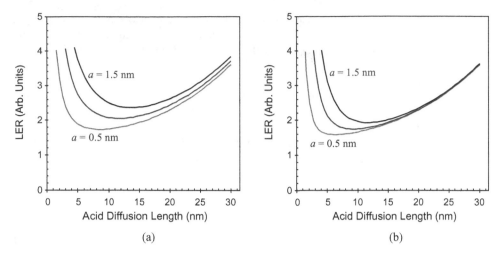

Figure 9.28 *Prediction of LER trends for a 45-nm feature using the generic conditions found in Equation (9.99) and using three values of the deblocking reaction capture range a (0.5, 1.0 and 1.5 nm): (a) assuming a 2D problem, and (b) for a 3D problem*

For $\sigma_D \ll L$ this becomes

$$LER \propto \left(1 + \frac{\pi^2 \sigma_D^2}{4L^2}\right)\sqrt{1 - (K_{\text{amp}} t_{\text{PEB}})\langle m^*\rangle \ln\langle m^*\rangle \left(\frac{2a}{\sigma_D}\right)^p \frac{\langle n_{0-\text{block}}\rangle}{\langle n_{0-\text{PAG}}\rangle}} \qquad (9.98)$$

To evaluate this expression, consider some typical values. Let $M_0 N_A = 1.2/\text{nm}^3$, $G_0 N_A = 0.042/\text{nm}^3$, $\langle h \rangle = 0.3$, $\langle m^* \rangle = 0.6$. Thus, the LER becomes

$$LER \propto \frac{\sigma_D^2}{1 - e^{-\pi^2 \sigma_D^2/2L^2}}\sqrt{1 + 15\left(\frac{2a}{\sigma_D}\right)^p} \qquad (9.99)$$

Figure 9.28 shows the trend of LER versus acid diffusion for a 45-nm feature for three different values of the deprotection capture range a, 0.5, 1.0 and 1.5 nm. In each case, there is a diffusion length that minimizes the LER. Below the optimum diffusion length, LER is limited by σ_{m^*} so that increasing the diffusion will improve LER. Above the optimum diffusion length the LER is gradient limited, so that increases in diffusion further degrade the gradient and worsen the LER.

9.9   Summary

The above discussion traces the aerial image log-slope through to its ultimate impact on exposure latitude (see Table 9.1). The resist contrast relates the original image contrast to the final development rate gradient, encompassing exposure, chemical amplification,

Table 9.1 *Summary of lithography process steps and their corresponding information metrics*

Process Step	Information	Error Sources	Information Metric
Design	Polygons, binary	(usually assumed perfect)	
Mask	Amplitude transmittance, $t_m(x,y)$	CD and registration errors, corner rounding, phase and transmittance	
Aerial Image	$I(x,y)$	Diffraction limitation, aberrations, defocus, flare, polarization	NILS
Image in Resist	$I(x,y,z)$	Substrate reflections/ thin film effects, polarization effects, defocus through the resist	NILS
Exposure	Latent Image $m(x,y,z)$ or $h(x,y,z)$ (before PEB)	Exposure dose errors	Latent image gradient
Post-exposure Bake	Latent Image $m^*(x,y,z)$ (after PEB)	Thermal dose errors, diffusion	Latent image gradient
Development	Development Rate $r(x,y,z)$ + Resist Profile (CD, sidewall angle, resist loss)	Finite contrast, r_{max}/r_{min}	Development rate log-slope, gamma + exposure latitude, CD error

diffusion and development. Since NILS degrades with focus in a predictable way, the final Equation (9.88) provides a simple means of capturing the impacts of both focus and exposure on final CD control. Obviously, this simple gradient approach has its limitations. In particular, the impact of isofocal bias on depth of focus is not captured by looking only at gradients. However, the gradient-based optimization embodied in the use of the normalized image log-slope has proven to be an extremely powerful theory-based guide to full lithographic optimization.

The chain rule of differentiation provides a simple approach for identifying the effects of each step in the process:

$$\frac{\partial \ln R}{\partial x} = \left(\frac{\partial \ln R}{\partial m^*} \right) \left(\frac{\partial m^*}{\partial m} \right) \left(\frac{\partial m}{\partial \ln E} \right) \left(\frac{\partial \ln I}{\partial x} \right) \tag{9.100}$$

Comparing this equation to the lithographic imaging Equation (9.53) shows that the photoresist contrast is a combined metric of exposure, PEB and development:

$$\gamma = \left(\frac{\partial \ln R}{\partial m^*} \right) \left(\frac{\partial m^*}{\partial m} \right) \left(\frac{\partial m}{\partial \ln E} \right) \tag{9.101}$$

Problems

9.1.   Using results from Chapter 3, show that the aerial image NILS falls off approximately quadratically with defocus.

9.2.   Consider the three-beam aerial image of lines and spaces:

$$I(x) = a_0^2 + 4a_0a_1 \cos(2\pi x/p)4a_1^2 \cos^2(2\pi x/p)$$

Assuming equal lines and spaces, plot $pILS(x)$ versus x/p for $0 \leq x \leq p/2$.

9.3.   A plot of DOF versus NA (for a given feature) looks something like

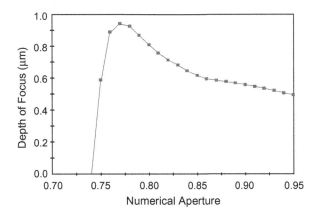

Explain qualitatively why there is an optimum NA that maximizes the depth of focus.

9.4.   Using the following approximation for exposure latitude,

$$EL = 20 \frac{\partial \ln E}{\partial \ln CD}$$

graphically compare Equations (9.6) and (9.91) [let $\gamma = 10$ and $D_{eff}/CD = 2$]. What are the best-fit values of α and β using NILS in the range from 1.5 to 3.5?

9.5.   Using Equations (9.20) and (9.21),

a)   Derive an equation for $\left.\dfrac{\partial m^*/\partial x}{\partial m/\partial x}\right|_{x=p/4}$ for $N = 4$.

b)   For this case, derive an expression for the approximate magnitude of the relative error in Equation (9.22).

9.6.   Derive Equation (9.44) [hint: first assume $m^* =$ constant, then assume h_{eff} is constant].

9.7.   Show that in the limit as $\eta \rightarrow 0$, Equation (9.39) becomes Equation (9.25).

9.8.   For the blocked polymer latent image when quencher is used, show that the maximum relative LIG becomes

$$\left.\frac{\partial m^*/\partial x}{ILS}\right|_{max} \approx \frac{1}{\eta}(1-q_0)\ln(1-q_0)$$

whenever $\eta < \eta_c$.

9.9. Show that at the optimum conditions (i.e. when the blocked polymer LIG is at its maximum),

$$\frac{1-e^{-\alpha_f \eta}}{\eta} = \frac{1}{h_{\text{eff}} + \eta}$$

9.10. Consider a resist obeying the original kinetic development model so that $n = 12$, $m_{\text{th}} = 0.65$, and $r_{\text{max}}/r_{\text{min}} = 2000$. If one desired to reformulate the resist to increase the maximum log-development rate versus m^* by 20%, how much would each one of these variables have to be individually changed?

9.11. In the limit of $\sigma_D \gg a$, derive an approximate proportionality for LER. In this regime, how does increasing diffusion affect LER?

References

1 Levenson, M.D., Goodman, D.S., Lindsey, S., Bayer, P.W. and Santini, H.A.E., 1984, The phase-shifting mask II: imaging simulations and submicrometer resist exposures, *IEEE Transactions on Electron Devices*, **ED-31**, 753–763.

2 Levinson, H.J. and Arnold, W.H., 1987, Focus: the critical parameter for submicron lithography, *Journal of Vacuum Science and Technology B*, **B5**, 293–298.

3 Mack, C.A., Understanding focus effects in submicron optical lithography, *Proceedings of SPIE: Optical/Laser Microlithography*, **922**, 135–148 (1988), and *Optical Engineering*, **27**, 1093–1100 (1988).

4 Mack, C.A., 1987, Photoresist process optimization, *KTI Microelectronics Seminar, Interface '87, Proceedings*, 153–167.

5 Smith, M.D. and Mack, C.A., 2001, Examination of a simplified reaction-diffusion model for post exposure bake of chemically amplified resists, *Proceedings of SPIE: Advances in Resist Technology and Processing XVIII*, **4345**, 1022–1036.

10
Resolution Enhancement Technologies

In general, the optimum image quality for an arbitrary object is obtained by using a classical imaging system with zero aberrations. Semiconductor lithography, however, uses a very limited set of objects (arrays of lines and spaces, isolated lines, small square holes called contacts, etc.) in limited arrangements and orientations. In many cases, the sizes of these objects are also limited to set values. Thus, lithography can pose a unique imaging problem – given a single object (say, an array of contact holes of a given size), is it possible to design a special imaging system which is better than a classical imaging system for the printing of this one object? The answer is yes.

The design of an imaging system optimized for a reduced class of objects has been called *wavefront engineering* by one of the earlier pioneers in this area, Marc Levenson.[1] Wavefront engineering can be thought of in a number of different ways, but essentially it refers to manipulating the optical wavefront exiting the projection lens to produce improved images for certain types of objects. There are three basic ways of modifying the wavefront – manipulating the object (the mask), adjusting the illumination of the object, or modifying the wavefront directly with a pupil filter. The use of a pupil filter is difficult with current lens designs since the pupil is not directly accessible within the lens. Modified illumination and masks, however, are widely used in semiconductor lithography to improve imaging.

In its simplest form, an imaging system uses an object which is identical to the desired image. However, knowing the limitations of the imaging system, one can design a new object which produces an image more like the desired pattern. By adjusting the transmittance of the mask, one can improve the exposure latitude and depth of focus of a given image. Ideally, the spatial variation of both the phase and the transmittance of the mask would be modified in whatever way needed to improve the image. From a practical point of view though, masks for semiconductor lithography are essentially devices with discrete levels of transmittance (the simplest mask is binary, either 100 or 0% transmittance).

Fundamental Principles of Optical Lithography: The Science of Microfabrication, Chris Mack.
© 2007 John Wiley & Sons, Ltd.

Each added level of transmittance adds considerably to the cost and complexity of the mask. For this reason, mask modification falls into two categories – mask shaping (also called *optical proximity correction*, or OPC) and *phase-shifting masks* (PSMs). Mask shaping adjusts the shape of the mask features to improve the shape of the resulting image (for example, by applying serifs to the corners of a rectangular pattern). The difficulty comes in predicting the needed mask shapes (a typical chip will require masks with hundreds of millions or billions of patterns that must be shaped). PSMs change (or possibly add to) the levels of transmittance on the mask to include phase differences between transmittance levels. Light shifted by 180° will interfere with unshifted light to produce well-controlled dark areas on the image. The difficulty again includes designing the mask, but also involves more complicated mask fabrication.

Modifying the illumination of the mask will also result in different wavefronts and thus different images. As we saw in Chapters 2 and 3, tilting the illumination can double the resolution of grating (line/space) patterns and can also improve the depth of focus of gratings of certain periods. This tilting, however, does not improve the performance of nonrepeating patterns. Thus, the direction (or directions) of the illumination can be customized for given mask features, but there is no one illumination direction which is best for all mask features. Modified illumination (also called *off-axis illumination* or OAI) can be combined with mask modification (OPC and possibly PSM) to produce results better than either approach alone, but with increasingly complicated processes.

Collectively, these three techniques, OPC, PSM and OAI, are known as *resolution enhancement technologies* (RETs). To see how they actually improve resolution, it is important to first carefully define what is meant by the term resolution.

10.1   Resolution

The resolution limit of optical lithography is not a simple function. In fact, resolution limits differ depending on the type of feature being printed. In general, however, there are two basic types of resolution: the smallest pitch that can be printed (the pitch resolution) and the smallest feature that can be printed (the feature resolution). While related, these two resolutions are limited differently by the physics of lithography, and have different implications in terms of final device performance. Pitch resolution, the smallest linewidth plus spacewidth pair that will print, determines how closely transistors can be packed together on one chip. This resolution has the greatest impact on cost per function and functions per chip. Feature size resolution determines the characteristics and performance of an individual transistor, and has the greatest impact on chip speed and power consumption. Obviously both are very important.

Pitch resolution is the classical resolution discussed in most optics textbooks and courses. It is governed by the wavelength of the light used to form the images and the numerical aperture of the imaging lens. The classical pitch resolution is given by a version of the Rayleigh resolution equation:

$$pitch\ resolution = k_{pitch}\frac{\lambda}{NA} \tag{10.1}$$

where λ is the vacuum wavelength of the lithographic imaging tool, NA is the numerical aperture, and k_{pitch} depends on the details of the imaging process. Ultimately, k_{pitch} can be as low as 0.5, but only with tremendous effort (as will be discussed below). Values of 0.7–1.0 are typical, with a trend toward decreasing values approaching the 0.5 limit.

While pitch resolution has a hard physical limit given by the Rayleigh equation, feature resolution is limited by the ability to control the critical dimension (CD) of the feature. As features are made smaller, control of the CD of that feature becomes harder. There is no hard cutoff, only a worsening of CD control as the feature size is reduced. Feature size control is governed by the magnitude of various process errors that inevitably occur in a manufacturing environment (such as focus and exposure errors), and the response of the process to those errors (see Chapter 8). In order to improve CD control, one must simultaneously reduce the sources of process errors and improve process latitude (the response of CD to an error). Interestingly, process latitude is similar to pitch resolution in that it also depends most strongly on the imaging wavelength and numerical aperture, though in more complicated ways.

Process latitude is an exceedingly general concept, but usually the most important process latitudes are the interrelated responses of CD to focus and exposure. Using the process window to characterize these responses (see Chapter 8), we define depth of focus (DOF) as *the range of focus which keeps the resist profile of a given feature within all specifications (linewidth and profile shape) over a specified exposure range*. One of the most insidious difficulties of lithography is that small features tend to have smaller depth of focus.

10.1.1 Defining Resolution

Resolution is, quite simply, the smallest feature that you are able to print (with a given process, tool set, etc.). The confusion comes from what is meant by 'able'. For a researcher investigating a new process, 'ability' might mean shooting a number of wafers, painstakingly searching many spots on each wafer, and finding the one place where a small feature looks somewhat properly imaged. For a production engineer, the *manufacturable* resolution might be the smallest feature size which provides adequate yield for a device designed to work at that size. We can define resolution, similar to our definition of DOF, in such a way that it can meet these varied needs.

Producing an adequately resolved feature in a realistic working environment means printing the feature within specifications (linewidth, sidewall angle, resist loss, line edge roughness, etc.) over some expected range of process variations. As we have seen before, the two most common process variations are focus and exposure. Since our definition of depth of focus includes meeting all profile specifications over a set exposure range, a simple definition of resolution emerges: *the smallest feature of a given type which can be printed with a specified depth of focus*. This definition is perfectly general. If the exposure latitude specification used in the DOF definition is set to zero and the DOF specification in the resolution definition is set to zero, the 'research' use of the term resolution is obtained (if it prints once, it is resolved). If the exposure latitude and DOF specifications are made sufficiently large to handle all normal process errors encountered in a manufacturing line, the 'manufacturing' use of the term resolution is obtained. As with the

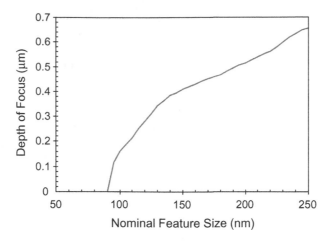

Figure 10.1   *Resolution can be defined as the smallest feature which meets a given DOF specification. Shown are results for equal lines and spaces, λ = 193 nm, NA = 0.9, σ = 0.7, typical resist on a nonreflective substrate*

definition of DOF, the choice of the specifications determines whether the resulting resolution is appropriate to a given application.

Figure 10.1 illustrates this concept of resolution. The depth of focus for a pattern of equal lines and spaces is shown as a function of feature size. (For this and subsequent figures, the DOF is based on profile specifications of CD ±10%, sidewall angle >80°, resist loss <10%, and an exposure latitude specification of 6%. All focus and exposure errors are assumed to be random. Each data point assumes that nominal exposure and focus were adjusted to give the best process window and thus the largest possible DOF. Mask linearity – the ability to print different feature sizes at the same time – is not considered here, but could easily be added as a constraint.) If zero DOF is required, the equal line/space resolution for this process would be about 80 nm. A requirement of 160-nm DOF would increase the minimum printable feature size to 100 nm, and a requirement of 250-nm DOF would degrade the resolution further to 115 nm. Obviously, a simple statement of the resolution without clearly stating the DOF requirement (and thus the profile and exposure latitude requirements) would be of little use.

Figure 10.2 illustrates how a given process, tool set, etc., does not have a single resolution for all feature types. The contact hole shows the worst resolution under these conditions, while the isolated line has the greatest resolution for small required DOF. Figure 10.3 illustrates how a careful definition of resolution can elucidate fundamental lithographic behavior, such as the role of numerical aperture. For larger features, lower NA gives more depth of focus. But for smaller features, the DOF falls off more quickly for the lower NA. This results in the well-known effect of an optimum NA to give the greatest DOF. But it also impacts resolution in an interesting way. If no DOF is required, the resolution (the point where each curve in Figure 10.3 hits the *x*-axis) follows the familiar trend of increased resolution with increased NA. If, however, a nonzero DOF is required, the behavior of resolution with NA becomes more complicated.

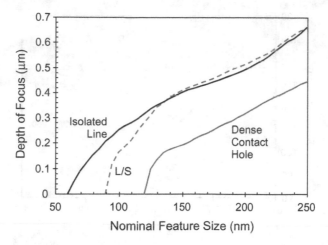

Figure 10.2 *Comparison of the resolution for different feature types (λ = 193 nm, NA = 0.9, σ = 0.7). Here, L/S means equal lines and spaces*

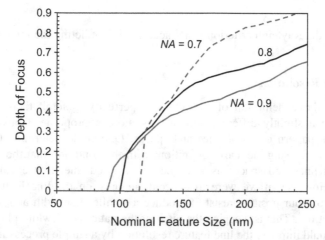

Figure 10.3 *The definition of resolution can be used to study fundamental lithographic trends, such as the impact of numerical aperture (NA) on resolution (λ = 193 nm, σ = 0.7, equal lines and spaces)*

Figure 10.4 expands on the results of Figure 10.3 and shows the resolution of equal line/space arrays as a function of numerical aperture for different DOF specifications.[2] For example, with a required DOF of 200 nm, the resolution reaches an optimum (a minimum in the curve at a feature size of 105 nm) at a numerical aperture of 0.84. Larger numerical apertures actually reduce the resolution! As the required DOF is reduced, the NA which gives maximum resolution moves out to higher values. Also shown on the graph is the Rayleigh resolution criterion ($R = k_1 \lambda/NA$) for comparison. For a nonzero

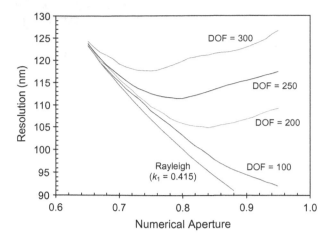

Figure 10.4 *Resolution as a function of numerical aperture is more complicated than Rayleigh's criterion would imply (λ = 193 nm, σ = 0.7, equal lines and spaces). Graphs show different resolution versus NA results for different minimum DOF specifications in nanometers*

required DOF, the Rayleigh criterion is not accurate in predicting the influence of NA on resolution.

10.1.2   Pitch Resolution

Although the above definition of resolution is perfectly general, there is a class of features where a slightly different approach is more appropriate. Consider a simple repeating mask pattern of equal lines and spaces. One could determine the resolution of this pattern type using the above definition, with the condition that the resist features must be equal lines and spaces as well (i.e. the desired linewidth equals the desired spacewidth). However, if we were to concentrate only on the line feature, we could easily overexpose our positive resist to produce a smaller linewidth and, most likely, a smaller 'resolution'. That is, keeping the pitch of the pattern (linewidth plus spacewidth) constant, we could improve the line feature resolution by a simple processing change, but only at the expense of the space feature resolution. Is this truly an improvement in resolution?

The answer depends on the application. If only the width of the line is critical, then resolution should be based only on the line feature. The electric performance of a device, for example, may be critically dependent on the linewidth of a given device structure, but only marginally affected by the accompanying space feature size. In most cases, however, the space feature is also critical (affecting capacitance at interconnect levels and strain at the gate level, for example). In fact, the ability to decrease both linewidth and space-width simultaneously allows manufacturers to shrink chip sizes, putting more chips on a wafer and providing a huge economic driver for the quest for better resolution. For such applications, where linewidth and spacewidth are both critical , one can modify the above definition of *feature* resolution to produce a definition for *pitch* resolution: *the smallest pitch of a given duty cycle which can be printed with a specified depth of focus*, where duty cycle is defined as the ratio of spacewidth to linewidth.

At first glance the difference between the feature resolution and the pitch resolution seems almost trivial – hardly worth the effort to propose a separate definition. However, use of these two different types of resolutions reveals that the physical limits to resolution can be quite different for each. Consider the simple case of forming an image of an equal line/space mask pattern illuminated with a single wavelength, normally incident plane wave (i.e. coherent illumination). For such a case, there will be a hard cutoff for the pitch resolution: when the pitch drops below λ/NA no image whatsoever is formed for any duty cycle. Regardless of the profile, exposure latitude and DOF specifications, no pitch below this limit can be imaged. For the case of an isolated line, there is no equivalent 'hard cutoff' of the feature resolution, which instead exhibits a gradual reduction in profile control as the feature size is decreased.

In general terms, the feature resolution is limited by photoresist profile control and is a complicated function of wavelength and numerical aperture. Ultimately, the pitch resolution is limited by the cutoff of discrete diffraction information passing through the objective lens and is a relatively simple function of wavelength and numerical aperture. To understand this simple functionality, one must understand that a single diffraction order passing through the objective lens produces a single plane wave of light striking the wafer (see Chapter 2). Two plane waves at the wafer (coming from two separate diffraction orders) will interfere with each other to produce a sinusoidal pattern of light and dark (giving spaces and lines). If the two plane waves strike the image plane (i.e. the wafer plane) with angles θ_1 and θ_2 with respect to the optical axis (that is, a normal to the image plane), then the period of the resulting image will be

$$period = \frac{\lambda}{\sin\theta_1 - \sin\theta_2} \qquad (10.2)$$

This general expression can be used to understand a variety of imaging situations. For the coherent illumination case described above, $\theta_2 = 0$ for the zero order and $\sin\theta_1$ has a maximum value of NA, giving the λ/NA resolution mentioned above. In other words, for three-beam imaging the minimum k_{pitch} is 1.0. For the special case of two symmetrical beams, $\theta_1 = -\theta_2 = \theta$ and the period becomes

$$period = \frac{\lambda}{2\sin\theta} \qquad (10.3)$$

The ultimate pitch resolution of an imaging tool is obtained when $\sin\theta$ becomes its maximum value, the NA:

$$ultimate\ pitch\ resolution = \frac{\lambda}{2NA} \qquad (10.4)$$

for all duty cycles. Thus, for two-beam imaging, the minimum k_{pitch} is 0.5. This ultimate pitch resolution uses two-beam imaging that can be obtained from an alternating phase-shifting mask or an optimized off-axis illumination scheme (see sections 10.3 and 10.4). However, the actual pitch resolution may not be this good if the resulting printed image does not meet the required DOF specifications (that is, if the patterns are limited by the feature resolution of the line or the space making up the pitch).

10.1.3   Natural Resolutions

Besides the two kinds of resolution discussed so far, feature resolution and pitch resolution, a third type of resolution emerges from a study of optical imaging: natural resolutions of an imaging system. Since a discussion of natural resolution will make use of many of the concepts that will be developed in the sections that follow, its treatment will be delayed until the end of this chapter, in section 10.5.

10.1.4   Improving Resolution

Considering both pitch resolution and feature resolution, the following approaches are commonly used to improve resolution:

- Reducing the exposure wavelength
- Increasing the imaging lens numerical aperture
- Reducing k_{pitch} by using two-beam imaging
- Increasing the focus–exposure process window size
- Reducing the magnitude of process errors such as focus and exposure errors.

Unfortunately, several of these factors work against each other. For a given feature, there is an optimum numerical aperture that gives the largest process window. Increasing the NA further will reduce the process window, making the feature resolution worse even as the pitch resolution is improved. Reducing wavelength is always good, but as a practical matter, it is extraordinarily difficult since we are limited by our ability to engineer materials with the proper optical properties at the lower wavelength. Of course, reducing the magnitude of process errors is a never ending quest with cost being the only possible downside. That leaves two final resolution enhancement approaches: reducing k_{pitch} with two-beam imaging and increasing the size of the process window.

By far the most effective and popular process window improvement approach has been improvements in the photoresist. Over the years, resist capabilities have undergone dramatic progress. While we are still far from being able to ignore the photoresist (that is, we do not have a diffusionless infinite contrast resist), resist performance today is high enough that even small improvements in optical imaging can be seen in the final patterns.

Attempts to improve the process window by optical means (sometimes called optical 'tricks') include:

- Optimization of the mask pattern shape (called optical proximity correction, OPC)
- Optimization of the angles of light illuminating the mask (called off-axis illumination, OAI)
- Adding phase information to the mask in addition to intensity information (called phase-shifting masks, PSM).

Collectively, these optical approaches (each of which will be described in detail in the sections that follow) are known as resolution enhancement technologies (see Figure 10.5). While some techniques improve feature resolution at the expense of pitch resolution,

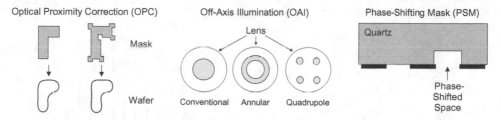

Figure 10.5 *Examples of the three most common resolution enhancement technologies*

many of the resolution enhancement technology (RET) approaches can improve pitch resolution and increase the process window simultaneously, a seemingly no-compromise path to resolution enhancement. However, the most promising RETs (especially the best PSM techniques) require a revolution in chip layout design that has yet to occur. Ultimately, a k_{pitch} as low as 0.5 is possible, but only for chips designed specifically to take advantage of these RETs.

10.2 Optical Proximity Correction (OPC)

Since, in general, the smallest features in a design are the hardest to print, lithography processes are optimized to best print those smallest features. Unfortunately, the larger features (which, while *larger*, cannot be considered large) are rarely faithfully reproduced using this same lithography process. The nonlinear nature of a high-resolution lithography process ensures that a 'what you see is what you get' imaging ideal, where all features on the mask are faithfully reproduced in resist, will never be achieved.

10.2.1 Proximity Effects

Proximity effects are the variations in the linewidth of a feature (or the shape for a 2D pattern) as a function of the proximity of other nearby features. The concept of proximity effects became prominent several decades ago when it was observed that electron beam lithography can exhibit extreme proximity effects (backscattered electrons can travel many microns, exposing photoresist at nearby features). Optical proximity effects refer to those proximity effects that occur during optical lithography (even though they may not be caused by optical phenomenon). The simplest example of an optical proximity effect is the difference in printed linewidth between an isolated line and a line in a dense array of equal lines and spaces, called the *iso-dense print bias*.

Although many factors may affect the iso-dense print bias, such as developer flow or PEB diffusion, in general this bias is fundamentally the result of optics – the aerial images for dense and isolated lines are different. For high-resolution features, the diffraction patterns from isolated and dense lines are significantly different (see Chapter 2), resulting in different aerial images, as shown in Figure 10.6. In this case, the isolated line will print wider than the dense line (assuming a positive photoresist), giving a positive iso-dense print bias. It is important to note that this result is not a 'failing' of the optical system,

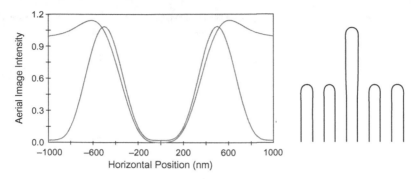

Figure 10.6 *The iso-dense print bias is fundamentally a result of the difference in the aerial images between isolated and dense lines. In this case, the isolated line is wider than the line in a dense array of equal lines and spaces (0.5-micron features, $\lambda = 365\,nm$, $NA = 0.52$, $\sigma = 0.5$)*

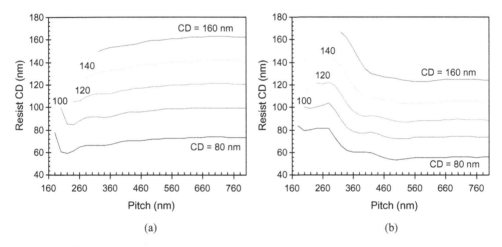

Figure 10.7 *Resist CD through pitch for different nominal feature sizes (used to fully characterize 1D proximity effects) can be very different as a function of the optical imaging parameters used: (a) conventional illumination, $\sigma = 0.7$, and (b) quadrupole illumination, center $\sigma = 0.8$ ($\lambda = 193\,nm$, $NA = 0.85$, binary mask, dose set to properly size the 100-nm line/space pattern)*

but a natural consequence of the physics of imaging. Also, aberrations in the optical system can change the magnitude of the bias, sometimes significantly.

The proximity effect is very feature size dependent. For large features, the diffraction patterns for isolated and dense lines are similar, giving very little differences in the aerial images. As feature size shrinks, the differences grow. Thus, to fully characterize one-dimensional proximity effects, it is common to measure resist CD versus pitch for a range of nominal feature sizes (Figure 10.7). It is very important to note that the resulting behavior (the CD-through-pitch curves) is very sensitive to the exact lithographic conditions used. For example, a switch from conventional to quadrupole illumination causes a dramatic change in the shape of the CD-through-pitch curves (Figure 10.7a vs. Figure 10.7b), even changing the sign of the iso-dense bias for some features.

The photoresist will also influence proximity effects. A systematic study using lithography simulation showed that the resist property that most significantly influences proximity effects is resist contrast.[3] In simulation terminology, it is the resist dissolution selectivity parameter n of the original kinetic development model (which is directly proportional to resist contrast, see Chapter 7) that influences proximity effects. As an example of how an optical proximity effect might change with resist contrast, Figure 10.8 shows the influence of the next nearest feature on the linewidth of a nominal 400-nm feature using i-line lithography and a conventional resist. No proximity effects would result in the flat line shown at 400-nm linewidth. Five curves are shown corresponding to five different resist contrasts (subjectively called low, medium, high, state-of-the-art and future contrast resists). These resists correspond to dissolution selectivity parameters of 4, 5.5, 7, 10 and 16, respectively. The infinite contrast resist corresponds to the width of the image-in-resist itself since an infinite contrast resist will reproduce the image-in-resist exactly (baring diffusion).

Obviously, resist contrast plays an important role in determining the actual printed proximity effect. For example, the iso-dense bias (the difference in linewidth between isolated and dense features) for the state-of-the-art resist is twice that of the low-contrast resist for the case shown in Figure 10.8. Also, the lower-contrast resists show a dip in linewidth as the spacewidth is initially increased, whereas the higher-contrast resists do not. Incidentally, the range of printed linewidths often exceeds the iso-dense bias quite significantly, as Figure 10.8 clearly shows.

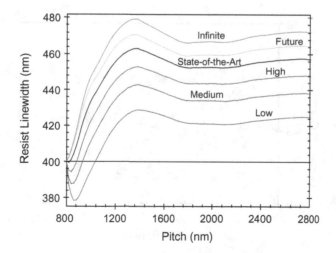

Figure 10.8 *Proximity effects for different resist contrasts (400-nm nominal features, NA = 0.52, σ = 0.5, i-line). The increasing pitch corresponds to increasing distance between 400-nm lines*

10.2.2   Proximity Correction – Rule Based

Since the difference between the desired and actual CD printed on the wafer as a function of feature size and pitch is, for the most part, a systematic error, it should be possible to correct for this error. The only practical means for correction is to change the CD on the mask to compensate for these proximity effects. This compensation is called *optical proximity correction* (OPC).

The initial goal of OPC, then, is to determine the optimal mask shape to get the desired resist shape – often called the 'inverse problem' in imaging. This simple question is more complicated than it appears. As with most such questions, the answer depends on what is meant by 'optimal'. Consider first the simple case of printing long lines. What is the optimal width of the mask line to get the desired resist line for many different linewidths and proximity to other lines? One simple definition of optimum could be just obtaining the right linewidth at the nominal exposure dose and focus. As an example, consider a 100-nm resolution process using the 100-nm equal line/space pattern as the baseline. Without any bias on the mask for this feature, we determine the exposure dose to properly size this feature. Our simple requirement, then, is that all other mask features must properly print at their correct size at this dose as well. Since proximity effects prevent this from occurring naturally, we can only obtain this result by changing the feature sizes on the mask to correct for these proximity effects.

Figure 10.9 shows one example of what the mask bias solution might look like. The mask correction (CD bias) is defined as the actual absorber width that produces the desired resist CD minus the nominal (unbiased) absorber width. Thus, a positive bias means the absorber has been made bigger. Each curve shows the amount of linewidth bias needed as a function of the nominal feature width. For this example, we have defined our starting point as zero bias for the most difficult feature, the 100-nm lines and spaces. An alternate

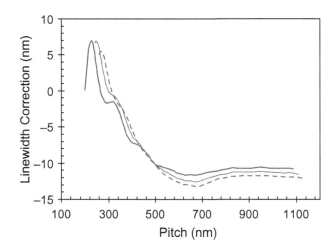

Figure 10.9 *Design curves of the mask linewidth bias (in wafer dimensions) required to make all of these features print at the nominal linewidth: 100 nm (thick line), 120 nm (thin line) and 140 nm (dashed line). Dose set to require no bias at 100-nm lines and spaces ($\lambda = 193$ nm, NA = 0.93, $\sigma = 0.7$, 6 % ESPM)*

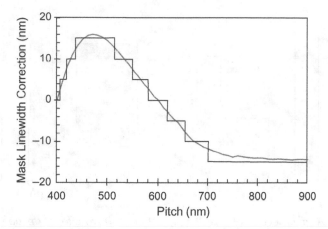

Figure 10.10 *Discretized design curve (the stair-step approximation to the actual smooth curve) appropriate for use in a design rule table (5-nm correction grid used)*

(though less common) approach would be to set zero bias for some large feature and adjust the bias of the smaller features to match the dose-to-size of that larger feature.

Our definition of the optimum mask bias so far is a simple one – the bias which gives the proper printed linewidth at the nominal process conditions. Other definitions may include the tolerance to variations in process conditions (for example, the maximum overlap of the focus–exposure process windows of the various features). Ultimately, the best solution is that which gives the tightest distribution of linewidths on the wafer for all of the features in the presence of typical process variations.

While the curves of Figure 10.9 show how mask correction is a continuous function of nominal CD and pitch, the actual implementation of these corrections on a real design is often snapped to a design grid in order to ease the design and mask-making burden. Typical design grids are 1–3 % of the minimum resist feature size. For example, a typical design curve is snapped to a 5-nm grid in Figure 10.10. When discretized in this way, the information in the design curve can be easily captured in a correction table, called an OPC rules table.

While all of the examples shown here have been symmetrical (the proximity of other features is identical on both sides of the main feature), in general this will not be the case. Thus, corrections are better described as edge movements rather than linewidth changes. A simple OPC rule might be something like this: 'If a feature is between A and B in width, and the next feature to the right is between X and Y away, move the right edge of the feature by Z'. Thus, the shape of the feature in the design is changed (its edge is moved) depending on the size of the feature and its proximity to its nearest neighbor. The values of the parameters of the rule (the amount of correction Z as a function of A, B, X and Y in this example) are empirically determined for a given lithography process. Since a large set of these parameters is typically required for each rule, the collection of these parameters is put into a table called a rules table. Figure 10.11 shows an extremely simple case where one-dimensional edge movements based on a rules table were applied.

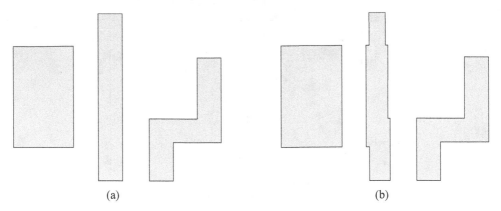

Figure 10.11 *A small section of a design (a) before, and (b) after correction of the middle feature with a simple 1D rule-based correction*

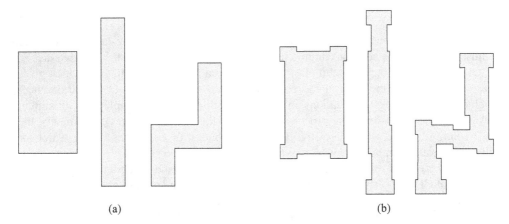

Figure 10.12 *A small section of a design (a) before, and (b) after the use of a simple 1.5D rule-based correction*

There are two key aspects of successfully implementing rule-based OPC: (1) having the right set of rules, and (2) having the right parameter values for those rules. While rule-based OPC is conceptually simple and easy to implement for one-dimensional corrections, the rules can get very cumbersome when considering two-dimensional effects. Corner rounding and line-end shortening (see Chapter 8) are two particular examples of important 2D proximity effects that should be addressed with OPC. An intermediate approach is to treat line-ends with their own set of rules, apply serifs to corners, and use 1D rules for all of the other edges. Sometimes called '1.5D' correction, an example of this approach is shown in Figure 10.12.

Rule-based OPC is fairly simple to implement. First, a set of CD-through-pitch curves are experimentally measured for a given process. From these data, a one-dimensional rules table is created by interpolating from these data. Optimum line-end treatments and corner serif sizes are also empirically determined for the given process. Corrections are made by implementing the rules as scripts for a design rule checker (DRC) software package. This approach to OPC became important when feature sizes dropped below 500 nm, and was widely used at the 250-nm technology generation. However, by the 130-nm technology generation, the accuracy and robustness of rules-based OPC began to falter. It was at this technology generation that most semiconductor manufacturers began to switch to model-based OPC.

10.2.3   Proximity Correction – Model Based

Rule-based OPC suffers from a difficult scalability problem: in order to increase the accuracy of the OPC, the number of rules must grow. This growth in rules is highly non-linear – a small increase in OPC accuracy may require a very large increase in the number of rules. Additionally, the required accuracy of OPC scales at least linearly with the technology node dimensions. Since each rule requires several experimentally derived parameters, the result is a dramatic increase in effort. For these reasons, most semiconductor manufacturers realized that *model-based OPC* was required at the 90-nm technology generation.

Model-based OPC replaces experimentally derived rules with a calibrated lithography model that predicts the proximity effects for the actual chip pattern. Edges of the features are then iteratively moved until the predicted resist shape matches the desired feature shape to within a preset tolerance. A well-designed model-based OPC procedure requires only very few rules to operate. Instead, considerable effort is put into the development and calibration of a simplified lithography model that will predict how each design pattern will print.

The constraints on the lithography models used for OPC are severe. Since an entire chip must be simulated (with its hundreds of millions or even billions of features), OPC models must be very fast and should also be parallelizable (to take advantage of distributed computing). At the same time, these models must be very accurate, predicting the printed shapes on the wafer for every pattern of the design accurately to within a few percent of the minimum design rule size. This combination of very fast speed with acceptable accuracy has resulted in the development of hybrid physical/empirical lithography models, sometimes called *compact models*.

OPC models combine a reasonably accurate and physically correct aerial imaging model with a very simple resist model. Etch effects are sometimes included, or separate etch models are applied, to predict the after-etch feature sizes and shapes. For the aerial image calculation, an approximate solution to the Hopkins imaging equations called *Sum of Coherent Sources* (SOCS) provides extremely fast computation times by pre-computing a small number of coherent convolution kernels for a given source and lens pupil (see Chapter 2). Diffusion effects in the resist are usually approximated by allowing the aerial image to diffuse, which can be conveniently calculated using a Gaussian diffusion kernel in the SOCS formulation. Resist development effects are approximated using a threshold plus bias or a variable threshold model (see Chapters 2 and 7). Alternately, resist effects can be captured as empirical kernels in the SOCS formulation. In either case, these empirical resist models require extensive calibration for a given resist process.

Model-based OPC begins by first dividing the original design into edge segments, each of which can be individually moved during the OPC process. Simulation of the design predicts a resist shape, which is then compared to the original design at various measurement points (for example, the midpoint of the segment) to calculate an edge placement error (EPE, see Chapter 8). This EPE is then used to guess how much the segment must be moved so that the resist shape will have zero EPE. Edge segments are iteratively moved and the resist shape resimulated until all of the EPEs are below a preset limit (typically requiring three to six iterations). As with rule-based OPC, edge positions are usually snapped to a design grid to reduce the design and mask complexity. An example of model-based OPC is shown in Figure 10.13. The aggressiveness of the OPC can be adjusted by controlling the segment size (smaller segments mean more aggressive OPC), the smallest movement allowed for an edge compared to its connecting edge (that is, the smallest allowed *jog* size), and the design grid, among other parameters. A more aggressive OPC results in a larger number of design vertices and greater mask-making cost.

In general, model-based OPC is used to find a mask shape that produces an acceptable resist shape for a given lithography process that is assumed to be operating at its nominal conditions, i.e. at best focus and exposure. Thus, while OPC can be used to extend the

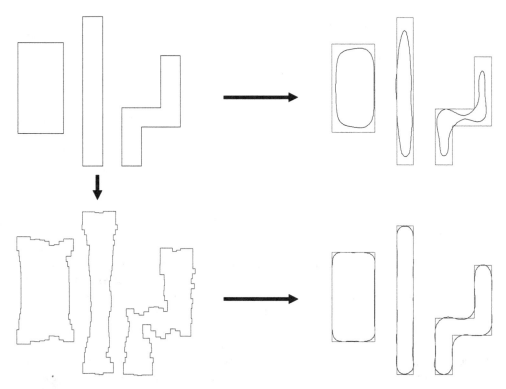

Figure 10.13 *Example of model-based OPC: the original design (upper left) prints very poorly (upper right). After aggressive model-based OPC, the resulting design (lower left) prints very close to the desired shape (lower right). OPC and simulations done using PROLITH*

linear resolution (the smallest features that can be printed while still printing all larger features acceptably), it does not address the issue of process window, and thus the true manufacturable resolution. One important technique used with OPC that addresses process window as well as sizing issues is the *subresolution assist feature*, as described next.

10.2.4 Subresolution Assist Features (SRAFs)

Scattering bars, also called subresolution assist features (SRAFs), are narrow lines or spaces placed adjacent to a primary feature in order to make a relatively isolated primary line behave lithographically more like a dense line.[4] The problem being solved is generically described as the problem of iso-dense bias. Isolated features will almost always print at a feature size significantly different than the same mask feature surrounded by other features. The pitch curves of printed CD versus pitch for various nominal mask dimensions show the problem (see Figure 10.7). While sizing the mask to give the correct CD on the wafer for all pitches certainly works (this is the conventional OPC approach), there is another isolated-versus-dense difference that is not addressed by this bias OPC.

The response of an isolated feature to focus and exposure errors is significantly different than the same-sized dense line. Figure 10.14 shows example focus–exposure matrices for dense and isolated lines after the isolated line has been sized to give the proper CD at the best focus and exposure needed by the dense features. The different shapes of the Bossung curves produce different shapes for the process windows, which limits the overlapping depth of focus even when the features nominally have the same best exposure dose.

Scattering bars are designed to reduce the difference in the focus response of an isolated feature compared to a dense feature by making the isolated feature seem more 'dense'. This becomes especially important when an off-axis illumination scheme is optimized for greatest depth of focus of the dense features (a topic that will be discussed extensively in section 10.3). The overlapping process window for the dense and isolated lines of Figure 10.14 is shown in Figure 10.15a. The curvature of the isolated process window severely limits the useable, overlapping DOF.

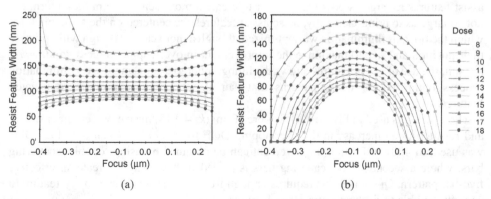

(a)   (b)

Figure 10.14 *Focus–exposure matrices (Bossung curves) for (a) dense and (b) isolated 130-nm features (isolated lines biased to give the proper linewidth at the best focus and exposure of the dense lines, λ = 248 nm, NA = 0.85, quadrupole illumination optimized for a 260-nm pitch)*

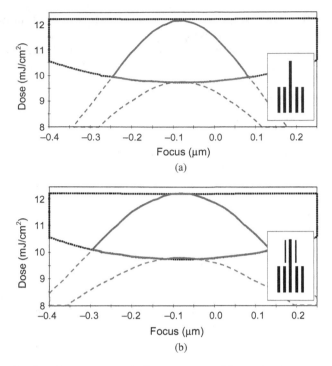

Figure 10.15 *Overlapping process windows generated from the focus–exposure matrices of dense and isolated lines for (a) isolated lines with bias OPC (overlapping DOF = 300 nm) and (b) isolated lines with scattering bars (overlapping DOF = 400 nm)*

An SRAF, as the name implies, is a subresolution feature that is not meant to print. In fact, it must be carefully adjusted in size so that it never prints over the needed process window. This determines the most important trade-off in scattering bar design: make the assist features as large as possible in order to create a more dense-like mask pattern, but not so large as to print. Generally, these assist features are centered on the same pitch for which the off-axis illumination was optimized, though a more careful design will optimize their position and size to maximize the improvement in overlapping process window. As a result, the use of assist features allows the lithographer to design an off-axis illumination process optimized for dense patterns that can also be used to print more isolated features.

The assist bars used in Figure 10.15b were 50 nm ($k_1 = 0.17$) in size (wafer dimensions), and resulted in an increase in the overlapping DOF from 300 nm, when only bias OPC was used, to 400 nm. Further improvement can be obtained by using 'double' scattering bars, where a second set of scattering bars is placed further away to create an effective five-bar pattern. Of course, this requires enough free space around the primary feature to actually be able to fit these extra assist features.

While the concept of using scattering bars to improve the DOF of isolated features is a simple one, its practical implementation is anything but simple. Unlike the idealized case of an isolated line, real patterns contain lines with a variety of pitches (i.e. nearby

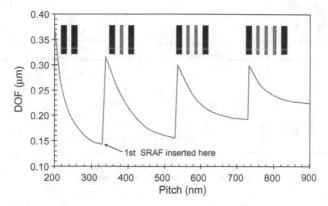

Figure 10.16 *Schematic diagram of SRAF placement showing the discontinuous effect of adding an SRAF as the pitch grows (main feature size is 100 nm)*

patterns with different distances away), each of which must be outfitted with an optimal assist feature or features, if one can fit. While bias OPC can be used on the intermediate cases where the space between two lines is not large enough to accommodate an assist feature, these intermediate pitches do not benefit from the DOF advantages of SRAFs (see Figure 10.16). And then, of course, there is the problem of what to do with line ends, corners, and other 2D patterns. Rule-based SRAF placement is quite common, but has difficulty with 2D placement. Model-based SRAF placement is difficult, but shows promise for complex 2D geometries. These issues can be resolved, however, and subresolution assist features are commonly used in many chip designs. Polysilicon gate and contact levels, in particular, have seen benefits from using SRAFs. For contacts and other dark-field mask levels, the SRAFs take the form of clear slots (spaces) rather than assist lines.

10.3   Off-Axis Illumination (OAI)

Off-axis illumination[5,6] (OAI) is one of the three major resolution enhancement technologies that have enabled optical lithography to push practical resolution limits far beyond what was once thought possible. In order to effectively use off-axis illumination, the shape and size of the illumination must be optimized for the specific mask pattern being printed. This section will describe how to optimize the most popular types of off-axis illumination – dipole, quadrupole and annular illumination – to maximize depth of focus for a given pitch.

Off-axis illumination refers to any illumination shape that significantly reduces or eliminates the 'on-axis' component of the illumination, that is, the light striking the mask at near-normal incidence. By tilting the illumination away from normal incidence, the diffraction pattern of the mask is shifted within the objective lens. For the case of a repeating pattern, the diffraction pattern is made up of discrete diffraction orders. If the pitch of the repeating pattern is small, only a few diffraction orders can actually make it

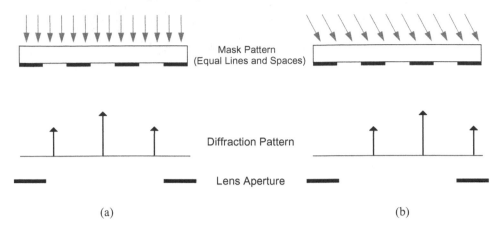

Figure 10.17 *Off-axis illumination modifies the conventional imaging of a binary mask shown in (a) by tilting the illumination, causing a shift in the diffraction pattern as shown in (b). By positioning the shifted diffraction orders to be evenly spaced about the center of the lens, optimum depth of focus is obtained*

through the finite size lens. As discussed in Chapter 3, placing those diffraction orders that make it through the lens evenly about the center of the lens leads to improved depth of focus (DOF). Thus, the main advantage of off-axis illumination is an increase in DOF (and thus the resolution) for small-pitch patterns. For these small pitch patterns, OAI changes the imaging from three-beam to two-beam imaging.

In spatial frequency terms, the distance between the zero and first diffracted orders is $1/p$, where p is the pitch. It will be convenient to convert this spatial frequency distance to 'sigma space', the spatial frequency normalized by λ/NA. In this normalized coordinate system, the maximum spatial frequency passing through the lens is 1.0. Thus, the distance between diffraction orders in sigma space is $\lambda/(pNA)$. To center the zero and first orders about the center of the lens, the zero order (found at the exact center of the lens for normally incident light) must be shifted by $\lambda/(2pNA)$ in sigma space. Thus, this becomes the optimum illumination tilt to give maximum DOF (Figure 10.17). Of course, tilting in the opposite direction [a $-\lambda/(2pNA)$ shift in sigma space] will produce the same effect. Combining both tilts into one illumination shape produces an illumination called *dipole illumination* that adds the required telecentricity behavior to the imaging. Note that the optimum illumination tilt is pitch dependent.

Real lithography, however, adds a significant complication to this otherwise simple picture. The line/space pattern shown in Figure 10.17 has a specific orientation (the lines are running into and out of the page) that results in an optimum tilt as shown in the figure. Most integrated circuit designs will contain many line and space-like features that are oriented both vertically and horizontally. A perspective plot of the same diffraction situation may make this point clearer, as shown in Figure 10.18. If the illumination is tilted by the amount discussed above, that tilt, in a specific direction, will only help the lines and spaces that are properly oriented with respect to that tilt. The other orientation of lines will not only not be improved by the illumination tilt, they are likely to

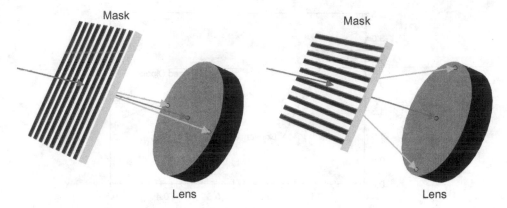

Figure 10.18 *The position within the lens of the diffracted orders from a pattern of lines and spaces is a function of the orientation of the lines and spaces on the mask*

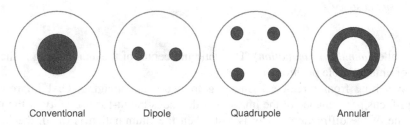

     Conventional   Dipole   Quadrupole   Annular

Figure 10.19 *Various shapes for conventional and off-axis illumination*

be significantly degraded in imaging performance. If both vertical and horizontal lines are to be imaged together on the same mask, an illumination shape must be used that provides optimum tilts for both geometries. The simplest shape that provides this optimum tilt for both horizontal and vertical line/space patterns is called *quadrupole illumination*.

Quadrupole illumination takes the optimum dipole generated for one orientation of lines and spaces, then shifts it both up and down in the other direction to create the proper angles for the other orientation of lines. The result is four poles evenly spaced about the center of the lens, as shown in Figure 10.19. In sigma space, the radial position of the center of each pole with respect to the center of the lens that gives optimum DOF is $\sqrt{2}\lambda/(2\,pNA)$. Note that this positioning of the quadrupoles gives the same horizontal and vertical spacing between poles as in the dipole case, but places them closer to the edge of the lens aperture.

While the quadrupole shape provides optimal performance for vertical and horizontal lines, other orientations (such as a line/space array oriented at 45°) will not be optimum. For any orientation of lines, the optimal dipole for that pattern will be spread in a direction perpendicular to the line orientation, and can be shifted parallel to the lines in any amount that keeps the dipoles within the lens. If the mask will contain arbitrary orientations of lines, many rotations of the dipoles will produce an annulus of illumination (and

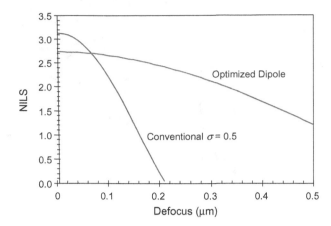

Figure 10.20   *The impact of off-axis illumination on the log-slope defocus curve (NA = 0.85, λ = 193 nm, binary chrome-on-glass mask, 120-nm lines and spaces). The dipole had a radius of 0.2 in sigma space*

thus is called *annular illumination*). The optimum center of the annular ring is the same as the optimum dipole position.

By switching from three-beam imaging to two-beam imaging, OAI improves the through-focus performance of the image, but degrades the in-focus quality of the image. Since one of the diffraction orders is lost when the illumination is tilted, the in-focus image will have reduced image log-slope (that is, reduced NILS). But if the tilt is optimized for the pitch being imaged, this reduced NILS will not change much as the image goes out of focus, as seen in Figure 10.20.

For each illumination shape discussed – dipole, quadrupole and annular illumination – there is one illumination size that maximizes the DOF for a given pitch. However, this illumination shape is only optimum for that one pitch. While pitches close to this optimum will get most of the benefit of the off-axis illumination, pitches sufficiently far away from the optimum will receive little or no benefit. In particular, isolated lines do not see the benefit of improved DOF when using off-axis illumination. Figures 10.21 and 10.22 show an example of how NILS and the DOF of a 100-nm line vary with pitch for quadrupole illumination nominally optimized for a 200-nm pitch. The DOF reaches a maximum when the pitch is near the designed-for pitch of 200 nm (the nonzero diameter of the poles means that the maximum DOF is obtained at a slightly larger pitch, in this case 230 nm). As the pitch increases, the DOF very quickly drops, leveling off at the DOF of an isolated line (which in this case is about half of the maximum DOF).

There is, however, a very convenient solution to the problem of lack of DOF for isolated lines: subresolution assist features. As discussed before, adding assist features around the more isolated lines will make them behave more like dense features. When using OAI, this means that the more isolated features can gain some of the benefit that OAI gives in terms of DOF. But, as we have seen before, SRAFs can be inserted between lines only if there is sufficient room to fit them. In general, SRAFs can be inserted when the pitch is greater than about 1.7 times the minimum pitch. For pitches less than about 1.3 times

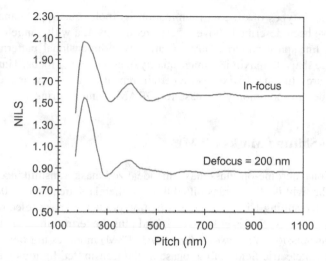

Figure 10.21   *Quadrupole illumination optimized for a pitch of 200 nm showing how NILS varies with pitch both in-focus and with a moderate amount of defocus (NA = 0.85, λ = 193 nm, 100-nm line, chrome-on-glass mask, quadrupole settings of 0.8/0.2)*

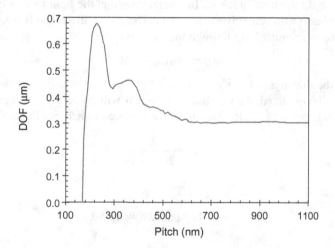

Figure 10.22   *Quadrupole illumination optimized for a pitch of 200 nm showing how isolated lines do not show improved DOF (NA = 0.85, λ = 193 nm, 100-nm line, chrome-on-glass mask, quadrupole settings of 0.8/0.2)*

the minimum pitch, the off-axis illumination provides sufficient DOF. Thus, a range of pitches emerges, roughly about 1.3–1.7 times the minimum pitch, where NILS and DOF are low and there is no possibility of using SRAFs to improve DOF. These pitches are sometimes called 'forbidden' pitches,[7] indicating the lithographer's desire that these pitches be avoided during circuit design. The forbidden pitch phenomenon significantly complicates the use of off-axis illumination since most chip designs will employ a wide range of pitches.

The three major types of off-axis illumination – dipole, quadrupole and annular illumination – have been described above. There are, however, a wide range of other shapes possible. Like line/space patterns, any repeating two-dimensional pattern will have an optimum source shape to maximize image quality over a range of focus. Thus, customized source shapes are often used whenever a certain repeating 2D shape is the most critical pattern in the design (commonly the case for DRAM manufacturing).

10.4   Phase-Shifting Masks (PSM)

For a conventional chrome-on-glass mask, the idealized mask transmittance is considered to be binary: the light is 100% transmitted through the glass areas and 100% blocked by the chrome. The resulting Kirchhoff approximation predicts a mask (electric field) transmittance function $t_m(x,y)$ that is either 0 or 1 (and thus the term 'binary' mask to describe this type of transmission). A *phase-shifting mask* (PSM) modifies not only the amplitude of the transmitted electric field but the phase of the transmitted light as well.

Consider the cross section of a mask structure shown in Figure 10.23. Two nearby regions of the mask transmit 100% of the monochromatic light, but experience different optical path lengths. In one region, the light passes through extra glass (or fused silica) of thickness d and refractive index n_g. In a nearby region, the light travels through air of the same thickness (with refractive index of 1). The phase difference between two plane waves traveling perpendicularly through the mask will be

$$\Delta\phi = 2\pi d(n_g - 1)/\lambda \qquad (10.5)$$

By adjusting the thickness of this extra layer of glass, any phase difference between the two waves can be obtained. As we shall soon see, it will be very advantageous to make this phase-shift exactly 180° (π). The thickness required to achieve this will be

$$d_{180} = \frac{\lambda}{2(n_g - 1)} \qquad (10.6)$$

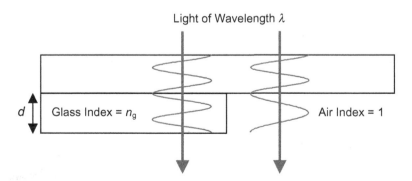

Figure 10.23   *Cross section of a mask showing how the phase of the light transmitted through one part of the mask can be shifted relative to the phase of light transmitted through a nearby part of the mask*

The fused silica most commonly used for mask blanks has a refractive index of 1.56 at 193 nm and 1.51 at 248 nm. Thus, the resulting glass thickness to give a π phase shift is slightly less than one wavelength.

The phase shift depicted in Figure 10.23 can be achieved by adding extra fused silica to the region on the left, or by etching away the same amount of mask blank material on the right. Alternately, some other material might be added to the mask blank so that both the phase and the amplitude transmittance of the light can be manipulated. There are an extremely large number of possible variations when both the amplitude and the phase of the transmitted light can be varied. However, practical mask-making considerations limit the number of different types of transmittance regions on the mask to a small number, typically two or three. And while many styles of phase-shifting masks have been investigated, only the two most common, alternating PSM and attenuated PSM, will be described in some detail below.

10.4.1 Alternating PSM

A conventional, binary chrome-on-glass mask of lines and spaces will produce a diffraction pattern of discrete diffraction orders at spatial frequencies that are multiples of one over the pitch (see Chapter 2). For a high-resolution pattern, only the zero and the plus and minus first diffraction orders pass through the lens (which has a spatial frequency cutoff of NA/λ), as seen in Figure 10.24a. In fact, it is the interference of the zero-order light with the first orders that produces the bright and dark image of the proper pitch. If the pitch is made too small, the first-order light diffracts at an angle too large to fit through the objective lens and no image is produced. The resolution limit, then, occurs when the

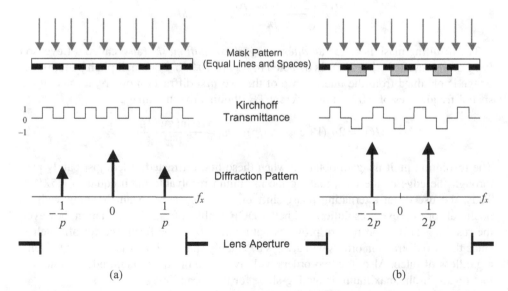

Figure 10.24 *A mask pattern of equal lines and spaces of pitch p showing the idealized amplitude transmittance function and diffraction pattern for: (a) binary chrome-on-glass mask; and (b) alternating phase-shift mask*

first diffracted order (spatial frequency of 1/pitch) lands exactly at the edge of the aperture (spatial frequency of NA/λ) so that the minimum resolvable pitch is equal to λ/NA (i.e. $k_{pitch} = 1.0$). Additionally, the use of partially coherent illumination can extend this classical resolution limit, but only at the expense of reduced image quality.

Consider now a repeating line/space pattern (with lithographic pitch p = linewidth + spacewidth) where every other space has been phase-shifted by $180°$. Letting w be the spacewidth, the diffraction pattern can be calculated as the superposition of one repeating space with pitch $2p$ and transmittance 1, plus another repeating space of pitch $2p$ and transmittance -1, shifted by p. Applying the superposition and the shifting theorems of the Fourier transform, the diffraction pattern will be

$$T_m(f_x) = \sum_{j=-\infty}^{\infty} \frac{\sin(j\pi w/2p)}{j\pi} \delta\left(f_x - \frac{j}{2p}\right) - \sum_{j=-\infty}^{\infty} e^{i2\pi f_x p} \frac{\sin(j\pi w/2p)}{j\pi} \delta\left(f_x - \frac{j}{2p}\right)$$

$$= \sum_{j=-\infty}^{\infty} \frac{\sin(j\pi w/2p)}{j\pi}[1-(-1)^j]\delta\left(f_x - \frac{j}{2p}\right)$$

(10.7)

When j is even, the diffraction order from the shifted space exactly cancels out the diffraction order from the unshifted space. Thus, only the odd orders survive, giving

$$T_m(f_x) = \sum_{j=-\infty}^{\infty} a_j\delta\left(f_x - \frac{j}{2p}\right)$$

(10.8)

where

$$a_j = \begin{cases} 0, & j = even \\ \dfrac{2\sin(j\pi w/2p)}{j\pi}, & j = odd \end{cases}$$

The resulting mask is called an *alternating phase-shift mask*[8] (also called a Levenson PSM or abbreviated Alt-PSM), as depicted in Figure 10.24b. For small pitch patterns, the image is obtained from the interference of the two first diffraction orders, located at the spatial frequencies of $\pm 1/2p$. For coherent TE illumination, the image is

$$I(x) = 2a_1^2(1+\cos(2\pi x/p)), \quad a_1 = \frac{2\sin(\pi w/2p)}{\pi}$$

(10.9)

The resolution limit is again obtained when these first diffracted orders just barely pass through the edge of the lens, making the minimum resolvable pitch equal to $0.5\lambda/NA$. Thus, the use of an alternating phase-shift mask leads to two-beam imaging with its potential for improved resolution and better DOF. Unlike off-axis illumination, however, the resulting diffraction pattern provides optimum defocus performance for all pitches, since the two diffraction orders will always be equally spaced about the center of the lens, regardless of pitch. Also, the two orders will always be of equal magnitude, a condition that results in the maximum image log-slope for two-beam imaging.

While the example given above is for a repeating pattern of lines and spaces, alternating PSM can be applied to the printing of any narrow, dark feature by shifting the phase of light on one side of the feature compared to the other side. Consider an isolated line of width w, where the clear region on the right side of the feature is phase-shifted by 180° relative to the transmittance on the left side of the feature (Figure 10.25). The diffraction pattern for this phase-shifted, isolated line is

$$T_m(f_x) = i \frac{\cos(\pi f_x w)}{\pi f_x}$$
(10.10)

For coherent TE illumination, the resulting aerial image can be expressed in terms of sine integrals:

$$I(x) = \frac{1}{\pi^2} \left[\mathrm{Si}\left(\frac{2\pi NA}{\lambda}(x + w/2) \right) + \mathrm{Si}\left(\frac{2\pi NA}{\lambda}(x - w/2) \right) \right]^2$$
(10.11)

where

$$\mathrm{Si}(\theta) = \int_0^\theta \frac{\sin z}{z} \, dz$$

It is interesting to examine the behavior of this image as w goes to zero. For an unshifted line, of course, the image would disappear as the chrome linewidth went to zero. But as Figure 10.26 shows, as the chrome width goes to zero the aerial image width approaches

Figure 10.25   *Cross section of a mask of an isolated phase-shifted line*

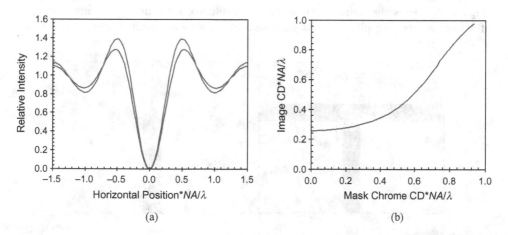

(a)

(b)

Figure 10.26   *Behavior of an isolated phase-shifted line as a function of the chrome line width: (a) coherent aerial images for w = 0 and 50 nm, and (b) the aerial image width (at an intensity threshold of 0.25) as a function of the mask chrome width*

a constant value of about $0.26\lambda/NA$ (see section 10.5.3). Setting $w = 0$ in Equation (10.11), we obtain the coherent TE aerial image for this isolated phase edge:[9]

$$I(x) = \frac{4}{\pi^2} \mathrm{Si}^2\left(\frac{2\pi NAx}{\lambda}\right) \tag{10.12}$$

As Figure 10.26 and Equation (10.12) show, a 0–180° phase edge will print as a narrow dark line (in fact, the class of *strong* phase-shifting masks can be defined as those masks that use a phase edge to print a dark line). While this can be very good if a narrow dark line is desired, it can be very bad if the phase edge occurs incidentally and a dark line is not desired at that point. As will be discussed in the following section, it can be very difficult to ensure that no unwanted phase edges appear in a phase-shifted layout.

10.4.2   Phase Conflicts

'Strong' phase-shift masks, such as alternating PSM, have seen only limited use in manufacturing, despite their potential for nearly doubling resolution and extending DOF by even more. Strong shifters invariably use etched quartz to create the phase shift and require two mask writing steps (including an alignment in between). While the much greater complexity and expense of making alternating PSMs have certainly been an impediment to their adoption, it is not the mask manufacturing that is the biggest problem, but the mask design. Alternating PSM works by shifting the phase of the clear region to one side of a small line by 180° relative to the phase of the light coming from the other side of the line. While simple in concept, attempting to phase-shift an arbitrary layout of lines will invariable lead to *phase conflicts*.

There are two basics types of phase conflicts, as seen in Figure 10.27: no phase shift where you want it, and a phase shift where you don't want it. The first type (Figure 10.27a) results in a lack of phase shift across a critical feature when there is an odd wrapping of phase assignments. This 'nonshifted' feature will not properly print. The second type (Figure 10.27b) is also called the termination problem since it usually occurs at the end of a line. Alternating phase across each side of a line will result in those two phases

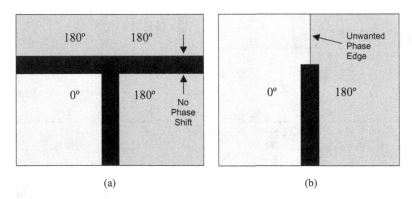

(a)   (b)

Figure 10.27   *Types of phase conflicts: (a) no phase shift across a critical pattern, and (b) the phase termination problem producing an unwanted phase edge*

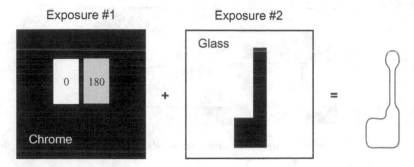

Figure 10.28 *Simple example of a double-exposure alternating phase-shift mask approach to gate-level patterning*

meeting at the line end. Whenever two opposing phases meet, a dark interference line is created causing a resist line to print. (In fact, as seen in the previous section, when printing small lines, the use of chrome is almost superfluous – it is the 0 to 180° phase transition that causes the dark lines to print.)

The phase conflict problems do not have an obvious solution. Taking an arbitrary layout and coloring the clear regions with two phase colors will lead to numerous conflicts in all but a few special cases. One solution, though not yet proven, is to force the original layout to be 'phase friendly', a layout where no phase conflicts can occur. It is unclear whether such a solution can in fact be developed and if so, what the design trade-offs will be. The alternative is costly but more practical: use two exposures from two masks. The phase-shift mask is used to define the critical features using a dark-field background to avoid phase conflicts. A second exposure uses a second bright-field mask that prints the noncritical features as well as the open areas not exposed with the first dark-field mask (Figure 10.28). Besides the obvious cost disadvantage due to the lower lithography tool throughput, this double-exposure approach also tends to have less-than-optimal transistor density (small lines can be printed, but not as close together as desired), and has not been proven to be practical on all critical levels. The double-exposure PSM process has been used successfully for polysilicon gate layers on microprocessors, where gate CD control has a profound effect on the value of the device.

10.4.3 Phase and Intensity Imbalance

The most obvious way of shifting the phase of light on a photomask is to change the thickness of quartz that one ray of light must pass through compared to another ray. This is most easily done by etching the quartz under one space by a set depth d while leaving the quartz of an adjacent space unetched (Figure 10.29). Equation (10.5) tells us the amount of phase shift as a function of the etch depth of this trench. This equation can also tell us how an error in etch depth turns into an error is phase. For 193-nm lithography, that translates into about 0.9° phase error per nanometer etch depth error.

These previous equations relating phase change to etch depth make an important assumption – that the light is traveling vertically. However, as light goes through the etched quartz hole, it begins to diffract at the bottom of the hole and its directions deviate

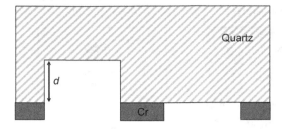

Figure 10.29   *Example of a simple alternating phase-shift mask manufacturing approach*

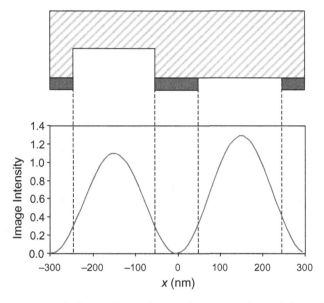

Figure 10.30   *Intensity imbalance shown for an alternating phase-shift mask of equal lines and spaces*

from a straight line. The smaller the hole, the greater this diffraction effect. This causes a difference in the actual phase of the light compared to the geometric 'straight line' approximation, and this difference is a function of the size of the etched space. Diffraction through the etched space causes a second problem: some of the diffracted light doesn't make it out of the hole. As a result, the etched quartz space appears dimmer than the unetched space, creating an intensity imbalance in addition to the phase error (Figure 10.30). As with the phase error, the degree of intensity imbalance is spacewidth dependent.

The real, physical effects of trying to create a 180° phase shift by etching the quartz as described above are detrimental to lithography. The intensity imbalance causes the two adjacent spaces to print as different linewidths, and the phase imbalances cause these spaces to have different best-focus settings. There are a number of possible ways to fix

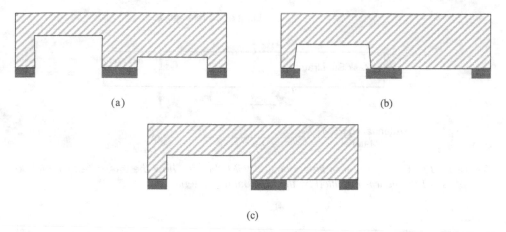

Figure 10.31 *Different approaches for fixing the phase error and intensity imbalance in alternating PSM: (a) dual trench, (b) undercut etch and (c) biased space*

these problems. In the *dual trench* method (Figure 10.31a), both the unshifted and shifted spaces are first etched to some depth, then the shifted space is further etched to create the desired phase shift. This approach can reduce both the phase errors and the intensity imbalance, though not perfectly for all space widths, and it complicates the mask-making process. In the *undercut etch* method (Figure 10.31b), the shifted space is etched some-what isotropically, causing some undercut of the chrome. This widening of the etched space reduces the intensity imbalance appreciably. However, to get the intensity imbal-ance to approach zero, the amount of undercut can become excessive, causing possible reliability problems with the overhanging chrome. Finally, the *biased space* approach (Figure 10.31c) uses an OPC tool to bias the etched space larger in order to eliminate the intensity imbalance. This bias can be adjusted as a function of the space size and pitch in order to eliminate intensity imbalance across the full range of features. In general, none of these approaches is perfect at eliminating the phase error as a function of size and pitch.

The most common approach to alternating PSM manufacturing is to combine the biased space approach with a small amount of undercut. While not perfect, it can provide very good intensity balance and minimum phase error across all pitches.

10.4.4 Attenuated PSM

Consider a general repeating line/space transmittance function (with pitch p) where the first feature (width w_1) has amplitude and phase transmittance of t_1 and ϕ_1, respectively, and the second feature (width $w_2 = p - w_1$) has amplitude and phase transmittance of t_2 and ϕ_2 (see Figure 10.32). The resulting diffraction pattern will be

$$T_m(f_x) = \sum_{j=-\infty}^{\infty} a_j \delta\left(f_x - \frac{j}{p} \right) \qquad (10.13)$$

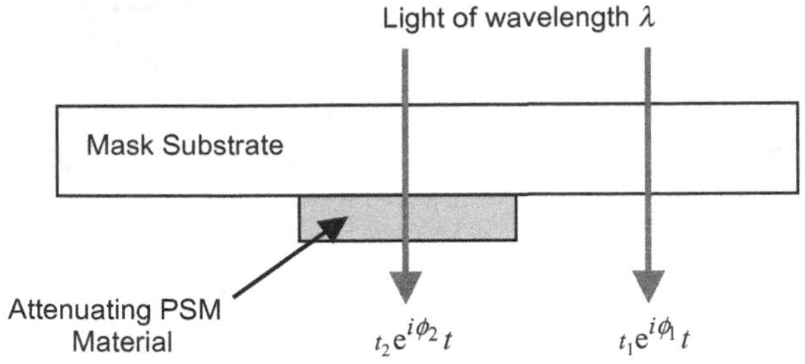

Figure 10.32   *Cross section of an attenuated PSM showing how the transmitted amplitude and phase of the light is modified by the attenuating material*

where

$$a_j = t_1 e^{i\phi_1}\left(\frac{\sin(j\pi w_1/p)}{j\pi}\right) + t_2 e^{i\phi_2}(-1)^j\left(\frac{\sin(j\pi w_2/p)}{j\pi}\right) \qquad (10.14)$$

or,

$$a_0 = (t_1 e^{i\phi_1} - t_2 e^{i\phi_2})\left(\frac{w_1}{p}\right) + t_2 e^{i\phi_2}$$

$$a_j = (t_1 e^{i\phi_1} - t_2 e^{i\phi_2})\left(\frac{\sin(j\pi w_1/p)}{j\pi}\right), \quad j \neq 0 \qquad (10.15)$$

Without loss of generality, it will be convenient to let $\phi_1 = 0$ so that ϕ_2 represents the phase difference between the two features. If $t_1 = 1$ and $t_2 = 0$, we have the conventional chrome-on-glass result. If we let $t_1 = 1$ but allow $t_2 > 0$, the phase of this second feature will now matter. Given our experience above, we will let this phase shift be $180°$. Thus, the diffraction orders are

$$a_0 = (1+t_2)\left(\frac{w_1}{p}\right) - t_2$$

$$a_j = a_{-j} = (1+t_2)\left(\frac{\sin(j\pi w_1/p)}{j\pi}\right), \quad j \neq 0 \qquad (10.16)$$

A mask of this type is called an *attenuated PSM* (also known as an embedded PSM, or EPSM, and sometimes called a half-tone PSM).

Equation (10.16) shows the impact of allowing the nominally dark feature to transmit some light that is phase-shifted compared to the bright feature: the zero order is decreased, the first order is increased. Consider the simple case of coherent three-beam imaging where the aerial image (for TE illumination) is

$$I(x) = a_0^2 + 2a_1^2 + 4a_0 a_1 \cos(2\pi x/p) + 2a_1^2 \cos(4\pi x/p) \qquad (10.17)$$

For the case of equal lines and spaces, the NILS at the nominal line edge is

$$NILS = 4\pi \frac{a_1}{a_0} \qquad (10.18)$$

For an attenuated PSM, using Equation (10.16) evaluated for equal lines and spaces, the NILS becomes

$$NILS = 8\left(\frac{1+t_2}{1-t_2}\right) \tag{10.19}$$

As the transmittance of the attenuated line increases, the NILS increases as well.

While the discussion above shows the advantages of an attenuated PSM for three-beam imaging, a high-resolution lithography process will often use off-axis illumination in order to improve depth of focus for small pitch patterns. For this case, a coherent two-beam image (for TE illumination) is

$$I(x) = a_0^2 + a_1^2 + 2a_0a_1\cos(2\pi x/p) \tag{10.20}$$

For the case of equal lines and spaces, the NILS at the nominal line edge is

$$NILS = 2\pi\frac{a_0a_1}{a_0^2 + a_1^2} \tag{10.21}$$

For the attenuated PSM case,

$$NILS = \frac{(1-t_2)(1+t_2)}{\dfrac{(1-t_2)^2}{4} + \dfrac{(1+t_2)^2}{\pi^2}} \tag{10.22}$$

For the three-beam imaging case, the best NILS is obtained by making the 0th order small and the 1st order big. For the two-beam case, maximum NILS comes from making the two orders equal in amplitude. To achieve this maximum,

$$\frac{(1-t_2)}{2} = \frac{(1+t_2)}{\pi} \quad \text{or} \quad t_2 = \frac{\pi-2}{\pi+2} \approx 0.222 \tag{10.23}$$

The above equations relate diffraction orders, images and the NILS to the electric field transmittance of the attenuated PSM material. It is common, however, to specify an attenuated PSM by its intensity transmittance, equal to $(t_2)^2$. Thus, for the two-beam imaging equal line/space patterns, the optimum intensity transmittance of the PSM is 4.93%.

Note that the transmittance used in the above equations was set to 1.0 for the transmittance of the mask substrate. Thus, the attenuated material transmittance is defined relative to the transmittance of the mask substrate. It is common, however, for the manufacturers of mask blanks to specify the transmittance of the EPSM blank in absolute terms. Letting t_{blank} be the transmittance of the EPSM coated substrate, and t_{glass} the transmittance of the glass or fused silica substrate without EPSM material, then

$$t_2 = \frac{t_{blank}}{t_{glass}} \tag{10.24}$$

For example, a common commercially available EPSM blank has an intensity transmittance of 6%. The fused silica substrate itself has a transmittance of approximate 92% at 248 nm and about 91% at 193 nm. Thus, the relative intensity transmittance of the EPSM features is actually 6.5 and 6.6% for 248- and 193-nm wavelengths, respectively.

Attenuated PSM has been very widely adopted for contact and via printing, and is fairly mainstream for other critical lithography layers as well. This type of PSM is generally called a 'weak' shifter – it uses phase-shifting to improve the image at the edge, but does not define the edge by a 0–180° phase transition. Its great advantage is the simplicity and low cost of replacing chrome-on-glass (COG) masks, the nonphase-shift alternative. Essentially, an existing design based on COG can be converted to an EPSM by simply recalibrating the optical proximity correction models used to apply OPC to the design. Mask manufacturing, while certainly more difficult than COG, is not dramatically different (the chrome is replaced by a more complex absorber such as molybdenum silicon) and only somewhat more costly. Attenuated PSM does not suffer from phase conflicts and exhibits much reduced phase and intensity variation as a function of feature size as compared to alternating PSM. When coupled with off-axis illumination and SRAFs, it can provide most of the benefit of the much more difficult and expensive alternating PSM alternative.

One problem with the use of attenuated PSM is the occurrence of sidelobes. Figure 10.33 shows an image of an isolated space ($k_1 = 0.48$) using a 6% attenuated PSM. Interference between light from the space and light from the background EPSM material cancels out near the space edge to make the intensity go to zero and sharpen the image near the edge (i.e. increase the NILS). However, away from the edge diffracted light from the unshifted space can add constructively with the EPSM transmission to give a bright region called a *sidelobe*. Even though the intensity transmittance in this attenuated area is only 6%, the brightness of this sidelobe is 17% (in this case). A higher level of EPSM transmittance is likely to cause the sidelobe to print, resulting in catastrophic failure. Sidelobes can be suppressed, however, using strategically placed assist slots. Since the phase of the sidelobe electric field is about 180° with respect to the unshifted transmittance of a space, placing an unshifted assist space (small enough in size so that it won't print) at the location of the sidelobe will have the paradoxical effect of reducing the intensity at that point.

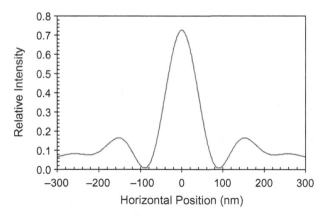

Figure 10.33   *An isolated space (100 nm, NA = 0.93, λ = 193 nm, σ = 0.5) imaged from a 6% attenuated PSM mask showing sidelobes*

10.4.5 Impact of Phase Errors

All of the analyses of phase-shifting masks above assume that the mask was designed and manufactured to give the ideal 180° phase shift. However, in practice there will invariably be phase errors. It will be important to understand the impact of phase errors on lithographic performance and the magnitude of these errors that can be tolerated. To investigate the impact of phase errors, the three analytical cases described in the previous sections (an isolated phase edge, an alternating PSM of lines and spaces, and an attenuated PSM pattern of lines and spaces) will be analyzed.

Consider an arbitrary isolated edge pattern oriented in the y-direction that has left and right sides with amplitude and phase transmittances of t_1, ϕ_1 and t_2, ϕ_2, respectively. Looking first at the simple case of coherent illumination, the resulting aerial image electric field will be

$$E_{edge}(x) = \frac{1}{2}(t_1 e^{i\phi_1} + t_2 e^{i\phi_2}) + \frac{1}{\pi}(t_1 e^{i\phi_1} - t_2 e^{i\phi_2}) \mathrm{Si}\left(\frac{2\pi NAx}{\lambda}\right) \tag{10.25}$$

where $x = 0$ is the position of the edge. For an ideal phase edge, $t_1 = t_2 = 1$, $\phi_1 = 0$, and $\phi_2 = \pi$ (180°). For a phase edge with phase error, we can let $\phi_2 = \pi + \Delta\phi$. Using these values in Equation (10.25) and squaring the magnitude of the electric field to obtain the intensity of the aerial image,

$$I_{phase-edge}(x) = \frac{1}{2}(1 - \cos\Delta\phi) + (1 + \cos\Delta\phi)\frac{2}{\pi^2}\mathrm{Si}^2\left(\frac{2\pi NAx}{\lambda}\right) \tag{10.26}$$

For small phase errors, the small angle approximation to the cosine can be used,

$$\cos\Delta\phi \approx 1 - \frac{\Delta\phi^2}{2} \tag{10.27}$$

giving

$$I_{phase-edge}(x) = \frac{\Delta\phi^2}{4} + \left(1 - \frac{\Delta\phi^2}{4}\right)\frac{4}{\pi^2}\mathrm{Si}^2\left(\frac{2\pi NAx}{\lambda}\right) \tag{10.28}$$

Looking closely at Equation (10.28), it is interesting to observe that the effect of the phase error is to take the ideal image (when $\Delta\phi = 0$) and add a uniform (d.c.) flare (see Chapter 3) in the amount of $\Delta\phi^2/4$. Like flare, a phase error for this feature will add a background dose, in this case caused by the incomplete cancellation of the out-of-phase light across the edge.

To put some numerical relevance to this relationship, an 11.5° phase error is equivalent to a 1% flare level. Since flare levels of 1% or more are commonly tolerated in lithographic projection tools, one would expect phase errors of 11.5° or more should be easily tolerated. While the above analysis assumes coherent illumination, simulation of partially coherent imaging of the same mask feature has shown the same relationship between phase error and flare (see Figure 10.34).

It is well known that phase errors also cause a change in the response of a phase-mask feature to focus errors. Again using simulation, the effect of phase errors on the process

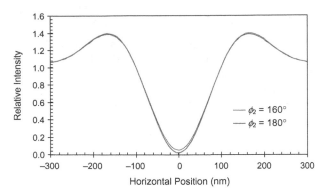

Figure 10.34 *Impact of a 20° phase error on the aerial image of an isolated phase edge* *(λ = 248 nm, NA = 0.75, σ = 0.5)*

window of an isolated phase edge is no different than the effect of flare on the process window. However, a new effect is introduced. When the phase error is not zero, going out of focus produces an asymmetry in the image that results in an effective image placement error. This image placement error varies linearly with defocus, acting just like a telecentricity error. Over the full range of focus within the process window, the impact on image placement error is about ±5 nm for a 10° phase error.

Consider now an alternating PSM pattern of lines and spaces with spacewidth w_s and linewidth w_l. Assuming that the lines have zero transmittance, we shall let two adjacent spaces have arbitrary transmittances and phases of t_1, ϕ_1 and t_2, ϕ_2, respectively. For high-resolution patterns, only the zero and first diffracted orders will pass through the lens and be used to generate the aerial image. Defining a coordinate system with $x = 0$ at the center of the first space and letting $p = w_s + w_l$ = the pitch, the amplitude of the zero and first diffraction orders will be given by

$$a_0 = \frac{w_s}{2p}(t_1 e^{i\phi_1} + t_2 e^{i\phi_2})$$

$$a_1 = a_{-1} = \frac{\sin(\pi w_s/2p)}{\pi}(t_1 e^{i\phi_1} - t_2 e^{i\phi_2})$$

(10.29)

The electric field of the aerial image (assuming coherent illumination for simplicity) will be

$$E(x) = a_0 + 2a_1 \cos(\pi x/p)$$

(10.30)

For an ideal alternating PSM, $t_1 = t_2 = 1$, $\phi_1 = 0$, and $\phi_2 = \pi$ (180°). For a mask with phase error, we can let $\phi_2 = \pi + \Delta\phi$. Using these values in Equation (10.29),

$$a_0 = \frac{w_s}{2p}(1 - e^{i\Delta\phi}) \approx -i\Delta\phi\left(\frac{w_s}{2p}\right)$$

$$a_1 = \frac{\sin(\pi w_s/2p)}{\pi}(1 + e^{i\Delta\phi}) \approx \frac{2\sin(\pi w_s/2p)}{\pi}\left(1 + \frac{i\Delta\phi}{2}\right)$$

(10.31)

where the approximate relations on the right make use of a small angle approximation. Using these values in Equation (10.30) and squaring the magnitude of the electric field to obtain the intensity of the aerial image,

$$I_{\text{Alt-PSM}}(x) \approx \left(1 - \frac{\Delta\phi^2}{4}\right)I_{\text{ideal}}(x) + \Delta\phi^2\left(\frac{w_s}{2p}\right)^2 \tag{10.32}$$

where I_{ideal} is the image of the alternating PSM for no phase error. Just as in the case of the phase edge discussed above, the impact of phase error on the image of an alternating PSM looks just like the addition of flare to the image. In this case, however, the amount of effective flare is reduced by the square of the shifted space duty ratio of the mask ($w_s/2p$). For equal lines and spaces, $w_s/2p = 1/4$ so that the effective flare for this Alt-PSM pattern is four times smaller than the effective flare for an isolated phase edge.

It would seem, then, that the wonderful properties of the alternating phase-shifting mask make them relatively immune to the small phase errors that will inevitably occur during mask fabrication. Unfortunately, a closer look reveals a more insidious problem. When the image goes out of focus, any phase error in the mask will interact with the defocus to cause an asymmetry between the shifted and unshifted spaces. The culprit lies in the zero order.

For no phase error, the zero order is completely missing ($a_0 = 0$). As a result, the image is formed completely by the interference of the two first orders. Since these orders are evenly spaced about the center of the lens, they have a natural immunity to focus errors, making improved depth of focus one of the most attractive qualities of the alternating phase mask. As the zero order grows (with increased phase error on the mask), this third beam ruins the natural symmetry of the first orders and adds a focus dependency. Again for the case of coherent illumination and assuming small phase errors, the aerial image when out of focus will be

$$I_{\text{out-of-focus}}(x,\delta) \approx I_{\text{in-focus}}(x) + 2\Delta\phi\left(\frac{w_s}{2p}\right)E_{\text{ideal}}(x)\sin(\pi\delta\lambda/p^2) \tag{10.33}$$

where E_{ideal} is the electric field for the no phase error, no defocus case [i.e. Equation (10.30) with $\Delta\phi = 0$].

Obviously, the interaction of phase error $\Delta\phi$ and defocus distance δ adds an error term to the in-focus aerial image as shown in Equation (10.33). The nature of this error term lies in the nature of the electric field E_{ideal}. Unlike the intensity, the electric field can be negative. In fact, the electric field will be negative under the phase-shifted spaces and will be positive under the unshifted spaces. Thus, for a positive focus error ($\delta > 0$) and phase error ($\Delta\phi > 0$), the error term in Equation (10.33) will make the shifted space dimmer and the unshifted space brighter. For a negative focus error, the opposite will be true. This effect is largest for smallest pitches p as can be seen from the argument of the sine term. This asymmetry between adjacent spaces through focus is the most detrimental effect of phase errors for an alternating phase-shifted mask and is illustrated in Figure 10.35.

The lithographic consequences of this intensity imbalance of adjacent spaces through focus are quite interesting. From the perspective of the space, the brighter space will print

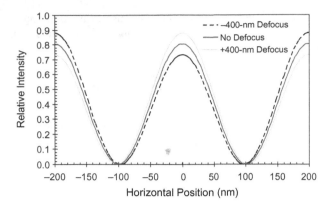

Figure 10.35   *Aerial images for an alternating phase-shifting mask with a 10° phase error for +400 nm defocus (dots), no defocus (solid), and –400 nm of defocus (dashed). (100-nm lines and spaces with* λ *= 248 nm, coherent illumination, 0.25 <* k_1 *< 0.5)*

wider (assuming a positive resist), so that one space will be wider on one side of focus while its neighbor will be wider on the other side of focus. Things look quite different from the perspective of the line, however. As the space to one side of the line gets wider with a focus error, the space to the other side of the line will get narrower. As a result, the linewidth remains about constant but its position shifts away from the brighter space and toward the dimmer space. Thus, the impact of the interaction of phase and defocus errors for an alternating PSM can be seen as an error in the placement of the line in the PSM array. When both transmission and phase errors are present, the two effects on space image brightness can cancel at some defocus. The resist image with the best fidelity would then appear out of focus.

While alternating aperture PSMs are important, the overwhelming majority of phase-shift masks used today are attenuated PSMs. How does a small phase error affect the lithographic performance of an attenuated PSM? Consider an equal line/space pattern where the line has an electric field amplitude and phase transmittance of t and ϕ, respectively. For a mask with phase error, we can let $\phi = \pi + \Delta\phi$. Using this value in Equation (10.15), and assuming that the phase error is small,

$$a_0 = \frac{1}{2}(1 - te^{i\Delta\phi}) = \frac{1}{2}(1 - t\cos(\Delta\phi)) - i\frac{1}{2}t\sin(\Delta\phi) \approx \frac{1}{2}\left(1 - t + \frac{t\Delta\phi^2}{2}\right) - i\frac{1}{2}t\Delta\phi$$

$$a_1 = a_{-1} = \frac{1}{\pi}(1 + te^{i\Delta\phi}) \approx \frac{1}{\pi}\left(1 + t - \frac{t\Delta\phi^2}{2}\right) + i\frac{1}{\pi}t\Delta\phi$$

(10.34)

Calculating the magnitude and the phase of each diffraction order,

$$|a_0| \approx |a_0|_{ideal} + \frac{\Delta\phi^2}{4}\left(\frac{t}{1-t}\right)$$

$$\angle a_0 \approx -\Delta\phi\left(\frac{t}{1-t}\right)$$

(10.35)

$$|a_1| \approx |a_1|_{\text{ideal}} - \frac{\Delta\phi^2}{2\pi}\left(\frac{t}{1+t}\right)$$
$$\angle a_1 \approx \Delta\phi\left(\frac{t}{1+t}\right)$$

(10.36)

Let's investigate the impact of the changes in the magnitude and phase of each diffracted order separately. As shown in Equations (10.35) and (10.36), the magnitudes of the orders vary as the phase error squared (and so should be quite small for small errors). As an example, for a 6% EPSM with a 10° phase error, the magnitudes of the zero and first orders change by only +0.7 and −0.2%, respectively. The resulting impact on the aerial image is quite small, less than 0.2% intensity difference in most cases.

The phase of the diffraction orders, on the other hand, varies directly as the EPSM phase error. In fact, the phase difference between the zero and first orders, which ideally would be zero, becomes in the presence of EPSM phase error

$$\angle a_1 - \angle a_0 \approx \frac{2\Delta\phi t}{1+t^2}$$

(10.37)

What is the impact of such a change in the phase difference between the diffraction orders? Focus also causes a phase difference between the zero and first orders. For the simple case of coherent illumination (that is, three-beam imaging), a defocus of δ causes a phase difference of

$$\angle a_1 - \angle a_0 \approx \frac{\pi\delta\lambda}{p^2}$$

(10.38)

Thus, the effect of the EPSM phase error will be to shift best focus by an amount given by

$$\delta_{\text{shift}} \approx \frac{2p^2\Delta\phi t}{\pi\lambda(1+t^2)}$$

(10.39)

Possibly a more useful expression for this effective focal shift is as a function of the depth of focus. Using the simple Rayleigh criterion for DOF,

$$DOF \approx k_2\frac{p^2}{\lambda}$$

(10.40)

where k_2 is some number less than 1 (0.7 is reasonable). Combining these expressions, and assuming that $t^2 \ll 1$,

$$\delta_{\text{shift}} \approx \frac{2}{\pi k_2}\Delta\phi t DOF \approx \Delta\phi t DOF$$

(10.41)

For a 6% EPSM, focus will shift by about 0.4% of the DOF per degree of EPSM phase error. Remembering that Equation (10.39) was derived under the simple assumption of coherent illumination, full image simulations show that the use of partial coherence can double or triple the focus shift compared to the coherent case. Thus, a 10° phase error

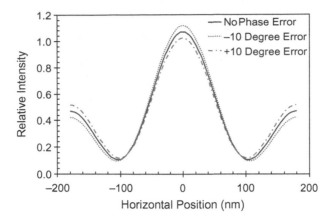

Figure 10.36   *A small phase error in an EPSM mask changes the aerial image in the same way as a small shift in focus. Here, ±10° phase error moves the image closer and farther away from best focus (wavelength = 248 nm, NA = 0.8, 180-nm lines/space pattern, coherent illumination, 150-nm defocus). For this case, a 10° phase error shifts best focus by about 14 nm*

might cause best focus to shift by 10 % of the DOF, potentially causing a 20 % loss in DOF. Off-axis illumination, however, tends to lower this effect since this illumination is specifically intended to minimize the impact of phase errors between the zero and first orders. Figure 10.36 illustrates this focus-shift effect for three-beam imaging, showing also that unlike alternating PSM, there is no pattern placement change through focus in the presence of an EPSM phase error.

In general, attenuated phase-shift masks are much less sensitive to phase errors than alternating phase-shift masks. The major impact of small phase errors is a focus shift, so the biggest worry would be a variation of EPSM phase across a reticle rather than a mean to target error. Since cleaning can cause phase errors in an EPSM mask, understanding the impact of these errors is quite important.

10.5   Natural Resolutions

The discussion above defined feature resolution with respect to CD control, (using depth-of-focus as the metric of CD control) and pitch resolution with respect to the frequency cutoff of the imaging lens. There are, however, three special purpose metrics of resolution that can also be used to define the ultimate capabilities of an imaging system. These special metrics, though defined for very specific mask features, have almost universal appeal as 'natural' metrics of imaging resolution.

10.5.1   Contact Holes and the Point Spread Function

The first metric concerns the smallest possible contact hole that can be printed. Consider a mask pattern of an isolated contact hole in a chrome (totally dark) background. Now let that hole shrink to an infinitesimal pinhole. If this small pinhole could be imaged in a positive resist, how big would the resist hole be? Of course, there is a practical problem with experimentally determining this 'resolution': as the pinhole shrinks, the intensity of light reaching the wafer becomes infinitesimally small, making the required exposure time for the resist grow to infinity. However, from a theoretical perspective, we can avoid this problem by assuming we are very patient and calculating the printed result.

Thinking first of just the imaging tool, what would be the aerial image resulting from this infinitely small pinhole? We will normalize our aerial image coming from the pinhole to have a peak intensity of 1.0 when in-focus for an ideal, aberration-free optical system. The aerial image of a pinhole, when normalized in this way, is called the *point spread function* (PSF) of the optical system (see Chapter 2). The PSF is a widely used metric of imaging quality for optical system design and manufacture, and is commonly calculated for lens designs and measured on fabricated lenses using special benchtop equipment. For classical imaging applications, the PSF can be calculated as the square of the magnitude of the Fourier transform of the imaging tool exit pupil. For an in-focus, aberration-free system of wavelength λ, the pupil function is just a circle whose radius is given by the numerical aperture NA, and the PSF becomes

$$PSF_{\text{ideal}} = \left| \frac{J_1(2\pi\rho)}{\pi\rho} \right|^2 \tag{10.42}$$

where J_1 is the Bessel function of the first kind, order one, and ρ is the radial distance from the center of the image normalized by multiplying by NA/λ.

How wide is the PSF? For large contact holes, the normalized intensity at a position corresponding to the mask edge (that is, at the desired contact hole width) is about 0.25–0.3. If we use this intensity range to measure the width of the PSF, the result is a contact hole between 0.66 and $0.70\lambda/NA$ wide. Thus, this width represents the smallest possible contact hole that could be imaged with a conventional chrome-on-glass (i.e. not phase-shifted) mask. If the contact hole size on the mask approaches or is made smaller than this value, the printed image is controlled by the PSF, not by the dimensions of the mask. Making the contact size on the mask smaller only reduces the intensity of the image peak. Thus, this width of the PSF, the ultimate resolution of a chrome-on-glass contact hole, is a natural resolution of the imaging system.

If instead of using a chrome-on-glass mask when making our conceptual pinhole an attenuated PSM is used, the width of the resulting PSM is decreased. Assuming that the mask background electric field transmittance is t and the phase shift is exactly $180°$, the resulting PSF is

$$PSF_{\text{ideal}} = \left| \frac{\frac{J_1(2\pi\rho)}{\pi\rho} - t}{1-t} \right|^2 \tag{10.43}$$

A graph of this PSF for $t = 0.245$ (a 6% intensity transmittance ESPM) compared to $t = 0$ (a chrome-on-glass mask) is shown in Figure 10.37. At an intensity threshold of

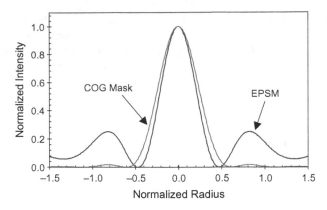

Figure 10.37   *Comparison of the ideal PSM from a chrome-on-glass (COG) mask and a 6%
intensity transmittance embedded phase-shifting mask (EPSM)*

0.25, the COG PSF has a width of about $0.705\lambda/NA$, whereas the EPSM mask produces
a PSF with a width of $0.595\lambda/NA$ (16% narrower). Note that the narrowing of the PSF
comes at the expense of increased sidelobes, which reach an intensity of nearly 0.25 for
this 6% EPSM mask. Thus, 6% intensity transmittance represents about the maximum
practical transmittance level when printing isolated contact holes unless some sort of
sidelobe suppression (in the form of strategically placed assist slots) is employed.

10.5.2   The Coherent Line Spread Function (LSF)

If instead of an infinitely small pinhole in a dark background an infinitely narrow space
is used, the resulting image is called the *line spread function* (LSF). Unlike the PSF,
however, the LSF image will depend on the illumination used. For coherent illumination,
the LSF can be calculated as an integral over one dimension of the coherent PSF:

$$LSF(x) = \int_{-\infty}^{\infty} \frac{J_1\left(\dfrac{2\pi NA}{\lambda}\sqrt{x^2+y^2}\right)}{\dfrac{\pi NA}{\lambda}\sqrt{x^2+y^2}}\,dy \tag{10.44}$$

For incoherent illumination,

$$LSF(x) = \int_{-\infty}^{\infty} \left|\frac{J_1\left(\dfrac{2\pi NA}{\lambda}\sqrt{x^2+y^2}\right)}{\dfrac{\pi NA}{\lambda}\sqrt{x^2+y^2}}\right|^2 dy \tag{10.45}$$

Alternately, the coherent LSF can be calculated using the image of an isolated space.
From Chapter 2, the TE coherent image of an isolated space of width w is

$$I(x) = \frac{1}{\pi^2}\left[\operatorname{Si}\left(\frac{2\pi NA}{\lambda}(x+w/2)\right) - \operatorname{Si}\left(\frac{2\pi NA}{\lambda}(x-w/2)\right)\right]^2 \tag{10.46}$$

Normalizing this image so that the peak intensity is 1,

$$\frac{I(x)}{I(0)} = \left[\frac{\mathrm{Si}\left(\dfrac{2\pi NA}{\lambda}(x+w/2)\right) - \mathrm{Si}\left(\dfrac{2\pi NA}{\lambda}(x-w/2)\right)}{2\mathrm{Si}\left(\dfrac{\pi NA}{\lambda}w\right)} \right]^2 \tag{10.47}$$

Taking the limit of Equation (10.47) as w goes to zero (by applying L'Hopital's rule and realizing that the derivative of the sine integral is just the sinc function),

$$LSF(x) = \lim_{w \to 0} \frac{I(x)}{I(0)} = \left[\frac{\sin(2\pi NAx/\lambda)}{2\pi NAx/\lambda} \right]^2 \tag{10.48}$$

Using an intensity threshold of 0.25, the coherent illumination LSF has a width of $0.603\lambda/NA$, and is thus 15 % narrower than the PSF. The LSF is a natural resolution limit of an isolated space.

10.5.3   The Isolated Phase Edge

Another special mask feature which exhibits a similar natural resolution behavior is the 180° phase edge, as described previously in section 10.4.1. Consider a chromeless mask with a large region shifted by 180° to produce a long, straight boundary between the 0° and the 180° regions. Since light transmitted to either side of this edge will have a 180° phase difference, the light that diffracts and interferes under the edge will cancel out, producing a dark line centered under the phase edge. As a result, this isolated 180° phase edge will print as a narrow line in a positive photoresist.

For coherent TE illumination, the in-focus image of an isolated phase edge was given in Equation (10.12). Figure 10.38 shows a graph of this equation. The width of this image

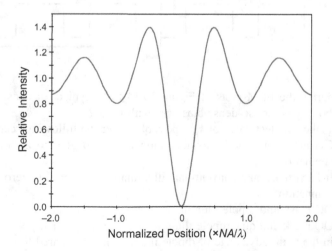

Figure 10.38   *The aerial image of an isolated 180° phase edge (shown here using coherent illumination) will produce a narrow line in a positive resist*

can be estimated in the same way as the PSF. Assuming an intensity level between 0.25 and 0.3, the width of the phase edge line is between 0.26 and $0.29\lambda/NA$. The use of partially coherent illumination does not appreciably change this width.

What controls the width of the image of an isolated phase edge? Obviously the edge itself does not have a 'width'. Like the PSF and the LSF, the width of the phase edge image is in fact an inherent property of the imaging system, controlled by its numerical aperture and wavelength. In fact, analogous to the LSF, the image of Equation (10.12) could be called the *phase edge line spread function*. The size of this tiny line is the 'natural' resolution of any 180° phase edge. But its importance goes beyond just the printing of small isolated lines. A phase-shifted mask can be thought of as a collection of 0–180° phase transitions. Each transition has a tendency to print at its natural linewidth of about $0.26\lambda/NA$. Much insight can be gained by approaching phase-shifting mask design with this natural linewidth in mind.

For the special types of mask features described here, the resolution of that feature becomes a function only of the properties of the imaging system and, strangely, becomes independent of the size of the feature on the mask. I call these resolution properties the 'natural' resolutions of the imaging tool.

Problems

10.1. Figure 10.1 shows an example of how the depth of focus varies with feature size. Generate an equivalent plot using the Rayleigh DOF criterion. Assume a 193-nm wavelength, equal lines and spaces, coherent three-beam imaging, and $k_2 = 0.8$.

10.2. Derive the diffraction pattern of a semi-dense line (width w, pitch p) with single SRAFs (width w_{SRAF}) placed in the center of the spaces:

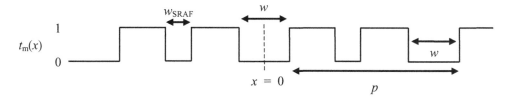

10.3. Considering the FEM plots of Figure 10.14 to be typical of dense and isolated lines, how does the iso-dense bias vary with focus?

10.4. What is the smallest value of k_{pitch} possible under the following scenarios:
   (a) COG mask and conventional illumination as σ goes to zero (i.e. coherent illumination)
   (b) Alt-PSM mask and conventional illumination as σ goes to zero (i.e. coherent illumination)
   (c) COG mask and dipole illumination with very small poles
   (d) COG mask and quadrupole illumination with very small poles

10.5. Explain the main advantage of dipole illumination compared to quadrupole illumination. Explain the main disadvantage of dipole illumination compared to quadrupole illumination.

10.6. For an optimized dipole illuminator, a two-beam aerial image results:

$$I(x) = a_0^2 + a_1^2 + 2a_0a_1 \cos(2\pi x/p)$$

Since the dipole can only print small-pitch patterns oriented in one direction, one possible solution is to add a second dipole rotated by 90° to obtain an illuminator called a *cross-quadrupole*:

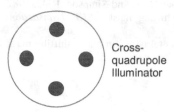

Cross-quadrupole Illuminator

When printing at the pitch for which the dipole and quadrupole are optimized, derive the resulting in-focus aerial image (assume TE illumination) for a cross-quadrupole.

10.7. The coherent two-beam aerial image for an alternating PSM mask is given in Equation (10.9). From this, derive the image for a chromeless repeating line/space pattern by letting the width of the chrome line in the alternating PSM mask go to zero. Qualitatively, what is the effect of reducing the chrome width?

10.8. Derive Equation (10.10), the diffraction pattern for an isolated phase-shifted line.

10.9. Derive Equation (10.11), the coherent aerial image for an isolated phase-shifted line.

10.10. Derive Equation (10.15) from Equation (10.14).

10.11. Equations (10.16) and (10.17) give the coherent (TE) image for an attenuated PSM pattern of lines and spaces. What happens to this image as t_2 goes to 1? (Such a mask is called a chromeless PSM.) What is the result for the case of equal lines and spaces?

10.12. Derive an equation for the coherent LSF when using an attenuated PSM. What is the maximum sidelobe intensity for the LSF for a 6% attenuating PSM?

10.13. Derive an expression for the normalized image log-slope of the isolated phase edge, assuming the nominal feature size is at an intensity threshold of 0.25.

References

1 Levenson, M.D., 1993, Wavefront engineering for photolithography, *Physics Today*, **46**, 28–36.

2 Fukuda, H., Imai, A., Terasawa, T. and Okazaki, S., 1991, New approach to resolution limit and advanced image formation techniques in optical lithography, *IEEE Transactions on Electron Devices*, **38**, 67–75.

3 Arthur, G. and Martin, B., 1995, Investigation of photoresist-specific optical proximity effect, *Micro- and Nano-Engineering '95*, Aix-en-Provence, France.

4 Chen, J.F., Laidig, T., Wampler, K. and Caldwell, R., 1997, Optical proximity correction for intermediate-pitch features using sub-resolution scattering bars, *Journal of Vacuum Science and Technology B*, **15**, 2426–2433.

5   Mack, C.A., 1989, Optimum stepper performance through image manipulation, *KTI Microelectronics Seminar, Proceedings*, 209–215.
6   Fehrs, D.L., Lovering, H.B. and Scruton, R.T., 1989, Illuminator modification of an optical aligner, *KTI Microelectronics Seminar, Proceedings*, 217–230.
7   Socha, R., Dusa, M., Capodieci, L., Finders, J., Fung Chen, J., Flagello, D. and Cummings, K., 2000, Forbidden pitches for 130-nm lithography and below, *Proceedings of SPIE: Optical Microlithography XIII*, **4000**, 1140–1155.
8   Levenson, M.D., Viswanathan, N.S. and Simpson, R.A., 1982, Improving resolution in photolithography with a phase-shifting mask, *IEEE Transactions on Electron Devices*, **ED-29**, 1828–1836.
9   Mack, C.A., 1991, Fundamental issues in phase-shifting mask technology, *KTI Microlithography Seminar Interface '91, Proceedings*, 23–35.

Appendix A

Glossary of Microlithographic Terms

'When I use a word, it means just what I choose it to mean – neither more nor less.'
Humpty Dumpty (Lewis Carroll's *Alice's Adventures in Wonderland*)

The following is a list of common words or phrases and their corresponding definitions as used in the field of semiconductor microlithography. For a field as rich and diverse as lithography, no list of terms could ever be complete. I hope, however, that the terms contained here are sufficiently representative that this glossary might be useful as a reference. In particular, people new to or outside of the field of lithography should benefit from its use. The definitions assume the reader has a general technical background and a basic familiarity with the semiconductor industry.

Format of Glossary Entries:
Glossary Item   Definition of the glossary item, in the context of semiconductor microlithography, is given here.
   Example: *An example sentence that illustrates the use of the* glossary item *appears here.*

The example sentence is meant to illustrate the use of the glossary item in the context of lithography, but is *not* a part of the definition of the item.

A
ABC **Parameters**   see Dill Parameters.

Aberrations, Lens   Any deviation of the real performance of an optical system (lens) from its ideal (Fourier optics) performance. Examples of lens aberrations include coma, spherical aberration, field curvature, astigmatism, distortion and chromatic aberration. One way to describe lens aberrations is through a Zernike polynomial fit to the wavefront error at the exit pupil of the lens for each field point.
   Example: *The aberrations of the objective lens caused a noticeable degradation in image quality.*

Fundamental Principles of Optical Lithography: The Science of Microfabrication, Chris Mack.
© 2007 John Wiley & Sons, Ltd.

Absorption Coefficient   The fractional decrease in the intensity of light traveling through a material per unit distance traveled. See also Extinction Coefficient.

Example: *The absorption coefficient of the ArF resist was so high that only a thin film of the resist could be used.*

Acid-Catalyzed Resist   A type of chemically amplified resist where an acid is the product of exposure and this acid serves as the catalyst for a thermal reaction which changes the solubility of the resist. See also Chemically Amplified Resist.

Example: Acid-catalyzed resists *are the most common type of resist for deep-UV lithography.*

Activation Energy   Defined by its role in the Arrhenius equation, the activation energy determines the temperature dependence of chemical reaction rate constants, diffusivities, and other temperature-dependent rate terms. High activation energies produce a large temperature dependence.

Example: *This low* activation energy *resist will begin its acid-catalyzed deblocking reaction at room temperature.*

Actinic Wavelength   The wavelength used to expose the photoresist in a lithographic system.

Example: *Measurement of the refractive index of the substrate at the* actinic wavelength *is necessary in order to design an optimal BARC.*

Additive Patterning   A process by which material is added in the places where the pattern is to be formed. Examples include lift-off and electroplating processes.

Example: *Due to our inability to plasma etch copper,* an additive patterning *approach was chosen instead.*

Adhesion Promoter   A chemical that is applied to the surface of a wafer in order to improve the adhesion of resist to the wafer, often by eliminating water from the wafer surface.

Example: *The HMDS, used as an* adhesion promoter, *was applied using the vapor prime unit.*

Advanced Process Control (APC)   The use of automated feedback and feed-forward loops to control a lithographic process.

Example: *Both rework rates and end-of-line yield were improved after turning on the overlay* APC *system.*

Aerial Image   An image of a mask pattern that is projected onto the photoresist-coated wafer by an optical system.

Example: *The* aerial image *of the isolated line was found to differ significantly from that of the dense line.*

Aligner   see Mask Aligner.

Alignment   The act of positioning the image of a specific point on a photomask (the alignment key) to a specific point on the wafer (the alignment target) to be printed. Alignment accuracy is the overlay measured at this alignment target.

Example: *The* alignment *system of the stepper used an advanced image recognition algorithm.*

Alignment Key   The pattern on a photomask used to perform alignment.

Example: *This mask had several* alignment keys *for use on several different steppers.*

Alignment Mark   see Alignment Key or Alignment Target.

Alignment Target   The pattern on a wafer used to perform alignment.

Example: *A mesa structure is used as the* alignment target *for the gate level of this device.*

Alternating PSM   A type of phase-shifting mask where the clear region to one side of a small chrome line is shifted in phase by 180° compared to the clear region to the other side of that same line. Also called alternating aperture PSM or Levenson PSM.

Example: *Although* alternating PSM *promises extreme resolution and good depth of focus, phase conflicts limit their use for general circuit patterns.*

Annular Illumination   A type of off-axis illumination where a doughnut-shaped (annular) ring of light is used as the source.

Example: *The use of* annular illumination *was found to give a noticeable improvement in depth of focus for these features.*

Antireflective Coating (ARC)   A coating that is placed on top or below the layer of resist to reduce the reflection of light, and hence, reduce the detrimental effects of standing waves or thin-film interference. See also Top Antireflective Coating and Bottom Antireflective Coating.

Example: *By optimizing the thickness of the* antireflective coating, *the swing curve amplitude was reduced to almost zero.*

APC   see Advanced Process Control.

Aperture, Numerical   see Numerical Aperture.

Aperture Stop   see Pupil, Lens.

ARC   see Antireflective Coating.

ArF   Argon Fluoride, a type of excimer laser used in optical lithography that emits light at about 193 nm.

Example: *Due to the difficulty in producing calcium fluoride lens components,* ArF *exposure tools are considerably more costly to manufacture.*

Arrhenius Coefficient   Defined by its role in the Arrhenius equation, the Arrhenius coefficient is the pre-exponential term in the equation that defines the temperature dependence of chemical reaction rate constants, diffusivities and other temperature-dependent rate terms.

Example: *The* Arrhenius coefficient *is often thought of as the extrapolation of the temperature-dependent rate constant to an infinitely high temperature.*

Arrhenius Equation   The temperature dependence of chemical reaction rate constants, diffusivities and other temperature-dependent rate terms as an exponential relationship with the inverse of absolute temperature.

Example: *The* Arrhenius equation *is used to determine how reaction rates and diffusion change with PEB temperature.*

Aspect Ratio   The ratio of a resist feature's height to its width.

Example: *The resist images suffered from pattern collapse whenever the* aspect ratio *exceeded about 3 : 1.*

Astigmatism   An aberration that results in a shift in best focus for radially oriented line patterns compared to tangentially oriented patterns.

Example: *Typically, a variation of H–V bias with focus is a sign of* astigmatism.

Attenuated PSM   A type of phase-shifting mask where the nominally dark region of the mask is allowed to transmit a fraction of the light (e.g. 6%) with a 180° phase shift from light transmitted through the clear regions of the mask.

Example: *Although alternating PSM provides better performance,* attenuated PSMs *have become very popular due to their ease of design and manufacture.*

Autofocus System   A part of a projection imaging tool that automatically places the top surface of the wafer a set distance from the focal plane.

Example: *Despite the sophistication of the scanner's* autofocus system, *the lithographer must still determine best focus manually by shooting a focus–exposure matrix.*

B

Bandwidth, Illumination   The range of wavelengths that is used to illuminate the mask, and thus to expose the resist.

Example: *The* illumination bandwidth *for a typical g-line stepper is about 10-nm FWHM.*

BARC   see Bottom Antireflective Coating.

Bias   see Mask Bias.

Binary Mask   A mask made up of opaque and transparent regions (for example, one composed of chrome and glass) such that the transmittance of the mask is either 0 or 1. Also called a binary intensity mask.

Example: *The needed resolution was not obtained using a conventional* binary mask.

Birefringence   A somewhat rare property of some materials (usually crystals) where the refractive index of the material is a function of the polarization of the light passing through the material.

Example: *The discovery that calcium fluoride exhibits significant intrinsic* birefringence *caused considerable consternation during the development of 157-nm lithography.*

Bleaching, Photoresist   The decrease in optical absorption of a photoresist due to the chemical changes that occur upon exposure to light.

Example: *Without* photoresist bleaching, *this resist could not be used effectively at thicknesses greater than about 2 μm.*

Bossung Curves   see Focus–Exposure Matrix (named after John Bossung, the engineer who first published these curves.).

Bottom Antireflective Coating (BARC)   An antireflective coating placed just below the photoresist to reduce reflections from the substrate.

Example: *The use of a* bottom antireflective coating *not only reduced the swing curve, but also nearly eliminated the effects of reflective notching.*

C

CAR   see Chemically Amplified Resist.

Catadioptric   An optical system made up of both refractive elements (lenses) and reflective elements (mirrors).
Example: *The* catadioptric *lens system was capable of accepting a much broader illumination bandwidth than conventional all-refractive lenses.*

Catoptric   An optical system made up of only reflective elements (mirrors).
Example: *The first Perkin–Elmer scanners used a unique* catoptric *lens design.*

Cauchy Coefficients   Coefficients of the Cauchy equation, which gives an empirical expression for the variation of the index of refraction of a material as a function of wavelength.
Example: *The* Cauchy coefficients *of the resist are needed in order to use the reflectance spectroscopy tool to measure resist thickness.*

CD   see Critical Dimension.

Characteristic Curve   see Contrast Curve.

Chemically Amplified Resist   A type of photoresist, most commonly used for deep-UV processes, which, upon pos-texposure bake, will multiply the number of chemical reactions through the use of chemical catalysis.
Example: *The* chemically amplified resist *exhibited a large sensitivity to airborne base contaminants.*

Chromatic Aberration   A change in the aberration behavior of a lens as a function of wavelength.
Example: *For KrF lithography tools, the main* chromatic aberration *is a linear shift in focus as a function of wavelength.*

Circular Definition   see Definition, Circular.

Clearing Dose (E_o)   see Dose-to-Clear.

Coater, Resist   Equipment used to perform resist coating. This equipment is often a part of a resist track or cluster tool.
Example: *This* resist coater *can be used at spin speeds from 1000 to 5000 rpm.*

Coating, Resist Spin   see Spin Coating.

Coherence Factor   see Partial Coherence.

Coherence, Spatial   The phase relationship of light at two different points in space at any instant in time. For mask illumination, the spatial coherence is determined by the range of angles incident on the mask.
Example: *For lithographic tools, the* spatial coherence *of the illumination is most easily described by the partial coherence factor.*

Coherent Illumination   A type of illumination resulting from a point source of light that illuminates the mask with light from only one direction. This is more correctly called *spatially* coherent illumination.

Example: *Although* coherent illumination *gave the best resolution performance for the phase-shifting mask, it resulted in very poor illumination uniformity.*

Coma   An aberration that is often seen as a difference in linewidth between the left and right lines in a group of five lines.

Example: Coma *also causes an asymmetry in resist profiles (right side versus left side) that changes as a function of focus.*

Condenser Lens   Lens system in an optical projection system that prepares light to illuminate the mask.

Example: *For Köhler illumination, the* condenser lens *forms an image of the source at the entrance pupil of the objective lens.*

Contact Printing   A lithographic method whereby a photomask is placed in direct contact with a photoresist-coated wafer and the pattern is transferred by exposing light through the photomask into the photoresist.

Example: *Although exhibiting good resolution,* contact printing *was limited by defect densities.*

Contrast, Image   see Image Contrast.

Contrast, Resist   see Photoresist Contrast.

Contrast Curve   see H–D curve.

Contrast Enhancement Layer (CEL)   A highly bleachable coating on top of the photoresist that serves to enhance the contrast of an aerial image projected through it.

Example: *The* contrast enhancement layer *resulted in improved resist sidewall angle, but at the cost of reduced throughput.*

Corner Rounding   The rounding of a nominally sharp, square corner of a printed lithographic feature due to the inherent resolution limits of the patterning process.

Example: *The* corner rounding *on the reticle resulted in a reduction of the total energy transmitted through the mask opening.*

Critical Dimension (CD)   The size (width) of a feature printed in resist, measured at a specific height above the substrate. Also called the linewidth or feature width. (Over time, the meaning of 'critical' has become vague, and it seems that any dimension worth measuring must be critical.).

Example: *The* critical dimension *specifications for this device are very tight.*

Critical Shape (CS)   An extension of the one-dimensional critical dimension to two-dimensional features, the critical shape is the polygon which defines the top-down (in the plane of the substrate) shape of a feature.

Example: *The line-end* critical shape *suffered from severe line-end shortening.*

Critical Shape Difference (CSD)   A statistical analysis (for example, the average magnitude) of a collection of vectors describing the difference (i.e. point-by-point distance) between two critical shapes.

Example: *The large* critical shape difference *between the two wafer patterns indicated a significant process problem.*

Critical Shape Error (CSE)   The critical shape difference between the pattern being measured and an ideal 'desired' critical shape.

Example: *A* critical shape error *of 20 nm was considered to be acceptable for this device pattern.*

D

Deep-Ultraviolet (DUV)   A common though vague term used to describe light of a wavelength in the range of about 150 to 300 nm. Also called deep-UV.

Example: *The transition of optical lithographic wavelengths from i-line to* deep-ultraviolet *accelerated as the industry dipped below the 350-nm resolution node.*

Deep-UV Lithography   Lithography using light of a wavelength in the range of about 150 to 300 nm, with about 250 nm being the most common.

Example: *Most lithographers agree that* deep-UV lithography *is required for device dimensions below 0.3 microns.*

Definition, Circular   see Circular Definition.

Defocus   The distance, measured along the optical axis (i.e. perpendicular to the plane of best focus) between the position of a resist-coated wafer and the position if the wafer were at best focus.

Example: *The amount of* defocus *cannot be determined without an adequate method of measuring best focus.*

Degree of Coherence   see Partial Coherence.

Dehydration Bake   A bake step performed on a wafer before coating with resist in order to remove water from the surface of the wafer.

Example: *The* dehydration bake *was only partially effective in removing water from the wafer surface.*

Depolarization   The change of light from being polarized to being unpolarized (that is, randomly polarized), generally as a result of scattering phenomenon.

Example: *Jones matrices cannot account for* depolarization *of light passing through the lens, though Mueller matrices can.*

Depth of Focus (DOF)   The total range of focus that can be tolerated, that is, the range of focus that keeps the resulting printed feature within a variety of specifications (such as linewidth, sidewall angle, resist loss and exposure latitude).

Example: *Optimizing the numerical aperture by finding the value that maximized the* DOF *of the critical feature was found to be very effective at improving CD control.*

Design Rule   A geometrical rule that defines minimum widths and/or spacings used when laying out a mask pattern.
Example: *Although the designer was not sure why the* design rule *forbade the use of this particular pitch, he reluctantly complied.*

Design Rule Checker (DRC)   A software package that checks a chip design for compliance with a set of design rules.
Example: *Since the* Design Rule Checker *tool had the capability to correct DRC violations, is was possible to program the tool to perform rule-based OPC.*

Developer   The chemical (typically a liquid) used to selectively dissolve resist as a function of its chemical composition.
Example: *Control of the temperature of the* developer *should be better than ±0.2 °C.*

Development   The process by which a liquid, called the developer, selectively dissolves a resist as a function of the exposure energy that the resist has received. Also called develop.
Example: *A puddle* development *process was used to reduce developer consumption.*

Development Rate   The rate (change in thickness per unit time) that the resist dissolves in developer for a given set of conditions.
Example: *The* development rate *was plotted as a function of exposure energy on a log-log scale.*

Development Rate Monitor (DRM)   An instrument used to measure the development rate of a resist by measuring the thickness of the resist *in situ* as the development proceeds.
Example: *The development rate as a function of exposure energy was characterized using a* development rate monitor.

Diattenuation   The difference in amplitude transmittance of a lens as a function of the polarization of the incident light.
Example: *At very large numerical apertures, the nonideal behavior of the lens antireflection coatings caused* diattenuation *at the highest spatial frequencies.*

Dichroism   The difference in the absorption of light by the lens as a function of the polarization of the incident light.
Example: *While diattenuation can be a concern for hyper-NA lenses,* dichroism *remains a very small problem.*

Die   A single, complete integrated circuit as printed on a wafer, possibly sliced but before packaging. Also called a chip.
Example: *Because of the size of the ASIC chip, the stepper could accommodate only one* die *in each exposure field.*

Diffraction   The propagation of light in the presence of boundaries. It is the property of light that causes the wavefront to bend as it passes an edge.
Example: *In an ideal imaging system, the quality of the aerial image is limited only by* diffraction.

Diffraction Limited   A description of a lens such that any aberrations in the lens are small enough as to be negligible. Theoretically, no lens can be perfect so that the term diffraction limited is always an approximation and the appropriateness of its use is situational.

Example: *In photographic systems and other imaging applications less stringent than lithography, lens are often described as* diffraction limited *when the RMS optical path deviation is less than a tenth of a wave.*

Diffraction Order   For a mask pattern that repeats indefinitely, the diffraction pattern becomes discrete, made up of regularly spaced points of light called diffraction orders.

Example: *In lithography, high-resolution line/space patterns are imaged with only the zero, and plus and minus first* diffraction orders *passing through the lens.*

Diffraction Pattern   The pattern of light entering the objective lens due to diffraction by a mask.

Example: *The* diffraction pattern *of a repeating pattern of lines and spaces is made up of discrete spots of light called diffraction orders.*

Diffusion Coefficient   A rate constant that defines the rate at which a particle will diffuse through a given medium for a given set of process conditions.

Example: *The* diffusion coefficient *of the acid in the chemically amplified resist was not constant during the post-exposure bake due to free volume generated by the amplification reaction.*

Diffusion Length   The average distance that a particle will diffuse for a given process.

Example: *The* diffusion length *of photoactive compound during PEB must be larger than the standing wave half period to be effective at removing standing waves from the resulting resist profile.*

Diffusivity   see Diffusion Coefficient.

Dill Parameters   Three parameters, named *A*, *B* and *C*, that are used in the Dill exposure model for photoresists. *A* and *B* represent the bleachable and nonbleachable absorption coefficients of the resist, respectively, and *C* represents the first-order kinetic rate constant of the exposure reaction. (Named for Frederick Dill, the first to publish this model.) Also called the photoresist *ABC* parameters.

Example: *The* Dill parameters *(A, B and C) were measured in a single optical transmittance experiment.*

Dioptric   An optical system made up of only refractive elements (lenses).

Example: Dioptric *lens systems require extensive effort to correct for the chromatic aberrations that are a natural part of all-refractive lenses.*

Dipole Illumination   A type of off-axis illumination where two circles or arcs of light are used as the source. These two circles are spaced evenly around the optical axis, either oriented vertically or horizontally.

Example: Dipole illumination *provides the greatest possible dense line resolution, but only for one orientation of lines and spaces.*

Direct-Write Lithography   A lithography method whereby the pattern is written directly on the wafer without the use of a mask.
Example: *Due to throughput limitations,* direct-write lithography *may never be practical for IC mass production.*

Dispersion   The variation of the index of refraction of a material as a function of wavelength.
Example: *Because of the* dispersion *of glass, lenses invariably suffer from chromatic aberration.*

Dissolution Inhibitor   A chemical which, when added to a photoresist, decreases the dissolution rate of the resist in developer. For many positive photoresists, the photoactive compound acts as a dissolution inhibitor.
Example: *If the* dissolution inhibitor *is bound directly to the novolac resin, diffusion during PEB does not occur.*

Dissolution Promoter   A chemical which, when added to a photoresist, increases the dissolution rate of the resist in developer. For many positive photoresists, the exposed photoactive compound acts as a dissolution promoter.
Example: *When exposed to light, the DNQ dissolution inhibitor becomes a mild* dissolution promoter.

Dissolution Rate   see Development Rate.

Distortion   An optical aberration that causes a variation in pattern placement error as a function of field position.
Example: *The variation of* distortion *from one stepper to another results in the need for lens matching when printing critical layers, or possibly even the use of a dedicated stepper.*

DOF   see Depth of Focus.

Dose   see Exposure Energy.

Dose-to-Clear (E_o)   The amount of exposure energy required to just clear the resist in a large clear area for a given process. Also called the clearing dose.
Example: *The* dose-to-clear *was measured once per shift and used as a process monitor.*

Dose-to-Size   The amount of exposure energy required to produce the proper dimension of the resist feature.
Example: *Changing the thickness of the photoresist resulted in a large change in the* dose-to-size *of the contact hole.*

DRM   see Development Rate Monitor.

DUV   see Deep Ultraviolet.

DUV Lithography   see Deep-UV Lithography.

Dyed Resist   A photoresist with an added nonphotosensitive chemical that absorbs light at the exposing wavelength.
Example: *Although the* dyed resist *was effective at reducing the swing curve, the resulting sidewall angle was unacceptably low.*

E

E-beam Lithography   see Electron Beam Lithography.

EBR   see Edge Bead Removal.

Edge Bead   A buildup of resist along the outer edge of a wafer caused by resist surface tension during the spin coat process.
Example: *If not removed, the* edge bead *causes contamination during subsequent wafer processing.*

Edge Bead Removal (EBR)   A process by which resist is removed from the outer edge of a resist-coated wafer in order to remove the thick 'bead' of resist that is usually formed along this edge during the spin coat process.
Example: *The spin coat module included both front and backside* edge bead removal *systems.*

Edge Placement Error (EPE)   A term used in optical proximity correction, this is a critical shape error where the distance vectors are constrained to be normal to the desired shape.
Example: *The model-based OPC system used the maximum* edge placement error *as the cost function of the optimization procedure.*

Electron Beam Lithography   Lithography performed by exposing resist with a beam of electrons. Also called e-beam lithography.
Example: Electron beam lithography *remains the most popular technique for producing high-resolution masks.*

Embedded PSM (EPSM)   see Attenuated PSM.

Entrance Pupil, Lens   The image of the pupil (also called the aperture stop) of an imaging lens when viewed from the entrance side of the lens.
Example: *The distance from the object to the* entrance pupil *of the lens is exactly equal to the distance from the exit pupil to the image times the reduction ratio of the lens.*

EPSM   see Attenuated PSM.

Etch Selectivity   The ratio of the vertical etch rate of the material that you wish to etch compared to the vertical etch rate of the material that you do not wish to etch (the masking material or the substrate material).
Example: *Sputtering is sometimes used because it is a very good anisotropic etching process, despite its lack of* etch selectivity.

Etching, Anisotropic   An etch process where the vertical etch rate within a given material is faster than the horizontal etch rate.
Example: *Sputtering is sometimes used because it is a very good* anisotropic etching *process, despite its lack of etch selectivity.*

Etching, Isotropic   An etch process where the etch rate within a given material is independent of position and direction.

Example: *While wet etch processes are simple and exhibit very good etch selectivity, their performance on fine patterns is limited by the fact that they are* isotropic etching *processes.*

EUV   see Extreme Ultraviolet.

EUV Lithography   Lithography using light of a wavelength in the range of about 5 to 50 nm, with about 13 nm being the most common. Also called soft x-ray lithography.

Example: *Although many problems remain,* EUV lithography *could potentially have both high resolution and large depth of focus.*

Excimer Laser   Laser using a gas or gases to create an excited dimer (e.g. KrF), usually resulting in pulsed deep-UV radiation.

Example: Excimer lasers *are used extensively in deep-UV lithography due to their extremely high output power.*

Exit Pupil, Lens   The image of the pupil (also called the aperture stop) of an imaging lens when viewed from the exit side of the lens.

Example: *The effective focal length of a lens is defined as the distance from the* exit pupil *to the image plane.*

Exposure   The process of subjecting a resist to light energy (or electron energy in the case of electron beam lithography) for the purpose of causing chemical change in the resist.

Example: *The chemically amplified resist was very sensitive to any delay between* exposure *and post-exposure bake.*

Exposure Dose   see Exposure Energy.

Exposure Energy   The amount of energy (per unit area) that the photoresist is subjected to upon exposure by a lithographic exposure system. For optical lithography, it is equal to the light intensity times the exposure time. Also called the exposure dose, or simply dose.

Example: *Accurate control of the* exposure energy *delivered to the resist is an important function of any lithographic exposure tool.*

Exposure Field   see Field, Exposure.

Exposure Latitude   The range of exposure energies (usually expressed as a percent variation from the nominal) that keeps the linewidth within specified limits.

Example: *A minimum* exposure latitude *of 10% is needed for this process in order to get adequate CD control.*

Exposure Margin   The ratio of the dose-to-size to the dose-to-clear.

Example: *In most cases, increasing* exposure margin *results in an increase in process latitude.*

Extinction Coefficient Another name for the absorption coefficient of a material, often using a base-10 definition.

Example: *The effectiveness of the dye was determined by measuring the* extinction coefficient *of the resist.*

Extreme Ultraviolet (EUV) A common though vague term used to describe light of a wavelength in the range of about 5 to 50 nm. Also called soft x-ray.

Example: *The historical progress of optical lithography toward ever smaller wavelengths has convinced some that* extreme ultraviolet *radiation will be the next logical step.*

F

Fab see Wafer Fab.

FE Matrix see Focus–Exposure Matrix.

Feature Size see Critical Dimension.

Field, Exposure The area of a wafer that is exposed at one time by the exposure tool.

Example: *An increase in the* exposure field *size allowed more die to be imaged per exposure, resulting in greater throughput.*

Field-By-Field Alignment A method of alignment whereby the mask is aligned to the wafer for each exposure field (as opposed to global alignment).

Example: *Although* field-by-field *alignment reduced throughput considerably, the improved overlay accuracy was worth the cost.*

Field Curvature An optical aberration that causes a variation in best focus as a function of field position.

Example: Field curvature *results in a systematic focus error that can only be partially corrected by a wafer tilt adjustment.*

Flare The unwanted light that reaches the photoresist as a result of scattering and reflection off surfaces in the optical system that are meant to transmit light. Also called background scattered intensity.

Example: *Contamination of the bottom surface of the lens resulted in a large increase in* flare.

Flood Exposure Exposure of the resist to blanket radiation with no pattern. For projection tools such as a stepper, this is also called an open-frame exposure (exposure with no mask or with a blank glass mask).

Example: *A* flood exposure *is the last step in the image reversal process.*

Focal Plane The plane of best focus of the optical system.

Example: *The best results typically come by placing the* focal plane *near the middle of the thickness of the resist.*

Focal Position   see Focus.

Focus   The position of the plane of best focus of the optical system relative to some reference plane, such as the top surface of the resist, measured along the optical axis (i.e. perpendicular to the plane of best focus).
Example: *The nonflatness of the wafer results in unavoidable* focus *errors.*

Focus–Exposure Matrix   The variation of linewidth (and possibly other parameters) as a function of both focus and exposure energy. The data are typically plotted as linewidth versus focus for different exposure energies and these plots are often referred to as smiley plots, spider plots, or Bossung curves.
Example: *The first step in measuring depth of focus is shooting a* focus–exposure matrix.

Fourier Optics   A mathematical description of imaging where diffraction is calculated as a Fourier transform, followed by multiplication by the pupil function, followed by a second Fourier transform to describe the focusing behavior of the imaging lens.
Example: *The* Fourier Optics *approach encourages a natural 'frequency domain' language for the description of imaging.*

G

G-Line   A line of the mercury spectrum corresponding to a wavelength of about 436 nm.
Example: G-line *steppers were the dominant lithography tools throughout the 1980s.*

GDS II   An industry standard file format for mask layout information.
Example: *When finished, the chip designer performed a final 'tape out', saving the mask layout data into* GDS II *format for transmittance to the mask shop.*

Glass Transition Temperature   The temperature (or the midpoint of the temperature range) at which a polymer makes a transition from behaving mostly like a solid to behaving mostly like a liquid.
Example: *After prebake, the* glass transition temperature *of the resulting resist film is approximately equal to the prebake temperature.*

Global Alignment   A method of alignment where the mask is aligned globally to the whole wafer (as opposed to field-by-field alignment).
Example: *Where applicable,* global alignment *is preferred due to its high throughput.*

H

H–D Curve   The standard form of the H–D or contrast curve is a plot of the relative thickness of resist remaining after exposure and development of a large clear area as a function of log-exposure energy. The theoretical H–D curve is a plot of log-development rate versus log-exposure energy. (H–D stands for Hurter–Driffield, the two scientists who first used a related curve in 1890.) Also called the photoresist contrast curve or characteristic curve.
Example: *The* H–D curve *was measured by exposing one wafer with a series of open-frame exposures of increasing energy.*

H-Line   A line of the mercury spectrum corresponding to a wavelength of about 405 nm.
Example: *The* h-line *of the mercury spectrum was essentially skipped as the industry moved from g-line directly to i-line steppers.*

H–V Bias   The difference in linewidth between horizontally and vertically oriented resist features that, other than orientation, should be identical.
Example: *The variation of* H–V bias *with focus was an indication of astigmatism in the stepper lens.*

Hard Bake   The process of heating the wafer after development of the resist in order to harden the resist patterns in preparation for subsequent pattern transfer. Also called postbake and postdevelop bake.
Example: *The* hard bake *step was necessary to ensure good etch resistance of the photoresist during plasma etching.*

Hurter–Driffield Curve   see H–D Curve.

Huygens' Principle   The idea that any wavefront can be decomposed into an array of spherically radiating point sources. The propagation of the wavefront can be calculated as the sum of the propagating point source spherical waves.
Example: Huygens' Principle, *when coupled with the concept of interference, can be used to derive a simple scalar diffraction theory.*

Hyper-NA   A euphemistic term to describe numerical apertures greater than 1.0.
Example: *Immersion lithography enables the design and manufacture of* hyper-NA *lenses.*

I
IC   see Integrated Circuit.

I-Line   A line of the mercury spectrum corresponding to a wavelength of about 365 nm.
Example: *Improved resolution made* i-line *steppers the lithography tool of choice since about 1990.*

Illumination, Köhler   see Köhler Illumination.

Illumination System   The light source and optical system designed to illuminate the mask for the purpose of forming an image on the wafer.
Example: *The* illumination system *in a modern stepper is more complicated than the entire stepper of 20 years ago.*

Image Contrast   A classic image metric useful for small equal line/space patterns only, the image contrast is defined as the difference between the maximum and minimum intensities in an image divided by their sum. Also known as the fringe visibility of two interfering plane waves.
Example: *Because of its limited usefulness,* image contrast *is not used in lithography as an image metric as often as the image log-slope.*

Image Log-Slope   The slope of the logarithm of the aerial image, usually defined at the nominal edge of the designed pattern.
Example: *A plot of the* image log-slope *versus defocus provides an excellent method of estimating depth of focus.*

Image Reversal   A chemical process by which a positive photoresist is made to behave like a negative photoresist.
Example: *The use of an* image reversal *process produced the reentrant profiles needed for metal lift-off.*

Immersion Lithography   A mode of optical lithography where an immersion fluid, with a refractive index greater than 1, fills the gap between the projection lens and the wafer.
Example: *The recent interest in* immersion lithography *is based on the hope of improved depth of focus at a constant resolution, or improved resolution at a (relatively) constant depth of focus.*

Imprint Lithography   A patterning method based on embossing where a topographic pattern on a mask is replicated as a topographic pattern in a polymer media by pressing the mask (called a template) directly into the polymer media. Also called nanoimprint lithography due to the high resolution possible.
Example: *The low-cost and high-resolution capabilities of* imprint lithography *make it a promising candidate for some niche applications.*

Incoherent Illumination   A type of illumination resulting from an infinitely large source of light that illuminates the mask with light from all possible directions. This is more correctly called *spatially* incoherent illumination.
Example: *In conventional photography, the available light exposes the subject to* incoherent illumination.

Index of Refraction   see Refractive Index.

Integrated Circuit (IC)   Many transistors, resistors, capacitors, etc., fabricated and connected together to make a circuit on one monolithic slab of semiconductor material.
Example: *Since the first* integrated circuit *was produced in the late 1950s, the number of transistors on a chip has grown exponentially.*

Intensity   A measure of the brightness of light that is defined either as the electromagnetic power per unit area or the electromagnetic power per unit solid angle, with the latter being the official (radiometry-based) definition. Physicists typically prefer the former definition, which is almost universally used by lithographers.
Example: *Exposure dose is the* intensity *of the light multiplied by the exposure time.*

Ion Beam Lithography   Lithography performed by exposing resist with a focused beam of ions.
Example: *The need for stencil masks has limited the acceptance of* ion beam lithography *outside of the research environment.*

Iso-Dense Print Bias   The difference between the dimensions of an isolated line and a dense line (a line inside an array of equal lines and spaces) holding all other parameters constant. Also called Iso-Dense Bias.

Example: *The* iso-dense print bias *is a strong function of feature size and partial coherence.*

Isofocal Bias   The difference between the isofocal linewidth and the desired resist feature width.

Example: *In general, the depth of focus is maximized when the* isofocal bias *becomes zero.*

Isofocal Dose   The dose at which the printed resist feature width equals the isofocal linewidth at best focus.

Example: *If possible, setting the process to use the* isofocal dose *can minimize the need for frequent focus adjustments.*

Isofocal Linewidth   The resist feature width (for a given mask width) that exhibits the maximum depth of focus (or the least sensitivity to focus variations).

Example: *The small isolated lines did not exhibit an* isofocal linewidth *over the range of exposures studied.*

J

Jones Pupil   see Pupil, Jones.

K

Köhler Illumination   A method of illuminating the mask in a projection imaging system whereby a condenser lens forms an image of the illumination source at the entrance pupil of the objective lens, and the mask is at the exit pupil of the condenser lens.

Example: *The use of* Köhler illumination *has become standard in projection lithography due to its superior uniformity.*

KrF   Krypton Fluoride, a type of excimer laser used in optical lithography that emits light at about 248 nm.

Example: KrF *exposure tools have been the most popular lithographic tools since the 250-nm resolution node.*

L

Latent Image   The reproduction of the aerial image in resist as a spatial variation of chemical species (for example, the variation of photoactive compound concentration).

Example: *The* latent image *was visible to the naked eye due to the change in the resist optical properties with exposure.*

LER   see Line Edge Roughness.

LES   see Line-End Shortening.

Levenson PSM   see Alternating PSM.

Lifting, Resist   The separation of the resist pattern from the substrate, either partially or completely, due to a loss of adhesion.
Example: Resist lifting *could not be avoided without the use of an adhesion promoter.*

Lift-Off Process   A lithographic process by which the pattern transfer takes place by coating a material over a patterned resist layer, then dissolving the resist to 'lift off' the material that is on top of the resist.
Example: *The* lift-off process *allowed the patterning of the metal without the use of an etch step.*

Line Edge Roughness (LER)   The deviation of a feature edge (as viewed top-down) from a smooth, ideal shape. That is, the edge deviations of a feature that occur on a dimensional scale much smaller than the resolution limit of the imaging tool that was used to print the feature.
Example: *One simple measure of* line edge roughness *is the RMS deviation of an edge from a best-fit straight line.*

Line-End Shortening (LES)   The reduction of the length of a line (where a line is defined here as any rectangular feature whose length is significantly greater than its width) as measured only at one end. Thus, the line-end shortening is characterized as the difference between the actual position of the end of a line and the intended (designed) position.
Example: *The amount of* line-end shortening *for the feature increased sharply when out of focus.*

Linear Resolution   The smallest feature that can be printed (using some agreed-upon criterion for resolution) while simultaneously allowing acceptable printing of all larger features.
Example: *While the use of OPC does not improve the ultimate resolution of a lithography process, its main benefit is in improving the* linear resolution.

Linearity   The variation of printed linewidth as a function of designed (or mask) line-width. In general, linearity is measured with a fixed duty cycle (equal lines and spaces, for example).
Example: *The use of OPC resulted in a marked improvement in* linearity *for both dense and isolated lines.*

Linewidth   see Critical Dimension.

Lithographer   1. A practitioner of lithography. 2. A harmless drudge.
Example: *The overworked and underappreciated* lithographer *paused for a moment and daydreamed, 'Will Moore's Law ever end?'.*

Lithography   A method of producing three-dimensional relief patterns on a substrate (from the Greek *lithos*, meaning stone, and *graphia*, meaning to write).
Example: *Although* lithography *is a centuries-old patterning technique, the small features used in integrated circuits make semiconductor lithography very challenging.*

LSI Large-Scale Integration, an integrated circuit made of hundreds to thousands of transistors.

Example: *As integrated circuits entered the* LSI *era, contact and proximity printing gave way to projection lithography.*

M

Mask A glass or quartz plate containing information (encoded as a variation in transmittance and/or phase) about the features to be printed. Also called a photomask or a reticle. (Historically, a photomask was the 1× mask used in contact or proximity printing, whereas the reticle was a higher magnification version of a single field used to make the photomask. Today, the terms photomask and reticle are used interchangeably for all masks used in optical lithography.).

Example: *Reduction projection printing significantly eases the burden of producing an acceptable* mask *compared to 1× lithography.*

Mask Aligner A tool that aligns a photomask to a resist-coated wafer and then exposes the pattern of the photomask into the resist.

Example: *The far superior throughput of* mask aligners *over direct-write lithography tools has made them the tools of choice for semiconductor manufacturing.*

Mask Biasing The process of changing the size or shape of the mask feature in order for the printed feature size to more closely match the nominal or desired feature size.

Example: *Although* mask biasing *complicates the design and mask-making process, the improvement in linewidth control that results could well be worth the effort.*

Mask Blank A blank mask substrate (e.g. quartz) coated with an absorber (e.g. chrome), and sometimes with resist, and used to make a mask.

Example: *The use of attenuated phase shifting masks greatly increases the cost of the* mask blank.

Mask Error Enhancement Factor The incremental change in the final resist feature size per unit change in the corresponding mask feature size (where the mask dimension is scaled to wafer size by the reduction ratio of the imaging tool). Abbreviated MEEF or MEF, a value of 1 implies a linear imaging of mask features to the wafer. Also called Mask Error Factor.

Example: *Although a linear imaging system produces a* mask error enhancement factor *of 1.0, near the resolution limit the MEEF often rises dramatically.*

Mask Error Factor see Mask Error Enhancement Factor.

Maskless Lithography Any one of a number of lithographic techniques (including direct-write lithography and programmable multimirror masks) that does not use a permanent, fixed mask to perform imaging.

Example: *For low-volume IC manufacturing,* maskless lithography *could offer a compelling cost of ownership advantage.*

Mask Linearity   The relationship of printed resist feature width to mask feature width for a given process.

Example: Mask linearity *is often used as a measure of the practical resolution of a process.*

Mercury Arc Lamp   A common light source used in lithographic exposure systems that produces intense radiation at the g-line, h-line and i-lines of the mercury spectrum.

Example: *The* mercury arc lamp *is the most common light source for optical lithography when the required resolution is greater than about 300 nm.*

MEEF   see Mask Error Enhancement Factor.

MEF   see Mask Error Enhancement Factor.

Metrology   The process of measuring structures on the wafer, such as the width of a printed resist feature or the overlay between two printed patterns.

Example: *Determination of practical linewidth control requirements must include* metrology *errors as well as process errors.*

Microlithography   Lithography involving the printing of very small features, typically on the order of micrometers or below in size.

Example: Microlithography *techniques are used extensively in semiconductor manufacturing as well as in compact disc mastering, thin-film head production, and many other advanced technologies.*

Mix-and-Match Lithography   A lithographic strategy whereby different types of lithographic imaging tools are used to print different layers of a given device.

Example: *The use of* mix-and-match lithography *allows for reduced equipment and process costs at the expense of more complicated overlay requirements.*

Model-Based OPC   An optical proximity correction technique that determines the level of correction (how much to move a design feature's edge) by iteratively simulating the lithographic result until the corrected design produces the desired resist pattern shape to within a preset tolerance.

Example: *At the 130-nm technology node, most semiconductor companies switch from using rule-based OPC to the more robust and accurate* model-based OPC.

Modeling   see Simulation.

Moore's Law   Named for Gordon Moore, one of the founders of Fairchild Semiconductor and Intel, the observation that the number of transistors on a typical chip doubles about 1–2 years. In lithography, this law has come to describe the exponential decrease in critical dimensions used in IC manufacturing over time.

Example: *By assuming that* Moore's Law *will continue to hold in the future, lithographic requirements can be predicted.*

Multilayer Resist (MLR)   A resist scheme by which the resist is made up of more than one layer, typically a thick conformal bottom layer under a thin imaging layer, possibly with a barrier layer in between.

Example: *The need for a very thin imaging layer can be met using a* multilayer resist *scheme.*

N

Nanolithography Lithography involving the printing of ultrasmall features, typically on the order of nanometers in size.

Example: Nanolithography *techniques are being used to research possible device technologies of the future.*

Negative Photoresist A photoresist whose chemical structure allows for the areas that are exposed to light to develop at a slower rate than those areas not exposed to light.

Example: *In theory, isolated lines or islands are best printed in* negative photoresist, *whereas spaces and contacts prefer a positive resist.*

NGL Next Generation Lithography, any potential successor to optical lithography for semiconductor manufacturing.

Example: *Despite billions of dollars invested in* NGL *technologies over the last two decades, optical lithography is still the only viable manufacturing technology for the foreseeable future.*

NILS see Normalized Image Log-Slope.

Normalized Image Log-Slope (NILS) The slope of the logarithm of an aerial image, measured at the desired photoresist edge position, normalized by multiplying by the nominal resist feature width. Generally, the sign of the slope is adjusted to be positive when the image is sloping in the correct direction. See also Image Log-Slope.

Example: *The* NILS *is a popular image metric because it is directly proportional to the feature's exposure latitude.*

Numerical Aperture (NA) The sine of the maximum half-angle of light that can make it through a lens, multiplied by the index of refraction of the media.

Example: *The* numerical aperture *of the lens can be adjusted over a specified range through the use of a motorized iris.*

O

OAI see Off-Axis Illumination.

Objective Lens The main imaging lens of a projection imaging system. Also called the projection lens, the imaging lens, or the reduction lens.

Example: *Weighing 500 kg, the stepper's* objective lens *can only be replaced using a specialized crane.*

Off-Axis Illumination (OAI) Illumination which has no on-axis component, i.e. which has no light that is normally incident on the mask. Examples of off-axis illumination include annular and quadrupole illumination.

Example: *Although a relatively old optical technique,* off-axis illumination *was only recently applied to the field of optical lithography.*

OPC see Optical Proximity Correction.

OPD see Optical Path Difference.

Optical Density   The base-10 logarithm of the intensity transmittance of a material of a given thickness.
Example: *The absorber material on the photomask had an* optical density *greater than 3.*

Optical Lithography   Lithography method that uses light to print a pattern in a photosensitive material. Also called photolithography.
Example: Optical lithography *will continue to be a workhorse of the semiconductor industry well into the 21st century.*

Optical Path Difference (OPD)   The difference in optical path (related to the difference in phase) between an actual wavefront emerging from a lens and the ideal wavefront, as a function of position on the wavefront.
Example: *The aberrations of the lens were determined by interferometrically measuring the wavefront* OPD.

Optical Proximity Correction (OPC)   A method of selectively changing the sizes and shapes of patterns on the mask in order to more exactly obtain the desired printed patterns on the wafer.
Example: *As minimum feature sizes are reduced below the imaging wavelength, some form of* optical proximity correction *is usually required.*

Optical Proximity Effect   Proximity effects that occur during optical lithography.
Example: Optical proximity effects *result in systematic linewidth variations across the chip.*

Overlay   A vector describing the positional accuracy with which a new lithographic pattern has been printed on top of an existing pattern on the wafer, measured at any point on the wafer. See also registration.
Example: *Improvements in* overlay *performance allowed the circuit designers to shrink the chip and reduce manufacturing costs.*

Overlay Correctables   Changes that can be made to the optical exposure tool (such as rotation or translation of the wafer stage or reticle stage) that would result in improved overlay if the same wafers were to be reworked and reprinted.
Example: *For each lot, a sample of wafers is measured and* overlay correctables *are automatically calculated and fed back to the stepper as part of the APC system.*

Overlay Mark   The target patterns printed on the wafer at two different lithography steps that allow the overlay between the two lithography patterns to be measured.
Example: *The* overlay mark *should be designed to minimize the impact of nonlithography process steps on overlay measurement accuracy and precision.*

Overlay Mark Fidelity   The variation in measured overlay due to (nonlithographically caused) local variations in the shape and structure of the overlay marks.
Example: *The* overlay mark fidelity *attempts to measure the susceptibility of the overlay marks to random, normal process variations.*

P

PAB   see Prebake.

PAC   see Photoactive Compound.

PAG   see Photoacid Generator.

Paraxial Approximation   The assumption that angles of light passing through a lens are small enough (close enough to the center axis of the lens) that spherical surfaces can be approximated as parabolic.
Example: *In the* paraxial approximation, *Snell's law becomes a linear function of incident and transmitted angles.*

Partial Coherence   Referring to the spatial coherence of light, the ratio of the sine of the maximum half-angle of illumination striking the mask to the numerical aperture of the objective lens. Also called the degree of coherence, the coherence factor, or the pupil filling function, this term is usually given the symbol σ.
Example: *Changes in the* partial coherence *of the projection tool result in significantly different imaging performance.*

Partially Coherent Illumination   A type of illumination resulting from a finite-sized source of light that illuminates the mask with light from a limited, nonzero range of directions.
Example: *All projection optical lithography tools in use today employ* partially coherent illumination.

Pattern Collapse   The mechanical failure of a resist feature such that the feature falls on its side. Pattern collapse is generally caused by unequal surface tension on the left and right sides of a tall photoresist line during drying after development.
Example: The phenomenon of pattern collapse *limits the aspect ratio for this resist to about 3.5.*

Pattern Placement Error   The difference between the position of the center of a resist pattern from the nominal (designed) center position. Pattern placement error is often used to describe pattern-dependent overlay. See also Overlay.
Example: *Besides reducing the resist linewidth control, lens aberrations can also result in* pattern placement errors.

Patterning   The processes of lithography (producing a pattern that covers portions of the substrate with resist) followed by etching (selective removal of material not covered by resist) or otherwise transferring the lithographic pattern into the substrate.
Example: *The repeated sequence of deposition followed by* patterning *allows for the complicated structures of an integrated circuit to be fabricated.*

PEB   see Post-exposure Bake.

Pellicle   A thin, transparent membrane placed above and/or below a photomask to protect the photomask from particulate contamination. Particles on the pellicle are significantly out of focus and thus have a much reduced chance of impacting image quality.
Example: *The mechanical strength of a* pellicle *is an important part of its practical use in manufacturing.*

Phase-Shifting Mask (PSM)   A mask that contains a designed spatial variation not only in intensity transmittance but phase transmittance as well.
Example: *Although complicated to design and make*, phase-shifting masks *offer significant improvements in resolution and depth of focus.*

Photoactive Compound (PAC)   The component of a photoresist that is sensitive to light. Also called a sensitizer.
Example: *The interaction of the* photoactive compound *with the resin is a controlling factor in resist performance.*

Photoacid Generator (PAG)   The light-sensitive component of a chemically amplified resist that generates an acid upon exposure to light.
Example: *The acid produced by the* photoacid generator *does not directly affect dissolution rate without the amplification reaction during the PEB.*

Photolithography   see Optical Lithography.

Photomask   A mask used in optical lithography.
Example: *The* photomask *industry changed considerably when the semiconductor industry switched from using 1× to 10× projection tools.*

Photoresist   A photosensitive material that forms a three-dimensional relief image by exposure to light and allows the transfer of the image into the underlying substrate (for example, by resisting an etch step).
Example: *The* photoresist *performs two functions: forming an image and resisting etch during pattern transfer.*

Photoresist Contrast   A measure of the resolving power of a photoresist, the photoresist contrast is defined in one of two ways. The measured contrast is the slope of the standard H–D curve as the thickness of resist approaches zero. The theoretical contrast is the maximum slope of a plot of log-development rate versus log-exposure energy (the theoretical H–D curve). The photoresist contrast is usually given the symbol γ.
Example: *The use of a material with a higher* photoresist contrast *resulted in improved sidewall angles and linewidth control.*

Pitch   The sum of the linewidth and spacewidth for a repeating pattern of long lines and spaces.
Example: *The optical proximity effects were characterized by measuring the change in resist linewidth as the* pitch *of the mask pattern was changed.*

Point Spread Function   The aerial image resulting from an infinitely small isolated pinhole on the mask. More correctly, it is the image resulting from a plane wave of light entering the entrance pupil of the lens.

Example: *Optical designers often use the* point spread function *as a means of characterizing the performance of a lens.*

Polarization   The orientation or direction of the electric field of a light wave.

Example: *By orienting the* polarization *of the illumination to be parallel to the line/space pattern, improved performance was obtained.*

Positive Photoresist   A photoresist whose chemical structure allows for the areas that are exposed to light to develop at a faster rate than those areas not exposed to light.

Example: Positive photoresists *remain the most common type of resist used in the semiconductor industry.*

Post-Apply Bake (PAB)   see Prebake.

Postbake   see Hard Bake.

Post-exposure Bake (PEB)   The process of heating the wafer immediately after exposure in order to stimulate diffusion of the PAC and reduce the effects of standing waves. For a chemically amplified resist, this bake also causes a catalyzed chemical reaction that changes the solubility of the resist.

Example: *Control of the temperature during the* post-exposure bake *is critical to linewidth control in most chemically amplified resists.*

Prebake   The process of heating the wafer after application (coating) of the resist in order to drive off the solvents in the resist. Also called softbake and post-apply bake.

Example: Prebake *is one of the least understood steps in resist processing.*

Process Latitude   The range over which a process parameter can be varied such that the lithographic results are still acceptable.

Example: *A large* process latitude *inevitably results in good linewidth control.*

Process Window   A window made by plotting contours that correspond to various specification limits, as a function of exposure and focus. One simple process window, called the CD process window, is a contour plot of the high and low CD specifications as a function of focus and exposure. Other typical process windows include sidewall angle and resist loss. Often, several process windows are plotted together to determine the overlap of the windows.

Example: *One of the most useful ways of characterizing the capabilities of a lithographic process is by examining the size of its focus–exposure* process window.

Projection Printing   A lithographic method whereby the image of a mask is projected onto a resist-coated wafer.

Example: *Since* projection printing *was first introduced in the early 1970s, its high-resolution and low-defect densities solved the problems of contact and proximity printing.*

Proximity Bake   A type of baking where the wafer is held in close proximity to a hot plate.

Example: Proximity baking *reduces the possibility of particle generation that can result from contact baking.*

Proximity Effect   A variation in the size or shape of a printed feature as a function of the sizes and positions of nearby features.

Example: *The coherence of the illumination determines the range of the* proximity effect *in optical imaging.*

Proximity Printing   A lithographic method whereby a photomask is placed in close proximity (but not in contact) with a photoresist-coated wafer and the pattern is transferred by exposing light through the photomask into the photoresist.

Example: *Although* proximity printing *reduced the defects inherent in contact printing, resolution was degraded due to greater diffraction.*

PSM   see Phase Shifting Mask.

Pupil, Jones   A mathematical description of the polarization-dependent transmission properties of a lens (named for R.C. Jones who invented the calculus used for polarization transmission descriptions in 1941).

Example: *As lens numerical apertures exceeded 1 (the so-called hyper-NA regime), scalar pupil descriptions of the lens had to be replaced with the more complete* Jones pupil description.

Pupil, Lens   The physical opening (somewhere within a lens) that constrains the range of angles than can pass through that lens. The size of a circular pupil is defined by its numerical aperture. Also called an aperture, stop, or aperture stop.

Example: *Ultimately, resolution is determined by the portion of the diffraction pattern that can pass through the entrance* pupil *of the objective lens.*

Pupil Filter   A device used to alter the amplitude and/or phase transmission of the light as it passes through the pupil of the objective lens.

Example: *Some researchers suggest that* pupil filters *might be able to improve resolution or depth of focus for specific mask features.*

Pupil Function   A mathematical function that describes the electric field transmission of the light as it passes through the pupil of the objective lens.

Example: *The* pupil function *of a lens includes the aperture (defined by its numerical aperture), the aberrations of the lens and any pupil filter that might be used.*

Q

Quadrupole Illumination   A type of off-axis illumination where four circles or arcs of light are used as the source. These four circles are spaced evenly around the optical axis.

Example: Quadrupole illumination *is especially useful for improving the depth of focus of small dense lines oriented in the x- or y-directions.*

Quantum Efficiency   Referring to photoresist exposure, the quantum efficiency is the average number of exposure reaction products produced when one photon is absorbed by the photoreactive species.

Example: *Reasonably good photoacid generators show* quantum efficiencies *of about 0.5.*

R

Raster Scan   A type of direct-write lithography where an exposing beam scans back and forth, covering the entire sample to be printed, while the beam is turned on and off to create the pattern.

Example: *Optical* raster scan *exposure tools are commonly used for cost-effective photomask production.*

Rayleigh Equations   Named for Lord Rayleigh, though modified for use in lithography, these equations relate resolution (R) and depth of focus (DOF) to the numerical aperture (NA) and wavelength (λ) of the imaging system.

$$R = k_1 \frac{\lambda}{NA} \quad DOF = k_2 \frac{\lambda}{NA^2}$$

The terms k_1 and k_2 are sometimes described as constants, but in reality are the scaled or dimensionless resolution and DOF, respectively. The DOF Rayleigh equation can also be corrected for high numerical aperture effects.

Example: *The* Rayleigh equations *are frequently misused by lithographers who do not understand their limitations.*

Reflective Notching   An unwanted notching or local feature size change in a photoresist pattern caused by the reflection of light off nearby topographic patterns on the wafer.

Example: *Although the* reflective notching *problem was reduced by using a dyed resist, only a bottom ARC could eliminate it.*

Reflectivity   The ratio of the reflected light intensity to the incident light intensity.

Example: *The amplitude of the swing curve is controlled by the* reflectivity *of the substrate.*

Refractive Index   The real part of the refractive index of a material is the ratio of the speed of light in vacuum to the speed of light in the material. The imaginary part of the refractive index is determined by the absorption coefficient of the material α and is given by $\alpha\lambda/4\pi$ where λ is the vacuum wavelength of the light.

Example: *The change in the* refractive index *of a material with wavelength is called dispersion.*

Registration   A vector describing the positional accuracy with which a lithographic pattern has been printed as compared to an absolute coordinate grid, measured at any point on the wafer. See also overlay.

Example: *Unlike wafers, where overlay is the most important measure, photomasks require* registration *specifications.*

Resin, Photoresist   A component of a photoresist that gives the resist its structural and etch-resistant qualities, and is not light-sensitive. The resin also interacts with the photoactive compound and/or its exposure products to affect the solubility of the resist in developer.
Example: *The most common* photoresist resin *used for typical g-line and i-line resists is a novolac resin.*

Resist   see Photoresist.

Resist Linewidth   see Critical Dimension.

Resist Gamma   see Photoresist Contrast.

Resist Reflectivity   The reflectivity of a photoresist-coated wafer. This reflectivity corresponds to the reflectivity that would be measured by bouncing light off of the resist-coated wafer. If a Top ARC or CEL is used, the reflectivity could include these films as well.
Example: *When coated on a reflective substrate, the* resist reflectivity *is a strong function of resist thickness due to thin-film interference effects.*

Resolution   The smallest feature of a given type that can be printed with acceptable quality and control. For example, resolution is often defined as the smallest feature of a given type that meets a given depth-of-focus requirement.
Example: *The traditional approaches to improving* resolution *are lower wavelength and higher numerical aperture.*

Resolution Enhancement Technologies (RETs)   A collection of techniques such as optical proximity correction, phase shifting masks and off-axis illumination, designed to improve the usable resolution of an optical lithography tool of a given numerical aperture and wavelength.
Example: *The widespread use of* resolution enhancement technologies *has enabled optical lithography to push to resolution limits thought impossible just a few years ago.*

RET   see Resolution Enhancement Technologies.

Retardance   The difference in the phase of the light transmitted through a lens as a function of its polarization.
Example: *A lens that exhibits* retardance *is not well described by a scalar pupil, but instead requires a Jones matrix.*

Reticle   see Mask.

Rule-Based OPC   An optical proximity correction technique that determines the level of correction (how much to move a design feature's edge) by applying empirically determined rules based on the proximity of that edge to other features.
Example: *The use of* rule-based OPC *began to wane as feature sizes dropped below 180 nm due to the exponential increase in the number of rules required for accurate correction.*

S

Saggital Lines   Line patterns oriented along the radial direction from the optical axis (i.e. the center) of an imaging system.
Example: *The* saggital lines *on the mask exhibited a different best focus than the tangential lines.*

Scalar Wave Theory   A simplified form of Maxwell's equations where the vector nature of light is ignored. In imaging applications, scalar theory will interfere two beams of light completely, regardless of their angle or polarization (i.e. regardless of the relative directions of the two electric field vectors).
Example: *As numerical apertures increase above 0.7,* scalar wave theory *becomes less and less adequate for predicting lithographic imaging phenomena.*

Scanner   A type of projection printing tool whereby the mask and the wafer are scanned past the small field of the optical system that is projecting the image of the mask onto the wafer.
Example: Scanners *offer the advantage of larger field size compared to steppers.*

Scanning Electron Microscope (SEM)   A machine that is used to inspect resist profiles and measure critical dimensions by bombarding the sample with electrons and detecting the backscattering of the electrons.
Example: *The critical dimension was measured in cross section using a* scanning electron microscope.

Scattering Bars   see Subresolution Assist Features.

SEM   see Scanning Electron Microscope.

Semiconductor Device   A transistor, resistor, capacitor, or integrated circuit made from a semiconductor material.
Example: *Advances in* semiconductor device *performance are typically driven by improvements in lithographic performance.*

Sensitizer   see Photoactive Compound.

Serif   A small ancillary pattern attached to the corners of the original pattern on a mask in order to improve the printing fidelity of the pattern.
Example: *The use of* serifs *can greatly reduce line-end shortening.*

Sidewall Angle   The angle that a resist profile makes with the substrate, usually estimated by modeling the resist profiles as a trapezoid.
Example: *After linewidth,* sidewall angle *is the most critical aspect of resist pattern quality.*

Simulation   The process of using physical models to predict the behavior of a complex process. These models are usually implemented as computer software.
Example: *Lithography* simulation *has become an essential tool for research, development and manufacturing.*

Smiley Plot   see Focus–Exposure Matrix.

Softbake   see Prebake.

Soft X-ray Lithography   see EUV Lithography.

Solvent, Photoresist   The solvent used to render a mixture of photoresist resin and photoactive compound or photoacid generator into a liquid form. This allows for spin coating of the resulting photoresist onto a wafer.
Example: *The* photoresist solvent *remaining after prebake has a significant impact on dissolution rates.*

Spatial Frequency   A scaled coordinate of the entrance or exit pupil of a lens, the spatial frequency refers to the Fourier transform used to calculate Fraunhoffer diffraction patterns. The center of the lens has a spatial frequency of zero and the edge of the lens is at the maximum spatial frequency, given by the numerical aperture divided by the wavelength.
Example: *The numerical aperture of a lens determines the maximum* spatial frequency *that can pass through the lens for a given imaging wavelength.*

Spherical Aberration   An aberration that often increases the asymmetric response of linewidth to positive versus negative focus errors.
Example: *Light traveling through the thickness of resist induces a small amount of* spherical aberration *in the resulting image.*

Spider Plot   see Focus–Exposure Matrix.

Spin Coating   The process of coating a thin layer of resist onto a substrate by pouring a liquid resist onto the substrate and then spinning the substrate to achieve a thin uniform coat.
Example: *Despite its apparent simplicity*, spin coating *can result in remarkably uniform photoresist films.*

SRAF   see Subresolution Assist Feature.

Standing Waves   A periodic variation of intensity as a function of depth into the resist that results from interference between a plane wave of light traveling down through the photoresist and one which is reflected up from the substrate.
Example: Standing waves *are reduced by lowering the reflectivity of the substrate, increasing the absorption in the resist, or by using broadband illumination.*

Standing Wave Effect   Caused by standing waves in the resist, the horizontal, periodic ridges formed along the sides of a resist profile.
Example: *The* standing wave effect *can be thought of as a loss in linewidth control.*

Step-and-Repeat Camera   see Stepper.

Step-and-Scan   A type of projection printing tool combining both the scanning motion of a scanner and the stepping motion of a stepper.
Example: Step-and-scan *systems combine the advantages of the scanner's larger field with the stepper's reduction capability.*

Stepper   A type of projection printing tool that exposes a small portion of a wafer at one time, and then steps the wafer to a new location to repeat the exposure. Also called a step-and-repeat camera.
Example: *Since their introduction in the late 1970s,* steppers *have dominated the lithographic market.*

Strehl Ratio   The ratio of the intensity at the peak of the actual point spread function of a lens to that at the peak of an ideal, aberration-free point spread function as formed by the same optical system.
Example: *Modern lithographic lenses have very low aberration levels, exhibiting* Strehl ratios *of 0.92–0.95.*

Stripping, Resist   Complete removal of the resist off the wafer after the lithographic and pattern transfer processes are finished.
Example: *Although often neglected, the ability to perform adequate* resist stripping *is an essential component in evaluating resist quality.*

Subresolution Assist Feature (SRAF)   Small features, usually in the form of parallel lines for a bright field pattern and parallel spaces for a dark field pattern, which are below the resolution limit of the imaging system but influence the lithographic behavior of the larger feature they are near. A common form of such subresolution assist features are often called scattering bars.
Example: *Sally discovered that the use of* subresolution assist features *in the form of two parallel lines running along either side of the main isolated line feature, of width equal to one-half the minimum design size and spaced one minimum design size away from the main feature, produced improved focus performance for the isolated line.*

Substrate   The film stack, including the wafer, on which the resist is coated.
Example: *The optical properties of the* substrate *can have a great impact on the lithography process.*

Substrate Reflectivity   The total reflectivity of the substrate beneath the resist. This is the reflectivity that light experiences after it passes through the resist and strikes the substrate.
Example: *Both the magnitude of the standing wave effect and the swing curve are determined by the* substrate reflectivity.

Subtractive Patterning   A process by which material is removed from the places where the pattern is not wanted. The standard sequence of deposition, lithography and etch is a subtractive patterning process.
Example: *The use of directional plasma etching enables very fine features to be formed in a* subtractive patterning *process.*

Surface Induction   see Surface Inhibition.

Surface Inhibition   A reduction of the development rate at the top surface of a resist relative to the bulk development rate. Also called surface induction.
Example: Surface inhibition *may improve the shape of the resist profile, though it may also result in reduced linewidth control.*

Surfactant   A 'surface-acting agent', a chemical that acts only on the surface of some material. For example, surfactants are commonly used in developers to reduce surface tension.
Example: *Surface inhibition can often be induced through the use of* surfactants *in the developer.*

Swing Curve   A sinusoidal variation of a parameter, such as linewidth or dose-to-clear, as a function of resist thickness caused by thin-film interference effects.
Example: *A* large swing curve *will make a lithographic process extremely sensitive to variations in resist thickness.*

Swing Ratio   Determined from the linewidth swing curve, the linewidths of the first two maximums are averaged together to give CD_{max}. Then using the linewidth at the minimum between these two maximums, called CD_{min}, the swing ratio is defined as:

$$SR = 2 * (CD_{max} - CD_{min})/(CD_{max} + CD_{min}) \times 100\%$$

Example: *By measuring the* swing ratio *as a function of ARC thickness, the optimum ARC thickness can be found.*

T

T-Top, Resist Profile   The T-shape of a resist profile caused by the formation of a low-solubility region at the top of a positive chemically amplified resist. This is usually caused by acid loss at the top of the resist due to atmospheric base contamination or acid evaporation.
Example: *The initial formation of a* resist T-top *determines the maximum permissible post-exposure delay.*

Tangential Lines   Line patterns oriented perpendicular to the radial direction from the optical axis (i.e. the center) of an imaging system.
Example: *The difference in best focus between* tangential lines *and saggital lines is called astigmatism.*

TARC   see Top Antireflective Coating.

TIS   see Tool-Induced Shift.

Tool-Induced Shift (TIS)   The difference in overlay measurements that results when the wafer is rotated by 180° and remeasured in the same overlay measurement tool.
Example: *While* tool-induced shift *can be easily measured and calibrated out of overlay measurements, variations in the* TIS *are more problematic.*

Top Antireflective Coating (TARC)   A thin film coated on top of the photoresist used to reduce reflections from the air–resist interface and thus reduce swing curves.
Example: *By optimizing the refractive index and the thickness of the* top antireflective coating, *the swing ratio was minimized.*

Top Surface Imaging   A resist imaging method whereby the chemical changes of exposure take place only in a very thin layer at the top of the resist.
Example: *The high absorption of polymers to EUV wavelength light means that* top surface imaging *may be required.*

U

ULSI Ultra Large-Scale Integration, an integrated circuit made of millions of transistors.

Example: *Today's* ULSI *circuits challenge every aspect of semiconductor manufacturing.*

UV Ultraviolet, the portion of the electromagnetic spectrum with wavelengths lower than can be seen by the human eye (typically taken to be wavelengths of about 400 nm and below).

Example: *Unlike g-line, the i-line wavelength of the mercury spectrum is in the* UV *and is not visible to the human eye.*

UV Cure A post-development process by which the resist patterns are exposed to deep-UV radiation (and often baked at the same time) in order to harden the resist patterns for subsequent pattern transfer. The UV cure is often a replacement for the hard bake step.

Example: *Without a UV curve, this resist would not hold up in the etch process.*

V

Vapor Prime A chemical treatment of a wafer to remove water from its surface in preparation for coating with resist in which the wafer is exposed to the vapor of an adhesion promoter.

Example: *The most effective method of applying adhesion promoter is the* vapor prime *method.*

Vector Scan A type of direct-write lithography where an exposing beam is not raster scanned but rather is moved directly to the area to be exposed before the beam is turned on and scanned over the exposure area.

Example: *For some mask patterns,* vector scan *exposure tools can show much greater throughput than traditional raster scan tools.*

Vector Wave Theory A complete and accurate treatment of imaging, based on Maxwell's equations, that accounts for the vector nature of light. In imaging applications, vector theory will interfere two beams based on the degree of overlap of their electric fields.

Example: *As numerical apertures increase above 0.7,* vector wave theory *becomes a requirement for accurately predicting lithographic imaging phenomena.*

VLSI Very Large-Scale Integration, an integrated circuit made of tens of thousands to hundreds of thousands of transistors.

Example: *The importance of lithography became obvious as the industry moved to* VLSI *circuits.*

W

Wafer A thin slice of semiconductor material on which semiconductor devices are made. Also called a slice or substrate.

Example: *The switch to a larger* wafer *size has greatly improved the economics of semiconductor production.*

Wafer Fab   The facility (building and equipment) in which semiconductor devices are fabricated. Also called a semiconductor fabrication facility.
Example: *The cost of a new* wafer fab *is dominated by the cost of the semiconductor fabrication equipment.*

Wavefront   Referring to the propagation of electromagnetic waves, any surface of constant phase.
Example: *Aberrations can be defined as the deviation of the actual* wavefront *emerging from the lens from the ideal* wavefront.

X

X-ray Lithography   Lithography using light of a wavelength in the range of about 0.1 to 5 nm, with about 1 nm being the most common. Usually takes the form of proximity printing.
Example: X-ray lithography *requires the use of proximity printing since focusing elements are difficult if not impossible to produce at these wavelengths.*

Y

Yield   The fraction of die (integrated circuits) began in a fab that work properly at the end of fabrication. Sometimes called die yield to distinguish from wafer yield, the fraction of wafer starts that finish production.
Example: *Ramping the* yield *during new technology introduction is critically important, with the goal of reaching 80%* yield *in 6 months.*

Z

Zernike Coefficients   The coefficients of the Zernike polynomial.
Example: *Knowledge of the* Zernike coefficients *across the field is essential to fully characterizing lens performance.*

Zernike Polynomial   A specific orthonormal polynomial, usually cut off at 36 terms, used to fit the wavefront error of a lens for a given field point. This polynomial characterizes the aberrations of the lens. (Named after Nobel prize-winner Frits Zernike.)
Example: *The* Zernike polynomial *not only provides a convenient function for fitting a measured wavefront error, but the individual terms of the polynomial have physical significance.*

For further glossary terms related to lithography, consult:

SEMI Standards M1-94, P5-94, P18-92, P19-92, P21-92, and P25-94

Appendix B
Curl, Divergence, Gradient, Laplacian

In the formulation of Maxwell's equations and the wave equation, some specialized notation is used to simplify the expression of derivatives. The symbol ∇ (usually pronounced 'del', though it is officially known by the name 'nabla', from the Greek for harp) represents a differential operator, but its meaning changes somewhat depending on its use. In particular, the same symbol is used in the symbolic representation of the curl ($\nabla \times$), the divergence ($\nabla \bullet$), the gradient (∇) and the Laplacian (∇^2). Definitions and some properties of these terms, as well as the cross and dot products of vectors, are given below.

B.1 Cross Product

In Cartesian coordinates, the cross product of two vectors U and V is given by

$$U \times V = (U_y V_z - U_z V_y)\hat{x} + (U_z V_x - U_x V_z)\hat{y} + (U_x V_y - U_y V_x)\hat{z} \qquad (B.1)$$

The cross product is sometimes called the vector product or the outer product. The cross product of two vectors is always perpendicular to both of those vectors – its direction is given by the 'right-hand rule' for the normal right-handed coordinate system. The magnitude of their cross product is

$$|U \times V| = |U||V|\sin\theta \qquad (B.2)$$

where θ is the angle between the two vectors. Thus, if the cross product of two nonzero vectors is zero, those vectors must be parallel to each other.

Lagrange's formula for repeated cross products can be useful:

$$A \times (B \times C) = B(A \bullet C) - C(A \bullet B) \qquad (B.3)$$

Fundamental Principles of Optical Lithography: The Science of Microfabrication, Chris Mack.
© 2007 John Wiley & Sons, Ltd.

Also note that the order of the cross product is important.

$$U \times V = -V \times U \tag{B.4}$$

B.2 Dot Product

In Cartesian coordinates, the dot product of two vectors U and V is given by

$$U \cdot V = |U||V|\cos\theta = U_x V_x + U_y V_y + U_z V_z \tag{B.5}$$

where θ is the angle between the two vectors. The dot product is sometimes called the scalar product or the inner product. It represents the product of the length of one vector with the portion of the length of the second vector that lies in the same direction as the first. If the dot product of two nonzero vectors is zero, then those vectors must be perpendicular to each other.

B.3 Curl

In Cartesian coordinates, the curl of a vector field F is defined as

$$\nabla \times F = \left(\frac{\partial F_z}{\partial y} - \frac{\partial F_y}{\partial z}\right)\hat{x} + \left(\frac{\partial F_x}{\partial z} - \frac{\partial F_z}{\partial x}\right)\hat{y} + \left(\frac{\partial F_y}{\partial x} - \frac{\partial F_x}{\partial y}\right)\hat{z} \tag{B.6}$$

(The designation of F as a *field* simply reflects the notion that the value of F varies as a function of position in space.) The curl of F is sometimes expressed as curl$F$. The curl is often thought of as showing the vector's rate of rotation at the point of evaluation. The curl of a vector field always points in a direction perpendicular to the vector at the point of evaluation.

B.4 Divergence

The divergence is an operator that produces a scalar measure of a vector field's tendency to originate from or converge upon a given point (the point at which the divergence is evaluated). In Cartesian coordinates, the divergence of a vector field F is defined as

$$\nabla \cdot F = \frac{\partial F_x}{\partial x} + \frac{\partial F_y}{\partial y} + \frac{\partial F_z}{\partial z} \tag{B.7}$$

The divergence of F is sometimes expressed as div$F$. Physically, the divergence is often thought of as the derivative of the net flow of the vector field out of the point at which the divergence is evaluated.

Divergence is a linear operator so that for two scalar constants a and b and two vectors U and V,

$$\nabla \cdot (aU + bV) = a\nabla \cdot U + b\nabla \cdot V \tag{B.8}$$

B.5 Gradient

In Cartesian coordinates, the gradient of a scalar field g is defined as

$$\nabla g = \frac{\partial g}{\partial x}\hat{x} + \frac{\partial g}{\partial y}\hat{y} + \frac{\partial g}{\partial z}\hat{z} \tag{B.9}$$

The gradient of g is sometimes expressed as grad$g$. It is interesting to note that the dot product of the gradient of a function with a unit vector gives the slope of the function in the direction of the unit vector. For example,

$$\hat{x} \cdot \nabla g = \frac{\partial g}{\partial x} \tag{B.10}$$

giving what is sometimes called the directional derivative.

B.6 Laplacian

The Laplacian operator, equal to the divergence of the gradient, operating on some scalar field g, is given in Cartesian coordinates as

$$\nabla^2 g = \nabla \cdot (\nabla g) = \frac{\partial^2 g}{\partial x^2} + \frac{\partial^2 g}{\partial y^2} + \frac{\partial^2 g}{\partial z^2} \tag{B.11}$$

The Laplacian is a second-order differential operator. The Laplacian can also operate on a vector field (such as F):

$$\nabla^2 F = \nabla^2 F_x \hat{x} + \nabla^2 F_y \hat{y} + \nabla^2 F_z \hat{z} \tag{B.12}$$

B.7 Some Useful Identities

For any scalar field g and any vector field F,

$$\nabla \cdot \nabla \times F = 0 \tag{B.13}$$

$$\nabla \times \nabla g = 0 \tag{B.14}$$

$$\nabla^2 F = \nabla(\nabla \cdot F) - \nabla \times (\nabla \times F) \tag{B.15}$$

$$\nabla \cdot (gF) = g\nabla \cdot F + \nabla g \cdot F \tag{B.16}$$

$$\nabla \times (gF) = g\nabla \times F + \nabla g \times F \tag{B.17}$$

B.8 Spherical Coordinates

The above definitions of the curl, divergence, gradient and Laplacian were all given in Cartesian coordinates. In spherical coordinates, vectors are defined by (r, θ, ϕ), where r is the length of the vector, θ is the angle with the positive z-axis, and ϕ is the angle with the x–z-plane. Spherical coordinates can be related to Cartesian coordinates by

$$r = \sqrt{x^2 + y^2 + z^2}$$
$$\theta = \cos^{-1}(z/r), \quad 0 \leq \theta \leq \pi \qquad \text{(B.18)}$$
$$\phi = \tan^{-1}(y/x), \quad 0 \leq \phi \leq 2\pi$$

or inversely by

$$x = r \sin\theta \cos\phi$$
$$y = r \sin\theta \sin\phi \qquad \text{(B.19)}$$
$$z = r \cos\theta$$

In spherical coordinates, the curl, divergence, gradient and Laplacian become:

Curl:

$$\nabla \times \boldsymbol{F} = \frac{1}{r\sin\theta} \left(\frac{\partial}{\partial\theta}(F_\theta \sin\theta) - \frac{\partial F_\phi}{\partial\phi} \right) \hat{r}$$
$$+ \left(\frac{1}{r\sin\theta} \frac{\partial F_r}{\partial\phi} - \frac{1}{r} \frac{\partial}{\partial r}(rF_\phi) \right) \hat{\theta} \qquad \text{(B.20)}$$
$$+ \left(\frac{1}{r} \frac{\partial}{\partial r}(rF_\theta) - \frac{1}{r} \frac{\partial F_r}{\partial\theta} \right) \hat{\phi}$$

Divergence:

$$\nabla \bullet \boldsymbol{F} = \frac{1}{r^2} \frac{\partial}{\partial r}(r^2 F_r) + \frac{1}{r\sin\theta} \frac{\partial}{\partial\theta}(F_\theta \sin\theta) + \frac{1}{r\sin\theta} \frac{\partial F_\phi}{\partial\phi} \qquad \text{(B.21)}$$

Gradient:

$$\nabla g = \frac{\partial g}{\partial r} \hat{r} + \frac{1}{r} \frac{\partial g}{\partial\theta} \hat{\theta} + \frac{1}{r\sin\theta} \frac{\partial g}{\partial\phi} \hat{\phi} \qquad \text{(B.22)}$$

Laplacian of a scalar:

$$\nabla^2 g = \frac{1}{r^2} \frac{\partial}{\partial r} \left(r^2 \frac{\partial g}{\partial r} \right) + \frac{1}{r^2 \sin\theta} \frac{\partial}{\partial\theta} \left(\sin\theta \frac{\partial g}{\partial\theta} \right) + \frac{1}{r^2 \sin^2\theta} \frac{\partial^2 g}{\partial\phi^2} \qquad \text{(B.23)}$$

Appendix C
The Dirac Delta Function

The Dirac delta function (also called the *unit impulse function*) is a mathematical abstraction which is often used to describe (i.e. approximate) some physical phenomenon. The main reason it is used has to do with some very convenient mathematical properties which will be described below. In optics, an idealized point source of light can be described using the delta function. Of course, real points of light will have finite width, but if the point is narrow enough, approximating it with a delta function can be very useful.

C.1 Definition

The Dirac delta function is in fact not a function at all, but a distribution (a generalized function, such as a probability distribution) that is also a measure (i.e. it assigns a value to a function) – terms that come from probability and set theory. However, for our purposes it will suffice to consider it a special function with infinite height, zero width and an area of 1. It can be considered the derivative of the Heaviside step function.

To help think about the Dirac delta function, consider a rectangle with one side along the x-axis centered about $x = x_0$ such that the area of the rectangle is 1 (this is equivalent to a uniform probability distribution). Obviously there are many such rectangles, as shown in Figure C.1. We can construct a Dirac delta function by starting with a square of height and width of 1. If we halve the width and double the height, the area will remain constant. We can repeat this process as many times as we wish. As the width goes to zero, the height will become infinite but the area will remain 1. Any unit area rectangle, centered at x_0, can be expressed as

Fundamental Principles of Optical Lithography: The Science of Microfabrication, Chris Mack.
© 2007 John Wiley & Sons, Ltd.

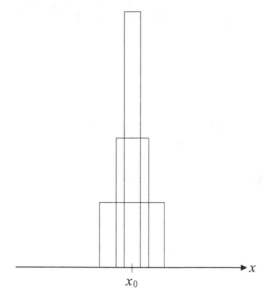

Figure C.1   *Geometrical construction of the Dirac delta function*

$$\delta_\varepsilon(x - x_0) = \begin{cases} 0, & x < x_0 - \dfrac{\varepsilon}{2} \\[2mm] \dfrac{1}{\varepsilon}, & x_0 - \dfrac{\varepsilon}{2} < x < x_0 + \dfrac{\varepsilon}{2} \\[2mm] 0, & x > x_0 + \dfrac{\varepsilon}{2} \end{cases} = \dfrac{1}{\varepsilon} rect\left[\dfrac{x - x_0}{\varepsilon}\right] \qquad (C.1)$$

where *rect* is the common rectangle function. The Dirac delta function, located at $x = x_0$, can be defined as the limiting case as ε goes to zero.

$$\delta(x - x_0) = \lim_{\varepsilon \to 0} \delta_\varepsilon(x - x_0) \qquad (C.2)$$

Although a rectangle is used here, in general the Dirac delta function is any pulse in the limit of zero width and unit area. Thus, the Dirac delta function can be defined by two properties:

$$\delta(x) = 0 \quad \text{when} \quad x \neq 0 \qquad (C.3)$$

$$\int_{-\infty}^{\infty} \delta(x) dx = 1 \qquad (C.4)$$

Any function which has these two properties is the Dirac delta function. A consequence of Equations (C.3) and (C.4) is that $\delta(0) = \infty$.

The function $\delta_\varepsilon(x)$ is called a 'nascent' delta function, becoming a true delta function in the limit as ε goes to zero. There are many nascent delta functions, for example, the

Gaussian pulse (a normal probability distribution, letting the standard deviation go to zero).

$$\delta(x) = \lim_{\varepsilon \to 0} \frac{1}{\varepsilon} e^{-\pi x^2/\varepsilon^2} \tag{C.5}$$

Extending this form to two dimensions,

$$\delta(x,y) = \lim_{\varepsilon \to 0} \frac{1}{\varepsilon^2} e^{-\pi(x^2+y^2)/\varepsilon^2} = \delta(x)\delta(y) \tag{C.6}$$

Generalizations to more dimensions are straightforward. Other nascent delta functions include the Airy disk function, the sinc function (see section C.2.4), and the Bessel function of order $1/\varepsilon$. In general, any probability density function with a scale parameter ε is a nascent delta function as ε goes to zero.

C.2 Properties and Theorems

The following sections will state some important identities and properties of the Dirac delta function, providing proofs for some of them.

C.2.1 Sifting Property

For any function $f(x)$ continuous at x_0,

$$\int_{-\infty}^{\infty} f(x)\delta(x - x_0)dx = f(x_0) \tag{C.7}$$

It is the sifting property of the Dirac delta function that gives it the sense of a measure – it measures the value of $f(x)$ at the point x_0.

Proof

Since the delta function is zero everywhere except at $x = x_0$, the range of the integration can be changed to some infinitesimally small range ε around x_0.

$$\int_{-\infty}^{\infty} f(x)\delta(x - x_0)dx = \int_{x_0-\varepsilon}^{x_0+\varepsilon} f(x)\delta(x - x_0)dx \tag{C.8}$$

Over this very small range of x, the function $f(x)$ can be thought to be constant and can be taken out of the integral.

$$\int_{x_0-\varepsilon}^{x_0+\varepsilon} f(x)\delta(x - x_0)dx = f(x_0) \int_{x_0-\varepsilon}^{x_0+\varepsilon} \delta(x - x_0)dx \tag{C.9}$$

From the definition of the Dirac delta function, the integral on the right-hand side will equal 1, thus proving the theorem. In fact, Equation (C.7) can be used as an alternate definition of the Dirac delta function. Any function $\delta(x-x_o)$ which satisfies the sifting property is the Dirac delta function.

C.2.2   Scaling Property

$$\delta(ax) = \frac{\delta(x)}{|a|} \tag{C.10}$$

C.2.3   Convolution Property

Convolution of a function f with a delta function at x_o is equivalent to shifting f by x_o.

$$f(x) * \delta(x - x_o) = f(x - x_o) \tag{C.11}$$

C.2.4   Identity 1

Another nascent delta function is the sinc function as the width of the sinc goes to zero:

$$\lim_{\varepsilon \to 0} \frac{\sin(x/\varepsilon)}{\pi x} = \lim_{a \to \infty} \frac{\sin ax}{\pi x} = \delta(x) \tag{C.12}$$

Proof

To prove identity 1, it is sufficient to show that this expression for the Dirac delta function satisfies sifting property:

$$\lim_{a \to \infty} \int_{-\infty}^{\infty} f(x) \frac{\sin ax}{\pi x} dx = f(0) \tag{C.13}$$

Breaking the integral into three sections, the outer two of which avoid the problem of dividing by zero at $x = 0$,

$$\int_{-\infty}^{\infty} f(x) \frac{\sin ax}{\pi x} dx = \int_{-\infty}^{-\varepsilon} + \int_{-\varepsilon}^{\varepsilon} + \int_{\varepsilon}^{\infty} \tag{C.14}$$

The first and last integral on the right-hand side are zero by the Riemann–Lebesgue lemma (an important theorem of the Fourier integral that will not be discussed here). The center integral can be evaluated by taking ε to be very small (but not zero). Over this very small range, $f(x)$ will be about constant:

$$\int_{-\varepsilon}^{\varepsilon} f(x) \frac{\sin ax}{\pi x} dx = f(0) \int_{-\varepsilon}^{\varepsilon} \frac{\sin ax}{\pi x} dx \tag{C.15}$$

Taking the limit as a goes to infinity,

$$\lim_{a \to \infty} \int_{-\varepsilon}^{\varepsilon} \frac{\sin ax}{\pi x} dx = \lim_{a \to \infty} \int_{-a\varepsilon}^{a\varepsilon} \frac{\sin x'}{\pi x'} dx' = \int_{-\infty}^{\infty} \frac{\sin x'}{\pi x'} dx' = 1 \tag{C.16}$$

Thus,

$$\lim_{a \to \infty} \int_{-\infty}^{\infty} f(x) \frac{\sin ax}{\pi x} dx = f(0) \tag{C.17}$$

C.2.5   Identity 2

$$\int_{-\infty}^{\infty} \cos(2\pi vx) dx = \delta(v) \tag{C.18}$$

Proof

The proof simply performs the integration and then applies identity 1.

$$\int_{-\infty}^{\infty} \cos(2\pi vx) dx = \lim_{a \to \infty} \int_{-a}^{a} \cos(2\pi vx) dx = \lim_{a \to \infty} \frac{\sin(2\pi va)}{\pi v} = \delta(v) \tag{C.19}$$

C.2.6   Identity 3 – $\mathcal{F}\{1\}$

The Fourier transform of one is the delta function:

$$\int_{-\infty}^{\infty} e^{-i2\pi vx} dx = \delta(v) \tag{C.20}$$

Proof

Changing the exponential into a sine and cosine,

$$\int_{-\infty}^{\infty} e^{-i2\pi vx} dx = \int_{-\infty}^{\infty} \cos(2\pi vx) dx - i \int_{-\infty}^{\infty} \sin(2\pi vx) dx \tag{C.21}$$

Since the sine is an odd function, the sine integral will vanish. Applying identity 2 to the cosine integral completes the proof.

C.2.7   Identity 4 – the Dirac Comb

The following identity is useful in the derivation of the diffraction pattern for a periodic line/space mask pattern with pitch p.

$$p \sum_{n=-\infty}^{\infty} e^{-i2\pi vnp} = \sum_{m=-\infty}^{\infty} \delta\left(v - \frac{m}{p}\right) \tag{C.22}$$

The function on the right-hand side of Equation (C.22) is called a *Dirac comb* of period p. This identity can be proved by recognizing that the Dirac comb is a periodic function that can be easily represented by a Fourier series. Direct calculation of the Fourier coefficients of the complex Fourier series produces Equation (C.22).

C.2.8   Relationship to the Heaviside Step Function

The Heaviside step function is defined as

$$u(x) = \begin{cases} 0, & x < 0 \\ 1 & x \geq 1 \end{cases} \tag{C.23}$$

The step function is related to the Dirac delta function by

$$\delta(x) = \frac{\mathrm{d}}{\mathrm{d}x} u(x) \quad \text{and} \quad u(x) = \int_{-\infty}^{x} \delta(t) \mathrm{d}t \tag{C.24}$$

Index